PLANT VITAMINS

Agronomic, Physiological, and Nutritional Aspects

A. Mozafar

CRC Press

Boca Raton   Ann Arbor   London   Tokyo

Library of Congress Cataloging-in-Publication Data

Mozafar, Ahmad.
    Plant vitamins: agronomic, physiological, and nutritional aspects
    / Ahmad Mozafar.
      p. cm.
    Includes bibliographical references and index.
    ISBN 0-8493-4734-3
    1. Plant vitamins. I. Title.
QK898.V5M63   1993
  581.19′26—dc20   93-8744
                                                   CIP

International Standard Book Number 0-8493-4734-3

Library of Congress Card Number 93-8744

Printed in the United States of America   3 4 5 6 7 8 9 0

Printed on acid-free paper

PREFACE

Vitamins are organic molecules that participate (directly or indirectly) in biological reactions in all living organisms. Organisms that cannot synthesize their own vitamins receive their vitamins through their foods or absorb some vitamins synthesized in their digestive tracts by various microorganisms. Humans depend on higher plants for most of their vitamin needs. Thus any factor that could reduce or increase the content of plant vitamins could eventually affect the vitamin intake by humans.

Increasing numbers of reports indicate that some vitamins, such as vitamins A, C, and E, because of their antioxidative properties, may also play some important nonvitamin roles in living organisms by reducing the risks of some forms of cancer. This has resulted in a new surge of interest by the public and scientific community in the roles of various vitamins in our overall health and thus in the vitamin content of our food.

There is a growing need to improve our knowledge of the agronomic and environmental stress factors that may alter the content of vitamins in the plant foods we consume. For example: (a) What are the natural variabilities in the content of vitamins in different varieties of the same plant? (b) Do plants vary in their vitamin content when grown in different geographical locations or in different years? (c) Does the use of chemical fertilizers, plant protection chemicals, and exposure of plants to various stress factors such as air and pollutants affect the plant vitamin contents? and (d) Are organically grown plants higher in vitamins?

The aim of this book is manifold: (a) to provide a concise synthesis of the available information on the factors acting on plants during the course of their growth that affect their vitamin contents at the time of their harvest, (b) to highlight those factors that have a negative or positive effect on plant vitamins so that the proper decision can be made to alter or avoid the conditions that may reduce the plant vitamins, and finally (c) to present the available information on the effects of various stress factors on the content of plant vitamins and the roles vitamins may play in the tolerance of plants to abiotic and biotic stress factors. It is justifiable to argue that, postharvest handling and losses being equal, a plant with a relatively higher vitamin content at the time of its harvest would also contain a proportionally higher amount of vitamin when it eventually reaches the dinner table. This may especially be true for vitamins such as ascorbic acid, especially in the socioeconomic situations where proper refrigeration of food products is not feasible. Thus any agronomic practices that could increase the content of this or other vitamins in plants at the time of their harvest may have a particular relevance for improving vitamin nutrition in cases where plant foods are consumed with the minimum amount of processing and facilities for proper storage are not available.

A relatively large portion of the literature on the factors affecting plant vitamins has been published in non-English (e.g., German, Russian, etc.) journals and has thus gone mostly unnoticed by English-speaking scientists. It was thus decided to cite all of the literature that could be collected irrespective of "age" (a) because there is no other literature available or (b) in order to be as comprehensive as possible. Since some of the classic works on agronomic factors influencing plant vitamins are published in non-English languages, it was decided to include a sample of the data in the form of tables or graphs so future readers should not need to go to the trouble of translating the cited works, some of which have appeared in journals hard to find in most libraries around the world.

I express my appreciations to the Hoffmann-La Roche of Basel (Switzerland) and the Swiss government grant for the advancement of industry (KWF) for providing assistance in my studies on the uptake of vitamins by the plant roots. I am grateful to Dr. Cyrus Abivardi for his valuable comments and encouragement during the course of writing this book. I am indebted to my family for their support and patience throughout this project.

Ahmad Mozafar

THE AUTHOR

Ahmad (Alex) Mozafar, Ph.D., is currently docent in the Division of Agronomy, Institute of Plant Sciences, at the Swiss Federal Institute of Technology in Zürich, Switzerland.

Dr. Mozafar obtained his B.S. degree in Horticulture (1964) from the University of Tehran, Iran, M.S. degree (1967) in Horticulture, and Ph.D. degree (1969) in Plant Physiology from the University of California, Riverside. He served as assistant and associate professor at the University of Rezaiyeh (presently Urmieh), Iran, from 1965 to 1978. He served as Chairman of the College of Sciences from 1976 to 1978 at Rezaiyeh University before taking up his post as Adjunct Professor at Texas Tech University in Lubbock from 1979 to 1980. In 1991, he joined the Swiss Federal Institute of Technology. His current research interests are the physicochemical and biological factors in the rhizosphere affecting the uptake of inorganic and organic compounds by the plant roots and the effects of agricultural practices on the quality and especially the vitamin content of plants.

Dr. Mozafar is a member of the American Society of Agronomy, Soil Science Society of America, Sigma Xi Society of America, and Swiss Society of Plant Physiology. He has authored two books in the Persian language and has authored numerous scientific papers.

to my family

CONTENTS

Contents   xi

Chapter 1

NUTRITIONAL ASPECTS

I. VITAMIN SOURCES FOR HUMAN NUTRITION

Vitamins are organic substances, present in minute amounts in natural foodstuffs, which are essential for normal metabolism. Their lack in the diet causes deficiency diseases.[1-6] At present, human beings are assumed to require at least 13 vitamins, most of which are received by consuming animal and plant products.[7] Vitamin D and niacin, both of which can be synthesized in the human body, are still classified as vitamins for historical reasons.[6]

The concentrations of vitamins in animal and plant tissues (products) are very different. Therefore, depending in the amount consumed, foods of animal or plant origin would contribute differently to the total vitamin intake by humans.[8] Furthermore, the year-round availability of foods, economic well-being and personal preference (vegetarian, etc.) of the individual, and the local customs could strongly affect the contribution of different foods to the total intake by different people.

Foods richest in a given vitamin do not necessarily contribute the most to our vitamin intake. For example, green pepper has about seven times more ascorbic acid than potato but, on the average, potato contributes more to the ascorbic acid intake of U.S. consumers than pepper because people consume more potato than pepper. On the other hand, tomato and citrus fruit, which, if compared with broccoli and spinach, are relatively low in their content of 10 vitamins and minerals, contribute the most to the supply of these vitamins and minerals to the diet of the U.S. population purely because they are consumed in relatively larger amounts.[7,9] Tomato, for example, has, on the average, 25 mg of ascorbic acid per 100 g. Thus only one small tomato (ca. 100 g) can supply ca. 40 and 67% and one cup (235 mL or 8 ounce) serving of tomato juice (containing ca. 35 mg of ascorbic acid) can supply 60–85%, of the Recommended Dietary Allowance for adults and children, respectively.[10]

Vegetables and fruits are the major source of ascorbic acid for human nutrition (Table 1). As people get to know new foods or develop new tastes and eating habits, however, the relative importance of each food item in supplying vitamins seems to change with time.[11] For example, in 1972 citrus (orange + grapefruit), potato, tomato, cabbage, green pepper, onion, and strawberry contributed 24.4, 19.7, 12.2, 5.1, 3.0, 1.81, and 1.8%, respectively, to the per capita availability of ascorbic acid to U.S. consumers.[12] The data of Karmas[13] published in

1

TABLE 1
Percent Contribution of Vitamins by Major Food Plants in the United States in 1983[13]

Food group	Ascorbic acid	Vitamin A	Thiamin	Riboflavin	Niacin	Pyridoxine	Folic acid
Citrus fruits	29.8	1.8	3.3	0.5	0.9	1.	11.9
Noncitrus fruits	11.9	5.8	1.9	1.8	1.9	7.3	4.3
Vegetables (dark green and deep yellow)	10.0	21.9	0.8	1.0	0.6	2.1	4.7
Other vegetables including tomato	25.9	13.5	6.0	4.4	5.1	10.6	21.5
Potato and sweet potato	13.5	5.3	4.8	1.3	6.1	9.6	5.2
Beans, peas, and nuts	0.0	0.0	5.1	1.9	6.6	4.8	10.5
Flour and cereal products	0.0	0.4	42.3	22.4	28.2	10.5	11.0
Total	91.9	48.7	64.2	33.3	49.4	46.5	69.1

1988, however, indicate that the shares of citrus fruits, potato, and sweet potato have undergone some changes so that these two food groups provided 29.8 and 13.5% of the total vitamin C in the diet of U.S. consumers at that time (Table 1). The latest information published in 1989 indicates that citrus fruit and potato provide 38 and 16%, respectively, of the total ascorbic acid in the U.S. diet.[14]

Depending on local climatic conditions, year-round availability of different fruits and vegetables strongly varies in different geographical locations around the world and is further influenced by the buying power of the people. These factors, along with the local habits, strongly affects the share of different plants in supplying different vitamins to the human diet. Thus in a country like Finland, where, other than potato, tomato, carrot, cucumber, and cabbage make up more than 70% of the fresh vegetables consumed (Heinonen et al., 1989), these vegetables may contribute a major portion of the vitamin C supply in the people's diet. In a tropical country like Venezuela, however, banana alone supplies 26% of the vitamin C in the people's diet.[15]

Vitamins are differently sensitive to destruction by heat, light, oxygen, and the pH of the medium they are kept in. During the past two decades the subject of vitamin loss during the transport, processing, storage, and preparation of foods prior to their consumption has been extensively investigated and reviewed.[16-20] Also, synthesis of vitamins by microorganisms and animals, and factors affecting these processes, have been extensively investigated.[3,5] For the plants, namely the living organisms on which we so strongly depend for the majority of our vitamins, however, little information is available on the site(s) of vitamin synthesis, their exact physiological role in the growth and development, and whether the functions attributed to different vitamins in microorganisms and animals are also valid for the plants themselves.[21-24]

There are some limited indications that plants may benefit from exogenous application of some vitamins under certain conditions.[25, 26] The main emphasis of this book, however, will be on the endogenous and exogenous factors that affect the content of plant vitamins. It is hoped that a better understanding of these relationships will also improve our understanding of the possible physiological role played by the vitamins in the plants themselves. The relationships between the content of vitamins in the plants and their response to diverse environmental stress factors, such as air and soil pollution and various pathogenic organisms and plant protection chemicals, are discussed in chapters 7–9.

In writing this book, it was not always possible to use one uniform terminology for the seemingly closely related plants. The case in point was species or subspecies of the gensu *Capsicum*, which may be referred to interchangeably as paprika, pepper, chilli (chili), or simply *Capsicum* in different English-speaking countries for the same types (or subspecies) of *C. annuum* or of *C. frutescens*. Some types of *C. annuum* are mild (sweet bell pepper type) and others are pungent.[27, 28] Furthermore in the non-English-speaking countries, the type called sweet (bell) pepper in the United States may be referred to as vegetables paprika (*Gemüsepaprika*) (in Germany), pepperoni (in Switzerland), paprica or pimento (in France), or chili (in India).[28, 29] Since in some of the reports the scientific names of the *Capsicum* plants used were not mentioned, it was decided to use the original terms, such as *Capsicum*, pepper, or chili, as used by the original authors rather than using one term, such as pepper, throughout the book.

II. TERMINOLOGY AND UNITS

A. Ascorbic Acid

In addition to their chemical names, vitamins may also be called by several other (sometimes synonymous) names or vitamers (Table 2). In this book these terms are used interchangeably, although in some cases they may not necessarily mean quite the identical chemical entities. For example, the terms vitamin C and ascorbic acid may not be identical since the dehydroascorbic acid, the oxidized form of ascorbic acid that could be present in plant tissues, also has vitamin C activity and thus the amount of vitamin C in a plant material may be more than the content of ascorbic acid in that tissue. The concentration of dehy-

TABLE 2
Commonly Used Names of Vitamins, Their Vitamers, and the
Recommended Dietary Allowance (RDA)[a]

Commonly used name	Vitamer, alternate, or related names (some now obsolete)	RDA (μg)
Vitamin C[b]	Ascorbic acid	60,000
Niacin	Nicotinic acid, nicotinamide, vitamin PP, B_3	19,000
Vitamin E	D-α-tocopherol, RRR-α-tocopherol, tocotrienols	10,000
Vitamin B_6	Pyridoxine, pyridoxal, pyridoxamine	2,000
Vitamin B_2	Riboflavin, vitamin G	1,700
Vitamin B_1	Thiamin, aneurine	1,500
Vitamin A	Retinol, β-carotene, carotenoid	1,000
Folic acid	Folacin, vitamin M	200
Vitamin K_1	Menadione, menaquinones, phylloquinone	80
Vitamin D	Ergocalciferol, cholecalciferol	5
Vitamin B_{12}	Cyanocobalamine	2
Biotin[c]	Vitamin H	—
Pantothenic acid[c]	—	—

[a] Compiled from references 3, 5, 6, and 14.
[b] arbitrary arranged based on their Recommended Dietary Allowances (RDA) for an adult man aged between 25 and 50, from the 1989 U.S. tables.
[c] Deficiencies of biotin and pantothenic acid are unknown and thus there is no evidence on which to base the RDA. Average daily intakes of 4000–7000 μg of pantothenic acid and 30–200 μg of biotin are apparently adequate.

droascorbic acid in fresh plant materials is, however, relatively low[3,30] and may vary from less than 5%[31] or 10%[32] in most fruits and vegetables. In some cases, however, 16% (in asparagus)[33] to 44% (in celery)[32] of the total vitamin C activity may be due to dehydroascorbic acid. Another source of variation may be due to changes in the relative amounts of reduced ascorbic acid and dehydroascorbic acid as fruits or vegetables mature or age. In tomato, for example, a large portion of total ascorbic acid is in the dehydroascorbic acid in the mature green fruits, a situation that fully reverses itself as the fruits mature and become red on the plants.[34]

By considering the reports, however, that in the fresh plant materials dehydroascorbic acid is not present at all.[35,36] and that the differences observed by some authors in the total vitamin C and the ascorbic acid (believed to be due to the dehydroascorbic acid) are due to some measurement artifacts,[5] the terms ascorbic acid and vitamin C are used interchangeably throughout this book. It is assumed that the error involved in the interchangeable use of these two terms should be negligible in most cases.

TABLE 3
Amount of β-Carotene as the Percentage of the Total
Carotenoids Present in Some Plants

Plant	β-carotene (% total)	Ref.
Red Pepper (paprika)	3–8	40
Guava	8	41
Tomato	10	42
Prune (Italian)	19	43
Green plant tissues (leaves)	20	42
Plum	16–34	44
Apricots	60	45
Carrots	44–88	46, 47, 48, 49
Sweet potato	86–90	50, 51

B. β-Carotene

Carotenoids account for the yellow, orange, or red color of maize, carrot, tomato, mango, sweet potato, apricot, and many other plant products. They frequently occur along with chlorophyll in all green plant tissues, and thus leafy vegetables such as spinach, cabbage, broccoli, and asparagus are good sources of carotenes, although in these plants the color of carotenes is masked by the chlorophyll molecules.

Among the 500–600 distinct natural carotenoids identified so far, only 50 of them could be converted to vitamin A in the animal body, and among them the all *trans-β* carotene has the most vitamin A activity.[37,38] Thus in this book the major emphasis is placed on the β-carotenes or vitamin A; in some cases, mostly in citing some older literature, however, only the total carotene was given by the authors and is thus reported as such.

β-carotene makes up a fraction of the total carotenes present in most plant tissues;[39] the size of this fraction, however, may vary from 8% in guava to 86–90% in sweet potatoes (Table 3). In *C. annuum lycopersiciforme flavum*, a yellow variety of paprika whose color never turns red upon ripening, the β-carotene was noted to constitute 4–20% of the total carotenoids present.[52]

Lycopene, a carotenoid without provitamin A activity, is the pigment responsible for the red color of many fruits and vegetables, may make up a major portion of the total carotenoids present in some fruits and vegetables: 83% in tomato[10] and 86% in guava.[41] In ripe tomato the ratio of lycopene to β-carotene is reported to be 1000 to 5 (80.77 to 0.42 μg/100 g, respectively).[53] This may, however, not be all in vain nutritionally since there is an indication that lycopene has a very powerful free-radical-scavenging property.[54] Since free-radical-scavenging compounds may have anticancer properties,[55–59] this observation, if substantiated by more research, would have a substantial impact on the nutritional values of lycopene-rich (most colorful) fruits and vegetables as compared to the dull-colored ones!

Carrot, tomato, sweet potato, cantaloupe, spinach, and orange contribute 13.9, 9.5, 5.6, 2.6, 2.2, and 1.3%, respectively, to the average per capita availability of vitamin A to U.S. consumers.[12] Carrot is such a good source of provitamin A for human nutrition that one dark-orange-colored carrot may supply 100% of the dietary adult requirement.[39] Not all regions of the world, however, get most of their carotene from carrot. For example, Thomas[15] cited a report by Jaffe and co-workers (originally published in 1963) that indicated that in the tropical countries such as Venezuela, banana supplied 48% of the vitamin A in the human diet at that time.

Carotene concentration is usually expressed as: μg carotene, IU (international unit) of vitamin A, a term that is now obsolete, or retinol equivalent (RE). One IU of vitamin A is equivalent to 0.6 μg of β-carotene and 1.2 μg of α-carotene and other carotenes with vitamin A activity.[51,60,61] One RE of vitamin A is equivalent to 6 μg of β-carotene and 12 μg of other carotenoids.[37,51] Thus the RE in a material can be calculated as:

$$RE = \frac{\mu g\ \beta\text{-carotene}}{6} + \frac{\mu g\ \text{other provitamin A carotenoids}}{12}$$

One IU is equal to 10.47 nmol of retinol or 0.3 μg of free retinol or 0.344 μg of retinyl acetate.[6]

C. Other Vitamins

Vitamin E and tocopherol may not be strictly synonymous.[5] Among the eight isomers (vitamers) of vitamin E present in nature and collectively called tocols, α-tocopherol has the highest vitamin E activity. Vegetable oils are very rich in vitamin E activity but α-tocopherol generally comprises less than 10% of their total tocopherol content. The tocopherol present in olive oil is exclusively in the form of α-tocopherol,[62] and in the foods of animal origin almost all of the vitamin E activity is due to their α-tocopherol content.[63] The use of international unit for the expression of vitamin E content of foods is now abandoned and is replaced by mg equivalent of (RRR)-α-tocopherol.[6]

The term niacin is used in the United States to specifically refer to nicotinic acid, and the amide form of this vitamin is referred to as niacinamide.[6] The concentration of niacin is expressed as mg niacin equivalent, which is the sum of preformed niacin plus 1/60 of the tryptophane based on the physiological equivalence of tryptophane and niacin.[6] In this book the concentrations of all vitamins are usually given as presented in the original publications. In some cases, however, the concentrations are converted into the presently used units.

III. NONVITAMIN ROLES

Do some vitamins play a role other than their classical role as vitamin? The answer to this question is a conditional yes in the light of recent evidence that some vitamins may play a role other than just preventing their typical clinical deficiency symptoms, i.e., a role that was not ascribed to them at the time of their discovery. For example, vitamins such as C and E and β-carotene have been implicated in playing a significant role in reducing the risks of some forms of cancer (e.g., cancers of the lung, stomach, and colon) and cardiovascular diseases because of their antioxidation properties.[51, 58, 59, 64–78] Ascorbic acid was recently noted to reduce the damage to DNA molecules in human sperm cell, caused by the endogenously generated oxygen radicals, and thus could play a significant role in reducing the risks of mutations in germ cells that may lead to birth defects.[79]

Although higher consumption of fruits and vegetables has been associated with a lower incidence of different cancers,[14, 80] the direct and unequivocal role of vitamin C in this regard has not been considered as proven yet.[14] The general belief is that the anticancer properties of fruits and vegetables may not be solely due to their vitamin A, C, and E content because fruits and vegetables may also contain numerous other micronutrients, such as Zn, Se, Cu, and Mn, which are essential for the proper functioning of antioxidant enzymes. Fruits and vegetables may also contain some not yet identified "nonnutritional compounds" that may inhibit carcinogenesis.[37, 69, 73, 80] No matter what there exact active constituents may eventually turn out to be, a balanced consumption of various vegetables is highly recommended for reducing the risks of some forms of cancer.[81]

The "back to nature" movement and the consumption of the foods produced by "organic" methods, supposedly free from undesirable chemicals, are gaining popularity.[11, 82] This has also resulted in a flourishing organic method of production in some parts of Europe and United States partly because of the claims made by their advocates that these products are of higher nutritional quality than the foods produced by the conventional methods. Thus a better knowledge of the factors that may decrease or increase the content of vitamins in plants is important not only for the scientist, but also for the general public. This is especially relevant since in recent years environmental pollution and the quality of foods have become the subject of ever-increasing debate by the general public and scientific community alike.

IV. PRESENCE IN PLANTS

All 13 compounds presently recognized as vitamins, with the exception of vitamin B_{12}, have been detected in plants, and thus plants supply human beings with a major portion of various vitamins. In the United States, for example, fruits and vegetables and various cereals and nuts supply ca. 92% of the ascorbic acid, 69% of the folic acid, 64% of the thiamin, 49% of the vitamin A, 49% of the niacin, 46% of the vitamin B_6, and 33% of riboflavin in the human diet (Table 1).[13]

Microorganisms are the major producers of vitamin B_{12} in nature.[3,5,22] A large portion of soil-inhabiting microorganisms are known to produce vitamin B_{12}[83-85] and other B vitamins.[86] In fact, soil is considered to be one of the richest sources of vitamin B_{12} in nature.[22,87] Ruminants (cattle, sheep, etc.) cover their need by absorbing the vitamin B_{12} synthesized in their forestomach by microorganisms. Part of this internally synthesized vitamin B_{12} is excreted in their feces and urine.[3,88,89] In sheep, for example, only 5% of the vitamin B_{12} produced in their rumen is absorbed and the rest (i.e., 95%) is excreted mostly through the feces.[90] The vitamin B_{12} excreted as such by the animal excrements would add to the soil's pool of this vitamin if these products were added to the soil directly or indirectly.[22]

Coprophagy (eating feces), a fairly common practice among most vertebrates, is one way some animals obtain their vitamin B_{12}. Another way is by consuming foods or water contaminated with soil.[89] Human beings receive most of their vitamin B_{12} from foods of animal origin since the vitamin B_{12} produced in their cecum and large gut cannot be effectively absorbed.[91] Vegetarian animals such as monkeys, however, receive their vitamin B_{12} when they consume "dirty" and soil-contaminated foods. In fact, if monkeys in captivity are fed food that is too clean, they may develop vitamin B_{12} deficiency![91] Also fruit bats have been reported to develop vitamin B_{12} deficiency if fed with washed fruits under laboratory conditions. It seems that in the wild these bats receive their vitamin B_{12} from that produced by the microorganisms naturally covering the fruit surfaces.[6]

Can humans also benefit from the vitamin B_{12} in the foods soiled with vitamin B_{12}-rich soil or manure? An interesting observation reported by Halsted[92] seems to indicate that this may be the case under certain conditions. He noted that in a developing country in the Middle East, although very poor farmers may practically live on bread, tea, and sugar, and some may rarely consume animal meat or eggs (which are rich in vitamin B_{12}), they rarely develop vitamin B_{12} deficiency (megaloblastic anemia) and their serum shows similar levels of this vitamin when compared with people consuming a diet adequate in this vitamin. One possible explanation given was that in these villages people and animals (donkey, cow, sheep, etc.) may often live in very close quarters and the yards are often littered with animal waste. It is thus not difficult to imagine that under these conditions, inhabitants might inadvertently ingest a certain quantity of manure and thus profit from its vitamin B_{12} content.

Herbert,[93] however, cited a work by Halsted indicating that villagers might get their vitamin B_{12} because they grow their vegetables in soil amended with human manure (*night soil*) and eat these vegetables without carefully washing them. It was thus concluded that the amount of vitamin B_{12} adhering to the dirty vegetables from the manure-rich soil was adequate to prevent the vitamin B_{12} deficiency in these people. An alternative explanation for these observations based on the absorption of vitamin B_{12} by the plant roots and its transport to the plant leaves is given below and in chapter 5.

With the exception of root nodules of legumes, which are rich in vitamin B_{12},[94-97] and one report indicating that plants may be able to synthesize this vitamin,[98] plants generally contain no vitamin B_{12} (Long and Wokes, 1968) or very little of it.[99-103] Some investigators are of the opinion that the vitamin B_{12} found in somes plants is due to experimental error and/or contamination of analyzed plant samples with soil. The recent observation that plants are capable of absorbing not only vitamin B_{12}, but also vitamin B_1 (thiamin) by their roots[104, 105] indicates that under certain conditions plants may in fact contain a low concentration of vitamin B_{12} inside their tissues. Since vitamin B_1 is also present in the soil,[106] the observation of Mozafar and Oertli[104, 105] raises the possibility that all of the vitamin B_{12}, and part of the vitamin B_1, found in the plants may have soil origin, and any factors that affect the content of these vitamins in the soil, such as addition of animal manure, may directly or indirectly affect the concentration of these vitamins in the plants (chapter 5).

V. LOSSES AFTER HARVEST

Concentration of vitamins in plants usually decreases after they are harvested; the rate of decline during storage or processing, however, strongly depends on the kind of vitamin, kind of plant or its variety, and the physicochemical conditions of the container or storage they are kept in.[7, 19, 20, 107-112] Therefore, the amount of a given vitamin in a given fruit and vegetable at the time of its harvest may not necessarily reflect the amount present in it by the time it is purchased by the consumer and, most important, by the time it is consumed as raw or cooked. For example, vitamin C loss in kale may amount to as much as 1.5% per hour. This means that just 24 hours after its harvest kale may already have lost ca. one-third of its original vitamin C.[109]

Bruising and wilting of plant materials during their handling and transport, by allowing the oxidizing enzymes access to the vitamin, can result in an accelerated vitamin loss.[109] For example, mustard plant (*Brassica juncea*) subjected to wilting conditions of wind with a speed of 24 km/h lost almost 92% of its vitamin C in 9 hours.[113] In addition, a four-hour exposure of Nigerian vegetables to sun at 35°C was found to drastically reduced the ascorbic acid content of okra (*Hibiscus esculentus*) by 63%, of water leaf (*Talinum triangulare*) by 71%, of fluted pumpkin leaf (*Telfairia occidentalis*) by 74%, and of African spinach (*Amaranthus hybridus*) by 82%.[114]

Shredding of cabbage can reduce its ascorbic acid content by 20%.[111] In addition, cucumber may lose up to 22% of its original ascorbic acid during slicing and 33–35% during standing for one hour. Cucumber salad, however, was noted to lose 22% of its ascorbic acid during preparation and an additional 8% and 11% during standing for one and two hours, respectively.[115]

The amount of vitamin loss that may occur between the field and the supermarket shelf, apart from its time component, may also be different in different produce. Available data, however, do not support the view that fruits from roadside stands are necessarily higher in vitamins than those purchased from the supermarkets. For example, Bushway et al.[116] noted that in the state of Maine, potato, cauliflower, and tomato purchased from roadside stands, which usually harvested their produce the night before, had significantly higher vitamin C than the same vegetables purchased from local supermarkets, which got their produce from all over the United States, taking 7–10 days from harvest to appearance on the supermarket shelf. The concentration of vitamin C in broccoli, cabbage, cantaloupe, green pepper, and spinach, however, was practically the same in the produce purchased from roadside stands as from supermarkets. The authors concluded that the modern methods of handling and storage of vegetables have been responsible for such a low rate of vitamin loss in the produce tested.[116]

The results of a more recent work done by Wu et al.,[117] however, tend to contradict the above findings of Bushway et al.[116] These investigators measured the ascorbic acid in green beans and broccoli under simulated conditions that took these products from the field to consumption: 3 days transport in a refrigerated truck to the supermarket, up to 7 days on the display shelf, and finally 3 days in the home refrigerator. Their results (Table 4) showed that green beans lost more than half of their original ascorbic acid during the 3 days they spent under simulated refrigerated transport conditions. The next 4 days on

TABLE 4
Ascorbic Acid and β-Carotene Contents in Green Beans and Broccoli
at Harvest and During Simulated Postharvest Transport, Marketing,
and Home Storage Periods[117]

Treatment[a]	Ascorbic acid (mg / 100 g FW)		β-carotene (mg / 100 g FW)	
	Green beans	Broccoli	Green beans	Broccoli
1 hour after harvest	16.97	107.18	0.29	0.41
5 hours after harvest	15.45	116.34	0.26	0.45
3 days in refrigerated truck	7.14	118.29	0.33	0.48
3 days on displays shelf	8.70	105.53	0.27	0.48
3 days on display + 3 days in the home refrigerator	6.16	104.47	0.30	0.36

[a] Simulated conditions of up to 5 hours at ambient temperature of 20–25°C, 3 days of transport in refrigerated truck at 4°C, and finally 3 days on display shelf of stores (at 10–16°C) and 3 days in the home refrigerator (4°C).

the display shelf did not cause any further decrease in vitamin C content. In contrast to green beans, the ascorbic acid content of broccoli increased significantly during the 3 days of simulated refrigerated transport. Even after an additional 4 days on the display shelf, the ascorbic acid content of broccoli was not statistically different from that measured one hour after its harvest. The β-carotene contents in green beans and broccoli were not affected by the transport and display conditions, so that even 4 days after being on the display shelf they had the same amounts of carotene as after their harvest. Wu et al.[117] also compared the vitamin contents of frozen green beans and broccoli with those of the fresh produce and reported that green beans frozen at −20°C for as long as 16 weeks had approximately twice the ascorbic acid as that in the fresh beans on the retail market's shelf. The ascorbic acid content of broccoli frozen at −20°C for 16 weeks, however, was about half of that in the fresh broccoli from the retail store's shelf. Some other examples of vitamin loss during the post harvest periods are shown in Table 5.

To make things even more complicated, one should note that not all vitamins in foodstuffs are equally available to the consuming organisms.[118] For example, only 20–30% of the biotin in sorghum is reported to be available to chicks (*Gallus gallus domesticus*), while the availability of biotin in wheat is zero.[119] Also, the niacin in cereal grains has a low biological availability. For example, in immature cereal grains as much as 70% of the niacin is not available,[14] and thus pellagra was noted to

TABLE 5
Loss of Some Vitamins During Food Processing[14, 110]

Vitamin	Type of processing	% loss
Pantothenic acid	Canned produce	85
	Frozen produce	37–58
Vitamin B_6	Milling of cereals	50–90
	Frozen produce	15–70

be endemic among jowar (sorghum) eaters in some parts of India.[119] Widdowson[120] cited a nutrition experiment according to which riboflavin and nicotinic acid in whole-meal flour were found to be less readily absorbed and utilized by the children than synthetic vitamins added to white flour.

REFERENCES

1. McDowell, L. R., *Vitamins in Animal Nutrition*, Academic Press, San Diego, 1989.
2. West, K. P., Jr., Howard, G. R., and Sommer, A., Vitamin A and infection: public health implications, *Annu. Rev. Nutr.*, 9, 63, 1989.
3. Friedrich, W., *Handbuch der Vitamine*, Urban & Schwarzenberg, München, 1987.
4. Gaby, S. K., Bendich, A., Singh, V. N., and Machlin, L. J., *Vitamin Intake and Health*, Marcel Dekker, New York 1991, 1.
5. Machlin, L. J., Ed., *Handbook of Vitamins*, 2nd ed., Marcel Dekker, New York, 1991.
6. Bender, D. A., *Nutritional Biochemistry of the Vitamins*, Cambridge University Press, Cambridge, 1992.
7. Salunkhe, D. K. and Desai, B. B., Effect of agricultural practice, handling, processing, and storage on vegetables, in *Nutritional Evaluation of Food Processing*, Karmas, E. and Harris, R. S., Eds., Van Nostrand Reinhold, New York, 1988, chap. 3.
8. Sauberlich, H. E., Kretsch, M. J., Johnson, H. L., and Nelson, R. A., Animal products as a source of vitamins, in *Animal Products in Human Nutrition*, Beitz, D. C. and Hansen, R. G., Eds., Academic Press, New York, 1982, 340.
9. Salunkhe, D. K., Bolin, H. R., and Reddy, N. R., *Storage, Processing, and Nutritional Quality of Fruits and Vegetables*, 2nd ed., Vol. 1, CRC Press, Boca Raton, Fla., 1991, 40.
10. Gould, W. A., *Tomato Production, Processing and Quality Evaluation*, 3rd ed., CTI Publications Inc., Baltimore, Md., 1992, 439.
11. Senauer, B., Asp, E., and Kinsey, J., *Food Trends and the Changing Consumer*, Eagan Press, St. Paul, Minn., 1991.
12. Senti, F. R. and Rizek, R. L., Nutrient levels in horticultural crops, *HortScience*, 10(3), 243, 1975.
13. Karmas, E., The major food groups, their nutrient content, and principles of food processing, in *Nutritional Evaluation of Food Processing*, Karmas, E. and Harris, R. S., Eds., Van Nostrand Reinhold, New York, 1988, chap. 2.

14. NRC (National Research Council), *Recommended Dietary Allowances 10th ed.*, *Report of the Committee on Dietary Allowances, Division of Biological Sciences, Assembly of Life Sciences, Food and Nutrition Board*, National Academy of Sciences, Washington, D.C., 1989.
15. Thomas, P., Radiation preservation of foods of plant origin. III. Tropical fruits: bananas, mangoes, and papayas, *CRC Crit. Rev. Food Sci. Nutr.*, 23, 147, 1986.
16. White, P. L. and Selvey, N., Eds., *Nutritional Quality of Fresh Fruits and Vegetables*, Futura Publ., Mount Kisco, N.Y., 1974.
17. Pattee, H. E., Ed., *Evaluation of Quality of Fruits and Vegetables*, AVI Publishing, Westport, Conn., 1985.
18. Karmas, E. and Harris, R. S., Eds., *Nutritional Evaluation of Food Processing*, Van Nostrand Reinhold, New York, 1988.
19. Eskin, N. A. M., Ed., *Quality and Preservation of Vegetables*, CRC Press, Boca Raton, Fla., 1989.
20. Eskin, N. A. M., *Quality and Preservation of Fruits*, CRC Press, Boca Raton, Fla., 1991.
21. Robbins, W. J., *Plants and Vitamins*, Chronica Botanica, Waltham, Mass., USA, 1943.
22. Sebrell, W. H., Jr. and Harris, R. S., Eds., *The Vitamins*, 2nd ed., Vol. 2, Academic Press, New York.
23. Lehninger, A. L., *Biochemistry*, 2nd ed., Worth Publishers, Inc, New York, 1977.
24. Stryer, L., *Biochemistry*, 3rd ed., Freeman and Co., San Francisco, 1988.
25. Chinoy, J. J., *The Role of Ascorbic Acid in Growth, Differentiation and Metabolism of Plants*, Nijhoff/Junk, The Hague, 1984.
26. Oertli, J. J., Exogenous application of vitamins as regulators for growth and development of plants—a review, *Z. Pflanzenernaehr. Bodenk.*, 150, 375, 1987.
27. Govindarajan, V. S., Capsicum-production, technology, chemistry, and quality. Part I: history, cultivation, and primary processing, *CRC Crit. Rev. Food Sci. Nutr.*, 22, 109, 1985.
28. Rylski, I., Pepper (*Capsicum*), in *CRC Handbook of Fruit Set and Development*, Monselise, S. P., Ed., CRC Press, Boca Raton, Fla., 1986, 341.
29. Franke, W., *Nutzpflanzenkunde*, Thieme Verlag, Stuttgart, 1985.
30. Erdman, J. W., Jr. and Klein, B. P., Harvesting, processing, and cooking influences on vitamin C in foods, in *Ascorbic Acid: Chemistry, Metabolism, and Uses*, Seib, P. A. and Tolbert, B. M., Eds., Adv. Chem. Series, no. 200, Am. Chem. Soc., Washington, D.C., 1982, 499.
31. Mapson, L. W., Metabolism of ascorbic acid in plants: part I. function, *Annu. Rev. Plant Physiol.*, 9, 119, 1958.
32. Wills, R. B. H., Wimalasiri, P., and Greenfield, H., Dehydroascorbic acid levels in fresh fruit and vegetables in relation to total vitamin C activity, *J. Agric. Food Chem.*, 32(4), 836, 1984.
33. Hudson, D. E., and Lachance, P. A., Ascorbic acid and riboflavin content of asparagus during marketing, *J. Food Qual.*, 9(4), 217, 1986.
34. Gonzalez, A. R. and Brecht, P. E., Total and reduced ascorbic acid levels in *Rin* and normal tomatoes, *J. Am. Soc. Hortic. Sci.*, 103(6), 756, 1978.
35. Lehnard, A., Anteil der L-Dehydroascorbinsäure am Gesamt-Vitamin-C-Gehalt pflanzlicher Produkte, *Z. Lebensm.-Unters. Forsch.*, 169, 82, 1979.
36. Tono, T., and Fujita, S., Determination of ascorbic acid by spectrophotometric method based on difference spectra. VI. Determination of vitamin C (ascorbic acid) in Satsuma mandrin fruit by difference spectral method and change in its content of the fruit during the development stage. *Nippon Shokuhin Kogyo Gakkaishi*, 32(4), 295, 1985; *Chem. Abst.*, 103, 138711j, 1985.

37. Ziegler, R. G. and Subar, A. F., Vegetables, fruits, and carotenoids and the risk of cancer, in *Micronutrients in Health and in Disease Prevention*, Bendich, A. and Butterworth, C. E., Jr., Eds., Marcel Dekker, New York, 1991, 97.

38. Ganguly, J., *Biochemistry of Vitamin A*, CRC Press, Inc., Boca Raton, Fla., 1989.

39. Gabelman, W. H. and Peters, S., Genetical and plant breeding possibilities for improving the quality of vegetables, *Acta Hortic.*, 93, 243, 1979.

40. Almela, L., López-Roca, J. M., Candela, M. E., and Alcázar, M. D., Carotenoid composition of new cultivars of red pepper for paprika, *J. Agric. Food Chem.*, 39(9), 1606, 1991.

41. Padula, M. and Rodriguez-Amaya, D. B., Characterization of carotenoids and assessment of the vitamin A value of Brasilian guavas (*P sidium guajava* L.), *Food Chem.*, 20, 11, 1986.

42. Simpson, K. L., Relative value of carotenoids as precursors of vitamin A, *Proc. Nutr. Soc.*, 42, 7, 1982.

43. Curl, A. L., The carotenoids of Italian prunes, *Food Sci.*, 28, 623, 1963.

44. Gross, J., Carotenoid pigments in three plum cultivars, *Gartenbauwissenschaft*, 49, 18, 1984.

45. Curl, A. L., Carotenoids of apricots, *Food Res.*, 25, 190, 1960; *Hortic. Abstr.*, 31, 333, 1961.

46. Simon, P. W. and Wolff, X. Y., Carotenes in typical and dark orange carrots, *J. Agric. Food Chem.*, 35, 1017, 1987.

47. Heinonen, M. I., Carotenoids and provitamin A activity of carrot (*Dawson carota* L.) cultivars, *J. Agric. Food Chem.*, 38, 609, 1990.

48. Lee, C. Y., Changes in carotenoid content of carrots during growth and post-harvest storage, *Food Chem.*, 20(4), 285, 1986.

49. Blattná, J., Krčová, and Blattný, C., Study of the composition of some carrot cultivars from different areas. I. Content of β-carotene, *Sbornik Vysoké Školy Chemicko-Technologické Praze, Potraviny*, E 45, 87, 1976; *Hortic. Abstr.*, 47, 11553, 1977.

50. Purcell, A. E. and Walter, W. M., Jr., Carotenoids of Centennial variety sweet potato, *Ipomea batatas* L., *J. Agric. Food Chem.*, 16(5), 769, 1968.

51. Woolfe, J. A., *Sweet Potato*, Cambridge University Press, Cambridge, 1992.

52. Matus, Z., Deli, J., and Szabolcs, J., Carotenoid composition of yellow pepper during ripening: isolation of β-cryptoxanthin 5,6-epoxide, *J. Agric. Food Chem.*, 39, 1907, 1991.

53. López-andréu, F. J., Lamela, A., Esteban, R. M., and Collado, J. G., Evolution of quality parameters in the maturation stage of tomato fruit, *Acta Hortic.*, 191, 387, 1986.

54. Di Mascio, P., Kaiser, S., and Sies, H., Lycopene as the most efficient biological carotenoid singlet oxygen quencher, *Arch. Biochem. Biophys.*, 274(2), 532, 1989.

55. Calabrese, E. J., *Nutrition and Environmental Health. The Influence of Nutritional Status on Pollutant Toxicity and Carcinogenicity, Vol. 1, The Vitamins*, Wiley-Interscience Publ., New York, 1980.

56. Calabrese, E. J., *Nutrition and Environmental Health. The Influence of Nutritional Status on Pollutant Toxicity and Carcinogenicity, Vol. 2, Minerals and Macronutrients*, Wiley-Interscience Publ., New York, 1981.

57. Calabrese, E. J., Does exposure to environmental pollutants increase the need for vitamin C? *J. Environ. Path. Toxicol. Oncol.*, 5(6), 81, 1985.

58. Thurnham, D. I., Anti-oxidant vitamins and cancer prevention, *J. Mironutr. Analy.*, 7, 279, 1990.

59. Byers, T. and Perry, G., Dietary carotenes, vitamin C, and vitamin E as protective antioxidants in human cancers, *Annu. Rev. Nutr.*, 12, 139, 1992.

60. Kays, S. J., *Postharvest Physiology of perishable Plant Products*, AVI Book, Van Nostrand Reinhold, New York, 1991.
61. Gerras, C., Golant, J., and E. Johan Hanna, E., *The Complete Book of Vitamins*, Rodale Press, Emmaus, Penn., 1977.
62. Gracián, J. and Arévalo, G., Tocopherols in vegetable oils, particularly olive oil, *Grasas Aceit.*, 16, 278, 1965; *Hortic. Abstr.*, 37, 7786, 1967.
63. Nikbin Meydani, S. and Blumberg, J. B., Vitamin E supplementation and enhancement of immune responsiveness in the aged, in *Micronutrients in Health and in Disease Prevention*, Bendich, A. and Butterworth, C. E. Jr., Eds., Marcel Dekker, New York, 1991, 289.
64. Doll, R., Nutrition and cancer: a review, *Nutrition and Cancer*, 1, 35, 1979.
65. Weisburger, J. H., Mechanism of action of diet as a carcinogen, *Nutrition and Cancer*, 1, 74, 1979.
66. Gormley, T. R., Downey, G., and O'Beirne, D., *Food, Health and the Consumer*, Elsevier Applied Sci., London, 1987.
67. Rensberger, B., Cancer, the new synthesis, *Science 84*, 5(7), 28, 1984.
68. Gey, K. F., Brubacher, G. B., and Stähelin, H. B., Plasma levels of antioxidant vitamins in relation to ischemic heart disease and cancer, *Am. J. Clin. Nutr.*, 45, 1368, 1987.
69. Marchand, L., Yoshizawa, C. N., Kolonel, L. N., Kankin, J. H., and Goodman, M. T., Vegetable consumption and the lung cancer risk: a population-based case-control study in Hawaii, *J. National Cancer Inst.*, 81, 1158, 1989.
70. Ziegler, R. G., A review of epidemiologic evidence that carotenoids reduce the risk of cancer, *J. Nutr.*, 119, 116, 1989.
71. Narbonne, J. F., Cassand, P., Daubeze, M., Colin, C., and Leveque, F., Chemical mutagenicity and cellular status in vitamin A, E and C, *Food Additives and Contaminants*, 7, 48, 1990.
72. Simon, P. W., Carrots and other horticultural crops as a source of provitamin A carotenes, *HortScience*, 25(12), 1495, 1990.
73. Butterworth, C. E., Jr. and A. Bendich, A., Introduction, in *Micronutrients in Health and in Disease Prevention*, Bendich, A. and Butterworth, C. E., Jr., Eds., Marcel Dekker, New York, 1991, 1.
74. Gaby, S. K. and Singh, V. N., β-carotene, in *Vitamin Intake and Health*, Gaby, S. K., Bendich, A., Singh, V. N., and Machlin, L. J., Eds., Marcel Dekker, New York, 1991, 29.
75. Gaby, S. K. and Singh, V. N., Vitamin C, in *Vitamin intake and Health*, Gaby, S. K., Bendich, A., Singh, V. N., and Machlin, L. J., Eds., Marcel Dekker, New York, 1991, 103.
76. Gaby, S. K. and Machlin, L. J., Vitamin E, in *Vitamin intake and Health*, Gaby, S. K., Bendich, A., Singh, V. N., and Machlin, L. J., Eds., Marcel Dekker, New York, 1991, 71.
77. Niki, E., Vitamin C as an antioxidant, in *Selected Vitamins, Minerals, and Functional Consequences of Maternal Malnutrition*, Simopoulos, A. P., Ed., World Rev. Nutr., Diet, Vol. 64, Karger, Basel, 1991, 1.
78. Gerster, H., Potential role of beta-carotene in the prevention of cardiovascular disease, *Int. J. Vit. Nutr. Res.*, 61, 277, 1991.
79. Fraga, C. G., Motchnik, P. A., Shigenaga, M. K., Helbock, H. J., Jacob, R. A., and Ames, B. N., Ascorbic acid protects against endogenous oxidative DNA damage in human sperm, *Proc. Natl. Acad. Sci. (USA)*, 88, 11003, 1991.
80. Birt, D. F. and Bresnick, E., Chemoprevention by nonnutrient components of vegetables and fruits, in *Cancer and Nutrition*, Alfin-Slater, R. B. and Kritchevsky, D., Eds., Plenum Press, New York, 1991, 221.

81. Knecht, P., Epidemiological studies of vitamin E and cancer risks, in *Micronutrients in Health and in Disease Prevention*, Bendich, A., and Butterworth, C. E., Jr., Eds., Marcel Dekker, New York, 1991, 141.

82. NRC (National Research Center), *Alternative agriculture*, National Academy Press, Washington, D.C., 1989, 116.

83. Lochhead, A. G. and Thexton, R. H., Vitamin B_{12} as a growth factor for soil bacteria, *Nature (London)*, 167, 1034, 1951.

84. Afrikyan, F. G., Vitamin B_{12} in soils, *Biol. Zh. Arm.*, 34(3), 253, 1981; *Chem. Abstr.*, 95, 41296f, 1981.

85. Avakyan, Z. G. and Afrikyan, F. G., Vitamin B_{12} in soils, *Biol. Zh. arm.*, 34(3), 253, 1981; *Chem. Abstr.*, 95, 41296f, 1981.

86. Mansurova, M. L. and Noskova, L. N., Ability of some bacteria isolated from Uzbek soils to produce B group vitamins, *Nauchn. Tr.–Tashk. Gos. Univ.*, 514, 1976; *Chem. Abstr.*, 88, 148667k, 1978.

87. Smith, E. L., *Vitamin B_{12}*, Methuen, London, 1960.

88. Maynard, L. A. and Loosli, J. K., *Animal Nutrition*, 6th ed., McGraw-Hill, New York, 1969, Chap. 9.

89. Pratt, J. M., *Inorganic Chemistry of Vitamin B_{12}*, Academic Press, London, 1972.

90. Friedrich, W., *Vitamin B_{12} and Verwandte Corrinoide*, George Thieme Verlag, Stuttgart, 1975.

91. Chanarin, I., *The Megaloblastic Anaemias*, Blackwell Sci. Publ., Oxford, 1969.

92. Halsted, J. A., Caroll, J., Dehghani, A., Loghmani, M., and Prasad, A. S., Serum vitamin B_{12} concentration in dietary deficiency, *Am. J. Clin. Nutr.*, 8, 374, 1960.

93. Herbert, H., Vitamin B-12: plant sources, requirements, and assay, *Am. J. Clin. Nutr.*, 48, 852, 1988.

94. Evans, H. J. and Kliewer, M., Vitamin B_{12} compounds in relation to the requirements of cobalt for higher plants and nitrogen fixing organisms, *Ann. N.Y. Acad. Sci.*, 112, 735, 1964.

95. Bond, G., Adams, J. F., and Kennedy, E. H., Vitamin B_{12} analogues in non-legume root nodules, *Nature (London)*, 207, 319, 1965.

96. Peive, Ya. V., Yagodin, B., Troitskaya, A., Compounds of B_{12} group of vitamins in nodules of leguminosae, *Field Crop Abstr.*, 23, 1371, 1970.

97. Troitskaya, G. N., Effect of conditions of soybean cultivation on vitamin B_{12} content in nodules, *Soviet Plant Physiol.*, 30, 127, 1983.

98. Fries, L., Vitamin B_{12} in Pisum sativum (L), *Physiol. Plant.*, 15, 566, 1962.

99. Long, A. G. and Wokes, F., Vitamins and minerals in plants, *Plant Foods Hum. Nutr.*, 1, 43, 1968.

100. Darken, M. A., Production of vitamin B_{12} by microorganisms and its occurrence in plant tissues, *Bot. Rev.*, 19(2), 99, 1953.

101. Gray, L. F. and Daniel, L. J., Studies of vitamin B_{12} in turnip greens, *J. Nutr.*, 67, 623, 1959.

102. Jathar, V. S., Deshpande, L. V., Kulkarni, P. R., Satoskar, R. S., and Rege, D. V., Vitamin B_{12}-like activity in leafy vegetables, *Indian J. Biochem. Biophys.*, 11, 71, 1974.

103. Petrosyan, A. P., Abramyan, L. A., and Sarkisyan, M. B., Dynamics of the accumulation of thiamine, biotin, and vitamin B_{12} in various parts of bean, I, *Vop. Mikorbiol.*, 4, 181, 1969, *Chem. Abstr.*, 73, 73880d, 1970.

104. Mozafar, A. and Oertli, J. J., Uptake and transport of thiamin (vitamin B_1) by the barley and soybean, *J. Plant Physiol.*, 139, 436, 1992.

105. Mozafar, A. and Oertli, J. J., Uptake of a microbially-produced vitamin (B_{12}) by soybean roots, *Plant Soil*, 139, 23, 1992.

106. Kononova, M. M., *Soil Organic Matter. Its Nature, its Role in Soil Formation and in Soil Fertility*, 2nd ed., Pergamon Press, London, 1966, 212.
107. Salunkhe, D. K., Pao, S. K., and Dull, G. G., Assessment of nutritive value, quality, and stability of cruciferous vegetables during storage and subsequent processing, *CRC Crit. Rev. Food Tech.*, 4, 1, 1973.
108. Fraiman, I. A., Comparative study of the vitamin composition of black currant varieties, *Sadovod., Vinograd. Vinodel. Mold.*, 31(5), 53, 1976; *Chem. Abstr.*, 85, 119559j, 1976.
109. Bender, A. E., *Food Processing and Nutrition*, Academic Press, London, 1978.
110. Sauberlich, H. E., Bioavailability of vitamins, *Prog. Food Sci. Nutr. Sci.*, 9, 1, 1985.
111. Kwiatkowska, C. A., Finglas, P. M., and Faulks, R. M., The vitamin content of retail vegetables in the UK, *J. Hum. Nutr. Diet.*, 2, 159, 1989.
112. Cotter, R. L., Macrae, E. A., Ferguson, A. R., McMath, K. L., and Brennan, C. J., A comparison of the ripening, storage and sensory qualities of seven cultivars of Kiwifruit, *J. Hortic. Sci.*, 66(3), 291, 1991.
113. Data, E. S. and Pantastico, E. B., Loss of ascorbic acid in fresh leafy vegetables subjected to light and different wind velocities, *Philippine Agriculturist*, 65(1), 75, 1982; *Food Sci. Technol. Abstr.*, 16(1), J118, 1984.
114. Akpapunam, M. A., Effect of wilting, blanching and storage temperatures on ascorbic acid and total carotenoids content of some Nigerian fresh vegetables, *Qual. Plant.–Plant Foods Hum. Nutr.*, 34, 177, 1984.
115. Lachance, P. A., Effects of preparation and service of food on nutrients, in *Nutritional Evaluation of Food Processing*, Harris, R. S. and Karmas, E., Eds., AVI Publishing, Westport, Conn., 1975, 463.
116. Bushway, R. J., Helper, P. R., King, J., Perkins, B., and Krishnan, M., Composition of ascorbic acid content of supermarkets versus roadside stand produce, *J. Food Qual.*, 12, 99, 1989.
117. Wu, Y., Perry, A. K., and Klein, B. P., Vitamin C and β-carotene in fresh and frozen green beans and broccoli in a simulated system, *J. Food Qual.*, 15, 87, 1992.
118. Yu, B. H. and Kies, C., Niacin, thiamin, and pantothenic acid bioavailability to humans from maize bran as affected by milling and particle size, *Plant Food Human Nutr.*, 43, 87, 1993.
119. Hoseney, R. C., Andrews, and Clark, H., Sorghum and pearl millet, in *Nutritional Quality of Cereal Grains*, Olson, R. A. and K. J. Frey, K. J., Eds., Agronomy 28, Am. Soc. Agronomy, Madison, Wisc., 1987, 397.
120. Widdowson, E. M., Extraction rates—Nutritional implications, in *Bread*, Spicer, A., Ed., Applied Science Publishers, London, 1975, 235.

Chapter 2

DISTRIBUTION OF VITAMINS IN DIFFERENT PLANT PARTS

I. LEAVES, ROOTS, AND FRUITS

Plant leaves, fruits, stems, and roots may strongly differ in their concentrations of vitamins (Tables 1 and 2).[1-15] Leaves of most vegetables may contain 2–6 times and leaves of spinach may even have 20 times higher ascorbic acid concentration than the stems.[16] Within each leaf, the leaf blade has higher ascorbic acid concentration than the leaf petiole[13,17,18] (Table 3). In broccoli, buds are much richer in ascorbic acid than the stalks.[16]

Higher concentrations of ascorbic acid and thiamin in the leaves than in the stems or roots presumably occur because these vitamins are both synthesized in the plant leaves.[19,20] The accumulation of thiamin in the cereal grains, for example, is believed to be brought about by the transport of this vitamin out of the leaves into the grains during the grains' maturity.[21,22]

In fruit plants, leaves may be higher in ascorbic acid content than the fruits. This has been noted in red raspberry[23] and tomato.[24] In black currant, although the leaves of Siberian cultivars were higher in ascorbic acid than their fruits, the reverse was true in the European cultivars.[25] No relationship could be found between the leaf and fruit content of ascorbic acid in the 19 apple and 10 pear cultivars tested.[26]

As compared with petiole, stem, and roots, leaf blade may also contain higher concentrations of other vitamins. This has been observed for carotene in sweet potato,[27] Swiss chard, spinach, and collards (Table 3); riboflavin and thiamin in tomato (Table 2); riboflavin in oats;[28] thiamin in parsley[29] and maize;[30] and tocopherol in spruce,[31,32] maize,[32] and alfalfa and lucerne.[33]

Information on the vitamin concentrations in the reproductive plant parts and on their possible role in the fertilization process is very limited, despite some reports that pollens are relatively rich in several vitamins. In maize, for example, pollen grains are higher in niacin and the tassels are higher in pantothenic acid content than the plant leaves (Table 4). Concentrations of vitamin C as high as 41–190 mg/100 g in the pollens of *Lilium candidum* and *Tulipa sp.*, respectively, and 402 mg/100 g in the spores of *Equisteum arvense* have been reported, which are much higher than what is usually found in most other plant tissues such as leaves.[35] In *Corylus avellana*, however, the vitamin C contents of pollens and green leaves were relatively similar, i.e., 272 and

TABLE 1
Concentration and Distribution of Vitamin C in Different Parts of a 17-Week-Old Potato Plant (Variety Doon Star)[14]

Plant part	Weight (g)	Vitamin C (mg/100 g)	Vitamin C (mg/part)[a]
Leaves	715	109	776
Stems	296	12	35
Old tubers	122	5	6
Roots	14	8	1
Stolon	15	9	1
New tubers	1045	39	407

[a] Calculated from the first two columns.

TABLE 2
Distribution of Vitamins (μg/g DW) in Different Parts of Tomato (Variety Bonny Best)[15]

Part	Riboflavin	Thiamin	Niacin
Fruit	6.8	12.0	104.7
Leaves	11.4	6.0	60.8
Roots	3.8	2.4	79.0

TABLE 3
Concentration of Carotene (CAR) (ppm DW) and Ascorbic Acid (AA) (mg/100 g DW) in Leaf Blade and Petiole of Different Vegetables[18]

Part	Swiss chard CAR	Swiss chard AA	Spinach CAR	Spinach AA	Collards CAR	Collards AA
Blade	516	425	650	720	511	796
Petiole	26	86	117	200	26	612

298 mg/100 g FW, respectively. Since plant leaves may contain some dead cells, it was concluded that the living cells of plant leaves may have a much higher concentration of vitamin C than that measured in the pollen cells.[35] Pollens of *Pinus massoniana* were noted to contain relatively high concentrations of vitamins A, B_1, B_2, B_6, C, nicotinic acid, folate, and pantothenic acid.[36]

TABLE 4
Concentration (μg / g DW) of Niacin and Pantothenic Acid in Different Tissues of Maize (Variety Ohio Gold)[34]

Part	Niacin	Pantothenic acid
Leaf and stalk	41[a]	10
Kernel	94	40
Tassel	35	18
Pollen	81	8

[a] Values rounded.

Whether vitamins play any role in the flowering and fertilization process is not clear. In *Citrus aurantium*, for example, the ascorbic acid content in the stigma and style was found to be higher during the prepollination stages and to decline following pollination.[37] This, along with the reports that (a) spraying of *Cucumis sativus* with ascorbic acid solution increased the production of male flower and at times changed the ratio of male/female flowers, and (b) spraying of biennial-bearing mango (cv. Langra) increased the flowering of the mango trees,[38] indicates that ascorbic acid may play some role in the reproduction process in some plants. Further research, however, is needed before any definite conclusion can be drawn.

II. LONGITUDINAL GRADIENTS

Vitamin concentrations in plants may also show a gradient along the longitudinal axis connecting the fruit or plant leaf to the mother plant; the differences, however, are relatively small. For example, the concentration of ascorbic acid was noted to be slightly higher in the stem halves than in the stylar halves of citrus fruits[39] and in the stem-end than the calyx-end of pears.[40] In pineapple[41] the concentration of ascorbic acid is progressively higher in the top, middle, and bottom sections of the fruit. In pepper, however, ascorbic acid is higher in the tip than in the base where the pepper is connected to the mother plant.[42]

In potato tuber, the concentrations of ascorbic acid[43–46] (Table 5) and thiamin[45, 47] are higher in the apical (rose, eye, or apex) end than in the basal (heel, stem, or attachment) end of the tuber. In sweet potato (anatomically a root tuber and not a stem tuber like potato), however, the tuber-end close to the mother plant was found to contain higher concentration of ascorbic acid[48] and 2–3 times higher concentration of carotene[48, 49] than the end close to the feeding roots. The concentration

TABLE 5
Concentration of Ascorbic Acid in Different Sections of Potato Tuber from the Rose (Flower End) to the Heel (Stem End)

Section	Ascorbic Acid (mg / 100 g)
1 (Heel)	4.3
2	5.5
3	6.8
4	7.1
5	7.6
6 (Rose)	8.1

From Smith, A. M. and Gillies, J., *Biochem. J.*, 34, 1312, 1940. With permission.

of α-tocopherol in sweet potato roots was found to be in the following order: center > upper part > lower part. The differences were, however, not significant.[27]

In carrot roots, a large vertical gradient exists in the ascorbic acid concentration—the highest being in the lower parts where cell division is more active and in the upper parts closest to the leaves; in the middle part, however, only a small amount of ascorbic acid is present.[50] In contrast, the carotene content of carrots was reported to be higher in the part located near the soil surface than in the bottom region.[51, 52] Higher concentration of ascorbic acid in the top-end (the part closer to the mother plant) than in the tip-end has been noted in parsnip[53] and in kohlrabi, celeriac, and radish.[54]

In asparagus[55, 56] and in bamboo shoots,[57] the concentration of ascorbic acid is higher in the stem tip than in the stem end (Table 6), and in cucumber,[58] the concentrations of riboflavin and niacin are noted to be higher in the flower end than in the stem end (Table 7).

Gradient in the vitamin concentration has also been observed along plant leaves. In monocotyledonous plants, the concentration of ascorbic acid decreases from the apex to the base of the leaves.[19] In the leaves of *Plantago lanceolata* (plantain) the following tocopherol concentrations were found in the 2.5-cm leaf intervals from the base to the leaf tip: 12, 17, 25, 40, 50, 25 ppm.[59] Even within a single needle of spruce, ascorbic acid concentration was noted to be slightly higher in the needle tip than in the needle base.[60]

By considering the fact that even some inorganic elements such as K, Ca, Mg, Mn, Fe, and B show gradients in their concentration along the plant leaves, with higher concentration being at the leaf tips,[61, 62] it may

TABLE 6

Distribution of Ascorbic Acid in the Tip (1 cm), Middle (4 cm), and Base (3–5 cm) Stem Sections of White and Blue Varieties of Asparagus[55]

Section	Ascorbic acid (mg / 100 g FW)	
	White	Blue
Tip	56	86
Middle	24	30
Base	16	17

TABLE 7

Concentration (μg / 100 g FW) of Riboflavin, Nicotinic Acid, Pantothenic Acid, and Biotin in Different Parts of Cucumber Fruit Grown in Sand Culture[58]

Cucumber part	Riboflavin	Niacin	Pantothenic acid	Biotin
Stem end with skin	7	100	200	0.7
Flower end with skin	13	140	190	1.3
Skin alone	18	180	350	1.8

be argued that the gradients observed in the concentration of different vitamins in different plant organs, such as in the leaves, fruits, stem-, or root-tubers reported above, may not be unique in this sense and may be a general phenomenon related to the plant anatomy and the network of phloem and xylem branches and their tributaries in different parts of the same organ.

III. PEEL VERSUS FLESH OF FRUITS: OUTER VERSUS INNER LEAVES OF VEGETABLES

A. Ascorbic Acid

Plant tissues (parts) located on the periphery (outside or more exposed positions) on plants usually contain higher concentrations of several vitamins, especially ascorbic acid (Table 8). This is not only true for leafy vegetables such as cabbage and lettuce, but also holds true for many different fruits and nonleafy vegetables and for vitamins other than ascorbic acid. In cucumber, for example, the skin not only contains a higher concentration of ascorbic acid (Table 8), it also contains higher

TABLE 8
Relative Concentration of Vitamins in Outer Versus Inner Tissues
(Skin or Peel vs. Flesh or Outer Leaves vs. Inner Leaves)
of Fruits and Vegetables

Plant	Difference	Ref.
	Ascorbic acid	
Apples	Peel > flesh	63–69
Asparagus	Peel > flesh	55
Cabbage	Outer leaves > inner leaves	18, 66, 71–75
Carrots	Peel > flesh	63
Carrots	Cortex > pith	75
Celeriac	Peel > pith	54
Cucumber	Peel > flesh	76
Cucumber	Cortex > flesh	77
Gooseberry	Peel > inside	78
Gourd	Peel > flesh	63
Guava	Peel > flesh	63
Guava	Pericarp > core	79
Kohlrabi	Peel > flesh	54, 63
Lettuce	Outer leaves > inner leaves	16, 80
Lettuce	Top parts > lower parts	77
Mango	Peel > flesh	81–85
Melon (Prince)	Cortex > flesh	78
Papaya	Skin > flesh	66
Passion fruit	Peel > flesh	86
Peaches	Peel > flesh	87
Pear	Peel > flesh	63
Pear	Peel > cortex	66
Pepper (sweet)	Pericarp > placenta	88, 89
Persimmon	Peel > flesh	90
Pimento	Pulp > placenta	91
Plum	Skin > flesh	92
Potato	Peel > flesh	16, 63, 64
Pumpkin	Peel > flesh	63
Radish	Peel > flesh	63
Spinach	Outer leaves > inner leaves	16
Strawberry	Outer part > inner part	79, 93
Strawberry	Pericarp > core	91
Strawberry	Cortex > medulla	66
Tamarind	Peel > flesh	63
Tomato	Peel > flesh	63, 94, 96
Tomato	Outer pericarp > inner pericarp	95
Tomato	Locular > pericarp	97
Turnip	Peel > flesh	63
Carrots	Peel = flesh	51, 98
Mango	Peel = whole fruit	99
Kiwifruit	Outer pericarp < inner pericarp	100
Okra	Flesh < placenta	78

TABLE 8   (continued)

Plant	Difference	Ref.
Pineapple	Near shell < near core	66
Potato	Cortex < pith	101–104
Potato	Peel < flesh	105
Spinach	Outer leaves < inner leaves	13
Tomato	Sacrocarp < placenta	91
Turnip root	Outer parts < inner parts	77

Biotin

Apples	Peel > flesh	106
Cucumber	Peel > flesh	58
Pear	Peel > flesh	106

Carotene

Cabbage	Outer leaves > inner leaves	18, 66, 107
Cabbage (chinese)	Outer leaves > inner leaves	108
Carrots	Phloem > xylem	51, 52, 98, 109–111
Cucurbita	Peel > pulp	112
Lettuce	Outer leaves > inner leaves	66, 113
Lettuce (iceberg)	Outer leaves > inner leaves	114

Folic acid

Mandarin	Peel > pulp	120
Potato	Peel > flesh	105

Niacin

Apples	Peel > flesh	66
Carrots	Peel > flesh	16, 115
Cucumber	Peel > flesh	58
Pear	Peel > flesh	66
Potato	Peel > flesh	66
Potato	Flesh > peel	105

Pantothenic acid

Cucumber	Peel > flesh	58

Pyridoxine

Potato	Peel = flesh	105

TABLE 8   (continued)

Plant	Difference	Ref.
Riboflavin		
Apples	Peel > flesh	66
Cabbage	Outer leaves > inner leaves	70
Carrots	Peel > flesh	16, 115
Cucumber	Peel > flesh	58
Pear	Peel > flesh	66
Potato	Peel > flesh	105
Tomato	Peel > flesh	66
Cabbage	Outer leaves = inner leaves	66
Carrots	Peel = flesh	51
Carrots	Phloem = xylem	98
Lettuce	Outer leaves = inner leaves	66
Sweet potato	Skin = flesh	116
Asparagus	Skin < flesh	117
Thiamin		
Cabbage	Outer leaves > inner leaves	70
Carrots	Peel > phloem	16, 51, 115
Sweet potato	Peel > flesh	116
Lettuce	Outer leaves = inner leaves	66
Cabbage	Outer leaves < heart	66
Pineapple	Near shell < core	60
Potato	Peel < flesh	47, 105
Tocopherol		
Apples	Peel > flesh	118
Cabbage	Outer leaves > inner leaves	118, 119
Cucumber	Peel > flesh	118
Lettuce	Outer leaves > inner leaves	118
Vitamin K$_1$		
Cabbage	Outer leaves > inner leaves	121

concentrations of riboflavin, niacin, pantothenic acid, biotin, and tocopherol as compared with the inner parts of the fruit (Table 7).

In the green plant tissues, ascorbic acid is highly concentrated in the chloroplast.[122] It has been estimated that 35–40% of the cell's total ascorbic acid may be present in its chloroplasts,[123] where its concentration may range from 13.5 mM[124] to 50 mM.[123] Since light intensity, to

TABLE 9
Ascorbic Acid in Different Parts of a Single Tomato Fruit
(Variety Fireball)

Fruit Part	Fresh weight (% total)	Ascorbic acid (mg / 100 g)	% Total
Epidermis (skin)	4	42[a]	6
Locule (juice and seeds)	23	28	23
Outer carpel wall	41	30	44
Inner carpel wall	32	32	27

[a] Values rounded.

From Ward, G. M., *Can. J. Plant Sci.*, 43, 206, 1963.

TABLE 10
Ascorbic Acid Concentration in Different Parts
of Pepper Fruit[89]

Fruit part	Ascorbic acid (mg / 100 g)
External portion adjacent to epidermis	102[a]
Pericarp	79
Placenta and seeds	17
Stem end of fruit	84
Central portion	92
Blossom end	102

[a] Values rounded.

which a given plant tissue is exposed, strongly affects the concentration of ascorbic acid in that tissue (see chapter 4), it is thus not surprising that in lettuce (Figure 9 in chapter 5), tomato (Table 9), pepper (Table 10), and apple (Table 11) the outer tissues, which are more exposed to light, also contain more ascorbic acid than the inner parts. This argument, however, does not hold true for the higher concentrations of carotenes, niacin, riboflavin, and thiamin in the outer tissues (peels) of underground plant parts, such as in carrots (Tables 8 and 12) (which are not normally exposed to much light). Also, no explanation could be given for the differences observed in the concentration of several other vitamins in different parts of other plants (Table 8) or the lower concentration of thiamin in potato peel as compared to its flesh (Table 13).

In citrus fruits the concentration of ascorbic acid in the peel is several times higher than that in the pulps or in the fruit juice, so much

TABLE 11
Concentration and Relative Distribution of Ascorbic Acid in Different Parts of Several Apple Varieties[65]

Variety	In the skin (mg / 100 g)	In the skin (% total[a])	In the flesh (mg / 100 g)	In the flesh (% total[a])	Average in the whole fruit (mg / 100 g)
Welschbrunner	34.3	56.4	1.1	43.6	2.43
Kronprinz Rudolf	26.6	39.4	1.7	60.6	2.69
Schöner aus Boskoop	22.0	14.3	5.5	85.7	6.16
London Pepping	61.8	21.1	9.6	78.9	11.68

[a] Percentage distribution is calculated based on the assumption that flesh and skin make up 96% and 4% of fruits weight, respectively.

TABLE 12
Concentration of Carotenes (mg / 100 g FW) and Thiamin (μg / 100 g FW) in Different Parts of Carrot Root[51]

Part	α-Carotene	β-Carotene	Thiamin
Peeling	1.7	5.0	120
Phloem (cortex)	2.2	7.7	50
Xylem	1.1	3.1	30

TABLE 13
Concentration of Thiamin (μg / 100 g FW) in the Skin Layer (Under the Corky Skin) and the Central Parts of Potato Varieties Majestic and King Edwards[47]

Tuber part	Majestic	King Edward
Skin layer	10	18
Tuber center	76	49

so that some 3/4 of the total ascorbic acid content of the fruit is located in its peel, which is normally discarded.[39, 125–129]

Higher concentrations of vitamins in the outer plant tissues are only of academic interest in cases where fruits or vegetables (such as strawberry, tomato, etc.) are usually consumed without peeling, or for those fruits such as citrus and mango, where peeling before consumption is inevitable. In other cases, however, such as milling of cereals, peeling or trimming of vegetables (e.g., lettuce, cabbage, cucumber,

TABLE 14
Concentration (μg / 100 g DW) of Vitamins in Different Fractions
of Four Sorghum Varieties[141]

Variety	Niacin	Pantothenic acid	Riboflavin	Biotin	Pyridoxine
		Endosperm			
Westland	4210	1090	110	11	470
Midland	4460	780	130	10	430
Cody	7050	950	90	11	370
Martin	2690	869	50	11	380
		Germ			
Westland	10550	5110	380	53	820
Midland	6950	2590	430	70	760
Cody	10830	2820	380	54	720
Martin	4980	2870	400	60	650
		Bran			
Westland	4340	1200	360	30	490
Midland	5380	840	480	37	510
Cody	6260	1130	470	34	400
Martin	2740	930	280	31	390

etc.) and apple, removal of outer plant parts prior to marketing or consumption is mainly a matter of marketing or cosmetic consideration. In these cases, a substantial loss of plant vitamins could be avoided if plants were consumed whole.

In apples, ascorbic acid is highly concentrated in and near the peel more than in pulp and the areas close to the core (pith).[130] Varieties may, however, strongly differ in this respect so that the ascorbic acid concentration in the apple peel may be 2–5 times[68, 106, 131] or even 31 times higher than in the apple flesh (34.1 vs. 1.1 mg/100 g in skin and flesh, respectively, in Welschbrunner) (Table 11). In the apple variety Oldenburg, the peel was noted to be 58 times higher in ascorbic acid concentration than the flesh close to the seeds (58 vs. 1 mg/100 g FW, respectively).[67] Thus the total amount of the apple's ascorbic acid located just in its peel may strongly differ in different apple varieties (Table 11). Therefore, depending on the size of the apple and the relative contribution of peel and flesh to the total weight of the apple (which may also differ considerably in different apple varieties),

TABLE 15
Distribution of Vitamins in Wheat Kernel

Fraction	Thiamin	Nicotinic acid	Riboflavin	Pantothenic acid
	(% in the whole kernel)			
Pericarp + testa	1	4	5	8
Aleurone	31	84	37	39
Embryo	2	1	12	4
Scutellum	63	1	14	4
Endosperm	3	12	32	41
	(μg / 100 g)			
Pericarp + testa	60	2570	100	780
Aleurone	1650	74100	1000	4510
Embryo	840	3850	1380	1710
Scutellum	15600	3820	1270	1410
Endosperm	13	850	70	390

From Hinton, J. J. C., Peers, F. G., and Shaw, B., *Nature* (*Lond.*), 172, 993, 1953. With permission.

8–17%,[16] 16–20%,[132] 33%,[133] or even 56% (in variety Welschbrunner) (Table 11) of the total ascorbic acid in apples may be just in their peel. In other words, peeling of apple may remove up to 50% of its total vitamin C content.[16, 134]

In potato, relative distribution of vitamin in different parts of the tuber is a matter of controversy. For example, ascorbic acid concentration in potato peel (cortex) is reported to be higher,[135] lower,[64, 102] or equal[14] to that of the rest (meat) of the tuber. It is probable that these apparent discrepancies may be due to the difference in the thickness of tissue sampled as "peel" by different authors. Loss of ascorbic acid by peeling of potato is estimated to range from 12 to 35%.[16]

B. Other Vitamins

Discarded (outer) leaves (parts) of vegetables may be much higher in carotene than the inner parts usually consumed. For example, outer leaves of cabbage and broccoli are reported to be 4–21[18, 136] or even 200 times[107] higher in carotene than the inner parts consumed. This, however, does not hold true in all cases. In asparagus, for example, the concentrations of ascorbic acid and riboflavin are 2–3 times higher in the edible portion than in the nonedible portion.[117]

In the cereals, vitamins are not evenly distributed in different parts of the kernel. Furthermore, varieties may differ in the relative concentration of various vitamins in different parts of their kernels (Table 14). In wheat kernels, for example, thiamin and niacin are highly concentrated in the scutellum or aleuron layer, respectively. Riboflavin and pantothenic acid, however, are distributed relatively more uniform in different parts of the kernel[137-139] (Table 15). In maize, nicotinic acid is highly concentrated in the aleurone layer and, in contrast to wheat, scutellum tissue contains a considerably higher percentage of the total seed vitamin.[138] One consequence of these differences in vitamin distribution between different parts of various cereal seeds is the differential loss of these vitamins during milling, as discussed by Pomeranz.[140]

Thiamin concentration was also noted to be higher (by twofold) in the tissues 2–3 mm under the skin of sweet potato roots than in the central part of the root, and in the skin of giant taro (*Alocasia macrorrhiza*) stem than in the stem center.[116]

Concentration of vitamin K_1 was noted to be higher by threefold in the outer than the inner leaves of cabbage grown in Montreal (Canada) (719 vs. 228 μg/100 g), and by sixfold in the same cabbage grown in Boston (U.S.A.) (449 vs. 72 μg/100 g).[121]

IV. SEEDS AND SPROUTS

Plant seeds contain a considerable amount of some (but not all) vitamins that are believed to be necessary for their viability and the germination process. During seed germination, some of the seed's original vitamins are mobilized. At the same time, a significant increase takes place in the total amount and the concentration of some vitamins in the young seedlings (sprouts), indicating that some vitamins may be synthesized during the germination process. Sprouts are thus believed to have an exceptionally high nutritional value and have become a very popular item in the daily diet of some countries.[142] Here, we will review the effect of seed germination on the content of vitamins in the sprout.

Effect of seed germination on the vitamin content of the emerging seedling or sprout has been observed by numerous investigators.[143-184] Increases of up to 29- to 86-fold in the ascorbic acid content of legumes (dry yellow peas, lentils, and faba beans) and a doubling of riboflavin in peas during a 4-day germination period have been documented.[185] Table 16 summarizes the changes taking place in the content of vitamins in different seeds as they germinate.

Environmental conditions such as temperature and light may affect vitamin synthesis during seed germination. Their effects, however, are strongly dependent on the kind of seed and vitamin under considera-

TABLE 16
Changes in the Vitamin Contents in the Seeds of Various Plants as They Germinate and Produce Sprouts[a]

Seed	Ascorbic	Biotin	Niacin	Pantothenic acid	Riboflavin	Thiamin
Alfalfa	+ [b]		+ +		+ +	0
Barley		+	+ +		+	+
Corn		+		+ +		+
Lentil	+ +		+ +		0	
Mungbean	+ +	+	+ +	+	+ +	0
Oats		+	+ +		+ +	0
Peas	+		+ +	+ +	+	0
Soybean	+		+	0	+	0
Wheat		+	+		+ +	+

[a] Compiled from various sources.
[b] 0 = no change; + = slight increase; + + = pronounced increase.

tion. For example, light was reported to have no effect on the concentration of thiamin, riboflavin, and nicotinic acid in several sprouts, especially if applied during the earlier periods of seed germination. The presence of light during seed germination, however, was noted to reduce the concentration of nicotinic acid in *Phaseolus mungo*, *Lens esculenta*, and *Vigna catiang*.[186] In barley sprouts, however, exposure of seeds to light during germination decreased the ascorbic acid and riboflavin in the sprouts produced.[161]

In soybean, although exposure to natural light was noted to decrease the ascorbic acid content of sprouts,[187] irradiation of sprouts with blue light (120 lux, 3 h/day) was found to increase vitamin C content by 31% as compared with sprouts grown in the dark.[188] Exposure of seeds to light was also noted to increase the provitamin A activity in mung bean sprouts[189] and the β-carotene concentration in the alfalfa sprouts.[181]

Not all vitamins increase in seeds as they germinate; the concentrations of some may not change or may even decrease.[169, 172, 179] For example, germination of *Phaseolus aureus*, *Pisum sativum*, and several varieties of soybean reduced the concentration of folic acid, did not affect that of thiamin, but slightly increased the concentration of pyridoxin, pantothenic acid, and biotin.[146] Nandi[186] reported that in *Phaseolus mungo*, *Lens esculenta*, *Vigna catiang*, and *Cicer arietinum*, the concentrations of thiamin, riboflavin, and nicotinic acid increased in all cases; the concentration of thiamin in *Vigna catiang* and *Cicer arietinum* and that of folic acid in *Phaseolus mungo* and *Lens esculenta*, however, decreased during seed germination. Also, in black-eyed peas,

Lima beans, and cottonseed the concentrations of nicotinic acid, riboflavin, pantothenic acid, pyridoxine, biotin, and inositol all increased, but the content of thiamin did not show a consistent rise in the seeds tested.[190] The concentration of niacin was noted to increase by 57–108% in oats and by 52% in rice but did not increase significantly in wheat, barley, and maize during seed germination.[191] These observations indicate that even the concentrations of a given vitamin may change differently in different seeds as they germinate.

Different varieties of the same plant may also have different amounts of vitamins in their sprouts. For example, Fordham et al.[192] noted a large variability in the ascorbic acid, tocopherol, and carotene content of sprouts of different varieties of peas and beans so much so that a 100-g serving of pea sprouts could deliver from 18.6 to 50.0 mg of ascorbic acid, and a 100-g portion of bean sprouts could deliver from 12.6 to 42.2 mg of ascorbic acid. Despite this large variability, the amount of vitamin C in sprouts is indeed considerable if one considers that the Recommended Dietary Allowance (RDA) for ascorbic acid is presently set at 60–100 mg/day.[193]

Thiamin concentrations in sprouts may range from 0.19 to 0.30 mg in pea sprouts and from 0.12 to 0.48 mg in bean sprouts; riboflavin may range from 0.11 to 0.38 mg in pea sprouts and from 0.11 to 0.29 mg in bean sprouts per each 100-g serving.[192] By considering the 1989 RDA for adult men for thiamin and riboflavin (1.5 and 1.7 mg/day, respectively),[193] it can be seen that a 100-g serving of these sprouts could contribute considerably to satisfying the needs for these vitamins.

REFERENCES

1. Carroll, G. H., The role of ascorbic acid in plant nutrition, *Bot. Rev.*, 9, 41, 1943.
2. Rodahl, K., Content of vitamin C (L-ascorbic acid) in arctic plants, *Trans. Bot. Soc. Edinburgh.*, 34, 205, 1944.
3. Gustafson, F. G., Distribution of thiamin and riboflavin in the tomato plant, *Plant Physiol.*, 22, 620, 1947.
4. Seybold, A. and Mehner, H., Über den Gehalt von Vitamin C in Pflanzen, Springer Verlag, Heidelberg, 1948.
5. Agarwala, S. C. and Hewitt, E. J., Molybdenum as a plant nutrient. IV. The interrelationships of molybdenum and nitrate supply in chlorophyll and ascorbic acid fractions in cauliflower plants grown in sand culture, *J. Hortic. Sci.*, 29, 291, 1954.
6. Hilbert, Tr., Betrachtungen über die Gladiole als Vitamin C-Trägerin und ihre praktische Verwertung, *Nahrung*, 1, 57, 1957.
7. Hoffman, I., Nowosad, F. S., and Cody, W. J., Ascorbic acid and carotene values of native eastern arctic plants, *Can. J. Bot.*, 45, 1859, 1967.
8. Achtzehn, M. K. and Hawat, H., Einfluss industrieller Vorbehandlung auf den Gehalt essentieller und nichtessentieller Inhaltsstoffe im Spinat, *Nahrung*, 15(5), 527, 1971.

9. Marciulionis, V., Biological and biochemical characteristics of promising silage plants. 2. Changes in the chemical composition of *Polygonum weyrichi* green mass in relation to growth and development of the plant, *Liet. TSR Mokslu Akad. Darb., Ser. C*, 3, 41, 1972; *Chem. Abstr.*, 78, 55411a, 1973.

10. Upadhyaya, P. P., Influence of radish mosaic virus infection on ascorbic acid content of radish (*Raphanus sativus*), *Curr. Sci.*, 46(18), 647, 1977.

11. Schneider, V., Über den Vitamin C-Gehalt von wildgemüsen und wildsalaten unter berücksichtigung jahreszeitlicher veränderungen und der unterschiedlichen Verteilung in den Pflanzenorganen, Ph. D. Thesis, Rheinischen Friedrich-Wilhelm University, Bonn, Germany, 1982.

12. Singh, V., Mathur, K., Sethia M., Bhojak, S., and Nag, T. N., Ascorbic acid content from some arid zone plants of Rajasthan, *Geobios*, 17, 35, 1990.

13. Hisaka, H., Ogura, N., Relationship of changes of components to that of appearance quality of spinach. II. Changes in the ascorbic acid contents at various parts od spinach leaves during storage, *Nippon Shokuhin Kogyo Gakkaishi*, 38(1), 41, 1991; *Chem. Abstr.*, 115, 27945s, 1991.

14. Lampitt, L. H., Baker, L. C., and Parkinson, T. L., Vitamin-C content of potatoes. I. Distribution in the potato plant, *J. Soc. Chem. Industry (Lond.)*, 64, 18, 1945.

15. Wilson, K. S. and Withner, C. L., Jr., Stock-scion relationships in tomatoes, *Am. J. Bot.*, 33, 796, 1946.

16. Lachance, P. A., Effects of preparation and service of food on nutrients, in *Nutritional Evaluation of Food Processing*, Harris R. S. and Karmas, E., Eds. AVI Publishing, Westport, Conn., 1975, 463.

17. Ijdo, J. B. H., The influence of fertilizers on the carotene and vitamin C content of plants, *Biochem J.*, 30, 2307, 1936.

18. Sheets, O, Leonard, O. A., and Gieger, M., Distribution of minerals and vitamins in different parts of leafy vegetables, *Food Res.*, 6, 553, 1941.

19. Åberg, B., Ascorbic acid, in *Handbuch der Pflanzenphysiologie*, Vol. 6, Ruhland, W., Ed., Springer Verlag, Berlin, 1958, 479.

20. Åberg, B., Vitamins as growth factors in higher plants, in *Handbuch der Pflanzenphysiologie*, Vol. 14, Ruhland, W., Ed., Springer Verlag, Berlin, 1961, 418.

21. Geddes, W. F. and Levine, M. N., The distribution of thiamin in the wheat plant at successive stages of kernel development, *Cereal Chem.*, 19, 547, 1942.

22. Kondo, H., Mitsuda, H., and Iwai, K., Thiamine synthesis in leaves of cereal crops, *J. Agr. Chem. Soc. Japan*, 24, 128, 1950–51; *Chem. Abstr.*, 46, 10305i, 1952.

23. Fejer, S. O., Johnston, F. B., Spangelo, L. P. S., and Hammill, M. M., Ascorbic acid in red raspberry fruit and leaves, *Can. J. Plant Sci.*, 50, 457, 1970.

24. Ananyan, A. A., Tarosova, E. O., Egiazaryan, A. G., and Avetisyan, S. A., Changes in vitamin C and carotene accumulation in tomato leaves and fruit, *Biol. Zh. Arm.*, 28(3), 93, 1975; *Hortic. Abstr.*, 46, 4721, 1976.

25. Gritsishin, Ī. S. and Snizhko, V. L., The vitamin C and sugar content in black currant berries and leaves, *Nauk. Pratsi Ukr. Sil'skogospod. Akad.*, 62, 109, 1973.; *Hortic. Abstr.*, 45, 7183, 1975.

26. Karamysheva, V. I. and Totubalina, G., Vitamins in apples and pears, *Sadovodstvo*, 4, 32, 1973; *Hortic. Abstr.*, 43, 7416, 1973.

27. Woolfe, J. A., *Sweet Potato*, Cambridge University Press, Cambridge, 1992.

28. Watson, S. A. and Noggle, G. R., Effect of mineral deficiencies upon the synthesis of riboflavin and ascorbic acid by the oat plant, *Plant Physiol.*, 22, 228, 1947.

29. Chaikelis, A. S., The thiamine content of herbs and medicinal plants, *J. Am. Pharm. Assoc.*, 35, 343, 1946.

30. Burkholder, P. R. and McVeigh, I., Studies on thiamin in green plants with the phycomyces assay method, *Am. J. Bot.*, 27, 853, 1940.

31. Franzen, J., Bausch, J., Glatzel, D., and Wagner, E., Distribution of vitamin E in spruce seedling and mature tree organs, and within the genus, *Phytochemistry*, 30(1), 147, 1991.
32. Franzen, J. and Haass, M. M., Vitamin E content during development of some seedlings, *Phytochemistry*, 30(9), 2911, 1991.
33. Khan, N. and Elahi, M., Studies on pigments and vitamin E at different stages of growth of some leguminose plants, *Pakistan J. Sci. Res.*, 20(4–5), 282, 1977.
34. Hunt, C. H., Rodrihuez, L. D., and Bethke, R. M., The effect of maturity on the niacin and pantothenic acid content of the stalks and leaves, tassels, and grain of four sweet corn varieties, *Cereal Chem.*, 27, 157, 1950.
35. Seybold, A. and Mehner, H., Über den Gehalt von Vitamin C in Pflanzen, Springer Verlag, Heidelberg, 1948.
36. He, Y., Development and use of *Pinus massoniana* pollens in the manufacture of carbonated beverages, *Shipin Kexue (Beijing)*, 84, 21, 1986; *Chem. Abstr.*, 106, 155045z, 1987.
37. Bhatia, D. S., Sidhu, T., Chander, P., and Malik, C. P., Post-pollination changes in enzymes and reserve metabolites in the stigma and style of *Citrus aurantium* L., *J. Plant Sci. Res.*, 21(1–4), 75, 1986; *Chem. Abstr.*, 107, 172575k, 1987.
38. Bauernfeind, J. C., Ascorbic acid technology in agriculture, pharmaceutical, food, and industrial applications, in *Ascorbic Acid: Chemistry, Metabolism, and Uses.*, Seib P. A., and Tolbert, B. M., Eds., Adv. Chem. Series, no. 200, Am. Chem. Soc., Washington D.C., 1982, 395.
39. Ting, S. V., Distribution of soluble components and quality factors in the edible portion of citrus fruits, *J. Am. Soc. Hortic. Sci.*, 94, 515, 1969.
40. Bal, J. S., Singh, S., and Sandhu, S. S., Quantitative variations in some metabolites in the different parts of two varieties of pear fruits, *J. Food Sci. Technol.*, 24(3), 144, 1987.
41. Miller, E. V. and Hall, G. D., Distribution of total soluble solids, ascorbic acid, total acid, and Bromelin activity in the fruit of the Natal pineapple (*Ananas comosus* L. Merr.), *Plant Physiol.*, 28, 532, 1953.
42. Simon, J., Ertrag und Vitamin C-Gehalt bei Paprika, *Bodenkultur*, 11, 208, 1960.
43. Smith, A. M. and Gillies, J., The distribution and concentration of ascorbic acid in the potato (*Solanum tuberosum*), *Biochem. J.*, 34, 1312, 1940.
44. Streighthoff, F., Munsell, H. E., Ben-dor, B. A., Orr, M. L., Cailleau, R., Leonard, M. H., Ezekiel, S. R., Kornblum, R., and Koch, F. G., Effect of large-scale methods of preparation and the vitamin content of food. I. Potatoes, *J. Am. Diet. Assoc.*, 22, 117, 1946.
45. Baird, E. A. and Howatt, J. L., Ascorbic acid in potatoes grown in New Brunswick, *Can. J. Res.*, 26C(4), 433, 1948.
46. Shekhar, V. C., Iritani, W. M., and Arteca, R., Changes in ascorbic acid content during growth and short-term storage of potato tubers (*Solanum tuberosum* L.), *Am. Potato J.*, 55, 663, 1978.
47. Meiklejohn, J., The Vitamin B_1 content of potatoes, *Biochem. J.*, 37, 349, 1943.
48. Speirs, M., Cochran, H. L., Peterson, W. J., Sherwood, F. W., and Weaver, J. G., The effects of fertilizer treatments, curing, storage, and cooking on the carotene and ascorbic acid content of sweetpotatoes, *South. Coop. Series Bull.*, 3, 5, 1945.
49. Ezell, B. D. and Wilcox, M. S., The ratio of carotene to carotenoid pigments in sweet-potato varieties, *Science*, 103, 193, 1946.
50. Kitagawa, Y., Distribution of vitamin C in the roots of root crops, *J. Jpn. Soc. Food Nutr.*, 24(5), 292, 1971; *Food Sci. Technol. Abstr.*, 4(6), J912, 1972.
51. Yamaguchi, M., Robinson, B., and MacGillivray, J. H., Some horticultural aspects of the food value of carrots, *Proc. Am. Soc. Hortic. Sci.*, 60, 351, 1952.

52. Michalik, H. and Bakowski, J., The content and distribution of carotenoids in carrots in relation to harvest time, *Biuletyn Warzywniczy*, 20, 417, 1977; *Hortic. Abstr.*, 49, 549, 1979.

53. Mayfield, H. L. and Richardson, J. E., Ascorbic acid content of parsnip, *Food Res.*, 5, 361, 1940.

54. Kröner, W. and Völksen, W., Über die Verteilung der Ascorbinsäure in einigen pflanzlichen Speicherorganen, *Biochem. Z.*, 314, 27, 1943.

55. Wolf, J., Untersuchungen an Spargel. I. Ascorbinsäure, *Gartenbauwissenschaft*, 15, 109, 1941.

56. Makus, D. J. and Gonzalez, A. R., Production of white asparagus using plastic row covers, in *Proc. Ann. Meetings—Arkansas St. Hortic. Soc.*, no. 110, Fayetteville, AR, 1989, 136; *Hortic. Abstr.*, 61, 8001, 1991.

57. Kitagawa, Y., Distribution of vitamin C in some stalk vegetables (brackens and bamboo shoots), *J. Jpn. Soc. Food Nutr.*, 24(8), 449, 1971; *Food Sci. Technol. Abstr.*, 5(8), J1158, 1973.

58. Haenel, H., Mikrobiologische Untersuchungen über den Gehalt von Faktoren des B-Komplexes in Lebens- und Futtermitteln, *Ernährungsforschung*, 1, 533, 1956.

59. Booth, V. H. and Hobson-Frohock, A., The α-tocopherol content of leaves as affected by growth rate, *J. Sci. Food Agric.*, 12, 251, 1961.

60. Osswald, W. F. and Elstner, E. F., Comparative studies on spruce diseases in sulfur dioxide exposed forest areas and unpolluted areas, *Spez. Ber. Kernforschungsanlage Juelich*, Juel-Spez-369, 54, 1986; *Chem. Abstr.*, 106, 143247d, 1987.

61. Nelson, P. V. and Boodley, J. W., Selection of a sampling area for tissue analysis of carnation, *Proc. Am. Soc. Hortic. Sci.*, 83, 745, 1963.

62. Benton Jones, J. Jr., Distribution of fifteen elements in corn leaves, *Soil Sci. Plant Analy.*, 1, 27, 1970.

63. Rudra, M. N., Distribution of vitamin C in different parts of common Indian foodstuffs, *Biochem. J.*, 30, 701, 1936.

64. Paech, K., Beitrag zur Kenntnis des Ascorbinsäure-(Vitamin C)-Gehaltes und dessen Veränderungen in pflanzlichen Lebensmitteln, *Forschungsdienst Sonderheft*, 16, 283, 1942.

65. Gross, E., Vitamin C-Untersuchungen an Äpfeln, *Gartenbauwissenschaft*, 17, 500, 1943.

66. Holman, W. I. M., The distribution of vitamins within the tissues of common foodstuffs, *Nutr. Abstr. Rev.*, 26(2), 277, 1956.

67. Matzner, F., Über den Gehalt und die Verteilung des Vitamin C in Äpfeln, *Erwerbsobstbau*, 4(2), 27, 1962.

68. Matzner, F., Über den Trockensubstanz-, Säure- und Vitamin-C-Gehalt in den Früchten der Sorten "Freiherr von Berlepsch" und "Roter Berlepsch," *Erwerbsobstbau*, 8(11), 208, 1966.

69. Martin, D., Vitamin C in apples, *N. Z. J. Agric.*, 116(6), 71, 1968; *Hortic. Abstr.*, 39, 313, 1969.

70. Wood, M. A., Collings, A. R., Stodola, V., Burgoin, A. M., and Fenton, F., Effect of large-scale food preparation on vitamin retention: cabbage, *J. Am. Diet. Assoc.*, 22, 677, 1946.

71. Branion, H. D., Roberts, J. S., Cameron, C. R., and McCready, A. M., The ascorbic acid content of cabbage, *J. Am. Diet. Assoc.*, 24, 101, 1948.

72. Meinken, M., Semiprocessed fruit and vegetable products, in *Nutritional Quality of Fresh Fruits and Vegetables*, White, P. L., and Selvey, N., Eds., Future Publ. Co. Mount Kisco, N.Y., 1974, 65.

73. Kitagawa, Y., Distribution of vitamin C in relation to the growth of edible vegetables, *J. Jpn. Soc. Food Nutr.*, 26(9), 551, 1973; *Food Sci. Technol. Abstr.*, 6(10), J1386, 1974.

74. Erdman, J. W., Jr. and Klein, B. P., Harvesting, processing, and cooking influences on vitamin C in foods, in *Ascorbic Acid: Chemistry, Metabolism, and Uses*, Seib, P. A. and Tolbert, B. M., Eds., Adv. Chem. Series, no. 200, Am. Chem. Soc., Washington, D.C., 1982, 499.

75. Oba, K., Changes in vitamin C contents and ascorbate oxidase activity of vegetables after cutting and washing, *Nippon Kasei Gakkaishi*, 41(8), 715, 1990; *Chem. Abstr.*, 114, 162712h, 1991.

76. McHenry, E. W. and Graham, M., Observations on the estimation of ascorbic acid by titration, *Biochem. J.*, 29, 2013, 1935.

77. Kitagawa, Y., Distribution of vitamin C in some vegetable fruits (cucumber, princemelon and okra), *J. Jpn. Soc. Food Nutr.*, 25(5), 436, 1972; *Food Sci. Technol. Abstr.*, 5(12), J1983, 1973.

78. Olliver, M., The ascorbic acid content of fruits and vegetables, *Analyst*, 63, 2, 1938.

79. Webber, H. J., The vitamin C content of guavas, *Proc. Am. Soc. Hortic. Sci.*, 45, 87, 1944.

80. Vogtmann, H., Kaeppel N., and von Fragstein, P., Nitrat- und Vitamin C-Gehalt bei verschiedenen Sorten von Kopfsalat und unterschiedlicher Düngung, *Ernährungs-Umschau*, 34, 12, 1987.

81. Mustard, M. J. and Lynch, S. J., Effect of various factors upon the ascorbic acid content of some Florida-grown mangos, *Fla. Agr. Expt. Stn. Bull.*, 406, 1, 1945; *Chem. Abstr.*, 42, 5580b, 1948.

82. Siddappa, G. S. and Bhatia, B. S., Tender green mangoes as a source of vitamin C, *Indian J. Hortic.*, 11, 104, 1954.

83. Spencer, J. L., Morris, M. P., and Kennard, W. C., Vitamin C concentration in developing and mature fruits of mango (*Mangifera indica* L.), *Plant Physiol.*, 31, 79, 1956.

84. Subramanyam, H., Krishnamurthy, S., and Parpia, H. A. B., Physiology and biochemistry of mango fruit, *Adv. Food Res.*, 21, 224, 1975.

85. Thomas, P. and Oke, M. S., Technical note: vitamin C content and distribution in mangoes during ripening, *J. Food Technol.*, 15, 669, 1980.

86. Pruthi, J. S., Physiology, chemistry, and technology of passion fruit, *Adv. Food Res.*, 12, 203, 1963.

87. Schroder, G. M., Satterfield, G. H., and Holmes, A. D., The influence of variety, size, and degree of ripeness upon the ascorbic acid content of peaches, *J. Nutr.*, 25, 503, 1943.

88. Beckley, V. A. and Notley, V. E., The ascorbic acid content of sweet peppers, *J. Soc. Chem. Ind.*, 62, 14, 1943.

89. Pepkowitz, L. P., Larson, R. E., Gardner, J., and Owens, G., The carotene and ascorbic acid concentration of vegetable varieties, *Plant Physiol.*, 19, 615, 1944.

90. Ito, S., The persimmon, in *The Biochemistry of Fruits and Their Products*, Vol. 2, Hulme, A. C., Ed., Academic Press, London, 1971, 281.

91. Kitagawa, Y., Distribution of vitamin C related to the growth of some fruits. II. Tomato, pimento and strawberry, *J. Jpn. Soc. Food Nutr.*, 26(2), 139, 1973; *Food Sci. Technol. Abstr.*, 6(3), J384, 1974.

92. Kel't, K., Nutrient accumulation in plum, *Sbornik Nauchnykh Trudov Estonskogo Nauchno-Issledovatel'skogo Instituta Zemledeliya Melioratsii*, *Plodovodstvo*, 50, 49, 1985; *Hortic. Abstr.*, 57, 4106, 1987.

93. Burkhart, L. and Lineberry, R. A., Determination of vitamin C and its sampling variation in strawberries, *Food Res.*, 7, 332, 1942.

94. Wokes, F. and Organ, J. G., Oxidizing enzymes and vitamin C in tomatoes, *Biochem. J.*, 37, 259, 1943.

95. McCollum, J. P., Effect of sunlight exposure on the quality constituents of tomato fruit, *Proc. Am. Soc. Hortic. Sci.*, 48, 413, 1946.

96. Ward, G. M., Ascorbic acid in tomatoes. I. Distribution and method of assay, *Can. J. Plant Sci.*, 43, 206, 1963.

97. Yoshida, K., Mori, S., Hasegawa, K., Nishizawa, N., and Kumazawa, K., On the ratios of the main components in locular and pericarp of tomato fruits, *Joshi Eiyo Daigaku Kiyo*, 16, 29, 1985; *Chem. Abstr.*, 105, 23300y, 1986.

98. Laferriere, L. and Gabelman, W. H., Inheritance of color, total carotenoids, alpha-carotene, and beta-carotene in carrot, *Daucus carota* L., *Proc. Am. Soc. Hortic. Sci.*, 93, 408, 1968.

99. van Lelyveld, L. J., Ascorbic acid content and enzyme activities during maturation of the mango fruit and their association with bacterial black spot, *Agroplantae*, 7, 51, 1975.

100. Selman, J. D., The vitamin C content of some kiwifruits (*Actinidia chenensis* Planch, variety Hayward), *Food Chem.*, 11, 63, 1983.

101. Mondy, N. I. and Ponnampalam, R., Potato quality as affected by source of magnesium fertilizer nitrogen, minerals, and ascorbic acid, *J. Food Sci.*, 51(2), 352, 1986.

102. Munshi, C. B. and Mondy, N. I., Ascorbic acid and protein content of potatoes in relation to tuber anatomy, *J. Food Sci.*, 54(1), 220, 1989.

103. Mondy, N. I. and Munshi, C. B., Chemical composition of potato as affected by the herbicide, metribuzin: enzymatic discoloration, phenols and ascorbic acid content, *J. Food Sci.*, 53(2), 475, 1988.

104. Mukerjee, D. and Chava, N. R., Storage response of potatoes after pre-harvest application of maleic hydrazide, *J. Indian Bot. Soc.*, 67, 103, 1988.

105. Augustin, J., Toma, R. B., True, R. H., Shaw, R. L., Teitzel, C., Johnson S. R., and Orr, P., Composition of raw and cooked potato peel and flesh: proximate and vitamin composition, *J. Food Sci.*, 44, 805, 1979.

106. Hulme, A. C., Some aspects of the biochemistry of apple and pear fruits, *Adv. Food Res.*, 8, 297, 1958.

107. Pirie, N. W., Optimal exploitation of leaf carotene, *Ecology Food Nutr.*, 22, 1, 1988.

108. Rinno, G., Die Beurteilung des ernährungsphysiologischen Wertes von Gemüse, *Arch. Gartenbau*, 13, 415, 1965.

109. Schlottmann, H., Fehler bei der einer Qualitätsanalyse vorausgehenden probeent-nahme und deren Einfluss auf die zuverlässigkeit der Befunde, *Qualt. Plant.*, 10, 301, 1963.

110. Toul, V. and Popíšilová, J., The contents of β-carotene and sugars in carrot varieties, *Bull. Vysk. Ust. Zelin.*, Olomouc, 7, 75, 1963; *Hortic. Abstr.*, 34, 2935, 1964.

111. Horváth-Mosonyi, M., Rigó, J., and Hegedüs-Völgyesi, E., Study of dietary fiber content and fiber components of carrots, *Acta Alimen.*, 12(3), 199, 1983.

112. Arima, H. K. and Rodriguez-Amaya, D. B., Carotenoid composition and vitamin A value of commercial Brazilian squashes and pumpkins, *J. Micronutr. Analy.*, 4, 177, 1988.

113. Simaan, F. S., Cowan, J. W., and Sabry, Z. I., Nutritive value of Middle Eastern foodstuffs. I. Composition of fruits and vegetables grown in Lebanon, *J. Sci. Food Agric.*, 15, 799, 1964.

114. Bureau, J. L. and Bushway, R. J., HPLC determination of carotenoids in fruits and vegetables in the United States, *J. Food Sci.*, 51(1), 128, 1986.

115. Streighthoff, F., Munsell, H. E., Ben-dor, B. A., Orr, M. L., Leonard, M. H., Ezekiel, and Koch, F. G., Effect of large-scale methods of preparation and the vitamin content of food. II. Carrots, *J. Am. Diet Assoc.*, 25, 511, 1946.

116. Bradbury, J. H. and Singh, U., Thiamin, riboflavin, and nicotinic acid contents of tropical root crops from the South Pacific, *J. Food Sci.*, 51(6), 1563, 1986.

117. Ournac, A., Vitamins C, B$_1$, and B$_2$ in asparagus. Variations during growth, storage and cooking, *Alimentation et la Vie*, 58(7/8/9), 164, 1970; *Food Sci. Technol. Abstr.*, 3(11), J1362, 1971.

118. Booth, V. H. and Bradford, M. P., Tocopherol contents of vegetables and fruits, *Br. J. Nutr.*, 17, 575, 1963.

119. Herting, D. C., Vitamin E. *Encyclo. Chem. Technol.*, 24, 214, 1984.

120. Pirtskhalaishvili, E. S., The effect of fertilizers on the folic acid content of leaves and fruit of Unshiu mandarin, *Subtrop. Kul't.*, 2, 161, 1971; *Hortic. Abstr.*, 42, 4879, 1972.

121. Ferland, G. and Sadowski, J. A., Vitamin K$_1$ (Phylloquinone) content of green vegetables: effect of plant maturation and geographical growth location, *J. Agric. Food Chem.*, 40, 1874, 1992.

122. Franke, W. and Heber, U., Über die quantitative Verteilung der Ascorbinsäure innerhalb der Pflanzenzelle, *Z. Naturforsch.*, 19b, 1146, 1964.

123. Gerhardt, E., Untersuchungen Über beziehungen zwischen Ascorbinsäure und Photosythese, *Planta*, 61, 101, 1964.

124. Law, M. Y., Charles, S. A., and Halliwell, B., Glutathion and ascorbic acid in spinach (*Spinacia oleracea*) chloroplasts, *Biochem. J.*, 210, 899, 1983.

125. Rauen, H. M., Devescovi., M., and Magnani, N., Der Vitamin C-Gehalt italianischer Orangen und daraus hergestellter Pulpen, *Z. Unters. Lebensmitt.*, 85, 257, 1943.

126. Cohen, A., The effect of different factors on the ascorbic acid content of citrus fruits. I. The dependence of the ascorbic acid content of the fruit on light intensity and on the area of assimilation, *Bull. Res. Council Israel*, 3, 159, 1953.

127. Eaks, I. L., Ascorbic acid content of citrus during growth and development, *Bot. Gaz. (Chicago)*, 125, 186, 1964.

128. Erickson, L. C., The general physiology of citrus, in *The Citrus Industry*, Vol. II, Reuther, W., Bachlor, L. D., and Webber, H. J., Eds., University of California. division of agricultural Science, Berkeley, Calif., 1968, 86.

129. Ting, S. V. and Attaway, J. A., Citrus fruits, in *The Biochemistry of Fruits and Their Products*, Vol. 2, Hulme, A. C., Ed., Academic Press, London, 1971, 107.

130. Murneek, A. E., Maharg, L., and Wittwer, S. H., Ascorbic acid (vitamin C) content of tomatoes and apples, *Univ. Missouri, Agric. Exp. Stn. Res. Bull.*, 568, 1954.

131. Mapson, L. W., Vitamin in fruits, in *The biochemistry of Fruits and Their Products*, Vol. 1, Hulme, A. C., Ed., Academic Press, London, 1970, 369.

132. Kessler, W., Über den Vitamin C-Gehalt deutscher Apfelsorten und seine Abhängigkeit von Herkunft, Lichtgenuss, Düngung, Dichte des Behanges und Lagerung, *Gartenbauwissenschaft*, 13, 619, 1939.

133. Martin, D., Ascorbic acid in Tasmanian apples, *Fld. Stat. Rec. CSIRO Div. Plant Ind.*, 5(1), 45, 1966; *Hortic. Abstr.*, 37, 4444, 1967.

134. Todhunter, E. N., Some factors influencing ascorbic-acid (vitamin-C) content of apples, *Food Res.*, 1, 435, 1936.

135. Lauersen, F. and Orth, W., Die Verteilung der Ascorbinsäure in der Kartoffelknolle, *Unters. Lebensmitt.*, 83(3), 193, 1942.

136. Massey, P. H., Jr., Eheart, J. F., and Young, R. W., Variety, year, and pruning effects on the dry matter, vitamin content, and yield of broccoli, *Proc. Am. Soc. Hortic. Sci.*, 68, 377, 1956.

137. Pollock, J. M. and Geddes, W. F., The distribution of thiamine and riboflavin in the wheat kernel at different stages of maturity, *Cereal Chem.*, 28, 289, 1951.

138. Heathcote, J. G., Hunton, J. J., C., and Shaw, B., The distribution of nicotinic acid in wheat and maize, *Proc. Roy. Soc. Ser. B*, 139, 276, 1952.

139. Hinton, J. J. C., Peers, F. G., and Shaw, B., The B-vitamins in wheat: the unique aleurone layer, *Nature* (*Lond.*), 172, 993, 1953.

140. Pomeranz, Y., Chemical composition of kernel structure, in *Wheat: Chemistry and Technology*, Vol. 1, Pomeranz, Y., Ed., Am. Association of Cereal Chemists. St. Paul, Minn., 1988, 97.

141. Hubbard, J. E., Hall, H. H., and Earle, F. R., Composition of the component parts of the sorghum kernel, *Cereal Chem.*, 27, 415, 1950.

142. Lorenz, K., Cereal sprouts: composition, nutritive value, food applications, *CRC Crit. Rev. Food Sci. Nutr.*, 13, 353, 1980.

143. Burkholder, P. R., Vitamins in dehydrated seeds and sprouts, *Science*, 97, 562, 1943.

144. McVeigh, L., Occurrence and distribution of thiamine, riboflavin, and niacin in Avena seedlings, *Bull. Torrey Bot. Club.*, 71(4), 438, 1944.

145. Burkholder, P. R. and McVeigh, I., The increase of the B vitamins in germinating seeds, *Proc. Natl. Acad. Sci. USA*, 28, 440, 1942.

146. Burkholder, P. R. and McVeigh, I., Vitamin content of some mature and germinated legume seeds, *Plant Physiol.*, 20, 301, 1945.

147. Nason, A., The distribution and biosynthesis of niacin in germinating corn, *Am. J. Bot.*, 37, 612, 1950.

148. Wai, K. N. T., Bishop, J. C., Mack, P. B., and Cotton, R. H., The vitamin content of soybeans and soybean sprouts as a function of germination time, *Plant Physiol.*, 22, 117, 1947.

149. Sreenivasan, A. and Wandrekar, S. D., Biosynthesis of vitamin C in germinating legumes, *Nature* (*Lond.*), 165, 765, 1950.

150. Wu, C. H. and Fenton, F., Effect of sprouting and cooking of soybeans on palatability, lysine, tryptophane, thiamine, and ascorbic acid, *Food Res.*, 18, 640, 1953.

151. Hall, G. S. and Laidman, D. L., Tocopherols and ubiquinone in the germinating wheat grain, *Biochem. J.*, 101, 5p. 1966.

152. Kylen, A. M. and McCready, R. M., Nutrients in seeds and sprouts of alfalfa, lentils, mung beans, and soybeans, *J. Food Sci.*, 40, 1008, 1975.

153. Segal, B., Segal, R., and Stoicescu, A., Biochemical transformations in the germination of the small pea, *Ind. Aliment. Agric.*, 95(1), 7, 1978; *Chem. Abstr.*, 88, 188410c, 1978.

154. Hamilton, M. J. and Vanderstoep, J., Germination and nutrient composition of alfalfa seeds, *J. Food Sci*, 44(2), 443, 1979.

155. Prudente, V. R. and Mabesa, L. B., Vitamin content of mung bean [*Vinga radiata* (L.) R. Wilcz.] sprouts, *Philipp. Agric.*, 64(4), 365; *Chem. Abstr.*, 97, 4873n, 1982.

156. Riaz, M., Qadri, R. B., and Warsi, S. A., Development of ascorbic acid and changes in the carbohydrate and nitrogenous fractions in sprouting legumes, *Proc. Pak. Akad. Sci.*, 17(1), 53, 1980; *Chem. Abstr.*, 96, 33559z, 1982.

157. Augustin, J., Cole, C. L., Fellman, J. K., Matthews, R. H., Tassinari, P. D., and Woo, H., Nutrient content of sprouted wheat and selected legumes, *Cereal Foods World*, 28(6), 358, 1983; *Chem. Abstr.*, 99, 69091s, 1983.

158. Kwon, J. H. and Yoon, H. S., Studies on nitrate and nitrite in foods. I. Changes of nitrate, nitrite and ascorbic acid contents during soybean germination, *Han'guk Yongyang Siklyong Hakhoechi*, 10(1), 39, 1981; *Chem. Abstr.* 98, 142115p, 1983.

159. Nasim, F. H., Khan, M. S., Mahmood, B. A., and Sheikh, A. S., Changes during the germination of some edible seeds, *J. Nat. Sci. Math.*, 22(1), 55, 1982; *Chem. Abstr.*, 98, 122943, 1983.

160. Abdullah, A. and Baldwin, R. E., Mineral and vitamin contents of seeds and sprouts of newly available small-seeded soybeans and market samples of mungbeans, *J. Food Sci.*, 49(2), 656, 1984.

161. Alexander, J. C., Gabriel, H. G., and Reichertz, J. L., Nutritional value of germinated barley, *Can. Inst. Food Sci. Technol. J.*, 17(4), 224, 1984.

162. El-Shimi, N. M. and Damir, A. A., Changes in some nutrients of fenugreek seeds during germination, *Food Chem.*, 14, 11, 1984.

163. Harrison, J. E. and Vanderstoep, J., Effect of germination environment on nutrient composition of alfalfa sprouts, *J. Food Sci.*, 49, 21, 1984.

164. Yamaguchi, Y., Composition of alfalfa sprouts. I. Contents of moisture, protein, fat, ash, carotene, vitamin C and vitamin B_1, *Kaseigaku Zasshi*, 34(10), 660, 1983; *Chem. Abstr.*, 100, 66878g, 1984.

165. Izumi, H., Tatsumi, Y., Maruta T., Vitamin C in fruits and vegetables. II. Effect of illumination intensity on the quality of radish seedlings, "kaiware daikon," *Nippon Shokuhin Kogyo Gakkaishi*, 31(11), 704, 1984; *Chem. Abstr.*, 102, 61019j, 1985.

166. Sawyer, C. A., DeVitto, A. K., and Zabik, M. E., Food service systems: comparison of production methods and storage time for alfalfa sprouts, *J. Food Sci.*, 50, 188, 1985.

167. Tajiri, T., Cultivation and keeping quality of bean sprouts. IX. Improvement of bean sprouts cultivation by combined treatment with artificial sunlight lamps, ethylene, and carbon dioxide, *Nippon Shokuhin Kogyo Gakkaishi*, 32(5), 317, 1985; *Chem. Abstr.*, 103, 138694f, 1985.

168. Taur, A. T., Pawar, V. D., and Ingle, U. M., Nutritional improvement of grain sorghum (*Sorghum bicolor* (L.) Moench) by germination, *Indian J. Nutr. Diet.*, 21(5), 168, 1984; *Chem. Abstr.*, 102, 130679p, 1985.

169. Harmuth-Hoene, A. E., Bognar, A. E., Kornemann, U., and Diehl, F. J., Der Einfluss der Keimung auf den Hährwert von Weizen, Mungbohnen und Kichererbsen, *Lebensm.-Unters. Forsch.*, 185, 386, 1987.

170. Meier-Ploeger, A. and Vogtmann, H., Keimlinge- eine sinnvolle Bereicherung des Gemüseangebotes, *Ernährungs-Umschau*, 33(12), 374, 1986.

171. Morinaga, T., Tsutsui, M, and Tanabe, A., Effects of germination conditions on the growth, contents of ascorbic acid and those of sugar in black matpe seeds, *Fukuoka Joshi Daigaku Kaseigakubu Kiyo*, 18, 7, 1987; *Chem. Abstr.*, 107, 20842g, 1987.

172. Kanbe, T., Kamei, M., Fujita, T., Odachi, J., and Sasaki, K., Changes in vitamin contents of kaiware daikon radish seedlings during cultivation, *Ann. Rep. Osaka City Inst. Public Health Environ. Sci.*, 49, 75, 1986; *Chem. Abstr.*, 108, 149102r, 1988.

173. Johnsson, H. and Oesterdahl, B. G., Content of vitamins in sprouts, *Vaar Foeda*, 41(1–2), 43, 1989; *Chem. Abstr.*, 111, 152360r, 1989.

174. Nnanna, I. A. and Phillips, R. D., Amino acid composition protein quality and water-soluble vitamin content of germinated cowpea (*Vinga unguiculata*), *Plant Foods Hum. Nutr.*, 39, 187, 1989.

175. Sattar, A., Durrani, S. K., Mahmood, F., Ahmad, A., and Khan, I., Effect of soaking and germination temperatures on selected nutrients and antinutrients of mungo bean, *Food Chem.*, 34(2), 111, 1989.

176. El, S. N. and Kavas, A., Nutritive value of lentil and mung bean sprout, *Doga: Turk Muhendislik Cevre Bilimleri Derg.*, 14(2), 240, 1990; *Chem. Abstr.*, 113, 170714e, 1990.

177. Colmenares de Ruiz, A. S. and Bressani, R., Effect of germination on the chemical composition and nutritive value of amaranth grain, *Cereal Chem.*, 67(6), 519, 1990.

178. Tommasi, F. and De Gara, L., Correlation between the presence of ascorbic acid and development of ascorbic peroxidase activity in *Avena sativa*, L. embryos, *Boll.-Soc. Ital. Biol. Sper.*, 66(4), 357, 1990; *Chem. Abstr.*, 113, 74980t, 1990.

179. Akinlosotu, A. and Akinyele, I. O., The effect of germination on the oligosaccharide and nutrient of cowpeas (*Vinga unguiculata*), *Food Chem.*, 39(2), 157, 1991.

180. Ghazali, H. M. and Cheng, S. C., The effect of germination on the physico-chemical properties of black gram (*Vinga mungo* L.), *Food Chem.*, 41(1), 99, 1991.

181. Nanmori, T., Mitsumatsu, A., Kohno, A., and Miyatake, K., Nutritional features of dry and germinating alfalfa seeds, *Nippon Eiyo, Shokuryo Gakkaishi*, 44(3), 221, 1991; *Chem. Abstr.*, 115, 230848e, 1991.
182. Sattar, A., Atta, S., Akhtar, M. A., Wahid, M., and Ahmad, B., Biosynthesis of ascorbic acid and riboflavin in radiated germinating chickpea, *Int. J. Vitam. Nutr. Res.*, 61, 149, 1991.
183. Steger, H. and Wallnöfer, P. R., Thiamingehalte in unterschiedlich behandelten Keimlingen von Weizen, Linsen und Sojabohnen, *Ernährungs-Umschau*, 38(1), 18, 1991.
184. Sattar, A., Neelofar, and Akhtar, M. A., Radiation effect on ascorbic acid and riboflavin biosynthesis in germinating soybean, *Plant Foods Hum. Nutr.*, 42, 305, 1992.
185. Hsu, D., Leung, H. K., Finney, P. L., and Morad, M. M., Effect of germination on nutritive value and baking properties of dry peas, lentils, and Faba beans, *J. Food Sci*, 45, 87, 1980.
186. Nandi, D. L., Effects of different light and darkness on the B-vitamin content of germinating pulses, *Food Res.*, 25, 88, 1960.
187. Kim, S. D., Jang, B. H., Kim, H. S., Ha, K. H., Kang, K. S., and Kim, D. H., Studies on the changes in chlorophyll, free amino acid and vitamin C content of soybean sprouts during light and dark periods, *Han'guk Yongyang Siklyong Hakhoechi*, 11(3), 57, 1982; *Chem. Abstr.*, 98, 142112k, 1983.
188. Kim, K. S., Kim, S. D., Kim, J. K., Kim, J. N., and Kim, K. J., Effect of blue light on the major components of soybean sprouts, *Han'guk Yongyang Siklyong Hakhoechi*, 11(4), 7, 1982; *Chem. Abstr.*, 99, 103943h, 1983.
189. Farhangi, M. and Valadon, L. R. G., Effect of light, acidified processing and storage on carbohydrates and other nutrients in mung bean sprouts, *J. Sci. Food Agric.*, 34(11), 1251, 1983.
190. Cheldelin, V. H. and Lane, R. L., B vitamins in germinating seeds, *Proc. Soc. Exp. Biol. Med.*, 54, 53, 1943.
191. Klatzkin, C., Norris F. W., and Wokes, F., Nicotinic acid in cereals. I. The effect of germination, *Biochem. J.*, 42, 414, 1948.
192. Fordham, J. R., Wells, C. E., and Chen, L. H., Sprouting of seeds and nutrient composition of seeds and sprouts, *J. Food Sci.*, 40, 552, 1975.
193. NRC (National Research Council), *Recommended Dietary Allowances, 10th ed.*, *Report of the Committee on Dietary Allowances, Division of Biological Sciences, Assembly of Life Sciences, Food and Nutrition Board*, National Academy of Sciences, Washington, D.C., 1989.

Chapter 3

GENETIC VARIABILITY

I. DIFFERENCES BETWEEN VARIETIES

Plants differ strongly in their vitamin content. This is true not only between different species, but also between different varieties of the same plant species. This has been documented by numerous authors and for many different kinds of fruits and vegetables grown in different parts of the world, the results of which have been summarized in Table 1. Here some selected cases are reviewed in greater detail.

A. Ascorbic Acid

The ascorbic acid content of different plant varieties may differ by a factor of 2–3 or higher in many fruits and vegetables (Table 1), some examples of which are graphically shown in Figure 1. Extreme differences of as much as 9–10 times in squash and peppers, 15 times in tomatoes, 20 times in muskmelons, 16–29 times in apples and pear varieties, and even 91 times in mangoes have also been noted.

In cabbage, ascorbic acid concentration may vary by a factor of 3.8-fold (180.9 vs. 48.0 mg/100 g).[30] Pointed-head varieties usually contain higher ascorbic acid than other varieties.[219] Franz et al.[220] compared three smooth-leaf and six curly parsley cultivars and noted that smooth-leaf cultivars were higher in vitamin C than the curly cultivars.

Early and late varieties of fruits and vegetables may have different vitamin contents. For example, spring-harvested cabbage varieties were found to be higher in ascorbic acid than the summer or fall-harvested varieties usually grown for the production of kraut or for storage,[30, 31, 218, 219] the differences in some cases being as much as five times.[221] Also, the early varieties of apples,[226] pear,[227] and cauliflower[228] are apparently higher in vitamin C than the late varieties. In contrast, late varieties of garlic,[222] olives,[223] *Ziziphus jujuba*,[224] and freestone peaches[96] were found to contain more vitamin C than the early varieties. Also late-ripening varieties of grapes were found to contain higher levels of vitamin B_1, B_2, and nicotonic acid than the early varieties.[225]

Are the most popular varieties of fruits and vegetables also richest in vitamins? The answer to this question seems to be no. For example, the most popular dessert apple varieties, such as Delicious, Jonathan, Spartan, and McIntosh, are among the varieties relatively low in ascorbic acid content when compared with some less popular varieties

TABLE 1
Range of Vitamin Concentrations in Different Varieties, Genotypes, or Strains of the Same Plant[a]

Plant	Number of varieties tested	Concentration range[b]	Factor[c] (max./min.)	Ref.
		Ascorbic acid (mg / 100 g FW)		
Apples	99	5–23	4.6	1
Apples	15	1–29	29.0	2
Apples	?	3–30	10.0	3
Apples	134	2–32	16.0	4
Apples	25	3–26	8.7	5
Apples	11	5–20	4.0	6
Apples	12	6–30	5.0	7
Apples	35	3–16	5.3	8
Apples	?	4–12	3.0	9
Apples	30	3–39	13.0	10
Apples	?	2–22	11.0	11
Apricots	2	7–16	2.3	12
Bananas (plantain)	5	20–24	1.2	13
Bananas	3	8–11	2.7	14
Bananas	5	1–9	9.0	15
Bananas (cooking)	4	6–16	2.7	16
Barberry[d]	?	12–53	4.4	17
Beans (snap)	19	16–32	2.0	18
Beans (snap)	171	10–29	2.9	19
Beans (snap)	22	9–16	1.8	20
Beans	15	11–27	2.4	21
Ber	35	70–165	2.3	22
Ber	5	86–113	1.3	23
Ber	7	60–101	1.7	24
Blackberry	20	12–28	2.3	25
Blueberry	7	12–21	1.7	26
Blueberry	10	3–9	3.0	27
Blueberry	6	4–8	2.0	28
Brussels sprout	4	117–152	1.3	29
Cabbage	30	48–181	3.8	30
Cabbage	7	33–53	1.6	31
Cabbage	12	44–119	2.7	32
Cabbage	6	35–55	1.7	33
Capsicum	15	126–218	1.7	34
Carrots	11	8–11	1.4	35
Cashew apple	16	144–274	1.9	36
Cassava[d]	4	120–150	1.2	37
Cassava	5	7–13	1.9	38
Cauliflower	15	48–84	1.7	39
Cauliflower	?	55–64	1.2	40

TABLE 1 (continued)

Plant	Number of varieties tested	Concentration range[b]	Factor[c] (max./ min.)	Ref.
Celeriac	6	10–12	1.2	41
Cherry	3	16–21	1.3	12
Cherry	37	13–55	4.2	42
Cherry (sour)	6	27–31	1.1	43
Cherry (sour)	12	9–29	3.2	44
Collards[d]	9	694–1088	1.6	45
Cranberry	10	27–38	1.4	46
Currant (black)	6	151–261	1.7	47
Currant (black)	?	88–266	3.0	3
Currant (black)	7	140–217	1.5	48
Currant (black)	?	93–340	3.7	49
Currant (black)	?	90–223	2.5	50
Currant (black)	11	80–290	3.6	51
Currant (black)	15	147–156	1.1	52
Currant (black)	24	139–355	1.5	53
Currant (black)	10	69–153	2.2	54
Currant (white)	7	10–80	8.0	51
Durian (*Durio zibethinus*)	?	32–58	1.8	55
Gram	40	2–6	3.0	56
Grape	11	2–4	2.0	57
Grape	32	3–4	1.3	58
Grape	31	1–3	3.0	59
Grape	8	5–13	2.6	60
Grape	?	6–18	3.0	61
Grapefruit	4	47–53	1.3	62
Guava	4	122–315	2.6	63
Guava	4	134–289	2.2	64
Guava	5	40–440	11.0	65
Guava	11	63–132	2.1	66
Guava	6	215–372	1.7	67
Guava	8	145–240	1.6	68
Guava	8	40–180	4.5	69
Kiwifruit	4	85–158	1.9	70
Kiwifruit	3	153–323	2.1	7
Kiwifruit	2	85–160	1.9	71
Kiwifruit	7	66–146	2.2	72
Kiwifruit	5	70–276	3.9	73
Lemon	4	37–45	1.2	74
Mango	30	9–72	8.0	75
Mango	?	5–70	14.0	76
Mango	?	13–178	13.7	77
Mango	?	5–70	14.0	78
Mango	4	4–20	5.0	79
Mango	4	40–143	3.6	80
Mango	30	1–91	91.0	81
Mango	9	14–57	4.1	82

TABLE 1   (continued)

Plant	Number of varieties tested	Concentration range[b]	Factor[c] (max./min.)	Ref.
Muskmelon	?	3–61	20.3	35
Muskmelon	10	20–34	1.7	83
Nectarine	2	11–13	1.2	12
Nectarine	27	3–14	4.7	84
Oranges	29	51–78	1.5	85
Oranges	9	40–55	1.4	86
Papaya	2	69–90	1.3	79
Papaya	12	46–126	2.7	87
Papaya	14	79–145	1.8	88
Peas	79	21–33	1.6	89
Peas	28	11–38	3.4	90
Peas	100	20–43	2.1	91
Peas	4	33–51	1.5	92
Peas	7	6–8	1.3	93
Peas[d]	4	101–157	1.3	94
Peas	6	12–21	1.7	40
Peaches	?	6–14	2.3	95
Peaches	8	6–16	2.7	12
Peaches	16	4–17	4.2	96
Pear	4	2–8	4.0	2
Pear	194	1–16	16.0	97
Persimmon	?	25–52	2.1	98
Persimmon	4	6–16	2.7	99
Persimmon	15	35–218	6.2	100
Pepper	20	147–225	1.5	101
Pepper (green)	10	104–173	1.7	102
Pepper (green)	11	95–170	1.8	103
Pepper (chilli)	12	34–91	2.7	104
Pepper (chilli)	11	29–300	10.3	105
Phyllanthus emblica	14	169–471	2.8	106
Plum	6	4–6	1.5	12
Pomegranate[e]	11	0.3–2.2	7.3	107
Potato	33	8–41	5.1	108
Potato	9	10–19	1.9	109
Potato	8	6–29	4.8	2
Potato	20	17–41	2.4	110
Potato	23	13–27	2.1	111
Potato	6	13–24	1.8	112
Potato	16	15–27	1.8	113
Potato	13	9–18	2.0	114
Potato	6	4–12	3.0	115
Potato[d]	8	57–120	2.1	116
Potato	6	12–22	1.8	117
Potato	?	15–25	1.7	118
Potato	7	12–18	1.5	119
Quince	2	16–127	7.9	2
Quince	10	20–76	3.8	120

TABLE 1 (continued)

Plant	Number of varieties tested	Concentration range[b]	Factor[c] (max./min.)	Ref.
Quince	11	15–63	5.2	121
Raspberry	27	18–37	2.1	25
Raspberry	30	21–33	1.6	122
Raspberry (leaf)	30	333–442	1.4	122
Raspberry	?	20–46	2.3	3
Raspberry	?	40–63	1.6	50
Rose hips[d]	5	2–55	27.5	123
Rutabaga	25	40–66	1.6	124
Soybean (immature seeds)	4	44–104	2.4	125
Spinach	13	28–41	1.5	126
Spinach	4	29–48	1.6	127
Squash (summer)	2	9–85	9.4	128
Squash (winter)	2	6–21	3.5	128
Strawberry	5	55–69	1.2	47
Strawberry	30	53–107	2.0	25
Strawberry	37	41–81	2.0	129
Strawberry	44	39–89	2.3	130
Strawberry	?	41–142	3.5	131
Strawberry	?	63–114	1.8	50
Strawberry	10	40–59	1.5	132
Strawberry	20	30–130	4.3	51
Strawberry	20	20–54	2.7	133
Strawberry	9	76–153	2.0	134
Sweet potato	15	21–30	1.4	135
Sweet potato	4	10–31	3.1	136
Sweet potato	10	32–73	2.3	137
Sweet potato	10	9–25	2.8	38
Strawberry	8	91–116	1.3	138
Strawberry	14	40–97	2.4	139
Strawberry	7	35–41	1.2	140
Sweet potato	20	20–54	2.7	141
Taro (*Colocasia esculenta*)	4	5–10	2.0	38
Taro (*Xanthosoma spp.*)	3	4–8	2.0	38
Taro (*Alocasia macrorrhiza*)	7	4–13	3.2	38
Taro (*Cyrtosperma chamissonis*)	4	6–10	1.7	38
Tomato	250	20–42	2.1	142
Tomato	24	11–23	1.2	35
Tomato	98	13–44	3.4	128
Tomato	14	12–33	2.7	143
Tomato	40	15–45	3.0	144
Tomato	24	13–24	1.8	145
Tomato	11	14–32	2.3	146
Tomato	3	15–18	1.2	147
Tomato	12	16–26	1.6	148
Tomato	?	15–22	1.5	149
Tomato	7	16–26	1.6	150
Tomato	5	8–18	2.2	151

TABLE 1   (continued)

Plant	Number of varieties tested	Concentration range[b]	Factor[c] (max./min.)	Ref.
Tomato	?	8–119	14.9	152
Tomato[e]	20	10–22	2.2	153
Tomato	3	26–34	1.3	154
Turnip greens	2	130–146	1.1	155
Yam (*Dioscorea alata*)	4	7–12	1.7	38
Yam (*D. esculenta*)	5	10–19	1.9	38

Biotin (μg / 100 g)

Apples	4	0.24–0.27	1.1	178
Cowpea	30	18–27	1.5	196
Pear	3	0.07–0.08	1.1	178
Sorghum	5	18–23	1.3	191

Folic acid (μg / 100 g)

Beans (classes)[d]	9	148–676	4.6	182
Peas[d]	4	266–576	2.2	94

β-carotene (μg / 100 g FW)

Apricots	8	1,210–3,490	2.9	128
Apricots[f]	2	110–285	2.6	12
Barberry[d]	?	160–1,510	4.4	17
Beans (snap)[f]	38	192–450	2.3	19
Blueberry	10	5–83	16.6	27
Capsicum	15	8,390–78,000	9.3	34
Carrots[f]	7	8,230–14,000	1.7	156
Carrots	4	6,830–10,620	1.5	157
Carrots	11	14,000–18,000	1.3	35
Carrots[f]	10	120–9,600	80.0	158
Carrots	23	850–8,500	10.0	159
Carrots	19	4,600–10,300	2.2	160
Cassava	8	10–1,130	113.0	161
Cassava[g]	5	3–14	4.7	162
Cherry	3	30–105	3.5	12
Clover (leaf)	3	3,810–4,010	1.0	163
Collards[f, d]	9	11,700–16,700	1.4	45
Grapefruit	5	420–960	2.3	164
Grapefruit[f]	5	1,210–11,170	9.3	164
Maize[f]	45	165–460	2.8	89
Maize[f]	34	210–610	2.9	165
Maize	125	30–730	24.3	166
Mango	5	661–2,545	3.8	167
Mango[g]	5	115–430	3.7	168
Nectarine	4	25–120	4.8	12
Orange (juice)[f]	9	360–2,452	6.8	169

TABLE 1 (continued)

Plant	Number of varieties tested	Concentration range[b]	Factor[c] (max./min.)	Ref.
Palm (oil)	4	20,210–102,610	5.1	170
Papaya	14	400–2300	5.7	88
Papaya	4	120–370	3.1	168
Peas[f]	79	125–539	4.3	98
Peas[f, d]	4	1,206–1,796	1.5	94
Peaches	8	30–180	6.0	12
Pepper (green)	10	340–460	1.3	102
Pepper (Mexican)	5	13–599	46.1	171
Pepper (paprika)[d]	7	35,000–156,000	4.5	172
Plum	6	90–225	2.5	12
Plum[f]	3	750–2,440	3.2	173
Sweet potato	6	130–5,180	39.8	174
Sweet potato[f, d]	15	500–44,600	89.2	135
Sweet potato[f]	11	186–15,800	84.9	175
Sweet potato	9	0–18,000	—	158
Sweet potato[f]	17	400–24,800	62.0	176
Sweet potato[f]	4	8,000–18,500	2.3	136
Tomato	24	490–800	1.6	35
Tomato	6	400–7,980	19.9	177
Tomato	10	400–573	1.4	146
Tomato	5	270–890	3.3	151
Watermelon	?	40–600	15.0	128

Niacin (μg / 100 g)

Plant	Number of varieties tested	Concentration range[b]	Factor[c] (max./min.)	Ref.
Apples	?	50–100	2.0	178
Apricots	2	1,100–1,400	1.3	12
Avocados	3	1,450–2,160	1.5	179
Barley[d]	3	8,990–9,840	1.1	180
Beans[d]	8	1,970–2,990	1.5	181
Beans (classes)[d]	9	850–3,210	3.8	182
Blueberry	10	1,000–1,700	1.7	27
Cherry	3	400–600	1.5	12
Cowpea	30	711–1,602	2.2	183
Grape	31	116–278	2.4	59
Maize	233	1,130–6,210	5.5	184
Maize	11	1,150–3,700	3.2	185
Maize	9	1,690–2,540	1.5	186
Mango	3	90–1,650	18.3	79
Millet (pearl)[d]	3	4,206–4,637	1.1	187
Nectarine	3	1,100–1,400	1.3	12
Oats	8	490–700	1.4	185
Papaya	2	330–770	2.3	79
Peas[d]	4	8,870–10,440	1.2	94
Peaches	8	900–1,100	1.2	12
Peanut[d]	12	17,500–26,300	1.5	188
Peanut	30	12,200–18,500	1.5	189

TABLE 1   (continued)

Plant	Number of varieties tested	Concentration range[b]	Factor[c] (max./min.)	Ref.
Pear	?	100–400	4.0	178
Pepper (green)	11	820–1,020	1.2	103
Plum	6	200–900	4.5	12
Potato	23	1,200–3,200	2.7	111
Potato	6	1,030–2,080	2.1	190
Rye	3	770–1,000	1.3	185
Sorghum[d]	5	2,810–6,800	2.4	191
Strawberry	2	470–630	1.3	79
Sweet potato	4	356–540	1.5	192
Sweet potato	3	259–887	3.4	193
Sweet potato	4	440–910	2.1	136
Taro (*Colocasia esculenta*)	4	268–1,310	4.9	193
Taro (*Xanthosoma spp.*)	3	711–1,078	1.5	193
Taro (*Alocasia macrorrhiza*)	4	220–769	3.5	193
Taro (*Cyrtosperma chamissonis*)	4	385–644	1.7	193
Wheat	10	5,100–6,300	1.2	194
Wheat[d]	2	594–716	1.2	180
Wheat (classes)	?	2,200–11,100	5.0	195
Yam (*Dioscorea alata*)	5	245–490	2.0	193
Yam (*D. esculenta*)	5	251–691	2.7	193

Pantothenic acid (μg / 100 g)

Plant	Number of varieties tested	Concentration range[b]	Factor[c] (max./min.)	Ref.
Apples	?	50–200	4.0	178
Avocados	3	900–11,400	12.7	179
Cowpea	30	1,730–2,240	1.3	196
Barley[d]	3	850–1,040	1.2	180
Maize	9	480–640	1.3	186
Pear	?	20–50	2.5	178
Sorghum[d]	5	900–1,550	1.7	191
Sweet potato	4	420–930	2.2	136
Wheat[d]	2	570–590	1.0	180
Wheat (classes)	?	719–1,990	2.6	195

Pyridoxine (μg / 100 g)

Plant	Number of varieties tested	Concentration range[b]	Factor[c] (max./min.)	Ref.
Avocados	3	39–62	1.6	179
Beans (classes)[d]	9	299–659	2.2	182
Cowpea	30	274–417	1.5	196
Gram (black)	6	26–28	1.1	197
Gram (Bengal)	5	31–47	1.5	197
Gram (Red)	6	26–36	1.4	197
Gram (green)	6	29–43	1.5	197
Peas[d]	4	518–666	1.3	94
Potato	3	20–30	1.5	198
Potato	6	13–42	3.2	190
Sorghum	5	40–62	1.5	191
Wheat (classes)	6	92–790	8.6	195

TABLE 1　(continued)

Plant	Number of varieties tested	Concentration range[b]	Factor[c] (max./min.)	Ref.
		Riboflavin (μg / 100 g)		
Apples	?	5–50	10.0	178
Apricots	2	30–40	1.3	12
Avocados	3	210–230	1.1	179
Beans (snap)	171	61–183	3.0	19
Beans (dry)	7	180–280	1.5	199
Beans (classes)[d]	9	112–411	3.7	182
Beans[d]	8	190–220	1.2	181
Blueberry	10	38–70	1.8	27
Cabbage	7	17–44	2.6	31
Cabbage	6	20–57	2.8	33
Carrots	11	11–61	5.5	35
Cauliflower	?	170–240	1.4	40
Cherry	3	20–30	1.5	12
Collards[d]	9	1,540–3,180	2.1	45
Cowpea	30	106–322	3.0	183
Durian (*Durio zibethinus*)	?	170–530	3.1	55
Gram (black)	6	210–300	1.4	197
Gram (bengal)	5	210–250	1.2	197
Gram (green)	6	220–280	1.3	197
Gram (red)	8	230–310	1.3	197
Grape	11	25–45	1.8	57
Grape	8	26–88	3.4	60
Mango	4	37–73	2.0	78
Mango	3	60–90	1.5	79
Millet (pearl)	3	245–339	1.4	187
Nectarine	4	30–40	1.3	12
Papaya	2	20–31	1.5	79
Peas	100	76–131	1.7	91
Peas[d]	4	516–653	1.3	94
Peaches	8	30–50	1.7	12
Pear	2	10–50	5.0	178
Peanut[d]	12	125–241	1.9	188
Plum	6	40–50	1.2	12
Potato	23	12–75	6.2	111
Sorghum[d]	5	80–200	2.5	191
Sweet potato	3	19–59	3.1	193
Sweet potato	4	300–440	1.5	136
Sweet potato	10	290–410	1.4	137
Taro (*Colocasia esculenta*)	4	16–40	2.5	193
Taro (*Xanthosoma spp.*)	3	14–29	2.1	193
Taro (*Alocasia macrorrhiza*)	4	12–29	2.4	193
Taro (*Cyrtosperma chamissonis*)	4	12–26	2.2	193
Wheat	15	94–141	1.5	200
Wheat	10	116–177	1.5	194
Wheat (classes)	?	60–310	5.2	195

TABLE 1   (continued)

Plant	Number of varieties tested	Concentration range[b]	Factor[c] (max./min.)	Ref.
Yam (*Dioscorea alata*)	5	15–53	3.9	193
Yam (*D. esculenta*)	5	18–44	3.0	193

Thiamin (µg / 100 g)

Plant	Number of varieties tested	Concentration range[b]	Factor[c] (max./min.)	Ref.
Apples	?	20–60	3.0	178
Apricots	2	20–30	1.5	12
Avocados	3	80–120	1.5	179
Barley[d]	80	308–728	2.3	201
Beans (snap)	171	43–117	2.7	19
Beans[d]	8	800–1,000	1.2	181
Beans (classes)[d]	9	810–1,320	1.6	182
Beans (dry)	7	390–730	1.9	199
Blueberry	10	20–27	1.3	27
Cabbage	7	56–83	1.5	31
Cabbage	6	40–100	2.5	33
Carrots	11	15–104	6.9	35
Cauliflower	?	160–230	1.4	40
Cherry	3	20–40	2.0	12
Cowpea	30	534–1,561	2.9	196
Durian (*Durio zibethinus*)	?	470–670	1.4	55
Grape	11	60–740	12.3	57
Grape	31	20–88	4.4	59
Grape	8	8–60	7.5	60
Linseed oil	9	910–1,120	1.2	202
Mango	4	57–60	1.0	78
Mango	3	50–90	1.8	79
Millet (pearl)	3	526–584	1.1	187
Nectarine	4	20–20	1.0	12
Oats[d]	24	380–630	1.7	203
Oats[d]	2	650–790	1.2	180
Oats[d]	45	389–707	1.8	201
Papaya	2	30–40	1.3	79
Peas	100	91–477	5.2	91
Peas[d]	4	1,182–1,458	1.2	94
Peaches	8	10–20	2.0	12
Peanut[d]	12	1,190–1,660	1.4	188
Pear	?	10–70	7.0	178
Plum	6	20–50	2.5	12
Potato	23	31–78	2.5	111
Sweet potato	3	43–123	2.9	193
Taro (*Colocasia esculenta*)	4	15–71	4.7	193
Taro (*Xanthosoma spp.*)	3	14–29	2.1	193
Taro (*Alocasia macrorrhiza*)	4	15–32	2.1	193
Taro (*Cyrtosperma chamissonis*)	4	12–59	4.9	197
Tomato	24	50–80	1.6	35

TABLE 1   (continued)

Plant	Number of varieties tested	Concentration range[b]	Factor[c] (max./min.)	Ref.
Wheat	15	400–690	1.7	200
Wheat	11	261–378	1.4	204
Wheat (summer)	10	360–510	1.4	205
Wheat (winter)	3	370–410	1.1	205
Wheat	10	328–624	1.9	194
Wheat (winter)	8	331–458	1.4	206
Wheat (durum)	90	234–443	1.9	206
Wheat (summer)	6	451–602	1.3	206
Wheat (summer)[d]	200	240–550	2.3	201
Wheat (classes)	?	126–990	7.9	195
Wheat (classes)	149	410–1,020	2.5	207
Yam (*Dioscorea alata*)	5	23–90	3.9	193
Yam (*D. esculenta*)	5	24–72	3.0	193

Tocopherols (μg / 100 g)

Plant	Number of varieties tested	Concentration range[b]	Factor[c] (max./min.)	Ref.
Clover (leaf)	3	2,850–3,170	1.1	163
Linseed oil	10	39,500–50,000	1.3	202
Maize[h]	42	5,500–22,400	4.1	208
Maize (embryo)[i]	100	0–40,930	—	209
Maize (embryo)[h]	100	0–13,820	—	209
Maize[h]	4	1,470–2,940	2.0	210
Maize[i]	4	3,860–7,000	1.8	210
Maize oil	125	30,000–330,000	11.0	166
Rapeseed oil	42	26,100–87,700	3.4	211
Pepper (bell)	10	450–8,240	18.3	212
Soybean oil	5	100,900–117,800	1.2	213
Sunflower oil	9	42,100–112,800	2.7	214
Wheat[h]	5	2,300–4,600	2.0	215
Wheat (classes)[h]	?	450–4,010	8.9	195
Wheat (classes)	?	490–14,010	28.6	195
Wheat (classes)[h]	8	1,430–2,600	1.8	216
Winged beans oil[i]	27	8,000–130,000	16.2	217

[a] Most of the data are from plants grown in the same geographical locations. In some cases, however, the data are from varieties grown in different geographical locations and thus the differences observed may partly be confounded by factors other than the genetic differences alone.

[b] For information as to whether the differences between different varieties were statistically significant or not, the reader is referred to the original publication. Some of the values are rounded off.

[c] The factor shows the ratio between the maximum and minimum concentration values.

[d] On the dry weight basis.

[e] Concentration based on 100 mL of juice.

[f] Total carotene.

[g] Retinol equivalent.

[h] α-Tocopherol.

[i] γ-Tocopherol.

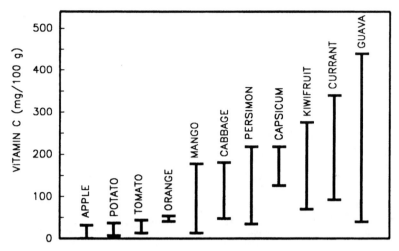

FIGURE 1. Range of vitamin C concentrations in different varieties of some fruits and vegetables. Compiled from different sources.

(Table 2). In extreme cases some very popular varieties of apples such as Granny Smith imported from Chile in May or Starking imported from Italy in June to Germany were found to contain no (zero) vitamin C.[10] In apples, the idea that older varieties are higher in vitamin C than the new varieties could not be ascertained, and taste and the acid content of apples do not seem to be related to its vitamin C content.[10] Also, the most popular varieties of tomato do not contain the highest vitamin C content.[152, 231, 232] The less popular cos-type (Romaine) lettuce variety may contain five times more ascorbic acid than the more popular crisp-head (iceberg) variety (24 vs. 5 mg/100 g, respectively).[128] Popular peach varieties (freestone) are on the average much lower in ascorbic acid than the clingstone varieties.[96]

The differences of 2–3 times or more in the ascorbic acid content of a large number of fruits and vegetables observed by so many authors and in different parts of the world (Table 1) are certainly high enough to warrant the genuine attention of the consumers and producers. There are cases reported in which the mere resistance of large commercial commodity distributors to accept a given vitamin-rich apple variety for distribution has been the main obstacle preventing it from becoming known to the consumers.[5] One would expect that a better awareness by consumers of the *invisible* differences in the vitamin content of different varieties of a given vegetable or fruit may eventually lead to higher sales of vitamin-richer products, which in turn would provide an incentive for plant breeders, producers, and distributors to accept high vitamin varieties.

B. β-Carotene

Varietal differences in carotene concentration may range from 3–4 times in apricots, cherry, maize, papaya, and peas; 5–9 times in grapefruits, oranges, peaches, and nectarines; 20 times in tomato; 46 times in Mexican peppers; 80 times in carrots; 89 times in sweet potato; and finally, up to 113 times in cassava (Table 1).

C. Other B Vitamins

Genetic differences have been observed in the concentrations of biotin, folic acid, niacin, pantothenic acid, pyridoxine, riboflavin, thiamin, and tocopherol in different fruits, vegetables, cereals, and legumes (Table 1). Varietal differences in thiamin concentrations may range from 2–3 times in barley, wheat, oats, beans, cherry, peaches, and potato varieties; to 5 times in varieties of peas; to 8 times in wheat classes (Table 1). Extreme differences of as large as 18 times have also been noted in the niacin content of mangoes, 13 times in the pantothenic content of avocados, 9 times in the pyridoxine content of wheat, and 10 times in the riboflavin content of different varieties of apples. In avocados, a fruit considered to be very rich in several B vitamins, the thiamin concentration in different varieties may range from 80 to 120 μg/100 g FW.[233] This is very close to the thiamin concentration in fish, milk, eggs, and all meats except pork.[179]

In maize, the niacin content of sugary hybrids (sweet corn) is found to be close to two times higher than that in the starchy hybrids.[185, 234, 235] There exists, however, a wide variation between different strains within the sugary and starchy hybrids.[234] A relationship between the sugary

TABLE 2
Concentration of Vitamin C in Different Apple Varieties[a]

High-vitamin varieties[b]	Vitamin C (mg / 100 g)	Low-vitamin varieties[b]	Vitamin C (mg / 100 g)
Weisser Winter-Kalvill	39	Belgolden	10
Ribston Pepping	31	Boiken	10
Undine	28	Cox Orange	10
Croncels	26	Jerseymac	10
Golden Noble (Carlisle Codlin)[c]	25	Albrechtapfel	9
Idared	24	Brettacher	9
Freiherr von Berlepsch	23	Jonathan	9
Maigold	22	Gascoyne's Scarlett	
Ananas Renette	21	Seedling	9
Habert	21	Winesap	9
Ontario	21	Jonathan	9
Northern Spy	19	Bohnapfel	8
Wegener	19	Golden Delicious	8
King of the Pippins (Goldparmäne)[c]	18	Gravensteiner	8

TABLE 2   (continued)

High-vitamin varieties[b]	Vitamin C (mg / 100 g)	Low-vitamin varieties[b]	Vitamin C (mg / 100 g)
Karmijn	18	Jacques lebel	8
Yellow Bellflower	18	Melrose	8
Kanadarenette (Paris Rambur)[c]	17	Roba	8
Boskoop	16	Auralia	7
Baumanns Renette	16	Clivia	7
Jonagold	16	Glockenapfel	7
Northern spy	16	Helios	7
White Transparant	15	James Grieve	7
Mustu	15	Richard	7
Kaiser Wilhelm	15	Singe Tillisch	7
Breuhahn	14	Alkmene	6
Gala	14	Dülmener	6
Juno	14	Gloster	6
Rheinischer Winterrambur	14	Landesberger Renette	6
Schweizer Orangenapfel	14	Laxtons Superb	6
Echter Altländer	14	Nordhausen	6
Zabergäu	14	Red Delicious	6
Zuccalmaglios Renette	14	Spartan	6
Blenheim	13	Starkrimson	6
Herma	13	Stayman	6
Lanes Prinz Albert	13	Watson Jonathan	6
Starking	13	Graham (Royal Jubille)[c]	5
Winter Bananas	13	Hammerstein	5
Golden Renette	12	Carola	4
Berner Rosen	11	Prizenapfel	4
Champagner Renette	11	Kalterer Böhmer	4
Elstar	11	Grany Smith	4
Early Blaze	11	Abbondanza	4
Glockenapfel	11	Morgenduft	4
Rome Beauty	11	Redspur	4
		Apollo	3
		Oldenburg	3
		Spartan	3
		Belford	3
		McIntosh	2
		Rote Stern Renette (Calville étilée)[c]	2

[a] Data are collected from apples grown in the different geographical locations around the world. Thus part of the difference might be due to the climatic conditions under which different varieties of apples were grown. Values are rounded off, and in some cases calculated average is based on the range of concentration given by the authors.

[b] Separation of varieties into vitamin-rich and vitamin-poor at the concentration level of 10 mg/100 g is arbitrary.

[c] The names in parenthesis are the synonym names according to reference 230.

Compiled from references 5, 8, 10, 128, 229, and 230.

endosperm and higher niacin concentration (as compared with starchy endosperm) has also been observed by other investigators.[236] Sugary kernels were shown to have a thicker niacin-rich aleuron tissue than starchy endosperms.[237] Other authors, however, are of the opinion that the thicker aleuron layer in itself is not an adequate explanation for the higher concentration of niacin in the sugary endosperms.[235]

The riboflavin concentration (μg/100 g) of different wheat classes are reported as: hard red, 110–140; soft white, 100–150; durum, 100–140.[195] Michela and Lorenz,[238] however, noted that among varieties of triticale, wheat, and rye, only some triticale varieties differed significantly in their concentration of riboflavin in the grain. Also, in pearl millet, little variation was observed in the riboflavin concentration in different varieties.[187]

In dry beans, although varieties may strongly differ in their thiamin content, the effect of environmental conditions (i.e., location of growth) on the thiamin content was considered to be much larger than the effect of variety.[181] In contrast, Kelly et al.[240] reported a threefold difference in the thiamin content of different varieties of beans and no difference in the thiamin content in plants grown in two different locations. Syltie and Dahnke[241] compared the vitamin B_1, B_2, B_6, and B_{12} content of two hard red spring wheats and found no significant differences between the varieties for their content of B_1, B_2, and B_6. Varieties, however, differed in their content of B_{12}, a situation that is hard to interpret since vitamin B_{12} is presently believed to be synthesized only by microorganisms. For a discussion on the possible uptake of vitamin B_{12} from the soil, refer to chapter 5.

No difference in the vitamin contents has also been noted in cereals. For example, Toepfer et al.[239] found no difference in the content of thiamin, riboflavin, niacin, and pyridoxine of hard red wheat and soft wheat. In Canada, Nik-khah et al.[180] did not find any significant difference in the riboflavin content of wheat, barley, and oat kernels grown for five years at 16 different locations in Saskatchewan. Nik-khah et al.[180] also did not find any significant difference in the thiamin content of two wheat varieties and three varieties of barley but found significant differences in the thiamin content of two varieties of oats grown in Saskatchewan.

Plant varieties may differ in their tocopherol contents (Table 1). In 1975, for example, U.S. durum wheat contained 2890 μg/100 g of tocopherol whereas soft white spring wheat contained 470 μg/100 g, i.e., a difference of more than sixfold. Avon (a soft white winter wheat) was noted to have 490 μg/100 g of total tocopherol while Coulee (a hard white winter wheat) contained as much as 14,010 μg/100 g, i.e., a difference of more than 28-fold![195]

In maize embryos, tocopherols may show a very wide range of concentrations: α-tocopherol from 0.0 to 138.2 and γ-tocopherol from

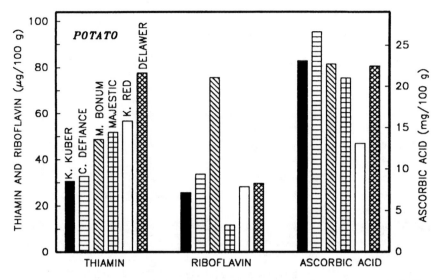

FIGURE 2. Concentrations of three vitamins in six varieties of potato grown in India. Varieties are arbitrarily arranged according to their thiamin content.[111]

0.0 to 40,930 μg/100 g. It was suggested that through selection, the total tocopherol of maize could be increased by as much as 35%.[209] In another study with 20 maize hybrids, the concentration of α-tocopherol in kernels was found to vary from 910 to 6440 μg/100 g (a difference of more than sevenfold) and that of γ-tocopherol, from 1,360 to 12,870 μg/100 g, or by a factor of more than nine times![235] Variety differences in the tocopherol concentration have also been found in alfalfa, corn, and paprika.[242]

II. RELATIONSHIPS BETWEEN VARIOUS VITAMINS

Is a given plant variety rich in one vitamin also rich in other vitamins? The answer to this question seems to be no. For example, potato tubers may differ in their thiamin concentration by a factor of more than 2 (30.6–77.6 μg/100 g FW), but those varieties with the highest thiamin concentration (Kufri Red and Delaware) were not necessarily those with the highest ascorbic acid or riboflavin (Figure 2).[111] Arbitrary arrangement of varieties of peas in the descending order of their thiamin content revealed that there exist no apparent relationships between the content of thiamin and that of ascorbic acid, carotene, or riboflavin in different varieties.[91] In sweet potato, the variety Centennial was highest in ascorbic acid, riboflavin, and carotenoids, while the variety Jasper was highest in niacin.[136]

III. NUTRITIONAL SIGNIFICANCE OF VARIETAL DIFFERENCES IN PLANT VITAMINS

Despite the large differences in the vitamin C content of various fruits and vegetables (Table 1), no single publication could be found in which the nutritional significance of varietal difference in vitamin C was addressed. One reason might be that up to now no one had collected such a large volume of data on varietal differences, and thus the scope of difference had gone unnoticed by most scientists. Without going into further detail, it suffices to mention that the large genetic variability in the concentration of vitamin C in different fruits and vegetables reported here needs to be taken seriously by scientists as well as by the general consuming public.

It is reasonable to expect that such huge variations in the vitamin content of different varieties could have real nutritional relevance. For example, Schuphan[5] used the case of high- and low-vitamin C varieties of apples such as Berlepsch and Golden Delicious (having 23 and 8 mg of ascorbic acid/100 g, respectively) to demonstrate this case. He proposed that if one were to cover the daily vitamin C need of a child (50 mg) with apples only, then one needed to give the child either one Berlepsch apple (230 g with 4% nonedible parts) or six Golden Delicious apples (710 g with 10% inedible parts)! It is, however, much easier to convince a child to eat one relatively large vitamin-rich apple rather than six relatively small and vitamin-poor apples a day.[5]

Do the large differences observed in the concentrations of other vitamins, such as carotene in carrots, peppers, guavas, and sweet potatoes, niacin and pyridoxine in wheat, and pantothenic acid in avocados, for example, have any nutritional significance? According to some authors, the difference in thiamin content of wheat varieties (classes) could be of considerable importance in the ordinary human nutrition if they are consumed without much milling.[91] If milled, however, the significance of genetic difference in thiamin content for the nutritional quality of cereal products is questionable since a considerable amount of thiamin is removed from the cereal grain during the milling process. Furthermore, there seems to be no relationship between the thiamin content in the wheat kernels and in its flour so that a wheat high in thiamin may not produce high-thiamin flour when compared with a wheat lower in thiamin.[243] Brown rice varieties may significantly differ in their content of riboflavin, nicotinic acid, and biotin. When milled, however, rice varieties differ in their content of pantothenic acid only.[244] Finally since the niacin present in cereal grains is mostly unavailable,[245] differences in the content of this vitamin in different cereals are suggested to be nutritionally inconsequential.

Although the above argument holds true for plant products polished, peeled, or milled prior to consumption, it is not valid for other products such as peas, beans, cabbage, cherry, grape, peaches, apples, and others, which are usually consumed whole. In these cases, any genuine difference in the vitamin content of different cultivars would probably translate into a higher intake of that particular vitamin by the consumer.

IV. COLOR AND VITAMIN CONTENT

Can the vitamin content of fruits be judged by their color or flavor? Although no scientific reason is known to us for such a relationship, some reports suggest that more colorful varieties may be higher in some vitamins. Not all the reports are, however, consistent.

A. Ascorbic Acid

In strawberry, for example, climatic conditions seem to affect the color of the fruits to a much lesser degree than their content of ascorbic acid, and therefore no close relationship between color and ascorbic acid content in this fruit could be observed.[141] Furthermore, those climatic conditions conducive to good color development do not appear to be most favorable for a high vitamin C content in strawberry. Since varieties respond differently to climatic conditions, it is not uncommon to find one variety having the best color but another having the highest vitamin C.[246] In contrast, Morris and Sistrunk (1991)[133] reported a highly significant correlation between the color and the ascorbic acid content among a large number of strawberry genotypes and noted that "...it appeared that there is a possibility of breeding and selecting for highly colored fruit with high ascorbic acid values."

Colored as compared with "white" varieties of some fruits and vegetables appear to be higher in their content of some vitamins. For example, red grapes were found to contain higher concentrations of thiamin, riboflavin, niacin, pyridoxine, pantothenic acid, and folic acid than white grapes.[247] Colored (and often high-sugar) varieties of grapes were also noted to be higher in ascorbic acid[248] as compared with white (and often low-sugar) varieties. Black currants have 4.5 times more ascorbic acid than the red or white varieties[249] (Table 1). Yellow varieties of passion fruit are noted to be slightly lower in ascorbic acid[250, 251] and considerably lower in vitamin A activity than the purple varieties.[252] Guavas grown in Brazil[67] contained ca. four times more ascorbic acid than those grown in Papua New Guinea,[253] but in both locations, white cultivars tended to have higher ascorbic acid than the red cultivars (in Brazil 372 vs. 215 and in New Guinea 68 vs. 52 mg/ 100 g in the white and red varieties, respectively).

Colored varieties of lettuce containing anthocyanin are higher in ascorbic acid than those lacking these pigments,[255] and the red cabbage

TABLE 3
Relationships Between the Color and the Carotene Concentration
(μg / 100 g) in the Carrot Roots[262]

Color of carrots	Total carotenoid	α-Carotene	β-Carotene
White	200	Trace	Trace
Yellow	1,430	30	210
Yellow-orange	2,020	100	770
Intermediate orange	4,070	460	1910
Light orange	6,750	1,750	3,640
Orange	10,390	3,210	5,650
Dark orange	13,110	4,670	6,910

varieties have higher ascorbic acid and choline content than the white varieties.[256] In asparagus, tips of a green variety (Mary Washington) were noted to contain more than twice the ascorbic acid than the two white varieties (51.2 vs. 18–23, respectively).[257] Also, covering the asparagus with black plastic produced light-colored spears, as compared with green spears produced under uncovered conditions, but also reduced their ascorbic acid by 50%.[258] Rose hips from red-dark pink and pure white species were noted to be higher in vitamins than the hips from yellow, cream, or light-pink flowering roses.[254] In *Capsicum*, as the fruit matures and its red color intensifies, not only does its β-carotene content increase tremendously (chapter 6), its ascorbic acid content also increases, so that red *Capsicum* may have 3–4 times more ascorbic acid than the green ones.[259, 260] In apples, colored portions were noted to have significantly higher ascorbic acid content than uncolored portions (see Table 5, chapter 4).

B. β-Carotene

Since most carotenoids are colorful pigments, there appears to be some relationship between the color and the β-carotene content in most fruits and vegetables studied. For example, carrots having orange and red colors are much higher in β-carotene than those varieties having yellow or white color (which in some cases may be fully devoid of carotenes)[158, 261, 262] (Table 3).

β-carotene is the principal yellow pigment in sweet potato, and thus there is a close relationship between the tuber's color and its β-carotene content. The β-carotene content, however, increases much more rapidly than the total pigments as the flesh of a variety becomes more deeply colored. Thus the sweet potato varieties with yellow-, orange-, or salmon-colored flesh contain a much higher concentration of

β-carotene than the varieties with white, creamy light-colored tubers, which may be fully devoid (have a zero concentration!) of β-carotene.[135, 158, 176, 263] The difference in the β-carotene of sweet potato varieties may be as high as 89 times (44,600 vs. 500 μg/100 g DW, respectively)[135] or much higher (Table 1).

In tomatoes, red and yellow varieties were found to contain somewhat similar ascorbic acid,[128] but the orange-colored variety (High Beta) may contain 25 times more β-carotene (30.6 vs. 1.4 ppm, respectively) than the red variety.[264] Thus pink- or white-colored varieties have much less vitamin A potency than the red-colored varieties.[149] In pumpkin, however, the orange-colored variety is higher in total carotenoids, and the yellow-colored variety is higher in vitamin A activity.[265]

In grapefruit the total carotenoid concentration and that of β-carotene do not vary proportionally in different varieties. Thus although varieties differed by a factor of more than ninefold in their total carotenoid (1210 to 11,170 μg/100 g), they differed by a factor of only 2.3-fold (420 to 960 μg/100 g) in their β-carotene content[164] (Table 1).

In the orange-fleshed papaya, the pigment β-cryptoxanthin, which has a 50% vitamin A activity, makes up to 62% of the total carotenoid present in the fruits. The consumers, however, prefer the red-colored variety Tailandia, whose color is mainly due to its high lycopene content (56–66% of the total carotenoids).[168]

Purple-colored passion fruit is higher in carotene and provitamin A activity than the yellow varieties.[250, 252] Red-colored watermelons are higher in total carotenoids and 14 times higher in β-carotene equivalent than the orange-colored varieties (1.4 vs. 0.1 mg/100 g FW, respectively), and red pepper is almost 11 times higher in β-carotene than green pepper (43.8 vs. 4.1 mg/100 g FW, respectively).[266]

Davidyuk and Vshivkova[267] compared the carotenoid content in the leaves and fruit of four yellow-fleshed and four white-fleshed varieties of peaches. They noted that the leaf carotenoids increased during the growing season and reached their maximum in late August. In autumn, the leaves of yellow-fleshed cultivars contained nearly twice as much carotenoids as the white-fleshed varieties. They concluded that a significant correlation exists between the accumulation of carotene in the leaves and in the flesh of peaches.

V. EFFECT OF ROOTSTOCKS ON THE FRUIT VITAMINS

Fruit trees are often grafted onto the roots of other varieties of the same plant or related plant species (rootstock) to improve the quality of the fruits and/or resistance of plants against various pathogenic organ-

TABLE 4
**Effect of Rootstocks on the Ascorbic Acid Concentration
in Sour Cherry Fruit[275]**

Rootstock	Ascorbic Acid (mg / 100 g DW)
P. avium	103[a]
P. mahaleb	79
Cer W8	108
Cer W13	117
Cer W15	94

[a] Values rounded.

TABLE 5
**Effect of Rootstocks on the Ascorbic Acid Concentration (mg / 100 mL)[a]
in the Fruit Juice of Four Orange Cultivars[292]**

	Rootstocks			
Cultivar	Sour orange	Sweet lime	Rough lemon	Egyptian lime
Bizzaria	68	60	58	57
Kinarti	66	62	56	61
Shamouti	56	53	50	46
Sigillata	63	68	62	57

[a] Values rounded.

isms.[268] There are numerous reports indicating that rootstocks can also affect the concentration of ascorbic acid in apricots,[269] apples,[270-274, 300] cherry[275-277] (Table 4), citrus fruits[278-292] (Table 5), grape,[293, 294] guava,[295] and *Acerola*,[296] that of carotene in grapefruit[297] and stasuma leaf,[298] and that of B vitamins in peaches.[299]

Effect of rootstocks on the vitamin C content of fruits seems to be altered under different environmental conditions,[284] and thus the effect of rootstocks on the vitamin C of citrus fruits, for example, appears to be dependent on the geographical location they are used in. For example, although sour orange rootstocks, as compared with other rootstocks, produced oranges in Florida[278] and pomelos in Spain[301] with relatively higher ascorbic acid, in Egypt, however, it was the sweet lemon rootstock that resulted in the highest ascorbic acid content in oranges.[287] In contrast to these, under Indian conditions rootstock did not affect the ascorbic acid content of oranges.[302, 303]

We note that rootstocks do not seem to affect the vitamin content and other quality characteristics of fruits in the same manner. For example, fruits of thorn-less Mexican lime grafted on *C. taiwanica* had higher ascorbic acid content than when grafted on other rootstocks but the trees produced less marketable fruits.[290] In sweet oranges, fruits grafted on sweet orange rootstock had the highest amount of juice and total soluble solids but those grafted on sour cherry had the highest vitamin C content.[281] Also, mandarins grafted on citrange rootstocks produced the more juicy fruits but their juice was lowest in ascorbic acid and carotenoids as compared with those grafted on Emperor mandarin, which had the opposite effect.[288] In sour cherry, trees grafted on *Prunus mahaleb*, compared with eight other rootstocks tested, produced the largest fruit yield but the fruits had the lowest quality with respect to sugar, total acids, and vitamin C content.[275]

In general, the effect of rootstocks on the vitamin C content is relatively small when compared with the effects discussed in other chapters. For example, the ascorbic acid concentrations in the oranges grown on two different rootstocks ranged from 37 to 42 mg/100 g.[289]

Rootstocks may also affect the rate of vitamin loss in the fruits after they have been harvested. For example, Arlt,[271] by comparing the ascorbic acid content of four varieties of apples grafted on 10 different rootstocks, noted that apples grown on certain rootstocks retained their ascorbic acid during storage much better than the same varieties grafted on other rootstocks.

VI. BREEDING PLANTS FOR HIGHER VITAMIN CONTENT

Up to now, the aim of most breeding efforts has been to increase yield; increase contents of starch, protein, or oil; improve sensory qualities such as flavor, color, aroma, and size; increase resistance to various pests; or decrease the content of some undesirable constituents.[338,339] Generally, relatively less attention has been given by scientists of the industrialized Western nations to increasing the vitamin content of food plants. Even in recent books on the criteria for improving the qualities of fruits and vegetables, the major emphasis has been placed on the sensory and cosmetic criteria (appearance, size, shape, color, texture, aroma, absence of defects, and tastes). As for the nutritional criteria, however, it is suggested that at the present time vitamin content of plants is rarely utilized by the consumers as a quality criterion in choosing different commodities. Also, in some recent congress proceedings one finds very few or no reports on the breeding efforts in increasing the vitamin content of fruits or vegetables or cereal grains.[11,133,339–345]

Although increasing the concentrations of some plant vitamins, such as ascorbic acid, through breeding is considered to be achievable,[292, 346] the lack of interest on the part of plant breeders to work in this area is surprising if one takes note of the large genetic difference so frequently observed in the concentrations of several vitamins in some important fruits and vegetables (Table 1).

That agronomic qualities such as resistance to various pests, keeping quality under storage, as well as taste, shape, and other culinary aspects are more important than pure vitamin content for the success of a given fruit or vegetable with the growers, distributors, and consumers can be illustrated using apples as an example. Thus, although newer varieties of apples may lack the flavor of older varieties, they are, however, considered more "foolproof" for the growers and distributors.[11] Furthermore, although some older apple varieties may have as high as 22 mg/100 g of ascorbic acid (Fisher & Kitson, 1991), most of the newer varieties are among those lowest in ascorbic acid content (Table 2)! The new varieties, however, because of the above-mentioned reasons, are grown by more and more farmers and thus are consumed more and more. In Switzerland, for example, a calculation based on the total tonnage of various apples produced shows that a relatively vitamin-poor apple variety such as Golden Delicious made up to 29.5% of all the apples produced, while the relatively vitamin-rich varieties such as Jonagold and Maigold made up only 8.8% and 5.2% of the total apples produced in this country in 1992.[347]

Most of the breeding efforts to improve plant vitamins have taken place in the former Soviet Union and Eastern European countries and are reported in non-English languages for which only English abstracts are available.[131, 304–337] Unfortunately, from the short abstract of most of these reports it is hard to judge whether these, in some cases very extensive breeding programs, have or have not led to any commercial use of the varieties high in vitamins. Following is a review of some of the breeding work already done for improving the vitamin content of fruits and vegetables.

A. Ascorbic Acid

Tomato is one of the major sources of ascorbic acid in human nutrition in many parts of the world. The breeding of tomato for high vitamin C content, however, has apparently not been very successful although Stevens and Rick[152] cite data showing that tomato varieties released in 1972 contained approximately 25% more ascorbic acid than those released in 1952.

Pospíšilová,[346] by crossing wild tomatoes (*Lycopersicon pimpinellifolim*) with cultivated types (*L. esculentum*), showed that the ascorbic acid content of the fruits of the hybrid plants was intermediate between those of the parents, and thus breeding was recommended for developing tomatoes with higher ascorbic acid content. According to other authors, however, there is no guarantee that once marketed, people would actually buy tomato varieties high in vitamin C since these varieties, along with their relatively lower yield, have a different shape, color, and taste than the varieties people are now used to.[152, 232, 349] In the words of others, people are more interested in the color of their vegetables and nobody asks about their vitamin content,[350] at least up to now!

According to others, the reason for the lack of success in breeding fruits or vegetables for high vitamin content is the economics of cost and return for this kind of undertaking and the fact that part of the plant vitamins are lost prior to their consumption in any case. For example, oranges and potatoes require several years for breeding and the efforts made so far have not resulted in any major change in the ascorbic acid content of the varieties released. In potato, there is a strong reduction in the ascorbic acid content during storage, and thus unless potato varieties could be found that retain their vitamins better during storage, a long-term attempt to increase the ascorbic acid content of potatoes and oranges is considered to be unrealistic.[349] If, however, one considers that different varieties may strongly differ in the rate at which they lose their vitamins during storage,[48] it seems logical to argue that selection for those varieties that lose their vitamins at a slower rate during food processing and storage may be worthwhile.

We note that although high ascorbic acid tomato (Double Rich) and cabbage (Huguenot) have been available to seed producers for a long time, no commercial production of these seeds has been undertaken![349] Also, in the former Soviet Union, despite a large number of published reports in breeding attempts for high-vitamin plants (see above), and although breeding for cucumbers, lettuce, celery, and parsley high in vitamin C has been reported, there are no indications whether any commercial varieties have been released.[340]

The only commercial success story known to this author is that of black currant. The reason for this success appears to be the enthusiasm of food companies marketing the black currant juice high in vitamin C, and thus they actively promoted the production of those black currant varieties high in ascorbic acid, a point that had very good consumers' appeal.[351]

Finally, the observation that duration and intensity of light available during plant growth and location of growth may alter the ascorbic acid content of plants by several fold (see chapter 4) makes any breeding program to improve the content of this vitamin in plants very difficult indeed. In other words, any newly bred high-vitamin C variety needs to maintain its higher vitamin content despite the strong effects of environmental factors imposed on it.

B. β-Carotene

1. Carrot

β-carotene, because of its colorful nature, would inevitably change the color of any plant part bred to be high in this compound. For example, high-β-carotene carrots are dark-orange in color while the yellow or white varieties are relatively low in this vitamin (Table 3). Available information indicates that breeding of carrots for high β-carotene has been successful without any adversely affecting other culinary qualities of carrots.[352] It is not known whether the high-carotene varieties of carrots with their "new" and "unfamiliar" color have gained the acceptance of the consumers or not. Based on other reports that, for example, local customs in some countries may even dictate the use of carotene-poor fruits and vegetables such as white-colored melons or sweet potatoes, one might suspect some difficulties in changing the habits of people toward any new variety with its unfamiliar or "undesired" color.[158, 262, 352–355]

2. Tomato

In tomato, a difference of more than 100 times (60 vs. 6750 μg/ 100 g) has been noted in the β-carotene content of progenies from crosses between the common red tomato (*L. esculentum*) and *L. hirsutum*.[128] As in the case of carrots, the customary habits and familiarity for color, shape, and taste seem to play a much stronger role than the vitamin concentration in the way consumers choose their tomatoes. Thus, demand for red tomatoes, which are high in lycopene but low in β-carotene, is much higher than for dark-orange-colored tomatoes, which are high in β-carotene.[158, 264, 352, 356] Therefore, although breeding of tomatoes for high carotene has resulted in Caro-Red and Caro-Rich varieties, which are 10 times richer in β-carotene than the conventional varieties, they have not been purchased enthusiastically by consumers because of their orange-red color! In a taste trial, participants reported that the carotene-rich tomato variety has a distinctly different flavor than the conventional variety. When color of the fruits was masked, however, participants could not detect any difference between the taste of the carotene-rich and carotene-poor varieties. This observation clearly indicates that the participants were biased for the red-colored tomatoes![128]

In a marketing trial, Munger[349] persuaded two growers to grow the Caro-Rich tomato. The growers selected were selling their vegetables directly to consumers in a high-income area whose residents were believed to be interested in buying products higher in nutritive values. In both cases, however, the consumers did not show much interest in the new tomato variety and growers refused to grow it for the second time! As a result of these and similar disappointing breeding attempts, breeders continue their work with good-looking (high-lycopene), low β-carotene tomatoes.[340]

3. Sweet Potato

Breeding has been successfully used to increase the β-carotene content of sweet potato,[176] and the new varieties have about three times higher β-carotene in their roots than the older varieties used some 40 to 50 years ago.[340,349] This, however, might not necessarily increase the share of sweet potato in the supply of carotene in the human diet since the per capita consumption of sweet potato has been declining, and therefore the average amount of provitamin A from sweet potato is lower now than in the past when lower-carotene varieties were cultivated.[349]

C. Other Vitamins

Breeding of cereals such as wheat for higher thiamin content is considered to be possible.[357] But because bran and germ tissues contain a major portion of the kernel's vitamins (chapter 2), and since commercial milling may remove ca. 68% of the thiamin, 58–65% of the riboflavin, and 85% of the pyridoxine contained in the wheat kernel,[358] there is little interest in selecting wheat and other cereals high in these vitamins,[359] only for them to be discarded by the milling process!

Are "older" varieties of cereals higher in vitamin than the "new" varieties? According to Witsch and Flügel,[360] who compared the thiamin content of old (less fertilizer-demanding) and newly selected (and highly fertilizer-demanding) varieties of wheat grown under similar soil and climatic conditions, kernels of older varieties were richer in thiamin than the newly bred varieties.

Niacin content of maize kernel could be increased through selection.[236,361,362] Breeding of high-niacin plants, however, is reported to be impractical since a major fraction of niacin in maize[235] and potato[363] is bound by phytin and is thus nutritionally unavailable. As for pantothenic acid, riboflavin, and thiamin, however, despite the genetic differences observed in different maize hybrids, the strong interactions between environmental and genetic factors have been one of the main reasons why up to now no major improvements have been achieved in increasing the concentration of these vitamins in maize.[235,361]

The above arguments seem to also hold true for the effect of fertilizers on plant vitamins, as discussed later (chapter 5). For example, top-dressing of rice with nitrogen fertilizer was found to significantly decrease the riboflavin but increase the niacin content of brown rice. After milling, however, no effect of fertilizer could be detected in the content of these vitamins in the milled rice[244] apparently because top-dressing with nitrogen fertilizer had only changed the vitamin content in those grain tissues (parts) that were later removed by the milling process!

According to some reports, scientists in the former Soviet Union have been involved in improving the niacin, pantothenic acid, thiamin, and pyridoxine contents in chili varieties through chemomutagenesis. There is no indication, however, whether they have succeeded in releasing commercial varieties high in these vitamins.[340]

In summary, although genetic variability in plant vitamins is available, successful cases of breeding and marketing of high-vitamin fruits or vegetables are rather limited. It is reasonable to assume that once consumers, through public awareness and education programs by nutritionists, distributors, or farmers, realize that, depending on the variety of produce they choose from the supermarket shelf, there might be a severalfold difference in the amount of vitamins, such as vitamin C, in the product they purchase, they may gradually change their habits and purchase the newly developed higher-vitamin varieties when and if they become available.

REFERENCES

1. Skard, O. and Weydahl, E., Ascorbic acid—vitamin C—in apple varieties, *Medl. Nor. Landbrukshoegsk.*, 30, 477, 1950.
2. Marx, Th., Über den Oxydationsschutz der L-Ascorbinsäure in frischem Pflanzenmaterial, *Z. Lebensm.-Unters. Forsch.*, 99(3), 180, 1954.
3. Patutina, L. G., Saumjan, K. V., and Sljapnikova, A. S., Vitamin-C-Gehalt im Obst, *Landwirt. Zentralbl.*, 19(11), 2857, 1974; *Hortic. Abstr.*, 45, 9325, 1975.
4. Schwerdtfeger, E., Wertgebende Inhaltsstoffe bei Obst und Gemüse in Abhängigkeit von Erbgut, Kulturmassnahmen und Verarbeitung, *Bericht der Landwirt.*, 54, 73, 1976.
5. Schuphan, W., *Mensch und Nahrungspflanze*, Dr. Junk Verlag, Den Haag, 1976.
6. Karklina, D., Chemical composition of new apple varieties, *LLA Raksti*, 196, 51, 1982; *Chem. Abstr.*, 98, 52159b, 1983.
7. Gersbach, K., Vitamin-C in Früchten (vor allem in Äpfeln), *Züri-Obst*, 45(10), 2, 1984.
8. Pätzold, G. and Finke, M., Ergebnisse analytischer Untersuchungen beim Apfel, *Gartenbau*, 31(2), 52, 1984.

9. Kovalenko, G. K., Shirko, T. S., Yaroshevich, I. V., Evdokimenko, V. M., and Yarokhovich, L. M., Chemical composition of promising regional (Belorussian) apple varieties, *Vestsi Akad. Navuk BSSR, Ser. Sel'skagaspad. Navuk*, 2, 79, 1984; *Chem. Abstr.*, 101, 71330x, 1984.

10. Hansen, H. and Bohling, H., Welche Bedeutung hat der Apfel für die Vitamin C-Versorgung, *Obstbau*, 9(5), 230, 1984.

11. Fisher, D. V. and Kitson, J. A., The Apple, in *Quality and Preservation of Fruits*, Eskin, N. A. M., Ed., CRC Press, Boca Raton, Fla., 1991, 46.

12. Wills, R. B. H., Scriven, F. M., and Greenfield, H., Nutrient composition of stone fruit (*Prunus* spp.) cultivars: apricot, cherry, nectarine, peach and plum, *J. Sci. Food Agric.*, 34, 1383, 1983.

13. Berre, S., Gallon, G., and Tabi, B., Vitamin C content in Cameroon tubers and plantain before and after cooking, *Annales de la Nutrition et de l'Alimentation*, 23(1), 31, 1969; *Food Sci. Technol. Abstr.*, 1(7), J556, 1969.

14. Thomas, P., Dharkar, S. D., and Sreenivasan, A., Effect of gamma irradiation on the postharvest physiology of five banana varieties grown in India, *J. Food Sci.*, 36, 243, 1971.

15. Jayaraman, K. S., Ramanuja, M. N., Dhakne, Y. S., and Vijayaraghavan, P. K., Enzymatic browning in some banana varieties as related to polyphenoloxidase activity and other endogenous factors, *J. Food Sci. Technol.*, 19, 181, 1982.

16. Seenappa, M., Laswai, H. S. M., and Fernando, S. P. F., Availability of L-ascorbic acid in Tanzanian banana, *J. Food Sci. Technol.*, 23(5), 293, 1986.

17. Shapiro, D. K., Anikhimovskaya, L. V., Narizhnaya, T. I., and Vereskovskii, V. V., Chemical characterization of fruit from several species of *Berberis* L. introduced into the *BSSR*, *Rastit. Resur.*, 19(1), 84, 1983; *Chem. Abstr.*, 98, 104347h, 1983.

18. Heinze, P. H., Kanapauk, M. S., Wade, B. I., Grimball, P. C., and Foster, R. L., Ascorbic acid content of 39 varieties of snap beans, *Food Res.*, 9, 19, 1944.

19. Hayden, F. R., Heinze, P. H., and Wade, B. L., Vitamin content of snap beans grown in South Carolina, *Food Res.*, 13, 143, 1948.

20. Niketić, G. and Sekulić, M., A study of the technological value of some varieties of snap bean, *Zborn. Rad. Poljopriv. Fak. Beograd*, 8(301), 1, 1960; *Hortic. Abstr.*, 32, 6581, 1962.

21. Marchesini, A., Majorino, G., Montuori, F., and Cagna, D., Changes in the ascorbic and dehydroascorbic acid contents of fresh and canned beans, *J. Food Sci.*, 40, 665, 1975.

22. Chadha, K. L., Gupta, M. R., and Bajwa, M. S., Performance of some grafted varieties of ber (*Zizyphus mauritiana* Lamk.) in Punjab, *Indian J. Hortic.*, 29, 137, 1972.

23. Singh, B. P., Singh, S. P., and Chauhan, K. S., Certain chemical changes and rate of respiration in different cultivars of ber during ripening, *Haryana Agric. Univ. J. Res.*, 11(1), 60, 1981; *Food Sci. Technol. Abstr.*, 13(11), J1782, 1981.

24. Kundi, A. H. K. Wazir, F. K., Ghafoor, A.; and Wazir, Z. D. K., Physico-chemical characteristics and organoleptic evaluation of different ber (*Ziziphus jujuba* Mill.) cultivars, *Sarhad J. Agric.* 5(2), 149, 1989; *Hortic. Abstr.*, 62, 884, 1992.

25. Hansen, E. and Waldo, G. F., Ascorbic acid content of small fruits in relation to genetic and environmental factors, *Food Res.*, 9, 453, 1944.

26. Matzner, F., Über den Vitamin-C-Gehalt der Kulturheidelbeeren, *Erwerbsobstbau*, 7(6), 105, 1965.

27. Bushway, R. J., McGann, D. F., Cook, W. P., and Bushway, A. A., Mineral and vitamin content of lowbush blueberries (*Vaccinium angustifolium* Ait.), *J. Food Sci.*, 48, 1878, 1983.

28. Lenartowicz, W., Zbroszczyk, J. and Plocharski, W., The quality of highbush blueberry fruit. Part. I. Fresh fruit quality of six highbush blueberry cultivars and their suitability for freezing, *Fruit Sci. Reports*, 17(2), 77, 1990; *Hortic. Abstr.*, 62, 3751, 1992.

29. Domagala, F. and Kmiecik, W., Comparison of the chemical composition of some cultivars of Brussels sprout with regard to suitability for the processing industry, *Acta Agrar. Silvestria, Ser. Agrar.*, 24, 119, 1985; *Chem. Abstr.*, 105, 59663h, 1986.

30. Burrell, R. C., Brown, H. D., and Ebright, V. R., Ascorbic acid content of cabbage as influenced by variety, season, and soil fertility, *Food Res.*, 5, 247, 1940.

31. Poole, C. F., Heinze, P. H., Welch, J. E., and Grimball, P. C., Differences in stability of thiamin, riboflavin, and ascorbic acid in cabbage varieties, *Proc. Am. Soc. Hortic. Sci.*, 45, 396, 1944.

32. Branion, H. D., Roberts, J. S., Cameron, C. R., and McCready, A. M., The ascorbic acid content of cabbage, *J. Am. Diet. Assoc.*, 24, 101, 1948.

33. Golubev, V. N. and Reut, O. V., Contents of biologically active substances in varieties of cabbage, *Iyv. Vyssh. Uchebn. Zaved., Pishch. Tekhnol*, 2, 131, 1988; *Chem. Abstr.*, 109, 53497t, 1988.

34. Rahman, F. M. M., Buckle, K. A., and Edwards, R. A., Changes in total solids, ascorbic acid and total pigment content of capsicum cultivars during maturation and ripening, *J. Food Technol.*, 13, 445, 1978.

35. Leveille, G. A., Bedford, C. L., Kraut, C. W., and Lee, Y. C., Nutrient composition of carrots, tomatoes and red tart cherries, *Fed. Proc.*, 33(11), 2264, 1974.

36. Nagaraja, K. V. and Krishnan Nampoothiri, V. M., Chemical characterization of high-yielding varieties of cashew (*Anacardium occidentale*), *Qual. Plant.—Plant Foods Hum. Nutr.*, 36, 210, 1986.

37. Ogunsua, A. O. and Adedeji, G. T., Effect of processing on ascorbic acid in different varieties of cassava (*Manihot esculenta*, Crantz), *J. Food Technol.*, 14, 69, 1979.

38. Bradbury, J. H. and Singh, U., Ascorbic acid and dehydroascorbic acid content of tropical root crops from the South Pacific, *J. Food Sci.*, 51(4), 975, 1986.

39. Jaiswal, S. P., Singh, J., Karus G., and Thakur, J. C., Some chemical constituents of cauliflower curd and their relationship with other plant characters, *Indian J. Agric. Sci.*, 44(11), 726, 1974.

40. Korobkina, Z. and Druzhinskaya, L. P., Vitamin value of frozen green peas and cauliflower, *Konservn. Ovoshchesush. Prom-st*, 11, 25, 1982; *Chem. Abstr.*, 98, 87878n, 1983.

41. Wedler, A., Overbeck, G., and Hentschel, H., Wertstoffanalysen und sensorische Bewertung von Knollensellerie-Sorten (*Apium graveolens var. rapaceum* L.), *Gartenbauwissenschaft*, 47(2), 85, 1982.

42. Jacquin, P., Vitamin C in cherries, *Ann. Technol. Agric.*, 14, 157, 1965; *Hortic. Abstr.*, 36, 403, 1966.

43. Stajic, N., Pomological characteristics and chemical composition of fruits in some important sour cherry cultivars, *Jugosl. Vocarstvo*, 18(69-70), 3, 1984; *Chem. Abstr.*, 103, 159340c, 1985.

44. Kostova, R., The chemical composition of sour cherries in relation to variety, precipitation and temperature, *Izv. Nauč.-izsled. Inst. Ovoštarstvo, Kostinbrod*, 2, 47, 1962; *Hortic. Abstr.*, 34, 413, 1964.

45. Sheets, O. A., Prementer, L., Wade, M., Gieger, M., Anderson, W. S., Peterson, W. J., Rigney, J. A., Wakeley, J. T., Cochran, F. D., Eheart, J. F., Young, R. W., and Massey, P. H., Jr., The nutritive value of collards, *South. Coop. Series Bull.*, 39, 5, 1954.

46. Matzner, F., Fruchtmerkmale und Inhaltsstoffe der Kulturpreiselbeeren (*Vaccinium macrocarpon* Ait.), *Erwerbsobstbau*, 13(7), 120, 1971.

47. Olliver, M., The ascorbic acid content of fruits and vegetables, *Analyst*, 63, 2, 1938.
48. Fraiman, I. A., Comparative study of the vitamin composition of black currant varieties, *Sadovod., Vinograd. Vinodel. Mold.*, 31(5), 53, 1976; *Chem. Abstr.*, 85, 119559j, 1976.
49. Shirko, T. S. and Yaroshevich, I. V., Change in the chemical composition of Belorussian black currants, *Konserv. Ovoshchesush. Prom-st.*, 6, 1983; *Chem. Abstr.*, 99, 69134h, 1983.
50. Shenshina, S. V., Untilova, A. E., and Levitskaya, E. S., Biochemical composition of strawberries, black currants and raspberries, *Sadovod. Vinograd. Vinodel. Mold.*, 38(2), 56, 1983; *Chem. Abstr.*, 98, 142130q, 1983.
51. Sojak, S. and Takac, J., The ascorbic acid content in some varieties of small fruits, *Vedecke Prace Vyskumneho Ustavu Ovocnych a Okrasnych Drevin v Bojniciach*, 5, 163, 1984; *Plant Breeding Abstr.*, 55, 1268, 1985.
52. Soskic, A. and Jasarevic, F., Major chemical components of fruits in some black currant cultivars, *Jugosl. Vocarstvo*, 20(75-76), 693, 1986; *Chem. Abstr.*, 107, 57691h, 1987.
53. Yankelevich, B. B., Melekhina, A. A., and Eglite, M., Vitamin content of black currants, *Tautsaimn. Derigo Augu Agroteh. Biol.*, Zinatne, P. T., Ed., Riga, USSR, 1986, 114; *Chem. Abstr.*, 107, 22197z, 1987.
54. Haffner, K. and Heiberg, N., Vitamin C in black currants, *Gartneryrket*, 76(5), 116, 1986, *Hortic. Abstr.* 58, 5536, 1988.
55. Woller, R., and Idsavas, A., Knowledge of exotic fruits, *Fluess. Obst.* 48(4a), 180, 1981; *Chem. Abstr.* 95, 39064k, 1981.
56. Chandra, S. and Arora, S. K., An estimation of protein, ascorbic acid and mineral matter content in some indigenous and exotic varieties of Gram (*Cicer arietinum* L.), *Curr. Sci.*, 37(8), 237, 1968.
57. Aizenberg, V. Ya. and Bagdasarayan, T. M., Vitamin content in several table wine varieties of grapes, *Izv. S-Kh. Nauk*, 7, 33, 174; *Chem. Abstr.* 83, 41450d, 1975.
58. L'yanova, Kh. Kh., Tekhneryadnova, R. T., and Pan'kova, Z. I., Vitamin content of grapes, *Vestn. S-kh. Nauki Kaz.*, 20(12), 47, 1977; *Chem. Abstr.*, 88, 73246m, 1978.
59. Pogosyan, S. A., Marutyan, S. A., and Adamyan, A. Kh., Some data on the vitamin content of table grapes with different ripening periods, *Vinodel. Vinograd. SSSR*, 2, 48, 1986; *Chem. Abstr.*, 105, 5386b, 1986.
60. Gamova, O. V., Dorokhov, B. L., Tsypko, M. V., Guzun, N. I., and Petrova, T. V., Vitamins in berries of new table grape varieties developed by the Moldavian Scientific Research Institute for Viticulture and Winemaking, *Sadovod. Vinograd. Mold.*, 11, 27, 1989; *Chem. Abstr.*, 112, 54018m, 1990.
61. Gamova, O. V., Dorokhov, B. L., Tsypko, M. V., Petrova, T. V., and Kharyuk, S. G., Ascorbic acid content of grapes, *Sadovod. Vinograd. Vinodel. Mold.*, 12, 24, 1990; *Chem. Abstr.*, 115, 90914w, 1991.
62. Akhbazava, D. M. and Tsilosani, M. V., Biochemical characteristics of fruit of different grapefruit varieties, *Subtrop. Kul't.*, 2, 111, 1990; *Chem. Abstr.*, 114, 41254b, 1991.
63. Rathore, D. S., Effect of season on the growth and chemical composition of guava (*Psidium guajava* L.) fruits, *J. Hortic. Sci.*, 51, 41, 1976.
64. Sharma, A. K. and Singh, S. N., Physicochemical characteristics of some guava (*Psidium guajava* L.) cultivars, *Food Farming Agric.*, 13(11-12), 191, 1981; *Chem. Abstr.*, 99, 4311u, 1983.
65. Wilson, C. W., Shaw, P. E., and Campbell, C. W., Determination of organic acids and sugars in guava (*Psidium guajava* L.) cultivars by high-performance liquid chromatography, *J. Sci. Food Agric.*, 33(8), 777, 1982.

66. Mitra, S. K., Ghosh, S. K., and Dhua, R. S., Ascorbic-acid content in different varieties of guava grown in West Bengal, *Sci. Cult.*, 50(7), 235, 1984.
67. Esteves, M. T. da C., Carvalho, V. D. de, Chitarra, M. I. F., Chitarra, A. B., and Paula, M. B. de, Characteristics of fruits of six guava (*Psidium guajava* L.) cultivars during ripening. II. Vitamin C and tannin contents, in *Anais do VII Congresso Brasileiro de Fruticultura*, *Vol. 2*, Florianópolis, Brazil, 1984, 490; *Hortic. Abstr.*, 55, 7323, 1985.
68. Du Preez, R. J. and Welgemoed, C. P., Variability in fruit characteristics of five guava selections, *Acta Hortic.*, 275(1), 351, 1990.
69. Salmah, Y., Physico-chemical characteristics of some guava varieties in Malaysia, *Acta Hortic.*, 269, 301, 1990.
70. Wehmeyer, A. S. and von Staden, D. F. A., The nutrient content of South African-grown kiwifruit, *Food Rev.*, 11(1), 80, 1984; *Food Sci. Technol. Abstr.*, 16(10), J1749, 1984.
71. Lombardi-Boccia, G., Cappelloni, M., and Lintas, C., Vitamin C content of kiwifruits: effect of ripening stage and post-harvest storage, *Riv. Soc. Ital. Sci. Aliment.*, 15(1-2), 45, 1986; *Chem. Abstr.* 105, 77763v, 1986.
72. Cotter, R. L., Macrae, E. A., Ferguson, A. R., McMath, K. L., and Brennan, C. J., A comparison of the ripening, storage and sensory qualities of seven cultivars of Kiwifruit, *J. Hortic. Sci.*, 66(3), 291, 1991.
73. Zhang, Y. P. and Long, Z. X., Seedling selections of *Actinidia deliciosa*, *J. Fruit Sci.*, 8(2), 124, 1991; *Hortic. Abstr.*, 62, 5610, 1992.
74. Ortiz, J. M., Garcia-Lidon, A., Tadeo, J. L., De Cordova, L. F., Martin, B., and Estelles, A., Comparative study of physical and chemical characteristics of four lemon cultivars, *J. Hortic. Sci.*, 61(2), 277, 1986.
75. Mustard, M. J. and Lynch, S. J., Effect of various factors upon the ascorbic acid content of some Florida-grown mangos, *Fla. Agr. Expt. Stn. Bull.*, 406, 1, 1945; *Chem. Abstr.*, 42, 5580b, 1948.
76. Chávez, J. F. and Jaffé, W. G., The nutritional importance of fruits in the Venezuelan diet, *Proc. Carib. Reg. Am. Soc. Hortic. Sci. 1964*, 8, 254, 1965; *Hortic. Abstr.*, 37, 1776, 1967.
77. Hulme, A. C., The mango, in *The Biochemistry of Fruits and Their Products*, Vol., 2, Hulme, A. C., Ed., Academic Press, London, 1971, 233.
78. Subramanyam, H., Krishnamurthy, S., and Parpia, H. A. B., Physiology and biochemistry of mango fruit, *Adv. Food Res.*, 21, 224, 1975.
79. Beyers, M., Thomas, A. C., and van Tonder, A. J., γ irradiation of subtropical fruits. 1. Compositional tables of mango, papaya, strawberry, litchi fruits at the edible-ripe stage, *J. Agric. Food Chem.*, 27(1), 37, 1979.
80. Thomas, P. and Oke, M. S., Technical note: vitamin C content and distribution in mangoes during ripening, *J. Food Technol.*, 15, 669, 1980.
81. Deb Sharma, D. and Biswas, P., Physico-chemical composition of some local cultivars of mango grown in West Bengal, *Indian Agric.*, 25(1), 7, 1981.
82. Kohli, K., Quadry, J. S., and Ali, M., Protein, amino acids and ascorbic acid in some cultivars of mango, *J. Sci. Food Agric.*, 39(3), 247, 1987.
83. Kaur, G., Lal, T., Nandpuri, K. S., and Sharma, S., Varietal-cum-seasonal variation in certain physicochemical constituents in muskmelon, *Indian J. Agric. Sci.*, 47(6), 285, 1977.
84. Zhivondov, A., Krusteva, M., and Grigorov, I., Chemical composition and flavor of nectarines of some introduced cultivars, *Grandinar. Lozar. Nauka*, 21(8), 3, 1984; *Chem. Abstr.*, 103, 52884f, 1985.

85. Nagy, S., Vitamin C contents of citrus fruit and their products: a review, *J. Agric. Food Chem.*, 28, 8, 1980.
86. Mohammad, I, Jilani, M. S., and Ghafoor, A., Physico-chemical characteristics of sweet orange (*Citrus sinensis* L.) cultivars grown in D. I. Khan, *Sarhad J. Agric.* 5(2), 145, 1989; *Hortic. Abstr.*, 62, 702, 1992.
87. Pal, D. K., Subramanyam, M. D., Divakar, N. G., Iyer, C. P. A., and Selvaraj, Y., Studies on the physico-chemical composition of fruits of twelve papaya varieties, *J. Food Sci. Technol.*, 17, 254, 1980.
88. Imungi, J. K. and Wabule, M. N., Some chemical characteristics and availability of vitamin A and vitamin C from Kenyan varieties of Papayas (*Carica papaya* L.), *Ecol. Food Nutr.*, 24, 115, 1990.
89. Scott, G. C. and Belkengren, R. O., Importance of breeding peas and corn for nutritional quality, *Food Res.*, 9, 371, 1944.
90. Jaiswal, S. P., Kaur, G., Kumar, J. C., Nandpuri, K. S., and Thakur, J. C., Chemical constituents of green pea and their relationships with some plant characters, *Indian J. Agric. Sci.*, 45(2), 47, 1975.
91. Heinze, P. H., Hayden, F. R., and Wade, B. L., Vitamin studies of varieties and strains of peas, *Plant Physiol.*, 22, 548, 1947.
92. Selman, J. D. and Rolfe, E. J., Studies on the vitamin C content of developing pea seeds, *J. Food Technol.*, 14, 157, 1979.
93. Pant, P. C., Rawat, P. S., and Joshi, M. C., A note on some biochemical constituents in different cultivars of peas under hilly conditions, *Sci. Cult.*, 47(7), 264, 1981; *Chem. Abstr.*, 96, 65691c, 1982.
94. Lee, C. Y., Massey, L. M., Jr., and Van Buren, J. P., Effects of post-harvest handling and processing on vitamin contents of peas, *J. Food Sci.*, 47, 961, 1982.
95. Schroder, G. M., Satterfield, G. H., and Holmes, A. D., The influence of variety, size, and degree of ripeness upon the ascorbic acid content of peaches, *J. Nutr.*, 25, 503, 1943.
96. Sistrunk, W. A., Peach quality assessment: fresh and processed, in *Evaluation of Quality of Fruits and Vegetables*, Pattee, H. E., Ed., AVI Publishing, Westport, Conn., 1985, 1.
97. Rādulescu, M. and Branişte, N., A biochemical study of 200 pear varieties and their utilization in the breeding, *Acta Hortic.*, 224, 237, 1987.
98. Ito, S., The persimmon, in *The Biochemistry of Fruits and Their Products*, Vol. 2, Hulme, A. C., Ed., Academic Press, London, 1971, 281.
99. Golubev, V. N., Khalilov, M. A., and Kostinskaya, L. I., Biochemical characterization of persimmons of Azerbaijan, *Subtrop. Kul't.*, 4, 145, 1987; *Chem. Abstr.*, 108, 130262w, 1988.
100. Homnava, A., Payne, J., Koehler, P., and Eitenmiller, R., Provitamin A (alpha-carotene, beta-carotene and beta-cryptoxanthin) and ascorbic acid content of Japanese and American persimmons, *J. Food Qual.*, 13, 85, 1990.
101. Penkeva, T., Chemical characteristics of some local pepper populations, *Capsicum Newsletter*, 4, 23, 1985; *Hortic. Abstr.*, 57, 1961, 1987.
102. Pepkowitz, L. P., Larson, R. E., Gardner, J., and Owens, G., The carotene and ascorbic acid concentration of vegetable varieties, *Plant Physiol.*, 19, 615, 1944.
103. Anikeenko, A. P. and Anikeenko, V. S., Chemical composition of technically ripe green peppers in the Adygei autonomous region, *Tr. Prikl. Bot. Genet. Sel.*, 66(3), 90, 1980; *Chem. Abstr.*, 95, 147150u, 1981.
104. Rampal, S., A note on vitamin C content in chillies, *Indian J. Hortic.*, 35, 373, 1978.
105. Khadi, B. M., Goud, J. V., and Patil, V. B., Variation in ascorbic acid and mineral content in fruits of some varieties of chilli (*Capsicum Annuum* L.), *Qual. Plant.—Plant Foods Hum. Nutr.*, 37(1), 9, 1987.

106. Lu, R., Wu, J., Zhan, X., and Zhuang, R., Changes of vitamin C content in fruits of *Phyllanthus emblica* and its products, *Shipin Kexue* (*Beijing*), 99, 46, 1988; *Chem. Abstr.*, 109, 21879c, 1988.

107. Al-Kahtani, H. A., Intercultivar differences in quality and postharvest life of pomegranates influenced by partial drying, *J. Am. Soc. Hortic. Sci.*, 117(1), 100, 1992.

108. Lampitt, L. H., Baker, L. C., and Parkinson, T. L., Vitamin-C content of potatoes. II. The effect of variety, soil, and storage. *J. Soc. Chem. Industry* (*Lond.*), 64, 22, 1945.

109. Baker, L. C., Parkinson, T. L., and Lampitt. L. H., The vitamin C content of potatoes grown on reclaimed land, *J. Soc. Chem. Ind.*, 65, 428, 1946.

110. Allison, R. M. and Driver, C. M., The effect of variety, storage and locality on the ascorbic acid content of the potato tuber, *J. Sci. Food Agric.*, 4, 386, 1953.

111. Swaminathan, K. and Pushkaranath, Nutritive value of indian potato varieties, *Indian Potato J.*, 4, 76, 1962.

112. Hyde, R. B., Variety and location effects on ascorbic acid in potatoes, *Food Sci.*, 27, 373, 1962.

113. Somogyi, J. C. and Schiele, K., Der Vitamin-C-gehalt verschiedener Kartoffelsorten und seine Abnahme während der Lagerung, *Int. Z. Vitaminforschung*, 36, 337, 1966.

114. Kemp, P. and Kemp, T. C., The ascorbic acid content of thirteen varieties of potato, *Proc. Nutr. Soc.*, 41, 6A, 1982.

115. Randhawa, K. S., Sandhu, K. S., and Kaur, G., Chemical evaluation of potato genotypes, *J. Res.* (*Punjab Agric. Univ.*), 21(3), 375, 1984.

116. Mozolewski, W., Rotkiewicz, W., and Czaplicka, G., The vitamin C and citric acid content of selected potato cultivars and lines and of crisps produced from them, *Acta Acad. Agric. Techn. Olsten. Techn. Alimen.*, 21, 43, 1987; *Plant Breeding Abstr.*, 57, 10948, 1987.

117. Haque, M. A. and Iqbal, M. T., Bio-chemical constituents and storage behavior of some important varieties of potato grown in Bangladesh, *Bangladesh J. Agric.*, 8(1/4), 1, 1983; *FSTA Abstr.*, 19(7), J133, 1987.

118. Burton, W. G., *The Potato*, Longman Sci. Tech., England, 1989.

119. Mullin, W. J., Jui, P. Y., Nadeau, L., and Smyrl, T. G., The vitamin C content of seven cultivars of potatoes grown across Canada, *Can. Inst. Food Sci. Technol. J.*, 24(3/4), 169, 1991.

120. Aliev, M. M. and Berezovskaya, N. N., Chemical composition of quince grown in Azerbaidzhan, *Konservnaya i Ovoshchesushil'naya Promyshlennost*, 4, 36, 1974; *Food Sci. Technol. Abstr.*, 7(6), J875, 1975.

121. Zhebentyaeva, T. N., and Khrolikova, A. Kh., Chemical and technological characteristics of new varieties and hybrids of quince, *Sadovod. Vinograd. Mold.*, 2, 31, 1989; *Chem. Abstr.*, 110, 211249r, 1989.

122. Fejer, S. O., Johnston, F. B., Spangelo, L. P. S., and Hammill, M. M., Ascorbic acid in red raspberry fruit and leaves, *Can. J. Plant Sci.*, 50, 457, 1970.

123. Rouhani, I., Khosh-kuhi, M., and Bassiri, A., Changes in ascorbic acid content of developing rose hips, *J. Hortic. Sci.*, 51, 375, 1976.

124. Nylund, R. E., Ascorbic acid content of twenty-five varieties of the rutabaga (*Brassica napobrassica*), *Proc. Am. Soc. Hortic. Sci.*, 54, 367, 1949.

125. Reddy, N. S. and Kumari, R. L., Effect of different stages of maturity on the total and available iron and ascorbic acid content of soybean, *Nutr. Rep. Internat.*, 37, 77, 1988.

126. Behr, U., Comparison of cultivars regarding the content of nitrate and other quality components in butterhead lettuce and spinach (in German), Ph.D. thesis, University of Hannover, 1988.

127. Russel, L. F., Mullin, W. J., and Wood, D. F., Vitamin C content of fresh spinach, *Nutr. Report Int.*, 28, 1148, 1983.

128. Stevens, M. A., Varietal influence of nutritional values, in *Nutritional Qualities of Fresh Fruits and Vegetables*, White, P. L., and Selvey, N., Eds., Futura Publ. Mount Kisco, N.Y., 1974, 87.

129. Slate, G. L. and Robinson, W. B., Ascorbic acid content of strawberry varieties and selections at Geneva, New York in 1945, *Proc. Am. Soc. Hortic. Sci.*, 47, 219, 1946.

130. Ezell, B. D., Darrow, G. M., Wilcox, M. S., and Scott, D. H., Ascorbic acid content of strawberries, *Food Res.*, 12, 510, 1947.

131. Anstey, T. H. and Wilcox, A. N., The breeding value of selecting inbred clones of strawberries with respect to their vitamin C content, *Sci. Agric.*, 30(9), 367, 1950.

132. Shirko, T. S. and Yaroshevich, I. V., Chemical composition of cultivated strawberries in central Belorussia, *Konservn. Ovoshchesush. Prom-st.*, 2, 14, 1984; *Chem. Abstr.*, 100, 155409x, 1984.

133. Morris, J. R. and Sistrunk, W. A., The Strawberry, in *Quality and Preservation of Fruits*, Eskin, N. A. M., Ed., CRC Press, Boca Raton, Fla., 1991, 182.

134. Manabe, T., Quality evaluation of strawberry for fresh fruit. Part 2. Effect of cultivars on general constituents, organic acid composition, volatile compounds, pectic substances and physical properties in the edible portion of strawberry, *Hiroshima Nogyo Tanki Daigaku Kenkyu Hokoku*, 8(4), 669, 1989; *Chem. Abstr.*, 114, 22633h, 1991.

135. Speirs, M., Dempsey, A. H., Miller, J., Peterson, W. J., Wakeley, J. T., Cochran, F. D., Reder, R., Cordner, H. B., Fieger, E. A., Hollinger, M., James, W. H., Lewis, H., Eheart, J. F., Eheart, M. S., Andrews, F. S., Young, R. W., Mitchell, J. H., Carrison, O. B., and McLean, F. T., The effect of variety, curing, storage, and time of planting and harvesting on the carotene, ascorbic acid, and moisture content of sweet potatoes, *South. Coop. Series Bull.*, 30, 3, 1953.

136. Lanier, J. J. and Sistrunk, W. A., Influence of cooking method on quality attributes and vitamin content of sweet potatoes, *J. Food Sci.*, 44(2), 374, 1979.

137. Villareal, R. L., Tsou, S. C. S., Lin, S. K., and Chiu, S. C., Use of sweet potato (*Ipomea batatas*) leaf tips as vegetables. II. Evaluation of yield and nutritive quality, *Exp. Agric.*, 15, 117, 1979.

138. Tafazoli, E. and Niknejad, M., Comparison of introduced and local strawberry varieties (*Fragaria ananassa* Duch.) in relation to yield, vitamin C content, soluble solids, and storage life, *Iran J. Agric. Res.*, 1(1), 45, 1971.

139. Dhaliwal. G. S. and Grewal, G. S., Yield and quality of strawberry fruits under subtropical conditions, *J. Res. Punjab Agric. University*, 21(3), 361, 1984.

140. Drogina, M. A., Biochemical composition of red raspberries and strawberries in the central zone of the Stavropol District (USSR), *Tr. Stravrop. Nauchno-Issled. Inst. S-Kh. Khoz.*, 103, 1981; *Chem. Abstr.*, 97, 123968g, 1982.

141. Sistrunk, W. A. and Morris, J. R., Strawberry quality: influence of culture and environmental factors, in *Evaluation of Quality of Fruits and Vegetables*, Pattee, H. E., Ed., AVI Publishing, Westport, Conn., 1985, 217.

142. Baudisch, W., Askorbinsäuregehalt von Mutanten der Kulturtomate *Lycopersicon esculentum* Mill., *Kulturpflanze*, 15, 105, 1967.

143. Burge, J., Mickelsen, O., Nicklow, C., and Marsh, G. L., Vitamin C in tomatoes: comparison of tomatoes developed for mechanical or hand harvesting, *Ecol. Food Nutr.*, 4, 27, 1975.

144. Fritz, D., Habben, F. J., Reuff, B., and Venter, F., Die Variabilität einiger qualitätsbestimmender Inhaltsstoffe von Tomaten, *Gartenbauwissenschaft*, 41(3), 104, 1976.

145. Lee, Y. C., Vitamin composition of tomato cultivars and computer simulation of ascorbic acid stability in canned tomato juice, Ph.D. Thesis, Michigan State University, 1976.

146. Watada, A. E., Aulenbach, B. B., and Worthington, J. T., Vitamins A and C in ripe tomatoes as affected by stage of ripeness at harvest and by supplementary ethylene, *J. Food Sci.*, 41, 856, 1976.

147. Drake, S. R. and Price, L. G., pH and quality of home-canned tomato-vegetable mixtures, *J. Food Qual.*, 5, 145, 1981.

148. Som, M. G. and Paria, N. C., Studies on the quality of tomato fruits of different varieties under West Bengal conditions, *Indian Agric.*, 27(4), 317, 1983.

149. Gould, W. A., *Tomato Production, Processing and Quality Evaluation*, 2nd Ed., AVI Publishing, Westport, Conn., 1983.

150. Balasubramanian, T., Studies on quality and nutritional aspects of tomato, *J. Food Sci. Technol.*, 21(6), 419, 1984.

151. Wann, E. V. and Jourdain, E. L., Effect of mutant genotypes $hp\ og^c$ and $dg\ og^c$ on tomato fruit quality, *J. Am. Hortic. Soc.*, 110(2), 212, 1985.

152. Stevens, M. A., and Rick, C. M., Genetics and breeding, in *The Tomato Crop*, Atherton, J. G., and Rudich, J., Eds)., Chapman and Hall, London, 1986, 35.

153. Bajaj, K. L., Kaur, G., Singh, S., and Nandpuri, K. S., Consistency analysis of important chemical constituents of various tomato (*Lycopersicon esculentum*) varieties, *Indian J. Agric. Sci.*, 58(6), 492, 1988.

154. Daood, H. G., Al-Qitt, M. A., Bshenah, K. A., and Bouragba, M., Varietal and chemical aspect of tomato processing, *Acta Alimen.*, 19(4), 347, 1990.

155. Reder, R., Speirs, M., Chochran, H. L., Hollinger, M. E., Farish, L. R., Geiger, M., McWhirter, L., Sheets, O. A., Eheart, J. F., Moore, R. C., and Carolus, R. L., The effects of maturity, nitrogen fertilization, storage and cooking, on the ascorbic acid content of two varieties of turnip greens. *South. Coop. Series Bull.*, 1, 1, 1943.

156. Hansen, E., Variations in the carotene content of carrots. *Proc. Am. Soc. Hortic. Sci.*, 46, 355, 1945.

157. Yamaguchi, M., Robinson, B., and McGillivray, J. H., Some horticultural aspects of the food value of carrots, *Proc. Am. Soc. Hortic. Sci.*, 60, 351, 1952.

158. Gabelman, W. H. and Peters, S., Genetical and plant breeding possibilities for improving the quality of vegetables, *Acta Hortic.*, 93, 243, 1979.

159. Bajaj, K. L., Kaur, G., and Sukhija, B. S., Chemical composition and some plant characteristics in relation to quality of some promising cultivars of carrot (*Daucus carota* L.), *Qualit. Plant.*, 30, 97, 1980.

160. Heinonen, M. I., Carotenoids and provitamin A activity of carrot (*Daucus carota* L.) cultivars, *J. Agric. Food Chem.*, 38, 609, 1990.

161. McDowell, I. and Oduro, K. A., Investigation of the β-carotene content of yellow varieties of cassava (*Manihot esculenta* Crantz), *J. Plant Foods*, 5, 169, 1983.

162. Penteado, M. de V. C., and De Almeida, L. B., Occurrence of carotenoids in roots of five cultivars of cassava (*Manihot esculenta*, Crantz) from Sao Paulo State, *Rev. Farm. Bioguim. Univ. São Paulo*, 24(1), 39, 1988; *Chem. Abstr.*, 110, 56297j, 1989.

163. Hjarde, W., Hellström, V., and Åkerberg, E., The content of tocopherol and carotene in red clover as dependent on variety, conditions of cultivation and stage of development, *Acta Agric. Scand.*, 13, 3, 1963.

164. Rouseff, R. L., Sadler, G. D., Putman, T. J., and Davis, J. E., Determination of β-carotene and other hydrocarbon carotenoids in red grapefruit cultivars, *J. Agric. Food Chem.*, 40, 47, 1992.

165. Grogan, C. O., Blessin, C. W., Dimler, R. J., and Campbell, C. M., Parental influence on xanthophyll and carotenes in corn, *Crop Sci.*, 3, 213, 1963.

166. Quackenbush, F. W., Firch, J. G., Brunson, A. M., and House, L. R., Carotenoid, oil, and tocopherol content of corn inbreds, *Cereal Chem.*, 40, 250, 1963.

167. Godoy, H. T. and Bodriguez-Amaya, D. B., Carotenoid composition of commercial mangoes from Brazil, *Lebensmitt. Wiss. Technol.*, 22, 100, 1989.

168. Kimura, M., Rodriguez-Amaya, D. B., and Yokoyama, S. M., Cultivar differences and geographical effects on the carotenoid composition and vitamin A value of papaya, *Lebensmitt. Wiss. Technol.*, 24, 415, 1991.

169. Miller, E. V. and Winston, J. R., Seasonal changes in the carotenoid pigments in the juice of Florida oranges, *Proc. Am. Soc. Hortic. Sci.*, 38, 219, 1941.

170. Trujillo-Quijano, J. A., Rodriguez-Amaya, D. B., Esteves, W., and Plonis, G. F., Carotenoid composition and vitamin A values of oils from four Brazilian palm fruits, *Fat Sci. Technol.*, 92(6), 222, 1990.

171. Mejia, L. A., Hudson, E., Gonzales de Mejia, E., and Vazquez, F., Carotenoid content and vitamin A activity of some common cultivars of Mexican peppers (*Capsicum annuum*) as determined by HPLC, *J. Food Sci.*, 53(5), 1448, 1988.

172. Almela, L., López-Roca, J. M., Candela, M. E., and Alcázar, M. D., Carotenoid composition of new cultivars of red pepper for paprika, *J. Agric. Food Chem.*, 39(9), 1606, 1991.

173. Gross, J., Carotenoid pigments in three plum cultivars, *Gartenbauwissenschaft*, 49, 18, 1984.

174. Ezell, B. D. and Wilcox, M. S., The ratio of carotene to carotenoid pigments in sweet-potato varieties, *Science*, 103, 193, 1946.

175. Hernández, T. P., Hernández, T. P., Constantin, R. J., and Miller, J. C., Inheritance of and method of rating flesh color in *Ipomoea Batatas*, *Proc. Am. Soc. Hortic. Sci.*, 87, 387, 1965.

176. Woolfe, J. A., *Sweet Potato*, Cambridge University Press, Cambridge, 1992.

177. Manuelyan, H., Yordanov, M., Yordanova, Z., and Ilieva, Z., Studies on β-carotene and lycopene content in the fruits of *Lycopersicon esculentum* Mill. X *L. chilense* Dun. Hybrids, *Qual. Plant.—Plant Foods Hum. Nutr.*, 25(2), 205, 1975.

178. Mapson, L. W., Vitamin in fruits, in *The Biochemistry of Fruits and Their Products*, Vol. 1., Hulme, A. C., Ed., Academic Press, London, 1970, 369.

179. Hall, A. P., Moore, J. G., and Morgan, A. F., B vitamin content of California-grown avocados, *Agric. Food Chem.*, 3, 250, 1955.

180. Nik-Khah, A., Hoppner, K. H., Sosulski, F. W., Owen, B. D., and Wu, K. K., Variation in proximate fractions and B-vitamins in Saskatchewan feed grains, *Can. J. Animal Sci.*, 52, 407, 1972.

181. Gough, H. W. and Lantz, E. M., Relation of variety and locality to niacin, thiamine, and riboflavin content of dried beans grown in three locations, *Food Res.*, 15, 308, 1950.

182. Augustin, J., Beck, C. B., Kalbfleish, G., Kagel, L. C., and Matthews, R. H., Variation in the vitamin and mineral content of raw and cooked commercial *Phaseolus vulgaris* classes, *J. Food Sci.*, 46, 1701, 1981.

183. Ogunmodede, B. K., and Oyenuga, V. A., Vitamin B content of cowpeas (*Vigna unguiculata* Walp). 1. Thiamine, riboflavin and niacin, *J. Sci. Food Agric.*, 20, 101, 1969.

184. Burkholder, P. R., McVeigh, I., and Moyer, D., Niacin in maize, *Yale J. Biol. Med.*, 16, 659, 1943.

185. Barton-Wright, E. C., The microbiological assay of nicotinic acid in cereals and other products, *Biochem. J.*, 38, 314, 1944.

186. Hunt, C. H., Ditzler, L., and Bethke, R. M., Niacin and pantothenic acid content of corn hybrids, *Cereal Chem.*, 24, 355, 1947.

187. Simwemba, C. G., Hoseney, R. C., Varriano-Martson, E., and Zeleznak, K., Certain B vitamins and phytic acid contents of pearl millet [*Pennisetum americanum* (L.) Leeke], *J. Agric. Food Chem.*, 32, 31, 1984.

188. Eheart, J. F., Young, R. W., and Allison, A. H., Variety, type, year, and location effects on the chemical composition of peanuts, *Food Res.*, 20(5), 497, 1955.

189. Cheema, P. S. and Ranhotra, G. S., The effect of habit of growth and environment on the tryptophan–nicotinic acid content of the groundnut (*Arachis hypogaea*) varieties, *Int. Nutr. Dietet.*, 5(2), 101, 1968; *Field Crop Abstr.*, 22, 2844, 1969.
190. Page, E. and Hanning, F. M., Vitamin B_6 and niacin in potatoes, *J. Am. Diet. Assoc.*, 42, 42, 1963.
191. Hubbard, J. E., Hall, H. H., and Earle, F. R., Composition of the component parts of the sorghum kernel, *Cereal Chem.*, 27, 415, 1950.
192. Pearson, P. B. and Luecke, R. W., The B vitamin content of raw and cooked sweet potatoes, *Food Res.*, 10, 325, 1945.
193. Bradbury, J. H. and Singh, U., Thiamin, riboflavin, and nicotinic acid contents of tropical root crops from the South Pacific, *J. Food Sci.*, 51(6), 1563, 1986.
194. Schuphan, W., Kling, M., and Overbeck, G., Einfluss geneticher und umweltbedingter Faktoren auf die Gehalte an Vitamin B_1, Vitamin B_2 und Niacin von Winter- und Sommerweizen, *Qualit. Plant.*, 15, 177, 1968.
195. Davis, K. R., Peters, L. J., and Le Tourneau, D., Variability of vitamin content in wheat, *Cereal Foods World*, 29(6), 364, 1984.
196. Ogunmodede, B. K. and Oyenuga, V. A., Vitamin B content of cowpea (*Vigna unguiculata* Walp). II. Pyridoxine, pantothenic acid, biotin and folic acid, *J. Sci. Food. Agric.*, 21, 87, 1970.
197. Raghunath, M. and Belavady, B., Riboflavin and total vitamin B_6 content of indian pulses: varietal differences and the effect of cooking, *J. Plant Foods*, 3, 205, 1979.
198. Blumenthal, A., Scheffieldt, P., Zum Vitamin-B_6-Gehalt Svhweizerischer Kartoffeln und deren Beitrag zur empfohlenen Tageszufuhr, *Mitt. Geb. Lebensmittelunters. Hyg.*, 77, 460, 1986.
199. Koehler, H. H. and Burke, D. W., Nutrient composition, sensory characteristics, and texture measurements of seven cultivars of dry beans, *J. Am. Soc. Hortic. Sci.*, 106(3), 313, 1981.
200. Conner, R. T. and Straub, G. J., The thiamin and riboflavin content of wheat and corn, *Cereal Chem.*, 18, 671, 1941.
201. Jahn-Deesbach, W., Neuere Untersuchungen über den Thiamin-(Vitamin-B_1)Gehalt des Getreidekorns, *Getreide Mehl und Brot*, 33, 256, 1979.
202. Marquard, R., Schuster, W., and Iran-Nejad, H., Untersuchungen über Tokopherol- und Thiamingehalt in Leinsaat aus weltweitem Anbau und aus dem Phytotron unter definierten Klimabedingungen, *Fette Seifen Anstrichmittel*, 79(7), 265, 1977.
203. Holman, W. I. M. and Godden, W., The aneurin (vitamin B_1) content of oats. I. The influence of variety and locality. II. Possible losses in milling, *J. Agric. Sci.*, 37, 51, 1947.
204. Greer, E. N., Ridyard, H. N., and Kent, N. L., The composition of British-grown winter wheat. I. Vitamin-B_1 content, *J. Sci. Food Agric.*, 3, 12, 1952.
205. Mijll Dekker, L. P. van der and H. de Miranda, H., The vitamin B_1 content of dutch wheat and the factors which determine this content, *Neth. J. Agric. Sci.*, 2, 27, 1954.
206. Jahn-Deesbach, W. and May, H., The effect of variety and additional late spring nitrogen application on the thiamin (vitamin B_1) content of the total wheat grain, various flour types, and secondary milling products, *Z. Acker-Pflanzenbau*, 135, 1, 1972.
207. Downs, D. E. and Cathcart, W. H., Thiamin content of commercial wheats of the 1940 crop, *Cereal Chem.*, 18, 796, 1941.
208. Combs, S. B. and Combs, G. F., Jr., Varietal differences in the vitamin E content of corn, *J. Agric. Food Chem.*, 33, 815, 1985.
209. Galliher, H. L., Alexander, D. E., and Weber, E. J., Genetic variability of alpha-tocopherol and gamma-tocopherol in corn embryos, *Crop Sci.*, 25, 547, 1985.

210. Grams, G. W., Blessin, C. W., and Inglett, G. E., Distribution of tocopherols within the corn kernel, *J. Am. Oil Chem. Soc.*, 47, 337, 1970.

211. Marquard, R., Der Einfluss von Sorte und Standort sowie einzelner definierter Klimafaktoren auf den Tokopherolgehalt im Rapsöl, *Fette Seifen Anstrichmittel*, 78(9), 341, 1976.

212. Feldheim, W., Der Vitamin E-Gehalt der Paprika-Früchte, *Z. Lebensm.-Unters. Forsch.*, 104, 24, 1956.

213. Marquard, R. and Schuster, W., Protein- und Fettgehalte des Kornes sowie Fettsäuremuster und Tokopherolgehalte des Öles bei Sojabohnensorten von stark differenzierten Standorten, *Fette Seifen Anstrichmittel*, 82(4), 137, 1980.

214. Marquard, R., Schuster, W., und Seibel, K. K., Fettsäuremuster und Tokopherolgehalt im Öl verschiedener Sonnenblumensorten aus weltweitem Anbau, *Fette Seifen Anstrichmittel*, 79(1), 137, 1977.

215. Cabell, C. A. and Ellis, N. R., The vitamin content of certain varieties of wheat, corn, grasses and legumes as determined by rat assay, *J. Nutr.*, 23, 633, 1942.

216. Davis, K. R., Litteneker, N., Le Tourneau, D., Cain, R. F., Peters, L. J., and McGinnis, J., Evaluation of the nutrient composition of wheat. I. Lipid composition, *Cereal Chem.*, 57, 178, 1980.

217. de Lumen, B. O. and Fiad, S., Tocopherols of winged bean (*Psophocarpus tetragonolobus*) oil, *J. Agric. Food Chem.*, 30, 50, 1982.

218. Smith, F. G. and Walker, J. C., Relation of environmental and heredity factors to ascorbic acid in cabbage, *Am. J. Bot.*, 33, 120, 1946.

219. Pritchard, M. K. and Becker, R. F., Cabbage, in Eskin, N. A. M., Ed., *Quality and Preservation of Vegetables*, CRC Press, Boca Raton, Fla., 1989, Chap. 9.

220. Franz, C., Fritz, D., and Eisenmann, J., Parsley for industrial processing. I. Varieties and sites, *Gemüse*, 18(11), 380, 1982; *Hortic. Abstr.*, 53, 5390, 1983.

221. Pyke, M., The vitamin content of vegetables, *J. Soc. Chem. Ind.*, 61, 149, 1942.

222. Bakuras, N. S., Zhila, E. D., and Mirbaizaev, S. M., Biochemical characteristics of local garlic varieties, *Selektsiya i Semenovodstvo*, 59, 80, 1985; *Plant Breeding Abstr.*, 57, 8381, 1987.

223. Golubov, V. N., Gusar, Z. D., and Mamedov, E. Sh., Cultivar-specific characteristics of olives grown in Azerbaijan, *Subtrop. Kul't.*, 6, 86, 1987; *Chem. Abstr.*, 109, 51732s, 1988.

224. Kuliev, A. A. and Akhundov, R. M., Changes in ascorbic acid and catechin contents of *Zizyphus jujuba* fruit during ripening, *Uchenyw Zap. Azerb. Un-t. Ser. Biol. Nauk*, 3/4, 54, 1975; *Hortic. Abstr.*, 47, 4141, 1977.

225. Pogosyan, S. A. and Khachatryan, S. S., Increasing the nutrient value of grapes, *Vinodel Vinograd. SSSR*, 5, 35, 1972; *Chem. Abstr.*, 77, 137463n, 1972.

226. Malishevskaya, M. F., Chemical composition of apples, in *Yuzhnoe Stepnoe Sadovodstvo, Dnepropetrovsk, Ukrainian SSR*, 1973, 257; *Hortic. Abstr.*, 44, 5330, 1974.

227. Snapyan, G. G., Palazyan, T. N., and Melkonyan, L. T., Biochemical and market quality characteristics of the fruit of autumn and winter pears in the Armenian SSR, *Biol. Zh. Arm.*, 37(3), 241, 1984; *Plant Breeding Abstr.*, 56, 10972, 1986.

228. Hasani, A., Islami, X., and Pasko, P., Study of sowing dates for early, midseason and late cauliflower in the field, *Buletni Shkencave Bujqësore*, 26(2), 13, 1987; *Hortic. Abstr.*, 58, 1445, 1988.

229. Matzner, F., Über den Gehalt und die Verteilung des Vitamin C in Äpfeln, *Erwerbsobstbau*, 4(2), 27, 1962.

230. Petzold, H., *Apfelsorten*, Neumann Verlag, Radebeul, Germany, 1990.

231. Reynard, G. B. Kanapaux, M. S., Ascorbic acid (vitamin C) content of some tomato varieties and species, *Proc. Am. Soc. Hortic. Sci.*, 41, 298, 1942.

232. Davies, J. N. and Hobson, G. E., The constituents of tomato fruit—the influence of environment, nutrition, and genotype, *CRC Crit. Rev. Food Sci. Nutr.*, 15, 205, 1981.

233. Biale, J. B. and Young, R. E., The avocado pear, in *The Biochemistry of Fruits and Their Products*, Vol. 2., Hulme, A. C., Ed., Academic Press, London, 1971, 2.

234. Mather, K. and Barton-Wright, E. C., Nicotinic acid in sugary and starchy maize, *Nature (Lond.)*, 157, 109, 1946.

235. Glover, D. V. and Mertz, E. T., Corn, in *Nutritional Quality of Cereal Grains*, Olson, R. A. and Frey, K. J., Eds., Agronomy 28, Am. Soc. Agron. Madison, Wis., 1987, 183.

236. Richey, F. D. and Dawson, R. F., Experiments on the inheritance of niacin in corn (Maize), *Plant Physiol.*, 26, 475, 1951.

237. Teas, H. J., A morphological basis for higher niacin in sugary maize, *Proc. Acad. Sci. (USA)*, 38, 817, 1952.

238. Michela, P. and Lorenz, K., The vitamin of triticale, wheat, and rye, *Cereal Chem.*, 53, 853, 1976.

239. Toepfer, E. W., Polansky, M. M., Eheart, J. F., Slover, H. T., and Morris, E. R., Nutrient composition of selected wheats and wheat products, *Cereal Chem.*, 49, 173, 1972.

240. Kelly, E., Dietrich, K. S., and Porter, T., Vitamin B_1 content of eight varieties of beans grown in two locations in Michigan, *Food Res.*, 5, 253, 1940.

241. Syltie, P. W. and Dahnke, W. C., The vitamin B_1, B_2, B_6, B_{12}, and E contents of hard red spring wheat as influenced by fertilization and cultivar, *Qual. Plant.—Plant Foods Hum. Nutr.*, 32, 51, 1983.

242. Bauernfeind, J. C., The tocopherol content of food and influencing factors, *CRC Crit. Rev. Food Sci. Nutr.*, 8, 337, 1977.

243. O'Donnell, W. W. and Bayfield, E. G., Effect of whether, variety, and location upon thiamin content of some Kansas-grown wheats, *Food Res.*, 12, 212, 1947.

244. Taira, H., Taira, H., Matsuzaki, A., and Matsushima, S., Effect of nitrogen topdressing on B-vitamin content of rice grain, *Nihon Sakumostsu Gakkai Kiji*, 45(1), 69, 1976; *Food Sci. Technol. Abstr.*, 11(2), M139, 1979.

245. National Research Council, *Recommended Dietary Allowances, 10. ed., Report of the Committee on Dietary Allowances, Division of Biological Sciences, Assembly of Life Sciences, Food and Nutrition Board*, National Academy of Sciences, Washington, D.C., 1989.

246. Lundergan, C. A. and Moore, J. N., Variability in vitamin C content and colour of strawberries in Arkansas, *Arkansas Farm Res.*, 24(1), 2, 1975; *Hortic. Abstr.*, 46, 2019, 1976.

247. Hall, A. P., Brinner, L., Amerine, M. A., and Morgan, A. F., The B vitamin content of grapes, musts, and wines, *Food Res.*, 21, 362, 1956.

248. Melkonyan, M. V., Astabatsyan, G. A., Belaryan, S. A., and Petrosyan, D. A., B group vitamins and ascorbic acid in the metabolism of grape hybrids contrasting in a combination of characters, *S-kh. Biol.*, 4, 60, 1984; *Hortic. Abstr.*, 56, 1639, 1986.

249. Tudor, T. A., Ceausescu, M. E., Murvai, M., and Cepoiu, N., An investigation of the physicochemical properties of some currant cultivars, *Lucrări Stiintifice, Institutul Agronomic "Nicolae Bălcesc," Bucuresti, Seria B, Horticultură*, 32(1), 75, 1989; *Hortic. Abstr.*, 62, 6442, 1992.

250. Pruthi, J. S., Physiology, chemistry, and technology of passion fruit, *Adv. Food Res.*, 12, 203, 1963.

251. Kikutani, N., Momma, K., Iguchi, M., Tomomatsu, T., Ozawa, K., Wada, M., Aoki, M., Tosaka, M., and Urano, M., Nutrient composition of passion fruits cultivated in Ogasawara Islands, *Ann. Rep. Tokyo Metropolitan Research Laboratory of Public Health*, 39, 164, 1988; *Hortic. Abstr.*, 61, 6481, 1991.

252. Homnava, A., Rogers, W., and Eitenmiller, R. R., Provitamin A activity of specialty fruit marketed in the United States, *J. Food Comp. Analy.*, 3, 119, 1990.
253. Baqar, M. R., Technical note: vitamin C content of some Papua New Guinean fruits, *J. Food Technol.*, 15, 459, 1980.
254. Gadzhieva, G. G., Vitamin level in the fruit of roses introduced on the Apsheron peninsula (botanical garden), *Izv. Akad. Nauk Az. SSR, Ser. Biol. Nauk*, 3, 23, 1978; *Chem. Abstr.*, 90, 118093u, 1979.
255. Pospíšilová, J., Toul, V., Problems of the relationship between L-ascorbic acid and plant pigments in vegetables, *Věd. Práce Vyzk. Ust. Zelin. Olomouci*, 2, 119, 1963; *Hortic. Abstr.*, 33, 4990, 1963.
256. Studentsov, O. V. and Ter-Manuel'yants, E. E., The chemical composition of different species and varieties of brassicas, *Tr. Prikl. Bot. Genet. Sel.*, 50(2), 47, 1973; *Hortic. Abstr.*, 44, 5633, 1974.
257. Amaro Lopez, M. A., Moreno Rojas, R., and Zurera Cosano, G., Vitamin C content of fresh asparagus, *Alimentria (Madrid)*, 29(234), 39, 1992; *Chem. Abstr.*, 117, 149770t, 1992.
258. Makus, D. J. and Gonzalez, A. R., Production of white asparagus using plastic row covers, in *Proc. Ann. Meetings—Arkansas St. Hortic. Soc.*, No. 110, Fayetteville, Arkansas, 1989, 136; *Hortic. Abstr.*, 61, 8001, 1991.
259. Saimbhi, M. S., Kaur, G., and Nandpuri, K. S., Chemical constituents in mature green and red fruits of some varieties of chili (*Capsicum annuum* L.), *Qualit. Plant.*, 27(2), 171, 1977; *Hortic. Abstr.*, 47, 10520, 1977.
260. Jiang, J., Wang, Z., and Wang, D., Studies on fruit growth and the accumulation of nutrients in sweet pepper, *Acta Agric. Univ. Pekinensis*, 11 (3), 333, 1985; *Hortic. Abstr.*, 56, 9733, 1986.
261. Mazza, G., Carrots, in *Quality and Preservation of Vegetables*, Eskin, N. A. M., Ed., CRC Press, Boca Raton, Fla., 1989, Chap. 3.
262. Laferriere, L. and Gabelman, W. H., Inheritance of color, total carotenoids, alpha-carotene, and beta-carotene in carrot, *Daucus carota* L., *Proc. Am. Soc. Hortic. Sci.*, 93, 408, 1968.
263. Purcell, A. E. and Sistrunk, W. A., Sweet potatoes: effect of cultivar and curing on sensory quality, in *Evaluation of Quality of Fruits and Vegetables*, Pattee, H. E., Ed., AVI Publishing, Westport, Conn., 1985, 257.
264. Stevens, M. A., Tomato flavor: effects of genotype, cultural practices, and maturity at picking, in *Evaluation of Quality of Fruits and Vegetables*, Pattee, H. E., Ed., AVI Publishing, Westport, Conn., 1985, 367.
265. Hidaka, T., Anno, T., and Nakatsu, S., The composition and vitamin A value of the carotenoids of pumpkins of different colors, *J. Food Biochem.*, 11, 59, 1987.
266. Simpson, K. L., Relative value of carotenoids as precursors of vitamin A, *Proc. Nutr. Soc.*, 42, 7, 1983.
267. Davidyuk, L. P. and Vshivkova, G. F., Comparative study of carotenoids in the leaves of white- and yellow-fleshed peach cultivars, *Tr. Gosudarstvennogo Nikitskogo Botanicheskogo Sada*, 83, 103, 1981; *Hortic. Abstr.*, 56, 841, 1986.
268. Rom, R. C. and Carlson, R. F., Eds., *Rootstocks for Fruit Crops*, Wiley Intersci. Publ., New York, 1987.
269. Serafimova, R., Effect of variety and rootstock on some qualitative characteristics of apricots in the Pomorie region, *Gradinar. Lozar. Nauka*, 14(8), 3, 1977; *Hortic. Abstr.*, 49, 3186, 1979.
270. Arlt, K., Die Beeinflussung des Vitamin-C-Gehaltes bei einigen Apfelsorten durch Unterlage und Lagerungsdauer, *Intensivobstbau*, 2, 90, 1962.
271. Arlt, K., Der Einfluss von Unterlage und Lagerdauer auf den Vitamin-C-Gehalt einiger Apfelsorten, *Arch. Gartenbau*, 14, 25, 1966; *Hortic. Abstr.*, 37, 438, 1967.

272. Šidová, E., Effect of rootstocks on the vitamin C content of apples, *Vedecké Práce Vyskumného Ustavu Ovocnych Okrasnych Drevín Bojniciach*, 2, 109, 1980; *Hortic. Abstr.*, 52, 3585, 1982.

273. Trunov, I. A., Effect of rootstock on photosynthetic productivity and chemical composition of apples, *Sbornik Nauchnykh Trudov, Vsesoyuznyi Nauchno-Issledovatel'skii Institut Sadovodstva Imeni I.V. Michurina*, 29, 26, 1979; *Hortic. Abstr.*, 52, 1350, 1982.

274. Karychev, K. G. and Talipov, A., Chemical composition of Jonathan apples grafted on different rootstocks, *Vestn. S-kh. Nauki Kaz.*, 6, 37, 1987; *Chem. Abstr.*, 107, 214837e, 1987.

275. Schmid, P. P. S., Scherf, K., and Schmidt, K., Inhaltstoffe von sauerkirsche der Sorte "Schattenmorelle" in Abhängigkeit von verschiedenen Unterlagen, *Erwerbsobstbau*, 24(7), 175, 1982.

276. Schmid, P. P. S., Schmidt, K., Kraus, A., Schimmelpfeng, H., Inhaltstoffe von Sauerkirschen in Abhängigkeit von Sorten und Unterlagen, *Erwerbsobstbau*, 25(8), 204, 1983.

277. Matzner, F., Vitamin-C-Gehalt in Früchten der "Schattenmorelle," *Erwerbsobstbau*, 18(6), 83, 1976.

278. Harding, P. L., Winston, J. R., and Fisher, D. F., Seasonal changes in the ascorbic acid content of juice of Florida oranges, *Proc. Am. Soc. Hortic. Sci.*, 36, 358, 1939.

279. Ali, N. and Rahim, A., Influence of different rootstocks on the seasonal changes in the vitamin "C" content (ascorbic acid) of the Valencia Late sweet orange, *Punjab Fruit J.*, 23(81), 10, 1960; *Hortic. Abstr.*, 31, 5134, 1961.

280. Elazzouni, M. M. and Elbarkouki, M. H., Rootstock effect on the physical and chemical composition of Jaffa orange fruit, *Ann. Agric. Sci. (Cairo)*, 6, 175, 1961; *Hortic. Abstr.*, 36, 7290, 1966.

281. Mann, S. S. and Naurial, J. P., Performance of pineapple sweet orange (*Citrus sinensis* Osb.) on different rootstocks, *Prog. Hortic.*, 10(1), 37, 1978; *Hortic. Abstr.*, 49, 2885, 1979.

282. Siddique, M., Khan, M. D., and Hamdard, M. S., Studies on the quality of Kinnow mandarin as influenced by various rootstocks, *W. Pakist. J. Agric. Res.*, 3(2/3), 129, 1965; *Hortic. Abstr.*, 37, 7692, 1967.

283. Ting, S. V. and Attaway, J. A., Citrus fruits, in *The Biochemistry of Fruits and Their Products*, Vol. 2, Hulme, A. C., Ed., Academic Press, London, 1971, 107.

284. Godfrey-Sam-Aggrey, W., A preliminary evaluation of the effects of environment and rootstock types on fruit quality of sweet orange cultivars in Ghana. II. Internal quality and maturity standards, *Ghana J. Agric. Sci.*, 6(2), 75, 1973; *Hortic. Abstr.*, 45, 8865, 1975.

285. Bhullar, J. S. and Nauriyal, J. P., Effect of different rootstocks on vigor, yield and fruit quality of blood red orange (*Citrus sinensis* Osbeck), *Indian J. Hortic.*, 32(1/2), 45, 1975.

286. Diamante de Zubrzycki, A. and Rodriguez, D. S., The influence of the rootstock on the physicochemical and organoleptic characteristics of Valencia Late orange, *Revista Agronómica del Noroeste Argentino*, 11(1/2), 109, 1974; *Hortic. Abstr.*, 46, 1558, 1976.

287. El-Azab, E. M., El-Gazzar, A., and Abd-El-Kader, H. M., Influence of four citrus rootstocks on growth, yield, fruit quality and foliage microelement composition of some citrus varieties, *Alexandria J. Agric. Res.*, 26(2), 423, 1978.

288. El-Zeftawi, B. M. and Thornton, I. R., Varietal and rootstock effects on mandarin quality, *Aust. J. Exp. Agric. Anim. Husb.*, 18, 597, 1978.

289. Baldry, J., Dougan, J., Shaked, A., and Bar-Akiva, A., Chemical analysis and taste panel evaluation of the fruit quality of Valencia Late oranges on two rootstocks, *J. Hortic. Sci.*, 57(2), 233, 1982.

290. Medina-Urrutia, V. M. and Valdez-Verduzco, J., Effect of rootstocks on seasonal changes of Mexican limes [*Citrus aurantifolia* (Christm.) Swingle] in Colima, Mexico, in *Proc. Int. Soc. Citriculture*, Vol. 1., Shimizu, Japan, Fruit Tree Research Station, 1982, 130; *Hortic. Abstr.*, 53, 8178, 1983.

291. Núñez, M. and Rodriguez, E., Influence of 9 rootstocks on the quality of Dancy mandarin fruits, *Cult. Trop.*, 3(3), 127, 1981; *Food Sci. Technol. Abstr.*, 16(8), J1324, 1984.

292. Bradley, G. A., Fruits and vegetables as world sources of vitamins A and C, *HortScience*, 7(2), 141, 1972.

293. Derendovskaya, A. I., Effect of stock variety on the accumulation of ascorbic acid in the leaves of grape scions, *Tr. Kishinev. S-Kh. Inst.*, 114, 116, 1974; *Chem. Abstr.*, 82, 54146p, 1975.

294. Mikeladze, È. G., Georgobiani, È. L., and Gvamichava, N. È., Studies on some biological indices of grapes in relation to grafting, *Soobshcheniya Akademii Nauk Gruzinskoi SSR*, 80, 153, 1975; *Hortic. Abstr.*, 46, 11196, 1976.

295. Teaotia, S. S. and Phogat, K. P. S., Effect of rootstocks on growth, yield and quality in guava (*Psidium guajava*), *Prog. Hortic.*, 2(4), 37, 1971; *Hortic. Abstr.*, 43, 5626, 1973.

296. Nakasone, H. Y., Miyashita, R. K., and Yamane, G. M., Factors affecting ascorbic acid content of the acerola (*Malpighia glabra* L.), *Proc. Am. Soc. Hortic. Sci.*, 89, 161, 1966.

297. Issa, J. and Mielke, E. A., Influence of certain citrus interstocks on beta-carotene and lycopene levels in 10-year-old "Redbush" grapefruit, *J. Am. Soc. Hortic. Sci.*, 105(6), 807, 1980.

298. Fishman, G. M., Chikovani, D. M., and Tatarishvili, A. N., Carotenoids of the leaves of satsuma trees in relation to rootstock, *Khimiya Prirodnykh Soedinenii*, 3, 459, 1988; *Hortic. Abstr.*, 58, 9268, 1988.

299. Vardzelashvili, M. G. and Gachechiladze, T. T., Effect of varietal features and the rootstock on the content of bioactive compounds in peaches grown in the piedmont zone of eastern Georgian SSR, *Tr. Gruz. S-kh. Inst.*, 2, 62, 1982; *Chem. Abstr.*, 101, 35925m, 1984.

300. Jasnowski, S. and Maćkowiak, S., The effect of interstock on the vitamin C content of apples, *Prace Instytutu Sadownictwa Skierniewicach*, 12, 49, 1968; *Hortic. Abstr.*, 42, 242, 1972.

301. Bello, L., Growth, production and fruit quality of two pomelo cultivars (*Citrus paradisi* MACF.) on two rootstocks, *Agrotecnia de Cuba*, 17(1), 25, 1985; *Food Sci. Technol. Abstr.*, 19(12), J45, 1987.

302. Mehrotra, N. K., Jawanda, J. S., and Vij, V. K., Evaluation of rootstocks for jaffa oranges (*C. sinensis* (L.) sbeck) under arid-irrigated conditions of Punjab, *Indian J. Hortic.*, 41, 29, 1984.

303. Chohan, G. S., Kumar, H., and Vij, V. K., Influence of rootstock on vigour, yield, fruit quality and tree health of sweet orange cv. Pineapple (*Citrus sinensis* Osbeck), *J. Res. Punjab Agric. Univ.*, 22(1), 43, 1985.

304. Brown, G. B. and Bohn, G. W., Ascorbic acid in fruits of tomato varieties and F_1 hybrids forced in the greenhouse, *Proc. Am. Soc. Hortic. Sci.*, 47, 255, 1946.

305. Savcenko, N. A., Selection of onions by the free intervarietal crossing method, *Agrobiologija*, 1, 68, 1961; *Hortic. Abstr.*, 31, 6386, 1961.

306. Khan, A. R., Some studies on the soluble salt contents and vitamins in important varieties of turnips and their hybrids, *W. Pakist. J. Agric. Res.*, 2(1/2), 8, 1964; *Hortic. Abstr.*, 35, 3327, 1965.

307. Rampal, S., Note on vitamin C content in diploid and tetraploid capsicum, *Curr. Sci.*, 34, 614, 1965; *Hortic. Abstr.*, 36, 3075, 1966.

308. Kiseleva, V. A., Directed variation in the biochemical characters in tomato plants when heterosis occurs, *Sborn. Trud. Aspir. Molod. Nauč. Sotrud.* (*Leningrad*), 15, 354, 1970; *Hortic. Abstr.*, 41, 6854, 1971.

309. Tuk, Ch. T., Comparative biochemical characteristics of diploid and tetraploid forms of Washington Naval oranges, *Subtrop. Kul't.*, 1, 105, 1973; *Hortic. Abstr.*, 44, 1163, 1974.

310. Nair, P. M. and George, M. K., Studies on four intervarietal crosses of *Capsicum annuum* with references to chemical constituents, *Agric. Res. J. Kerala*, 11(1), 61, 1973; *Hortic. Abstr.*, 45, 4999, 1975.

311. Milewski, J., Hybridization of roses in order to obtain a high vitamin C content in the fruits, *Prace Instytutu Badawczego Leśnictwa*, 470/475, 127, 1974; *Hortic. Abstr.*, 46, 1045, 1976.

312. Manzyuk, S. G. and Omel'chenko, I. E., Physiologically active substances of group B vitamins in inbred and parental form corn plants, *Geterozis S-kh. Rast. Ego Fiziol.-Biokhim. Biofiz. Osn.*, Turbin, N. V., Ed., Kolos, Moscow, 1975, 42; *Chem. Abstr.*, 87, 114810r, 1977.

313. Manzyuk, S. G. and Zakrevskaya, L. E., Content of B-group vitamins in seeds of corn and sunflower parental forms and hybrids in relation to the crossing capacity of the lines, in *Geterozis S-kh. Rast. Ego Fiziol.-Biokhim. Biofiz. Osn.*, Turbin, N. V., Ed., Kolos, Moscow, 1975, 36; *Chem. Abstr.*, 87, 98968r, 1977.

314. Gevorkyan, L. A. and Snkhchyan, G. L., Heredity with reference to group B vitamin levels in hybrid grape generations, *Biol. Zh. Arm.*, 31(1), 57, 1978; *Chem. Abstr.*, 89, 56551k, 1978.

315. Premachanra, B. R., Vasantharajan, V. N., and Cama, H. R., Improvement in the nutritive value of tomatoes: vitamin A-potent tomatoes and genetic aspects of the regulation of β-carotene biosynthesis, *Indian J. Exp. Biol.*, 16(4), 468, 1978; *Chem. Abstr.*, 89, 20453r, 1978.

316. Lakshminarayana, S. and Moreno-Rivera, M. A., Promising Mexican guava selections rich in vitamin C, *Proc. Fla. St. Hortic. Soc.*, 92, 300, 1979; *Food Sci. Technol. Abstr.*, 13(11), J1781, 1981.

317. Ludilov, V. A., Possibilities of improving the quality of peppers with regard to content of ascorbic acid and P-active substances, *Kach. Ovoshch. Bakhchevyh Kul't.*, 50, 1981; *Chem. Abstr.*, 97, 88921n, 1982.

318. Valuzneu, A. R., Zazulina, N. A., New Belorussian black currant varieties as initial material for breeding varieties of the intensive type, *Vestsi AN BSSR Ser. sel'skagaspad.*, 3, 75, 1981; *Plant Breeding Abstr.*, 54, 388, 1984.

319. Holden, J. H. W., The contribution of breeding to the improvement of potato quality, in *8th. Triennal Conf. European Assoc. Potato Res.*, Midlothian, United Kingdom, 1981; *Food Sci. Technol. Abstr.*, 15(2), J295, 1983.

320. Uzo, J. O., Inheritance of "Nsukka Yellow" aroma, ascorbic acid and carotene in *Capsicum annuum* L., *Crop Res.*, 22(2), 77, 1982; *Chem. Abstr.*, 99, 50437n, 1983.

321. Bassett, M. J., Strandberg, J. O., and White, J. M., Orlando Gold. A fresh market F_1 hybrid carrot for Florida, *Circular*, *Agric. Exp. Stn.*, *Univ. Florida*, S-296, 1982; *Hortic. Abstr.*, 54, 178, 1984.

322. Heneberg, R., Momirovic-Culjat, J., and Sebecic, B., Thiamin, riboflavin and niacin contents of seeds of some soybean lines derived from M. Goldsoy x M_{14} crossing. *Poljopr. Znan. Smotra*, 60, 27, 1983; *Chem. Abstr.*, 100, 117955q, 1984.

323. Khalil, R. M. and Omran, A. F., Total soluble solids and ascorbic acid inheritance in pepper fruits "*Capsicum annuum* L.," *Minufiya J. Agric. Res.*, 5, 363, 1982; *Chem. Abstr.*, 100, 82898u, 1984.

324. Kvasnikov, B. V. and Zhidkova, N. I., Breeding carrot for increased carotene content, *Kachestvo Ovoshch i Bakhch kul'tur.*, 60, 1981; *Plant Breeding Abstr.*, 54, 8470, 1984.

325. Rieksta, D., Ozola, L., Biologically active substances in the fruits of *Rosa rugosa* Thunb. hybrids, *Tautsaimnieciba Derigo Augu Agrotehnika un Biokimija*, 1981, 99, 1981; *Plant Breeding Abstr.*, 54, 5378, 1984.

326. Gupta, K. R. and Dahiya, B. S., Genetics of protein and ascorbic acid contents in peas (*Pisum sativum* L.), *Indian J. Hortic.*, 42(1-2), 101, 1985.

327. Leshchenko, A. K. and Mikhailov, V. G., Heterosis as a genetic basis for breeding soybean for high yield, *Doklady Vsesoyuznoi Ordena Lenina i Ordena Trudovogo Krasnogo Znameni Akademii Sel'skokhozyaistvennykh Nauk Imeni V. I. Lenina*, 11, 13, 1982; *Plant Breeding Abstr.*, 55, 7402, 1985.

328. Popov, P. S. and Kozhevnikova, V. N., Heritability of vitamin E in sunflower, *Selektsiya i Semenovodstvo USSR*, 10, 22, 1982; *Plant Breeding Abstr.*, 55, 7108, 1985.

329. Luk'yanenko, A. N., Luk'yanenko, E. Kh., and Glushchenko, E. Ya., Initial material for breeding tomato for improved chemical composition, *Byulleten' Vsesoyuznogo Ordena Lenina i Ordena Druzhby Narodov Nauchno issledovatel'skogo Instituta Rastenievodstva Imeni N. I. Vavilova*, 121, 37, 1982; *Plant Breeding Abstr.*, 55, 8260, 1985.

330. Vartapetyan, V. V., Breeding apple for increased content of biologically active substances, *Selektsiya yabloni v SSSR*, 142, 1981; *Plant Breeding Abstr.*, 55, 9862, 1985.

331. Melkonyan, M. V., Possibility of achieving heterosis for content of group B vitamins in grape berries by means of synthetic hybridization, *Doklady Vsesoyuznoi Ordena Lenina i Ordena Trudovogo Krasnogo Znameni Akademii Sel'skokhozyaistvennykh Nauk Imeni V. I. Lenina*, 5, 20, 1983; *Plant Breeding Abstr.*, 56, 2223, 1986.

332. Shirko, T. S. and Yaroshevich, I. V., Source material for apple breeding, *Vestsi Akad. Navuk BSSR, Ser. Sel'skagaspad. Navuk*, 1, 67, 1987; *Chem. Abstr.*, 107, 36685u, 1987.

333. Yadav, R. D. S., Gupta, C. R., and Singh, R. B., Genetics of vitamin C content in chilli (*Capsicum annuum* Linn.), *Proc. Natl. Acad. Sci., India, Sect. B*, 57(4), 411, 1987; *Chem. Abstr.*, 109, 89897n, 1988.

334. Khachatryan, S. S., Marutyan, S. A., and Adamyan, A. Kh., Vitamin content of the fruit of the table grape varieties of diverse origin, *Plant Breeding Abstr.*, 58, 2614, 1988.

335. Bakhvalova, A. G. and Lisovskii, G. M., The possbility of early selection in black currant seedlings for improved vitamin C content of the fruit, *Sel'skokhozyaistvennaya Biologiya*, 9, 36, 1987; *Plant Breeding Abstr.*, 58, 6939, 1988.

336. Hassan, A. A., Abdel-Fattah, M. A., and Ali, K. E., Inheritance of total soluble solids and ascorbic acid content in tomato, *Egypt. J. Hortic.*, 14(2), 155, 1987; *Chem. Abstr.*, 112, 52433u, 1990.

337. Vartapetyan, V. V. and Kocheshkova, T. V., Use of polyploid apple varieties in selection. III. The inheritance of vitamin C content in fruits from crosses of apple varieties with different ploidy, *Biol. Nauki (Moscow)*, 3, 130, 1990; *Chem. Abstr.*, 113, 208494y, 1990.

338. Olson, R. A., and Frey, K. J., *Nutritional Quality of Cereal Grains: Genetic and Agronomic Improvement*, Agronomy 28, Am. Soc. Agron. Madison, Wisconsin, 1987.

339. Pattee, H. E., Ed., *Evaluation of Quality of Fruits and Vegetables*, AVI Publishing, Westport, Conn., 1985.

340. Bittenbender, H. C. and Kelly, J. F., Improving the nutritional quality of vegetables through plant breeding, in *Nutritional Evaluation of Food Processing*, Karmas, E. and Harris, R. S., Eds., Van Nostrand Reinhold, New York, 1988, chap. 24.

341. Robinson, J. C., Ed., International Symposium on the Culture of Subtropical and Tropical Fruits and Crops, Vol. 1 & 2, *Acta Hortic.*, 275, 1990.

342. Giacometti, D. C., Ed., Symposium on tropical and subtropical fruit breeding, *Acta Hortic.*, 196, 1, 1987.

343. Eskin, N. A. M., Ed., *Quality and Preservation of Vegetables*, CRC Press, Boca Raton, Fla., 1989.
344. Eskin, N. A. M., Ed., *Quality and Preservation of Fruits*, CRC Press, Boca Raton, Fla., 1991.
345. Kays, S. J., *Postharvest Physiology of Perishable Plant Products*, AVI Book, Van Nostrand Reinhold, New York, 1991.
346. Jungk, A., Beeinflussung des Vitamin- und Mineralstoffgehaltes von Pflanzen durch Züchtung und Anbaumassnahmen, *Landwirt. Forsch. Sonderh.*, 32, 222, 1975.
347. Anonymous, *Ertrag der Apfelkulturen und der Birnenkulturen des Schweiz 1992*, Eidgenösische Alkoholverwaltung, Berne, Switzerland, 1993.
348. Pospíšilová, J., Inhalt von L-Ascorbinsäure in der F_1-Generation bei der Hybridisation von Tomaten, *Arch. Gartenbau*, 17(2), 85, 1969.
349. Munger, H. M., The potential of breeding fruits and vegetables for human nutrition, *HortScience*, 14(3), 247, 1979.
350. Gabelman, W. H., The prospects of genetic engineering to improve nutritional values, in *Nutritional Quality of Fresh Fruits and Vegetables*, White, P. L. and N. Selvey, N., Eds., Futura Publ. Co., Mount Kisco, N.Y., 1974, 147.
351. Williams, W., Genetic means and cultural methods for improving nutritional values of crops, *Qualit. Plant.—Plant Foods Hum. Nutr.*, 29(1-2), 197, 1979.
352. Simon, P. W., Carrots and other horticultural crops as a source of provitamin A carotenes, *HortScience*, 25(12), 1495, 1990.
353. Bradley, G. A. and Rhodes, B. B., Carotene, Xanthophyll, and color in carrot varieties and lines as affected by growing temperatures, *J. Am. Soc. Hortic. Sci.*, 94, 63, 1969.
354. Simon, P. W. and Wolff, X. Y., Carotenes in typical and dark orange carrots, *J. Agric. Food Chem.*, 35, 1017, 1987.
355. Simon, P. W., Wolff, X. Y., Peterson, C. E., Kammerlohr, D. S., Rubatzky, V. E., Strandberg, J. O., Bassett, M. J., and White, J. M., High carotene mass carrot population, *HortScience*, 24(1), 174, 1989.
356. Premachanra, B. R., Vasantharajan, V. N., and Cama, H. R., Improvement of the nutritive value of tomatoes, *Curr. Sci.*, 45(2), 56, 1976.
357. Whiteside, A. G. O. and Jackson, S. H., The thiamin content of canadian hard red spring wheat varieties, *Cereal Chem.*, 20, 542, 1943.
358. Betschart, A., Nutritional quality of wheat and wheat foods, in *Wheat: Chemistry and Technology*, Vol. II, Pomermilleranz, Y., Ed., Am. Assoc. Cereal Chemists Inc., St. Paul, Minn., 91, 1988.
359. Johnson, V. A. and Mattern, P. J., Wheat, rye, and triticale, in *Nutritional Quality of Cereal Grains*, Olson, R. A., and Frey, K. J., Eds., Agronomy 28, Am. Soc. Agronomy, Madison, Wis., 1987, 133.
360. Witsch, H. V. and Flügel, A., Untersuchungen über den Aneurinhaushalt höherer Pflanzen, *Ber. Dtsch. Bot. Ges.*, 64, 107, 1951.
361. Ditzler, L., Hunt, C. H., and Bethke, R. M., Effect of heredity on the niacin and pantothenic acid content of corn, *Cereal Chem.*, 25, 273, 1948.
362. Richey, F. D. and Dawson, R. F., A survey of the possibilities and methods of breeding high-niacin corn (maize), *Plant Physiol.*, 23, 238, 1948.
363. Kadam, S. S., Dhumal, S. S., and Jambhale, N. D., Structure, nutritional composition, and quality, in *Potato: Production, Processing, and Products*, Salunkhe, D. K., Kadam, S. S., and Jadhav, S. J., Eds., CRC Press, Boca Raton, Fla., 1991, 9.

Chapter 4

CLIMATE AND PLANT VITAMINS

I. LIGHT DURATION AND INTENSITY

A. Ascorbic Acid

1. Ascorbic Acid and Plant Metabolism

In the green plant tissues, a major portion of ascorbic acid is located in the chloroplast.[1-3] Whether ascorbic acid plays any direct role in the photosynthetic process or whether it functions solely as a scavenger of free radicals produced during the photosynthetic process is not very clear.[1,2,4,5] Although light intensity has been known to increase the concentration of ascorbic acid in the chloroplast[6] and the synthesis of glucose, the primary precursor of ascorbic acid, is strongly dependent on the light intensity, the conversion of glucose to ascorbic acid apparently does not depend on the light intensity.[7]

The exact site of ascorbic acid synthesis in the plant cells is not clear, despite numerous reports on the synthesis of ascorbic acid[5,8] and factors affecting its content in the plant. The observations that (a) albino (chlorophyll free) leaves contain less ascorbic acid than their green leaves and (b) variegated leaves have less ascorbic acid than the fully green leaves[9] may indicate a relationship between the chlorophyll and ascorbic acid content in the leaves. However, the fact that fully "white" leaves are not completely devoid of ascorbic acid, and may in fact contain a considerable amount of ascorbic acid (319 vs. 269 mg/100 g in the green and white leaves of *Daphne mesereum*, respectively), raises the question as to (a) the site of synthesis of ascorbic acid in the cells (chloroplasts, cytoplasm, etc.), (b) the origin of ascorbic acid found in the chlorophyll-free leaves, and (c) whether ascorbic acid could have been transported from other (green) plant leaves to the chlorophyll-free parts.[9]

That ascorbic acid can be transported from plant leaves to other plants parts has been reported.[10] Also, the presence of ascorbic acid in the underground plant parts, such as potato tubers, and in the photosynthetically less active plant parts such as fruits, point to the possible transport of this vitamin from the leaves to these organs rather than in situ synthesis of this vitamin in them. Other observations such as (a) lower ascorbic acid content in the citrus fruits, apples and tomatoes, located in the inner branches (i.e., more shaded locations) than those located on the outside (exposed) branches and (b) covering of only tomato fruits with a paper bag can reduce their ascorbic acid content considerably (page 93) indicate that at least part of the ascorbic acid present in the fruits is synthesized in situ.

In general, the concentration of ascorbic acid in the tubers and fruits is believed to be very closely associated with the carbohydrate metabolism and is thus strongly influenced by the light intensity under which plants are grown. This despite the fact that, as mentioned, the exact mechanism of ascorbic acid accumulation in these organs is not fully understood. With this in mind, the following discussion on the effect of light on plant vitamins will be limited to the general effect of light on vitamin content without any specific comments on whether the effects observed were due to a higher synthesis of vitamin in the leaves, in the edible part under consideration, or both.

2. Effect of Light

Increased duration of sunny periods or light intensity has been observed to increase the ascorbic acid content in acerola,[11] apples,[12,13] *Amaranthus hybridus* (African spinach),[14] asparagus,[15] blueberry,[16] cantaloupe,[17] *Capsicum*,[18] cherry,[13,19] citrus fruits,[20] green beans,[21] lettuce,[22-27] mandarin,[28] parsley,[22] pineapple,[29] potato,[17] spinach,[9,22] tomato,[22,30-40] tender-green mustard,[41] turnip greens,[42-44] several edible plants,[45] and citrus leaf. In the following section, some selected reports on the effect of light are reviewed in detail.

Early investigations on the effect of light on the ascorbic acid content of plants were conducted by comparing the ascorbic acid content in the plants differently exposed to sun under natural field conditions or by placing them under various light intensities under laboratory conditions. Åberg,[30] in a very extensive and pioneering work, noted a strong effect of light intensity and duration (photoperiod) on the ascorbic acid content of tomato leaf (Figure 1). By placing the attached or detached leaves of parsley, spinach, and lettuce in the dark and at different temperatures, it was also shown that the rate of ascorbic acid loss in the dark was accelerated when the temperature was increased.[22]

In tomatoes, the effect of light on the ascorbic acid content is very pronounced. For example, ascorbic acid concentration was found to be higher in the exposed fruits than in the shaded fruits, it was also higher in the stem end (top, exposed) side than in the lower (shaded) side, and a short-term period of cloudy weather could reduce the ascorbic acid content of greenhouse-grown fruits (Table 1). A short period of cloudy weather can also reduce the ascorbic acid concentration in the field-grown tomatoes; the effect is, however, much less pronounced than for the greenhouse-grown fruits. For example, Murneek et al.[47] reported that a period of cloudy weather prior to harvest decreased the ascorbic acid content in the greenhouse-grown tomatoes from 27.1 to 22.1 mg/100 g and in the field-grown tomatoes from 35.2 to 34.5 mg/100 g.

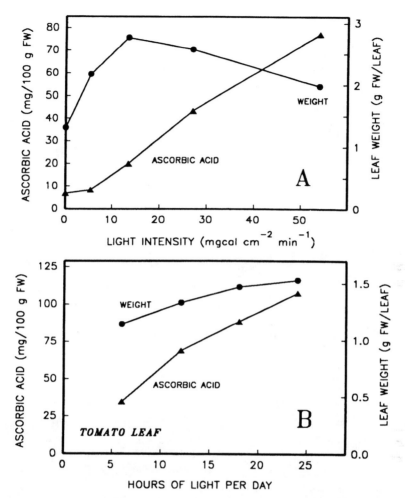

FIGURE 1. Effects of light intensity (A) and duration (B) on weight and ascorbic acid concentration in tomato leaves. (A) Tomato plants subjected for 10 days to different light intensities at 16 h/day photoperiod and under temperature-controlled conditions. Control plants (0 light intensity) were kept in complete darkness. (B) Plants were grown for 16 days with 33 mg cal/cm^2/min at 6–24 h of light per day. Data are for leaf five from tables 5 and 6 (for A) and for leaf three from table 12 (for B) of Åberg, B., *Ann. Roy. Agric. Coll. Sweden*, 13, 239, 1946.

TABLE 1

Effect of Periods of Cloudy or Sunny Weathers and the Degree of Exposure to Sunlight on the Ascorbic Acid Concentration in Tomato Fruits Grown in Greenhouse During June and July [47]

Stage of ripening and exposure	Ascorbic acid (mg / 100 g)
After a period of cloudy weather (*total light for the* 10 *days preceding day of harvest*: 3602 *g cal/cm²*)	
Whole fruit	22.0
Ripe, exposed side	21.0
Ripe, shaded side	16.2
After a period of bright sunny weather (*total light for the* 10 *days preceding day of harvest*: 6172 *g cal/cm²*)	
Ripe, exposed side	25.1
Ripe, shaded side	21.6
Fruits in direct sun	24.4
Fruits shaded by the leaves, same plant	19.8
Exposed (stem end half)	24.0
Shaded (lower) half	19.0

For further discussions on the ascorbic acid content of vegetables grown in greenhouse or in the field, see chapter 6. Also, by transferring tomato plants, at different stages of growth, to full sunshine or to shady (25% sunlight intensity) areas, Hamner et al.[32] showed that tomatoes ripened on the vine under full sunshine contained considerably higher ascorbic acid content than fruits ripened under shady conditions. Under experimental conditions, shading tomato fruits (by bagging them) and/or shading the whole plants, it has been shown that although general shading of the whole plants reduces the ascorbic acid content in the fruits, placing just the fruits in the dark can also reduce their ascorbic acid content, which indicates that direct exposure of tomato fruits to light is necessary to obtain the maximum ascorbic acid in the fruits[48,49] (Table 2).

Hassan and NcCollum[49] studied the effect of stage of fruit growth when it is exposed to low-light intensity on the final ascorbic acid concentration in ripe tomato. They placed small (2-3-cm diameter) green tomatoes 15 days after flowering in aluminum-covered bags and removed them 10 or 20 days later and let them grow till maturity. Other fruits were kept in the bags till they were ripe (i.e., 39 days).

TABLE 2

Effect of Subjecting the Tomato Fruits from 15 Days after Flowering to Darkness (by Bagging them) and/or Shading the Whole Plant with Cheesecloth (from July 8) on the Ascorbic Acid Concentration in Ripe Fruits Harvested from August to September [49]

Treatment	Ascorbic acid (mg / 100 g)
Unshaded plants, fruits bagged	17.5
Shaded plants, fruits bagged	17.3
Whole plant shaded	25.1
Unshaded plant, fruits shaded by foliage	24.2
Unshaded plant, fruits in direct sun (control)	33.5

TABLE 3

Effect of Bagging the Green Tomatoes[a] for Different Lengths of Time and Then Letting Them Ripen under Normal Sunlight Conditions on the Ascorbic Acid Concentration in the Ripe Tomato Varieties Illinois T19 and Garden State [49]

Treatment	Ascorbic acid (mg / 100 g)	
	Illinois T19	Garden State
Bagged for 10 days	25.0	22.4
Bagged for 20 days	23.2	20.1
Bagged for 39 days	16.8	13.5
Control (exposed)	30.4	26.8

[a] From the time they were 2–3 cm in the diameter (ca. 15 days old).

Comparison of ascorbic acid content of these tomatoes with that of tomatoes exposed to sun all through their growth showed that even 10 days of shading at the early stages of fruit development can significantly reduce the final ascorbic acid in the ripe fruits at the time of their harvest. Fruits that were kept in the dark from the 2-3-cm stage until maturity contained almost 50% less ascorbic acid than those fully exposed (Table 3). Even covering the pink-red fruits for a period of 10–12 days was noted to reduce the ascorbic acid content of tomatoes.[47]

Effect of degree of exposure to light on ascorbic acid content of the tomato is so strong that tomatoes grown on vines supported by poles were found to be significantly higher in ascorbic acid than those not supported by polês, which were thus shaded by the leaves.[50] Also,

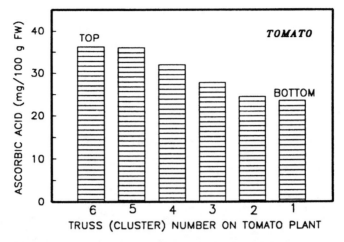

FIGURE 2. Ascorbic acid concentration in the red-ripe tomatoes taken from different clusters grown on tomato plant.[35]

Fryer et al.[35] reported that when tomato plants were grown as stacked plants in the greenhouse with supplemented illumination, red-ripe fruits harvested from higher clusters (trusses) contained significantly higher ascorbic acid concentration than those harvested from the lower clusters, again presumably because the latter fruits received less light (Figure 2). Krynska[51] reported that under wet and cold weather conditions tomatoes grown on higher slopes contained higher concentrations of ascorbic acid than those grown on lower ground. Whether this was related to a higher amount of light on higher slopes was not, however, specified.

Kondo[52] recently reported that shading and bagging of apple fruits significantly reduced the ascorbic acid concentration in their peels and pulp. Shaded fruits contained ca. four times less ascorbic acid in their peels (11.7 vs. 46.5 mg/100 g FW) and ca. six times less ascorbic acid in their pulp (0.6 vs. 3.8 mg/100 g FW) then the control plants. Also, the colored portions of fruits (presumably the side receiving more direct sunlight) contained significantly more ascorbic acid than the uncolored portions. It is noteworthy that in the bagged apples, just 15 days of exposure to light was enough to significantly increase the ascorbic acid concentration in the apple peels and pulp to a levels close to that of control fruits (Table 4).

In strawberries, the ascorbic acid content in the fruits may increase considerably when the light intensity or duration is increased.[53-57] For example, Hansen and Waldo,[56] in a pioneer work, noted that berries on plants fully exposed to light were much higher in ascorbic acid than

TABLE 4

Effect of Shading on the Ascorbic Acid Concentration in the Peel and Pulp of Apple Fruits (Variety Senshu)[52]

Treatment[a]	Ascorbic acid (mg / 100 g FW)	
	Pulp[b]	Peel[b]
Control[c]	3.8z	46.5z
Shaded[d]	0.6x	11.7x
Bagged until harvest[e]	1.3xy	6.3x
Bagged except for the last 15 days prior to harvest[e]	3.1yz	24.8y
Colored portion[f]	2.4[h]	40.4[i]
Colorless[g] portion	1.5	8.8

[a] Information given under footnotes c–g was obtained by communication with the author.

[b] Values in this column followed by different letters are significantly different at the 5% level according to Duncan's multiple range test.

[c] Fruits from unshaded outside branches.

[d] Fruits grown on branches shaded with cheesecloth, which reduced the natural light to 49% of its original amount.

[e] Fruits placed in bags that reduced the natural light to 28% of its original amount.

[f] Fruit side with the red color.

[g] Fruit side with the yellow color.

[h] Difference between the colored and colorless pulp is not statistically significant.

[i] Difference between the colored and colorless peel is significant at the 1% level.

those ripened under reduced light intensities. Shading of the entire plant (i.e., leaves as well as fruits) reduced the ascorbic acid concentration of fruits much more than when berries alone were shaded (Table 5). The fact that shading of the strawberry fruit alone reduced their ascorbic acid content to a slight extent indicates that a large part of the ascorbic acid is transported as such from the leaves to the fruits.[58] In other words, only part of the total ascorbic acid seems to be synthesized in the strawberry fruits themselves.

The effect of short-term low-light conditions on the ascorbic acid content in some fruits is so pronounced that strawberries harvested on the sunny days may contain much more ascorbic acid than those harvested on cloudy days.[58] It therefore seems safe to speculate that a relatively long period of foggy weather (days and weeks!), as is common in certain parts of Europe (like lowland areas of Switzerland), could reduce the ascorbic acid concentration in fruits and vegetables grown and harvested during these times. It is obvious that a recommendation to the commercial farmers not to harvest their strawberries, for example, immediately after a cloudy period would not be practical and

TABLE 5
Ascorbic Acid Concentration in Strawberry Fruits
after 21 Days of Shading the Whole Plant or the
Fruits Alone with One (1X) or Two (2X) Layers
of Muslin Cloth[56]

Treatment	Ascorbic acid (mg / 100 g FW)
Whole plant in sunshine	80
Berries only shaded 1 ×	71
Whole plant shaded 1 ×	64
Whole plant shaded 2 ×	48

From Hansen, E. and Waldo, G. F., *Food Res.*, 9, 453, 1944.
With permission.

realistic; under home gardening conditions, however, this would be a reasonable thing to do to avoid a product of lower nutritional quality.

In potato, although shading of the whole plants from the middle of May until their harvest was found to reduce the ascorbic acid content in the leaves considerably (from 90.1 to 66.3 mg/100 g in the variety Bliss Triumph), it did not affect the ascorbic acid content in the harvested tubers (from 37.8 to 37.6 mg/100 g, respectively), which clearly indicates a much stronger effect of light intensity on the synthesis of ascorbic acid in the leaves than on its amount in the storage organs (potato tubers).[17] This is in agreement with a recent report by Takebe et al.,[59] who showed that a 50% shading of sweet potatoes for one week reduced the ascorbic acid concentration in the plant leaves by 62–81% without affecting its concentration in the tubers.

The effect of light on the ascorbic acid content of pineapple is contradictory. Thus although Hamner and Nightingale[29] reported that experimental reduction of natural light by 50% increases the ascorbic acid content of fruits, Singleton and Gortner[60] noted a very close relationship between the short-term sunlight conditions and the fruits' ascorbic acid content so much so that just one week of sunny weather prior to harvest could increase the ascorbic acid content in the fruits by sixfold.[60] It appears that in pineapple, air temperature, in conjunction with sunlight, may also play an important role in the fruit's ascorbic acid content since Hamner and Nightingale[29] noted that the highest ascorbic acid levels were found in the fruits from those fields where the air temperature was the lowest during the six weeks previous to harvest.

Izumi et al.[28] grew mandarin trees for three months (July to October) under full sunlight (100% light) or under 50, 20, and 5% of full

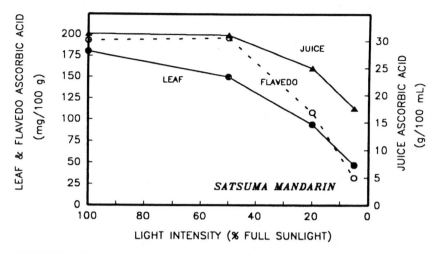

FIGURE 3. Effect of reduced intensity of natural light (by shading the trees with different types of cloths) from July 19 to October 19 on the ascorbic acid concentrations in the leaf and flavedo and juice of Satsuma mandarin grown in Japan.[28]

sunlight by covering the trees with different cloths and noted a significant decrease in the ascorbic acid content in the leaf and in the fruits' flavedo and its juice. This effect of light became especially pronounced when the light intensity was reduced below the 50% level (Figure 3).

An interesting circumstance where increased light was noted to affect the ascorbic acid in plants was reported for the berries of some *Vacciniaceae* plants. Mironov[61] reported that weak and medium-size fires in the lowland area of the former U.S.S.R. increased the ascorbic acid content in the berries produced after the fire. This was considered to be due to the improved physiochemical properties of the soil and that of the surviving trees, which were now exposed to more sunlight.

In some cases, plants, or part of them, are deprived of light in order to produce a product preferred by some consumer groups. This is the case for asparagus production, where plants are sometimes covered to produce white spears, which may have more consumer appeal. This practice, although it does not seem to affect other quality factors in asparagus, may reduce the ascorbic acid content by 50% when compared with uncovered green spears.[15]

B. Carotene

Evidence for the direct effect of light on the β-carotene content of fruits and vegetables is limited. This is partly because the effect of light is confounded with those of temperature and/or rainfall so much so that a decisive conclusion as to the single effect of each is very hard to make.[62] In most cases, however, it appears that despite the lack of a proven direct effect of light for carotenoid synthesis,[63] increased duration and intensity of light seem to promote the content of carotenes in several fruits and vegetables.

In tomato fruit, although light is not considered to be a requirement for the synthesis of carotenoid,[63, 64] increased light intensities could increase the carotene content of the fruits[65, 66] provided the temperature of fruits does not increase to high levels. In fact, it is well known that in warm geographical locations, tomatoes growing on the well-exposed positions on the vine often ripen to a yellow-red color instead of a deep red color[64] owing to the inhibitory effect of high temperature on lycopene synthesis (see page 109).

Apart from the above-mentioned confounding effect of temperature and light intensity, increased light intensity increases the carotene content of tomatoes, and thus immature tomatoes growing in the dark (bagged in aluminum foil) will have much less carotene than the control fruit exposed to light.[62] Pigment formation in tomato fruits was also noted to be inhibited when fruits were placed in dark bags (and additionally covered with aluminum foil to prevent the buildup of temperature within the bags). When the dark-grown fruits were ripened in light, however, their carotenoid content was similar to that of light-grown control fruits. Very little β-carotene was detected in the ripe but dark-grown tomato fruits.[67]

In the yellow varieties of peaches, the content of vitamin A (measured by bioassay) increased slightly in the fruits bagged from the blossom stage as compared with the control fruits. In apricots and yellow nectarines, however, vitamin A activity slightly decreased when fruits were bagged.[68] Kays[63] also notes that in peaches as well as in apricots the color development is greater in the absence of light.

Evidence for the effect of light on the carotene content of leafy vegetables or leafy crops is controversial. In lettuce, for example, shading of plants during the winter months (October to February)[23] or covering the plants with perforated plastic sheets[69] was noted to reduce the carotene content of the leaves. In turnip greens, however, reducing the light intensity by one-half 10 days prior to their harvest was found to increase their carotene content by 9.1%.[70] Also, in fodder plants and vegetables, the content of carotene in the leaves was generally higher during winter than summer months, although winter months were usually rainy and summer months were dry and hot,[71] indicating the possible confounding effects of light and temperature on the carotene content of plant leaves. Also, in the marine algae *Dunaliella bardawil*, accumulation of β-carotene is strongly dependent on the light intensity[72] but independent of the light quality.[73]

C. Other Vitamins

Light appears to increase the synthesis of thiamin[74] and niacin[75, 76] in the plant leaves. In turnip greens higher light intensity was noted to increase the thiamin concentration, especially when plants were grown on soils having lower pH (4.8 to 5.1) but not when soil pH was higher (6.6 to 6.9).[70]

The riboflavin content of turnip greens was not affected by the light intensity.[70] In tomatoes, however, riboflavin content was significantly affected by the position of plant pots in the greenhouse, which was attributed to the environmental difference in the greenhouse.[77] Gustafson[74] noted that leaves of tomato, potato, beans, peas, corn, and New Zealand spinach (*Tetragonia expansa*) exposed to higher light intensities contained slightly higher concentration of riboflavin than plants exposed to low light intensity.

II. LIGHT QUALITY

A. Ascorbic Acid

Information on the effect of light quality (wavelength spectrum) on plant vitamins is relatively rare and that which is available is mostly about ascorbic acid. For example, Åberg[30] noted that under controlled temperature conditions, the ascorbic acid content in kale and tomato leaves was not significantly affected by various combinations of sodium and mercury lamps. In the tomatoes grown in the greenhouse, however, supplementing the natural greenhouse light with blue fluorescent light (3400–4000 angstrom) or sodium vapor lamps (5800–5900 angstrom) increased the ascorbic acid concentration in the fruits considerably as compared with the control fruits (grown with natural sunlight entering the greenhouse). Sodium lamps, however, tended to be relatively more effective in this respect than fluorescent lamps.[36] Erner[46] grew sour orange seedlings for six months under plastic sheets transmitting different light wavelengths under equalized light intensities and noted that cutting the light spectrum above or below 500 nm reduced the ascorbic acid content in the plant leaves. Ascorbic acid content of leaves grown under blue light was 72% of that of the control plants grown under the same intensity of white light.[46] Finally, irradiation of tomato plants with UV-B radiation was noted to slightly decrease the concentration of ascorbic acid and carotene in the plants.[78]

Grimstad[24, 79] compared the effect of high-pressure Na(SON/T), metal halide (HPI/T), and warm-white fluorescent lamps on the chlorophyll and vitamin C content of two lettuce varieties and noted that lettuce grown with high-pressure Na lamps was lower in chlorophyll but higher in ascorbic acid than that grown with warm-white fluorescent lamps. The varieties, however, reacted differently to the kind of lamps used. The difference in lamps could not be explained by differences in their photosynthetically active radiation (PAR) since the fluorescent lamps that resulted in lower ascorbic acid produced higher PAR than the other two lamps.

In the wheat seedlings, red light was most active in stimulating the synthesis of vitamin C, and blue/violet in combination with green light was least effective in this respect.[80] Blue light from a halogen arc lamp was also reported to increase the ascorbic acid in radish roots as compared with those grown under a xenon lamp.[81] Effect of light quality on ascorbic acid of radish may also depend on the intensity of light used. For example, Zolotukhin et al.[82] noted that in radish leaves and roots the highest ascorbic acid content occurred when plants were exposed to short-wavelength (blue) radiation at intensities $> 100 \ \mathrm{W/m^2}$. At low-light intensity of $50 \ \mathrm{W/m^2}$, however, ascorbic acid content was higher when plants were radiated with long-wavelength (red) light.

B. Other Vitamins

Reports on the effect of light quality on the carotene content of tomatoes are inconsistent.[66, 83, 84–87] For example, Jen[86, 87] exposed the mature green tomatoes to the same quantity of radiant energy (by controlling the interfering effect of temperature) and noted that red light increased the rate of chlorophyll degradation whereas the blue portion of the spectrum stimulated the synthesis of carotenoids in the tomatoes ripened under different-colored lights.

Red and far-red lights may affect the concentration of different carotenoids in tomatoes by different extent so that exposure of fruits to red light was noted to simulate, but exposure to far-red light to inhibit, the synthesis of lycopene content of the fruits.[85, 88] The β-carotene content of tomatoes, however, is not affected by the exposure of fruits to red or far red light (Figure 4).[85] It thus appears that the synthesis of at least some carotenoids in the plants is mediated by the phytochrome system.[85, 89]

The effect of light quality on the carotenoid synthesis in *Capsicum* seems to be quite different from its effect on tomato discussed above. For example, when *Capsicum* fruits were wrapped in variously colored cellophane filters for 30 days, the total amount of carotenoids was higher in the fruits exposed to red light (transmitted through the blue filter) and was lower in the fruits exposed to green and blue lights (transmitted through the red and yellow filters, respectively) as compared with that in the fruits wrapped in white filter.[90] Leaves of red kidney beans contained maximum or minimum amounts of carotene when plants were grown with green or red light, respectively.[91]

Reports on the effect of light quality on vitamins other than ascorbic acid and carotene are very limited. Based on the only report known to

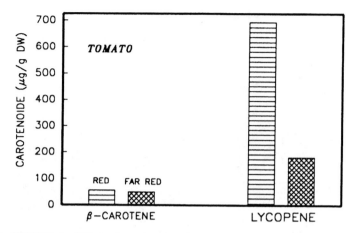

FIGURE 4. Effect of treating mature green tomatoes with red light (for 8 days at 14 h photoperiod) or with far-red light (for 30 min and kept in the dark for the remainder of the time) on the β-carotene and lycopene concentrations in the ripening tomatoes.[85]

me, the synthesis of lipoquinones in etiolated radish seedlings is apparently altered by the light quality, so that the far-red light stimulates the formation of more plastoquinone-9 than total α-tocopherol and enhances the formation of vitamin K_1.[92]

III. POSITION OF FRUIT ON THE PLANT

Effect of light on ascorbic acid is so strong that fruits growing on the outside or better-exposed branches of the trees or vine may contain more ascorbic acid than those growing on the inside or less-exposed and shaded branches. This has been observed in apples (Figure 5),[93] Brussels sprouts,[94] *Capsicum*,[95] oranges (Figure 6),[20, 96-104] and tomatoes (Tables 1 and 2).[35, 95, 105, 106] Citrus fruits located on the outside branches of the trees were noted to contain up to ca. 21% (9 mg/100 ml of juice) more ascorbic acid than the fruits grown on the inside locations.[20] In tomatoes this difference is even larger since fruits fully exposed to the sun contained 41–56% more ascorbic acid than those shaded by the plant leaves.[49]

Also, in a single apple fruit, the exposed part (i.e., the side facing the light and thus having a better color) may have higher ascorbic acid concentration than the shaded part (i.e., the side facing the dark or inner side of the tree)[12, 52, 93, 107, 108] (Table 6). This difference was found to be true not only in the highly colored varieties such as Winesap, but also in the less pigmented Stayman or in varieties without any red color

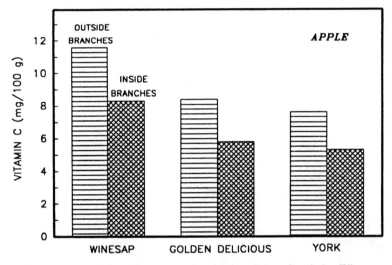

FIGURE 5. Ascorbic acid concentration in the apples of varieties Winesap, Golden Delicious, and York grown on the outside or inside branches of the same tree.[93]

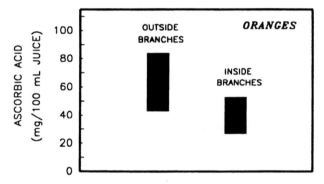

FIGURE 6. Range of ascorbic acid concentration in the juice of Valencia oranges (on sour orange rootstocks) grown on exposed (outside) or shaded (inside) branches of the same tree.[96]

such as Golden Delicious. Even after the skin was removed, a significant difference was noted in the ascorbic acid concentration in the exposed and shaded side of the apple flesh (Table 7). In the Golden Delicious apples, the "sunny" half of unpeeled apples contained 75% more ascorbic acid than the "shady" half (7.85 vs. 4.49 mg/100 g, respectively).[109] Also, in a single pepper, the side facing the sun, which had turned yellow, was noted to contain 50% more ascorbic acid than

TABLE 6
Ascorbic Acid Concentration in the Exposed and Shade Parts of
the Same Fruit Sampled from the Inside or Outside Branches
of Different Varieties of Apples[93]

Variety and tree condition	Ascorbic acid (mg / 100 g)	
	Exposed part	Shaded part
Winesap, old tree, outside fruit	11.2	9.5
Winesap, old tree, inside fruit	8.9	8.4
Stayman, old tree, outside fruit	9.1	7.0
York, old tree, outside fruit	8.5	5.8
Golden Delicious, young tree	3.8	3.1

TABLE 7
Ascorbic Acid Concentration[a] in the Skin and the Underlying Flesh
in the Red and Yellow Sides of a Single Apple Fruit
(Variety Wilshire)[107]

	(mg / 100 FW)	(mg / 100 mg protein)
Red skin	51	58
Flesh beneath the red skin	7	41
Yellow skin	29	33
Flesh beneath the yellow skin	4	26

[a] Values rounded.

the shady side, which was still green (255 vs. 166 mg/100 g, respectively).[110]

In tomato fruit, sections facing south (and thus more exposed to solar radiation) contain higher ascorbic acid than sections facing north.[40] Cherry fruits grown on the south side of a tree contain more ascorbic acid than fruits grown on the north side of the same tree.[13]

Also, leaves of several forest tress,[111] needles of spruces[112] (Table 8) and pine[113] and leaves of *Berberis virescens*[114] and Satsuma mandarin,[115] located on the south side of the plants were found to be much higher in ascorbic acid than those located on the north side.

The effect of position on ascorbic acid content of leaves has also been studied in leaves located at different heights on a tree and thus differently exposed to solar radiation. For example, spruce needles located at a 20-m height contained 401 mg/100 g whereas those located at 2 m above the ground contained much less, i.e., only 273 mg/100 g ascorbic acid.[9] Also, leaves of apples taken from the upper tier were

TABLE 8

Effect of Needle Age and Position of the Tree (South vs. North Side) on the Ascorbic Acid Concentration (ppm FW) in the Needles of Sprude (*Picea excelsa*) Measured in 1976[112]

	Year needles formed				
Tree side	1972	1973	1974	1975	1976
North side	950	950	1,300	1,050	800
South side	1,490	1,950	2,770	2,510	1,360

noted to contain higher ascorbic acid content than those from lower tiers.[116]

The position of fruits on the mother plant seems to also affect the rate at which fruits lose their ascorbic acid after they have been harvested. For example, Izumi et al.[104] noted that Satsuma mandarins grown on the exterior of the tree canopy maintained a higher ascorbic acid content during their storage than those grown in the interior positions on the tree.

IV. TEMPERATURE

Studies on the effect of temperature per se on the vitamin content of plants are relatively rare. Based on comparative studies on the vitamin content of plants grown in different geographical and thus diverse climatic conditions and at different times of year, it appears, however, that temperature could have a pronounced effect on the content of some vitamins in some plants. The effect of geographical location and the seasonal fluctuations in plant vitamins are treated on pages 126 and 112, respectively. Here, only those reports in which quantitative effect of temperature on plant vitamins has been studied are reviewed.

A. Ascorbic Acid

Air temperature seems to have very different effects on the ascorbic acid concentration of different fruits and vegetables. In the majority of cases, however, relatively lower temperatures during plant growth appear to increase the ascorbic acid in plants.

Vicia faba seedlings grown at 26°C contained two times more ascorbic acid than in those grown at 14°C. The situation in *Pisum sativum* seedlings, however, was the reverse, and in *Avena sativa* seedlings, the concentration of ascorbic acid was equal in the seedlings grown at 26 and 14°C.[45] In turnip tubers, no difference in ascorbic acid content could be detected when plants were grown with a daytime temperature of 25°C and subjected to 12°C for two nights, to 4°C for three nights, and to 0°C for six nights prior to their harvest as compared with the control plants, which were not subjected to the nightly low-temperature treatment. Also, storage of roots for 2–4 weeks at 0°C did not affect their vitamin C content.[117]

In tomatoes, ascorbic acid content seems to be increased when plants are grown at relatively higher temperature ranges. For example, Hammer et al.,[32] by growing tomatoes in growth chambers to maturity under two different temperatures regimes, noted that tomatoes from plants grown at a very cold temperature of 17°C (a temperature just above the lowest temperature that tomato can still grow and produce fruits) contained less ascorbic acid than those from plants grown at a warmer temperature of 25°C (15.6 vs. 18.5 mg/100 g FW, respectively). Also, Liptay et al.[118] reported that warmer temperatures improve the ascorbic acid content in the tomatoes. They reached this conclusion after growing tomatoes in Canada at different times of the year in greenhouses that did not have any supplemental light. They noted that tomatoes grown in the latter part of October, when the greenhouses were not yet heated and there were 158 h of sunshine per month, had a very low ascorbic acid concentration (7 mg/100 g FW). Tomatoes harvested in the month of March, however, when there were only 108 h of sunshine per month but the greenhouses were heated (31/22°C day/night temperature), contained more than twice the ascorbic acid concentration (16 mg/100 g FW). Under similar conditions of light and nighttime temperature, tomatoes grown at a daytime temperature of 31°C contained more ascorbic acid than those grown at 24°C (16 vs. 12 mg/100 g FW, respectively). On the other hand, Gutiev[119] noted that in tomato leaves, the concentration of ascorbic acid decreases if plants are grown at day temperatures above 30°C and night temperatures below 10°C, indicating that both night and day temperature may play an important role in the final ascorbic acid content in the tomato plant.

In contrast to tomatoes, in the leafy vegetables, lower temperature ranges seemed to increase their ascorbic acid content in most cases studied. For example, Rosenfeld[120] compared the effect of air temperatures on the ascorbic acid content of several vegetables and reported that in most cases, with the exception of turnip leaves, ascorbic acid concentration was highest when plants were grown at lower temperatures of 12–15°C and decreased if the air temperature was raised to 24°C. In turnip leaves, however, ascorbic acid concentration gradually increased with increase in temperature from 12 to 24°C (Figure 7).

Also, in apples, relatively lower temperatures during their growth seem to increase their ascorbic acid content. In Norway, for example, Skard and Weydahl[121] noted that the average ascorbic acid concentra-

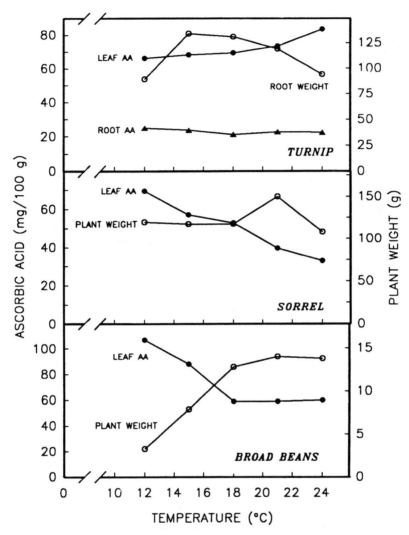

FIGURE 7. Effect of (air) temperature on yield and ascorbic acid (AA) concentration in different vegetables grown under natural light conditions in Norway.[120]

tion in nine apple varieties was 64.2 and 78.7 mg/100 g in two years with average summer temperatures of 16.25 and 13.5°C, respectively. They also noted that the ascorbic acid concentration of the same varieties of apples grown in another location with a relatively cold summer (average temperature of 12.3°C) was as high as 157.2 mg/100 g! Based on these observations, they concluded that low summer temperature increases the ascorbic acid concentration in apples. Efimova,[122] based on measurement of "active" temperatures recorded between 1969 to 1984 in the former Soviet Union, reported that a low sum of active temperatures accompanied by wet weather conditions resulted in higher ascorbic acid and low sugar content in eight varieties of apples. A negative relationship between temperature and the ascorbic acid content of apples was also noted by Shabalina.[123]

In agreement with the above, lower temperatures during growth were also noted to increase the vitamin C content in black currants,[124–126] parsley,[127] and pineapple.[29] Säkö and Laurinen,[126] for example, noted that in Finland black currants contained higher ascorbic acid in cold and cloudy years than in warm and sunny years. These authors, based on some cited reports, indicated that the low temperature during the cloudy periods apparently favors the synthesis of sugars and ascorbic acid and at the same time decreases the rate of ascorbic acid oxidation. Nilsson,[128] based on growth chamber studies, noted that the highest ascorbic acid content in black currant occurs if the plants are subjected to high light intensities and low night temperatures.

Part of the reason for the positive effect of relatively low temperatures in increasing the ascorbic acid in some plants may be a reduced rate of ascorbic acid degradation, and not necessarily a higher ascorbic acid synthesis, in the plants grown at lower temperatures. For example, Åberg[30] noted that leaves of tomato plants transferred (for a period of 13 days) from 23 to 15.5°C had 30% more ascorbic acid than those continuously kept at 23°C. Also, when detached leaves of kale, lettuce, parsley, spinach, and tomato plants were placed in the dark, the rate of ascorbic acid loss was much slower at 5°C than at 15°C. Somers et al.[43] investigated the effect of temperature (10–30°C) on the ascorbic acid in the discs of turnip greens and noted that increasing the temperature accelerated the loss of ascorbic acid if the leaf discs were kept in the dark. When discs were placed in the light, however, the rate of ascorbic acid accumulation increased as the temperature was increased from 14 to 30°C.

Arndt[129] tried to correlate the effect of natural air and soil temperature during the winter months (December to mid-March) with the ascorbic acid content of Brussels sprouts and reported that the ascorbic acid content of these vegetables underwent large fluctuations during this period and appeared to be related to the air temperature prevailing 4–5 days prior to the sampling date. Thus a relatively long and continuous period of air temperatures below 0°C reduced the ascorbic acid content of Brussels sprouts. A later increase in air temperature to about 2–4°C resulted in an increase in ascorbic acid. If the temperature went higher than 2–4°C, however, the content of ascorbic acid in Brussels sprouts decreased.

At the Polar Experiment Station in Murmansk (in the former Soviet Union), Kulikova[130] studied the ascorbic acid content of some 200 plant varieties during a 21-year period 1963–1984 and noted that vegetables grown in the open had higher ascorbic acid content than those grown in the greenhouse. If we assume (because it is not specified in the abstract of their report) that plants grown in the open in the polar regions were subjected to lower temperature and perhaps more light than their counterparts grown in the greenhouse, and if we further assume that the authors drew the above conclusion by comparing the ascorbic acid concentration in the plants grown during the same part of year (e.g., polar summer) in both the open and in the greenhouse, then this report is a very important finding indicating the relative importance of light and temperature to the ascorbic acid content under extreme climatic conditions. It thus appears that higher light intensity and lower temperature in the open were both favorable to higher ascorbic acid content in the plants investigated.

In summary, it appears that in most leafy and some fruit crops, a relatively lower temperature during their growth increases their ascorbic acid content. This, however, does not seem to hold true for all plants, for example, for tomatoes.

B. Carotene

Warmer temperatures during growth have been shown to increase the carotene content in carrots,[131–133] sweet potato,[134] and papaya.[135] In Valencia oranges, however, changing the soil temperature from 15 to 25°C did not have any effect on the carotenoid content of the fruits.[136]

Carrots grown in warmer locations often ripen faster and generally have a better taste and higher carotene content than those grown in cooler locations.[131] In Finland, vegetables grown in the south of the country were found to regularly contain more carotenes than those grown in the north.[137, 138] The observation that the concentration of carotene in the carrots grown during winter months may be 1/2 to 1/3 of those grown in spring and summer months[62, 132] points to the important positive role of temperature in carotene content in this vegetable. Maximum carotene content in carrots is reported to occur at a temperature range of 15.6–21°C and is reduced if the air temperature is lower or higher than the above range.[62, 132]

Effects of temperature on total carotene, β-carotene, and the color of carrots do not seem to go hand in hand. For example, Bradley et al.[139] reported that in carrots the best color is produced when the average soil temperature stays below 18°C for at least several weeks prior to harvest. In a different report, Bradley and Dyck[140] showed that higher preharvest temperature, normally associated with summer harvests, resulted in carrots with the highest total carotene content, lowest β- to α-carotene ratio, and generally poorest color. In comparison, carrots harvested in the fall (field temperatures of below 15°C) had the best color, lower total carotene, and higher β- to α-carotene ratio.

In tomatoes, optimum temperature for plant growth, development of best color (high lycopene content), and vitamin C concentration in the fruits do not seem to coincide. For example, Went et al.[141] noted that whenever tomatoes were grown under conditions favorable for best fruit development, the lycopene and carotene contents of fruits were low, and whenever the conditions for fruit growth were poor, the fruits had higher lycopene content. These authors noted that the β-carotene content of tomatoes, however, was very little influenced by the external temperature. Under field conditions tomatoes located on the well-exposed locations on vine often ripen to a yellow-red color and may not develop a deep red color, which is attributed to the high temperature in the tomatoes exposed to full sun.[65] Optimum temperature for the synthesis of red-colored lycopene pigment in tomato is 25°C, and its synthesis is virtually inhibited if the temperature exceeds 32°C.[144] The synthesis of β-carotene, however, remains unimpaired even if the temperature exceeds 30°C.[142, 144] In agreement with these is the report that ripening of breaker tomatoes would come to a virtual standstill if the fruits were held at 33°C, as is evident from the suppressed C_2H_2 evolution, color development, and softening.[145]

Sayre et al.[143] grew tomato plants at 24–27/18°C day/night temperatures until fruits started to turn color and then transferred the whole plants, or detached fruits, to three day/night temperature regimes and noted that the ascorbic acid concentration in the fruits was highest at 15/7°C, but the concentration of lycopene was highest at 24/15°C. Fruits grown at the higher temperature of 38/29°C developed very little color and had the lowest ascorbic acid and carotene content. In a more refined experiment, the authors noted that 27/18°C day/night temperature was the optimum temperature for both color development and carotene production in tomato fruits ripened on the vine or artificially ripened.

The positive effect of higher temperatures on carotene content has also been observed for some tropical fruits such as guava and papaya. In Brazil, for example, red-fleshed guavas grown in the warm northeastern parts of the country were higher in β-carotene content than those grown in the state of São Paulo, where the climate is relatively moderate.[146] Also, papayas (cultivar Formosa) grown in the hot region of Bahia were found to be higher in β-carotene than those grown in the São Paulo area.[147]

C. Other Vitamins

In general, environmental conditions inducive of better plant growth seem to also increase the content of some vitamins in plants. For example, concentrations of thiamin and riboflavin in endive and those

TABLE 9
Relative Vitamin Concentration in Plants Grown under
Low (10–15°C = L) and High (27–30°C = H)
Temperatures[148]

Plant[a]	Thiamin	Riboflavin	Nicotinic acid
Beans	L < H	L < H	—
Soybean	L < H	L < H	L = H[b]
Tomato	L < H	L = H	L < H
Clover	L = H	L > H	L > H
Peas	L = H	L > H	L = H
Spinach	L = H	L > H	L > H
Wheat	L = H	L > H	L > H
Broccoli	L > H	L > H	L > H
Cabbage	L > H	L > H	L = H
Lupines	L > H	L > H	L = H

[a] Plants are arbitrarily grouped according to the effect of temperature
on thiamin content.
[b] An equal sign (=) means little (less than 10%) or no difference.

of ascorbic acid and riboflavin in lettuce were found to be higher in the
plants grown in locations where conditions for plant growth were more
favorable, i.e., had more light and warmer temperatures, as compared
with plants grown in locations with less favorable climates.[133]

Gustafson[148] studied the effect of temperature on the growth and
vitamin content of several warm- and cold-season plants and reported
that neither does every plant respond to the same extent to the
temperature, nor are all vitamins affected in the same way. For example, those plant species making their best growth during the warmer
parts of the summer had higher vitamin content at relatively higher
temperatures, whereas plants like cabbage, clover, and peas, which
grow better during the cooler part of summer and in spring and fall,
tend to have higher vitamin content when grown at relatively lower
temperature ranges. The thiamin concentration is likely to be higher at
higher temperature and riboflavin lower at higher temperatures. Table
9 shows a summary of the work by Gustafson.[148]

In Kansas, periods of cool and rainy weather after flowering were
noted to increase the thiamin content of wheat.[149]. In contrast, wheat
grown in Holland was noted to contain less thiamin in the years with
cold and rainy summers than in the years with dry and warm summers.[150]

Marquard[151, 152] studied the factorial effects of temperature and
photoperiod on the tocopherol content of seeds of three mustards
(*Sinapsis sp.*) and rapeseed (*Brassica napus*). When plants were grown
under warm (24.5°C day/18°C night) temperatures, their seeds con-

tained a lower percentage of oil but their oil contained a higher concentration of tocopherol than seed of plants grown under cool (16.5/10°C day/night) temperatures. The length of the photoperiod did not appear to have any affect on the tocopherol content of the oil. Beringer and Saxena[153] reported that when oats, sunflower, and linseed were grown at 12 and 28°C daytime temperatures and a constant night temperature of 12°C, the tocopherol concentration in the oil of their seed oils was higher when plants were grown at warmer (28°C) than at colder (12°C) temperatures. In contrast to these, exposure of the algae *Euglena gracillis* to low-temperature stress was found to increase their content of α-tocopherol by six- to sevenfold.[154]

V. DIURNAL FLUCTUATIONS

Concentrations of vitamins in plants may undergo relatively rapid diurnal fluctuations. This has been observed for the ascorbic acid concentration in the leaves of *Arisarum vulgare*,[155] cowpea,[156] cotton,[157] barley,[158] broad beans,[158] chinese cabbage,[158] lettuce,[27, 159] parsley,[160] peas,[155] soybean,[161] spinach,[158] wheat,[158] potato,[162] *Rosa rugosa*,[163] peanut,[164] *Cannabis sativa, Cichorium endivia, Helianthus tuberosus, Helianthus autumnale, Phaseolus multiflorus*,[159] Satsuma mandarin,[115] turnip-rooted parsley,[165] tomato,[166, 167] and some forest trees,[45, 113, 168] and in the fruits of black currant, raspberry, and strawberry,[169] In most cases the maximum amount of ascorbic acid occurs shortly before or after midday and the minimum occurs during the early predawn hours.[45] In parsley, for example, the ascorbic acid concentration is often lower at 6:00 A.M. than at 6:00 P.M.[160] In black currant, strawberry, and raspberry, the ascorbic acid content was higher at midday than in the morning or in the evening.[169]

Although the above observations on the diurnal fluctuation in ascorbic acid seem to confirm the known effect of light on the ascorbic acid content in plants (page 90), the diurnal fluctuation in ascorbic acid seems to be also controlled by some internal biological clock. For example, Hagan[159] noted that even after three weeks of constant illumination, some plants still showed some fluctuations in their ascorbic acid content during a 24-period. The observation made by Izumi et al.,[115] however, that during sunny periods the ascorbic acid content in the leaves of Satsuma mandarin reaches it minimum during the early morning hours, but during cloudy periods its concentration is relatively higher during the night hours than during the daytime, may indicate that the diurnal fluctuation in the ascorbic acid in the leaves may be controlled by both photoperiod and temperature conditions.

β-carotene concentration in the plant leaves may also show a diurnal fluctuation; its lowest concentrations, however, occur mostly during the midday hours. During dull weather conditions, however, the difference between the carotene content at different times of the day is much less striking.[170] It was thus recommended that on sunny days, when such a fluctuation in β-carotene is most pronounced, if possible harvesting of the dark-green leafy vegetables during the midday hours should be avoided.

In turnip greens, riboflavin[171, 172] and thiamin[171] concentrations may significantly vary from one day to the next. In any given day, riboflavin content (expressed on a dry weight basis) was highest at 8:00 A.M. and lowest at 4:00 P.M.[172]

VI. SEASONAL FLUCTUATIONS

The term seasonal fluctuations as used here refers to the natural variations that have been observed in the vitamin concentration in freshly harvested plants, depending on the time of year they are planted and/or harvested. In this chapter the seasonal variations taking place in the vitamin content of stored products such as apples, potatoes, and so forth marketed at different times of the year are not being treated because this would be outside the scope and aim of the book. Some selected examples of this type of seasonal fluctuation in plant vitamins are given below just for those readers not familiar with the subject.

Apples and potatoes stored for different periods of time and marketed at different times of the year may have a very different ascorbic acid content. For example, Wills and El-Hgetany[173] noted that in Australia, fresh Delicious apples purchased in the 9th week of the year (late February, late summer in Australia) contained 14 mg/100 g vitamin C whereas those purchased in the 51st week (after a long period of storage) contained only 4 mg/100 g, i.e., a decrease of more than threefold. Also, potatoes sampled every two weeks over a 12-month period from retail stores in Australia showed a considerable seasonal fluctuation in their vitamin C content, which was at its peak during December to January (midsummer) and at its lowest during August to September (late winter to early spring). Potatoes purchased in summer contained about twice the vitamin C as the same varieties purchased in winter (23 vs. 10 mg/100 g in the Pontiac, for example).[174]

The reports discussed hereafter are mainly on the natural differences in the vitamin content of fresh produce harvested at different times of the year. We assume that a major reason for the seasonal variations in plant vitamins, as discussed below, is the composite effects of various environmental factors such as planting data, light, temperature, precipitation, and so forth, and is in some cases confounded with the stage of plant maturity. Since stage of maturity (chapter 6) and light and temperature (see pages 90 and 104, respectively) could all affect the content of different plant vitamins, the following discussion on the effects of seasonal variation on vitamin content must be judged with the above in mind.

A. Ascorbic Acid

Seasonal variations in ascorbic acid have been reported in cabbage,[175] grapefruits,[176] guava,[177] lettuce,[178] mango buds,[179] oranges,[180-182] parsley,[183-184] Persian lime,[185] pineapple,[186] prickly pears,[187] rose hips,[189] Stasuma mandarin,[190] spinach,[178, 191, 192] tomato,[47, 49, 193-196] and needles of forest trees such as *Larix* and *Aesculus*[197] and spruce.[198] Here some selected cases will be reviewed in more detail.

Fruits such as citrus have a relatively wide harvesting period, and some fruits may be harvested some months later than others. Fruits harvested late in the season, however, may be different in their ascorbic acid than those harvested early. For example, in Florida, early and midseason-harvested oranges (during November to February) were much higher in ascorbic acid than those harvested later (March to June). Although the fruits harvested toward the end of the harvesting season are generally more mature and are regarded to be of better quality, they are usually lower in their ascorbic acid content.[181, 199] Also, grapefruits harvested towards the end of the harvesting season were noted to be close to 50% lower in ascorbic acid than those harvested earlier (from 62 to 35 mg/100 ml). In tangerines, the ascorbic acid content was found to decline by more than 70% in the overmature fruits as compared with those harvested early in the season (35 vs. 10 mg/100 ml juice, respectively).[200] Stasuma mandarin cultivated in Japan was noted to show seasonal variations not only in ascorbic acid, but also in its β-carotene content. The ascorbic acid in Satsumas was highest in March and lowest in September (37 vs. 22 mg/100 g, respectively).[201]

Snap bean pods harvested in spring[202, 203] or midsummer[204] were noted to have a much a higher ascorbic acid content than pods harvested in the fall (Table 10). No consistent variations were noted in the ascorbic acid content in broccoli, chard, collards, kale, and lettuce harvested in different seasons.[205]

In some cases, plants grown during the colder months of the year may contain higher vitamin C than those grown during the warmer seasons. For example, Poole et al.[206] reported that in three out of five strains of cabbage tested, fresh samples taken in February contained significantly higher concentrations of ascorbic acid than samples of the same strains taken in June. Burrell et al.,[175] however, noted that depending on the variety tested, the ascorbic acid of cabbage may be higher or lower in the heads harvested in November than in those harvested in July (Table 11).

TABLE 10
Effects of Harvesting Season and Year on the Ascorbic Acid, Thiamin, and Riboflavin Concentrations[a] in Snap Beans

Year	Season	Ascorbic acid (mg / 100 g FW)	Thiamin (μg / 100 g FW)	Riboflavin
1943	Spring	21	75	94
	Fall	28	65	82
1944	Spring	21	93	89
	Fall	23	67	97
1945	Spring	23	103	119
	Fall	23	75	115

[a] Average and rounded data from table 6 of Hayden, F. R., Heinze, P. H., and Wade, B. L., *Food Res.*, 13, 143, 1948.

TABLE 11
Ascorbic Acid Concentration (mg / 100 g FW) in Cabbage Varieties in Relation to the Season They Were Harvested[175]

| Harvesting season | Ohio station varieties | | | |
	No. 5	No. 7	No. 8	No. 12
July	55	61	170	63
November	50	61	100	73

Seasonal variations in the ascorbic acid content have been noted in spinach grown in different parts of the world. For example, in 13 varieties of spinach grown in New York State, the average ascorbic acid concentration was one-third higher in the fall-grown crop than in the spring crop (99 vs. 60 mg/100 g, respectively).[207] Also spinach grown in Germany[191] or in the northern foothills of the Caucasus[178] was higher in ascorbic acid if grown during the fall and winter months than if grown during other seasons. In Egypt, however, comparisons of ascorbic acid in spinach sown on September 15, October 15, November 15, and December 15 showed that spinach planted at later dates had more ascorbic acid than that planted earlier.[192]

In the industrialized nations, tomatoes are available throughout the year. Tomatoes marketed during the winter months may be locally produced in greenhouses or shipped in from warmer areas. In most cases, however, tomatoes marketed during the winter in the northern hemisphere are lower in ascorbic acid than those marketed during summer. For example, in Massachusetts (U.S.A.), tomatoes sold in the supermarkets during late-winter months contained only one-third as much ascorbic acid as those offered during the summer months.[193] Also, in Ontario (Canada), comparisons between the ascorbic acid content of locally produced or imported tomatoes (as offered by the retail outlets) during the course of three years showed a yearly cycle in the ascorbic acid content, which was at its maximum during the summer months and reached its minimum during the winter months.[196]

Tomatoes cultivated in the same location may contain different concentrations of ascorbic acid depending on the time of year they are grown. In the Adana area of Turkey, for example, the ascorbic acid concentration in tomatoes grown on a find sandy loam soil was 180% higher in the spring-grown than in the fall-grown fruits (15.56 vs. 5.44 mg/100 ml juice).[208] In the Giza area of Egypt, tomatoes grown during the summer months were higher in ascorbic acid than those grown during the fall, winter, or spring months.[209]

The vitamin content of different varieties may not change in a similar manner during different times of year. For example, Shinohara et al.[195] noted that in Japan, the tomato cultivar FTVNR-3 grown during the winter months contained only one-half the ascorbic acid content of the same cultivar grown during the summer months. In some other cultivars tested, however, the ascorbic acid content of winter- and summer-grown crops was the same.

A relatively short span of time between date of harvest may also exert a strong effect on the content of ascorbic acid in some plants. For example, ascorbic acid content in the roots of radish (*Raphanus sativus*) harvested in Ano (Japan) decreased from 24–41 mg/100 g on June 6 to 15–32 mg/100 g in the roots harvested 14 days later (June 20). Whether short-term weather conditions such as cloudy periods were responsible for these changes was not mentioned.[210] In India, guava fruits harvested during the winter months were noted to contain more than twice the ascorbic acid than those harvested during the spring[211] or rainy period (summer months).[177, 211, 212]

Plant leaves may also show seasonal variations in their ascorbic acid content. This has been observed in the leaves of citrus trees[213, 214] and forest trees,[215–222] and in the leaves of plants grown in the arctic region of Greenland.[223] In the case of forest leaves, however, reports on the seasonal variations in ascorbic acid are not consistent. For example, Kazakova[216] noted that needles of conifer trees were lowest in ascorbic acid content during the spring months and the highest (3–4 times higher than the minimum) at the onset of winter.[216] In contrast, Liu et al.[218] reported that in the leaves of two evergreen plants (*Pinus tabulaeformis* and *Platycladus orientalis*) maximum content of ascorbic acid occurred in the spring months and gradually decreased thereafter. Leaves of three deciduous species (*Robinia pseudoacacia*, *Populus*

TABLE 12
Carotene Concentration (mg / 100 g) in Four Varieties of Carrots in Two Growing Seasons[132]

Season	Imperator	Tendersweet	Chantenay	Scarlet nantes
Summer[a]	14[c]	11	10	9
Winter[b]	4	4	5	4

[a] Carrots planted in May and harvested on October.
[b] Carrots planted in August and harvested in March of next year.
[c] Values rounded.

canadensis, and *Quercus variabilis*), however, retained high concentrations of ascorbic acid during the growing season, which decreased only during the autumn.

Plant roots may also undergo some seasonal fluctuations in their content of vitamins. For example, roots of dandelion plants were noted to contain higher concentrations of vitamins C and E and β-carotene in spring than in winter.[224] In addition roots of some tomato hybrids were found to contain higher ascorbic acid concentration during flower-bud formation (17.23 mg/100 g), while at the fruit-set stage the root's concentration of this vitamin was reduced to 4–5 mg/100 g.[225]

B. Carotene

Seasonal variations in carotene content have been noted in cabbage,[226] carrots,[62, 132, 227–229] kale,[147] lettuce,[26] snap beans,[204] and turnip greens.[230]

Carrots planted in the late spring and harvested in the summer or fall were found to contain higher carotene than those maturing during the winter months, which points to the strong temperature dependence of carotene production in this plant (see page 108). This has been observed for carrots grown in England,[227] Florida,[62] California,[228] and Oregon (Table 12). Usually carrots do not develop much carotenoids if sown too late, as compared with spring-sown carrots, even if they are allowed to grow through the winter months and are harvested during the next spring.[227] Carrots grown during the warmer season in California's Monterey County were found to be higher in total α- and β-carotenes and also to have a higher ratio of α/β-carotenes than those grown during the cooler season in Imperial County.[228] In contrast to the above, Krylov and Baranova[231] reported that carrots sown during autumn and winter months contained more carotene than those sown during spring. Also, in the polar regions, carrots sown in autumn were found to be higher in carotene than those sown in the spring.[229]

In contrast to the above, the carotene content in leafy and green vegetables seems to be lower during the warmer than in the colder seasons. For example, snap beans harvested in midsummer were found to be lower in carotene than those harvested in the fall. This difference was most pronounced if beans were harvested at their early stages of maturity when their carotene content was the highest.[204] Also, turnip greens grown in summer were noted be much lower in carotene than those grown in the winter in the greenhouse.[230] Hansen[205] studied the carotene content of broccoli, chard, collards, and kale harvested at intervals from July to April of the next year in Oregon and noted that with the exception of chard, carotene concentration in plants decreased during the fall and winter months and increased slightly during the early-spring months. In agreement with these reports, Mercadante and Rodriguez-Amaya[147] noted that in Brazil the β-carotene concentration in kale (cv. Manteiga) was significantly lower in the summer than in the winter.

Seasonal variations in carotenes have also been observed in the leaves of Scots pine[219] and in the leaves of several forage plants.[62] With the exception of alfalfa and bluegrass, generally forage harvested in midsummer contained lower carotene than that harvested in the spring or fall. In agreement with this is the observation that carotene concentration in several forage and vegetable plants is much lower in the plants cultivated in the dry and hot summer months than in plants cultivated during the rainy winter months.[71] In Florida, total carotene content in forage plants was found to be significantly lower during the dry season (February to March) than during the wet season (September to October) (1.2 vs. 17.6 ppm, respectively), and this difference was reflected in the vitamin A content in the livers of beef cattle grazing on forage, so much so that liver vitamin A ranged from 231 to 471 ppm in dry and wet seasons, respectively.[232]

According to some authors, seasonal variations in carotene content of different plants seem to have no overall bearing on the vitamin A availability to consumers. For example, Bureau and Bushway[233] studied the vitamin A content of 22 fruits and vegetables sampled from wholesale distributors from five major locations throughout the United States (Los Angeles, Dallas, Chicago, Miami, and Boston) three times during a year (November, March, and July) and noted large variations between the vitamin A activity in some fruits and vegetables depending on the time of year they were purchased. For example, on the average, the vitamin A content of spinach was relatively similar during the three times of the year tested. Asparagus, green beans, okra, sweet potato, and swiss chard, however, contained in some cases close to twice as much vitamin A activity when purchased in March as compared with November or July; carrots, oranges, and peaches contained much higher vitamin A when purchased in July as compared with March or November; and, finally, grapefruit, iceberg lettuce and summer squash had higher vitamin A when purchased in November as compared with March or July. In some cases, the difference was more than seven times, as in lettuce (122.9 vs. 16.5 RE/100 g in November and July, respectively). The ranges of differences observed in the samples collected during each month were, however, very large, as is shown for the lettuce purchased in November in which the values ranged from 16.2 to 579.9, i.e., a difference of close to 36-fold, which may be partly due to the fact that samples used for analysis were from different varieties, grown in different locations, and subjected to different postharvest handling conditions. Based on these findings, the authors noted no statistical differences among the locations or months of analyses on the plant vitamin A and concluded that " . . . for the total vitamin A active carotenoids, it does not matter where you obtain your fruits and vegetables or during what time of the year."[233]

C. Other Vitamins

Hayden et al.[203] measured ascorbic acid, thiamin, and riboflavin in some 200 varieties of snap beans during several years and noted that, on the average of all strains and different years, beans harvested in spring were consistently higher only in thiamin than those harvested in the fall (Table 10). In cabbage, a significantly higher concentration of thiamin was measured during the month of June than in February in three out of five strains tested.[206] In snap beans, higher niacin but lower riboflavin was noted in the beans harvested during midsummer than in the fall.[204] Highest thiamin content in the leaves of chives was noted to occur in April, in parsley in May-June, and in lemon balm (*Melissa officinalis*) in October.[184] In citrus leaves, the concentration of thiamin an riboflavin was maximum during the flowering and declined to a minimum during fruit ripening.[234]

The tocopherol content of forage crops may change during the growing season (Figure 8). In lucerne, for example, α-tocopherol rises from 35 ppm (DW basis) in March to 240 ppm in November. In grasses, however, α-tocopherol fell to a minimum in late spring, then rose and reached its maximum during summer, and finally fell in early autumn.[235] In the leaves of tea grown in Japan, α-tocopherol content in the green branches gradually increased from July to October, decreased in December, and then increases in spring as the new shoots appeared.[236]

No seasonal fluctuation was found in the folate content of various Nigerian foods.[237] In the leaves of Unshiu mandarins, however, folic acid was noted to show two peaks during summer and autumn whereas

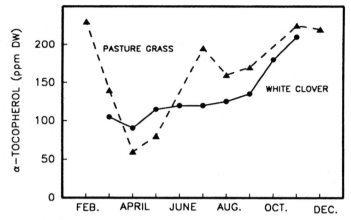

FIGURE 8. α-tocopherol concentration in fodder plants at different times of year.[235]

its concentration in the fruits continued to rise to a maximum in November.[238]

In summary, the concentration of vitamins in plants (fruits, vegetables, forage plants, forest trees, etc.) can vary strongly depending on the time of year they are grown and/or harvested. Not all vitamins, however, change in a parallel manner. This is especially true when different plants are compared with each other. Whether these variations, cumulated over all the vitamins in the food plants, would have any nutritional consequences for the vitamin nutrition of consumers during different months of year remains to be seen. In this respect, the situation may be very different in societies where plant products grown in faraway places with completely different climates are imported and are available to the consumers all year round as compared with those societies where the majority of plant foods are produced locally and may undergo enormous variations in kind and availability during different months of the year.

VII. YEARLY FLUCTUATIONS

Vitamin content in plants may also vary to a great extent from one year to the next (Table 13). Although the exact reasons for such large yearly fluctuations are not known, it is reasonable to assume that variation in factors such as the amount and distribution of sunny and cloudy periods and the magnitude and distribution of temperature and rainfall from one year to the next may be among the most important factors in this respect. Thus the smaller number of observations where

TABLE 13
Variations in Vitamin Concentrations in the Edible Parts of Plants in Different Years

Plant	Concentration range	Factor (max. / min.)	Ref.
Ascorbic acid (mg / 100 g)			
Apples	3.0–9.1	3.0	121
Apples[a]	52–88	1.7	239
Apples (Signe)	8.6–21.8	2.5	240
Apples	12.1–13.4	1.1	242
Apples (McIntosh)	0.6–6.5	10.8	241
Apples (Spartan)	1.7–17.5	10.3	241
Apples (Juno)	8.0–19.7	2.5	241
Apricots	3–4	1.3	243
Apricots	4.7–5.3	1.1	242
Asparagus	14–17	1.2	243
Beans (snap)[a]	89.7–120.0	1.3	244
Beans (green)[c]	9–23	2.5	245
Blueberry	8.3–8.9	1.1	246
Broccoli[a]	770–980	1.3	247
Cabbage	60–70	1.2	248
Cauliflower	55.6–109.6	2.0	138
Cherry[a]	82–177	2.2	239
Cherry[a]	58–98	1.7	249
Cherry (sour)	12.8–26.5	2.1	13
Cherry	4.0–8.9	2.2	242
Currant (red)[a]	91–204	2.2	249
Currant (black)[a]	619–824	1.3	249
Gooseberry[a]	81–204	2.5	239
Gooseberry[a]	89–139	1.6	249
Grape	2.3–6.4	2.8	249
Leek	16.1–22.3	1.4	138
Peas[c]	32–53	1.6	245
Peaches	6.2–17.4	2.8	242
Peaches	2.7–2.9	1.1	243
Pear	2.2–4.5	2.0	242
Plum	3.0–4.7	1.6	242
Raspberry	31.1–53.2	1.7	246
Spinach[c]	27–85	3.1	245
Strawberry[a]	147–545	3.7	239
Strawberry[a]	77–557	7.2	249
Strawberry	40.9–65.7	1.6	242
Strawberry	40–57	1.4	250
Strawberry	51.1–62.3	1.2	246
Swede	35.6–47.8	1.3	138
Sweet potato[a]	45–67	1.5	251
Tomato[c]	67–142	2.1	245
Tomato	9–12	1.3	243
Tomato	25.4–36.0	1.4	252

TABLE 13   (continued)

Plant	Concentration range	Factor (max./min.)	Ref.
Carotene (IU / 100 g)			
Asparagus	461–529	1.1	243
Apricots	1,489–1,755	1.2	243
Broccoli[a,b]	11,820–15,950	1.3	247
Broccoli[b]	620–800	1.3	253
Carrots[b]	44,900–57,700	1.3	254
Kale[b]	2,360–5,900	2.5	147
Peaches	303–558	1.8	243
Sweet potato[a,b]	13,500–21,300	1.6	255
Tomato	399–497	1.2	243
Niacin (μg / 100 g)			
Barley	8,650–10,000	1.2	256
Benas	2,110–2,480	1.2	257
Maize	2,090–2,160	1.1	262
Maize	2,130–2,440	1.4	259
Oats	1,290–1,760	1.4	256
Wheat	5,420–7,460		256
Pantothenic acid (μg / 100 g)			
Barley	750–1,120	1.5	256
Maize	490–530	1.1	262
Maize	560–680	1.2	259
Oats	660–950	1.4	256
Wheat	780–1,520	1.9	256
Pyridoxine (μg / 100 g)			
Barley	600–910	1.5	256
Oats	220–420	1.9	256
Wheat	380–520	1.4	263
Wheat	480–760	1.6	256
Riboflavin (μg / 100 g)			
Barley	170–220	1.3	256
Maize	99–112	1.1	259
Oats	140–200	1.4	256
Peanut	130–162	1.2	260
Wheat	75–94	1.2	259
Wheat	140–190	1.4	256

TABLE 13   (continued)

Plant	Concentration range	Factor (max./ min.)	Ref.
Thiamin (μg / 100 g)			
Barley	470–560	1.2	256
Beans	840–930	1.1	257
Cowpea[a]	1,080–1,650	1.5	258
Maize	370–530	1.4	259
Oats	680–870	1.1	256
Oats	420–610	1.2	259
Peanut	1,300–1,420	1.1	260
Wheat[c]	688–728	1.1	261
Wheat	330–390	1.2	259
Wheat (summer)	320–510	1.6	150
Tocopherol (μg / 100 g)			
Sunflower oil	66,000–69,700	1.1	264
Pumpkin seed oil	4,120–4,620	1.1	265
Wheat	490–4,010	8.2	266
Wheat[d]	730–2,150	2.9	266

[a] Dry weight basis.
[b] μg/100 g.
[c] International units per ounce.
[d] α-tocopherol.

no yearly fluctuations in plant vitamins have been observed[267] might have been due to the very uniform climate in the experiment site or to an unusually stable climatic condition in the years experiments were conducted.

Among various vitamins, ascorbic acid seems to show relatively larger yearly fluctuations in different plants than other vitamins and may range from ca. twofold in cherries, up to sevenfold in strawberries, and up to 10-fold in apples (Table 13). This is not surprising because ascorbic acid is one of the vitamins whose concentration is more strongly affected by light than that of any other vitamin (page 90). Yearly variations in the concentrations of other vitamins are relatively small and mostly in the range of less than a factor of 1.5, or a difference of less than 150%.

Different varieties may be differently susceptible to the yearly variation in their vitamin content. For example, Skard and Weydahl[121] noted that out of 46 varieties of apples, 30 showed their highest ascorbic acid in 1944, 11 in 1945, and five in 1943. Thus although the variety Starr

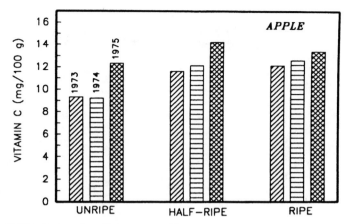

FIGURE 9. Vitamin C concentration in apples at three different stages of ripening in three different years. Data are average of five varieties.[242]

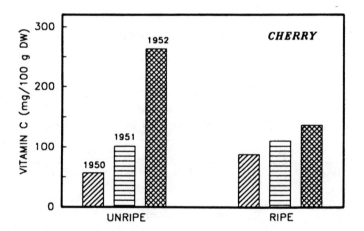

FIGURE 10. Vitamin C concentration in unripe and ripe sweet cherry in three years.[239]

showed very similar ascorbic acid concentration in the years 1943, 1944, and 1945 (9.5, 8.5, and 11.5 mg/100 g, respectively), the variety Lord Suffield showed its highest ascorbic acid in the year 1945 and the highest ascorbic acid content of the variety Emnet Early occurred in the year 1943. The yearly variations in the ascorbic acid content of fruits could also be very different in ripe and unripe fruits (Figures 9 and 10).

Putnam[268] noted significant yearly differences in the content of thiamin, riboflavin, biotin, and α-tocopherol in wheat and barley grown in England during the years 1975–1978.

TABLE 14

Yearly Variations in the Vitamin A and Ascorbic Acid Concentrations in Tomato, Asparagus, Apricot, and Peach[243]

Year	Tomato juice[a]	Asparagus spears	Apricot halves	Peach halves
		Vitamin A (IU/100 g)		
1972	400	529	—	—
1973	588	461	1489	547
1974	516	522	1755	558
1075	497	—	1496	303
		Ascorbic acid (mg/100 g)		
1972	12.2	17.0	—	—
1973	11.6	13.9	4.2	2.9
1974	9.5	16.7	3.3	2.9
1975	9.1	—	2.9	2.7

[a] With the exception of apricots, data are from the California products.

TABLE 15

Yearly Variations in the Nicotinic Acid and Pantothenic Acid Concentrations (mg / kg DW) in Wheat Kernels in Saskatchewan[256]

Year	Nicotinic acid			Pantothenic acid		
	Wheat	Barley	Oats	Wheat	Barley	Oats
1964	67.0b[a]	99.0a	17.6a	15.2a	10.0b	7.5bc
1965	68.4b	100.0a	14.8b	10.5b	11.2a	7.9b
1966	63.2c	90.4b	12.9c	7.8c	7.5c	7.4bc
1967	74.4a	97.1a	16.6a	10.2b	7.5c	6.6c
1968	54.2d	86.5c	13.6c	11.6b	10.3b	9.5a

[a] In each column means followed by the same letter are not significantly different at the 0.05 level.

Environmental factors that affect the content of each vitamin may not vary in a parallel manner from one year to the next, and thus different vitamins may fluctuate differently from one year to the next, a situation that may also be different in different plants (Tables 13–15) and geographical locations. Thus Eheart,[269] for example, noted that among the four locations tested (Georgia, Mississippi, South Carolina, and Virginia), lima beams showed a significant yearly difference in

TABLE 16
Summary of Analysis of Variance for the Effects of Year, Hybrid,
and Location of Growth on the Niacin and Pantothenic Acid
Concentrations in the Maize Kernels[262]

Source	Niacin	Pantothenic acid
Year	*[a]	**
Location	**	**
Hybrid	**	**
Year × location	**	**
Year × hybrid	**	**
Location × hybrid	**	NS

[a] NS, *, and ** indicate not significant, significant, and highly significant, respectively.

thiamin content in South Carolina, in riboflavin in Mississippi, and in ascorbic acid in Virginia. This indicates that not all vitamins are equally affected by the sum of environmental factors collectively referred to as "year-effect."

Carotene concentration in carrots[227] and in broccoli[247] showed significant yearly variations (Table 13) so that in some years the carotene concentration in carrots was extremely low and in other years very high.[227] For further examples of the effect of year on the concentration of vitamins in plants, see also Tables 21 and 24 in chapter 5, Table 3 and Figure 1 in chapter 6, and Table 1 in chapter 7.

The relative impact of fluctuations in the vitamin content of different varieties (variety-effect), on the one hand (chapter 3), and the year-to-year variation in the vitamin content of a given plant variety (year-effect), on the other, on our supply of various vitamins throughout the year may be a matter of nutritional significance. Ideally one would like to search for those high-vitamin varieties whose vitamin content is less sensitive to the year-to-year fluctuations. This might not be such a hard task if the variety-effect was not much larger than the year-effect. Although in some cases, the variety-effect on plant vitamins was found to be 3–4 times higher than the year-effect,[260] other workers present the view that the year-effect is far greater than that of variety and that different vitamins may not be equally affected by the year or variety effect.[257] Finally, the significant interactions observed between the effects of year, variety, and geographical location on the content of some plant vitamins (Table 16) make any attempt at breeding high-vitamin plants a very challenging job. As for the relative impact of year-effect and fertilizer-effect (chapter 5) on plant vitamins, some reports indicate that the year-effect may be 25 times higher than the average of effect of various fertilizer treatments.[270] This is easy to understand in view of the relatively smaller effect of most fertilizers on plant vitamins (chapter 5).

VIII. GEOGRAPHICAL VARIATIONS

A given plant grown in different geographical locations (different parts of the world, different locations in a given country or a state) may have very different vitamin contents (Table 17). From the reports at hand, it is, however, not certain what factor(s) bring about this "geographical-effect." Since most of the studies reported here were conducted with the same plant variety, it can be assumed that differences in light (intensity and duration), temperature, precipitation, and the chemical and physical properties of the soils in different geographical locations may be responsible for the major part of the effects observed. However, when the vitamin contents of plants grown in locations with extreme difference in climate are compared, factors such as length of the day, angle of the sun, light quality (such as the relative amounts of red and blue wavelengths) should also be taken into consideration.[287] Some selected reports are reviewed here.

A. Ascorbic Acid

The ascorbic acid content of fruits and vegetables may strongly vary depending on the location where a given plant is grown (Table 17). Differences as much as twofold in the ascorbic acid content of oranges (Tables 17, 18), turnip greens, and strawberry, fourfold in the carotene content of cabbage, fivefold in the riboflavin and niacin content of wheat, and eight- to ninefold in the thiamin, pyridoxine, folic acid, and tocopherol content of wheat have also been reported (Table 17). Guavas grown in Brazil[304] may contain ca. four times more ascorbic acid than those grown in Papua New Guinea[305] (372 vs. 68 mg/100 g in white varieties). In the United States, ascorbic acid concentration in the "Irish Cobbler" potatoes grown in Kentucky was less than one-third of that in the same potato variety grown in Maine.[311]

In the northern hemisphere, plants grown in the more northerly locations seem to contain more ascorbic acid than those grown in more southern places. Also, plant phenotypes adjusted (acclimated) to more northern regions seem to have higher ascorbic acid than their counterparts acclimated to more southern latitudes. This has been reported for apples[306–309] and Chinese radish grown in different parts of the former Soviet Union[311] and for apples grown in different parts of Norway.[121]

Skard and Waydahl[121] noted a 25-fold (24.8 vs. 0.8 mg/100 g) difference in the ascorbic acid concentration in the Wealthy apples grown in northern latitudes than in the southern locations in Norway

TABLE 17

**Variations in the Vitamin Concentrations in Plants Grown in Different
Geographical Locations within a Country, Continent,
or Around the World**[a]

Plant	Location	Concentration range	Factor (max./min.)	Ref.
Ascorbic acid (mg / 100 g)				
Apples	Norway	1–25[b]	25.0	121
Asparagus	Spain	20–85	4.2	271
Beans (snap)	Florida	13–20	1.5	272
Beans (snap)[c]	U.S.A.	79–107	1.3	244
Beans (lima) (DW)	U.S.A.	79–99	1.2	273
Beans (lima)	U.S.A.	23–34	1.5	269
Broccoli	Florida	56–98	1.7	274
Cabbage	Florida	39–65	1.7	272
Cabbage	U.S.A.	35–52	1.5	275
Carrots	U.S.A.	7–12	1.7	276
Cauliflower	Sweden	67–92	1.4	138
Celery	U.S.A.	5–8	1.6	275
Collards	Florida	52–105	2.0	274
Grapefruit	Sierra Leone	40–48	2.5	277
Leek	Sweden	16–32	2.0	138
Maize	U.S.A.	5–8	1.6	275
Onion	U.S.A.	5–10	2.0	275
Oranges	Iran	30–64	2.1	279
Oranges	Sierra Leone	32–81	2.5	277
Parsley	Sweden	141–163	1.2	138
Potato	Canada	6–21	3.5	280
Potato	Germany	13–20	1.5	281
Potato	New York	9–20	2.2	282
Potato[c]	U.S.A.	77–151	2.0	283
Potato	New Zealand	17–25	1.5	284
Sea buckthorn	Former U.S.S.R.	151–964	6.4	278
Strawberry	U.S.A.	21–80	3.8	250
Strawberry	U.S.A.	49–65	1.3	285
Swede	Sweden	34–56	1.6	138
Sweet potato	U.S.A.	45–97	2.1	255
Tangerine[g]	Sierra Leone	21–49	2.3	273
Tomato	U.S.A.	10–25	2.5	275
Turnip greens	U.S.A.	97–393	4.0	286
Carotene (µg / 100 g)				
Beans (snap)	Florida	319–533	1.7	272
Beans (snap)[c]	U.S.A.	2,430–3,250	1.3	244
Broccoli	Florida	810–1,170	1.4	273
Cabbage[d]	U.S.A.	57–213	3.7	275
Carrots	Florida	4,560–11,600	2.5	273
Carrots[f]	Finland	3,540–7,020	2.0	287
Carrots[e]	U.S.A.	1,300–21,000	1.6	276

TABLE 17   (continued)

Plant	Location	Concentration range	Factor (max./ min.)	Ref.
Carrots[d]	U.S.A.	12,228–18,538	1.5	275
Celery[d]	U.S.A.	60–213	2.5	275
Collards	Florida	3,280–4,640	1.4	273
Corn[d]	U.S.A.	99–351	3.5	275
Leek	Finland	1,400–3,800	2.7	137
Lettuce	Finland	1,700–2,300	1.3	137
Papaya[e]	Brazil	99–193	1.9	135
Parsley[f]	Finland	1,670–3,160	1.9	287
Spinach	Finland	2,000–3,200	1.6	137
Sweet pepper	Finland	1,320–2,130	1.6	287
Sweet potato	U.S.A.	7,000–2,2000	3.1	255
Tomato	U.S.S.R.	699–1,186	1.7	288
Tomato[d]	U.S.A.	530–1,040	2.0	275

Folic acid (μg / 100 g)

Plant	Location	Concentration range	Factor (max./ min.)	Ref.
Wheat[g]	Worldwide	16–78	8.6	293

Niacin (μg / 100 g)

Plant	Location	Concentration range	Factor (max./ min.)	Ref.
Beans	New Mexico	2,110–3,220	1.5	257
Maize	U.S.A.	1,990–2,220	1.1	262
Millet (pearl)	U.S.A.	3,189–4,258	1.3	295
Taro	South Pacific	671–932	1.4	291
Wheat[g]	Worldwide	2,200–11,100	5.0	291

Pantothenic acid

Plant	Location	Concentration range	Factor (max./ min.)	Ref.
Maize	U.S.A.	500–570	1.1	262

Pyridoxine (μg / 100 g)

Plant	Location	Concentration range	Factor (max./ min.)	Ref.
Asparagus	Spain	50–70	1.4	271

Riboflavin (μg / 100 g)

Plant	Location	Concentration range	Factor (max./ min.)	Ref.
Asparagus	Spain	90–120	1.3	271
Beans	New Mexico	200–240	1.2	257
Beans (lima)	U.S.A.	208–286	1.4	273
Beans (lima)	U.S.A.	91–119	1.3	269
Carrots	U.S.A.	48–80	1.7	276
Gram (red)	India	150–200	2.5	289
Peas	U.S.A.	870–2,400	2.8	290
Peanut	Virginia (U.S.A.)	128–158	1.2	260
Taro	South Pacific	17–34	2.0	291
Wheat	Germany	38–95	2.5	292
Wheat[g]	Worldwide	60–310	5.2	293

TABLE 17   (continued)

Plant	Location	Concentration range	Factor (max./ min.)	Ref.
Thiamin (μg / 100 g)				
Asparagus	Spain	90–120	1.3	271
Beans	New Mexico	930–1,330	1.4	257
Beans (lima)	U.S.A.	713–1,018	1.4	273
Beans (lima)	U.S.A.	272–324	1.2	269
Carrots	U.S.A.	76–157	2.1	276
Linseed oil	Worldwide	760–1,300	1.7	294
Millet (pearl)	U.S.A.	433–844	1.9	295
Peas	Australia	894–1,135	1.3	296
Peanut	Virginia (U.S.A.)	1,100–1,380	1.2	260
Taro	South Pacific	25–37	1.5	291
Wheat	Australia	378–449	1.2	297
Wheat	Canada	444–461	1.0	298
Wheat	Kansas	335–614	1.8	149
Wheat	Germany	371–476	1.3	299
Wheat	Germany	396–870	2.2	292
Wheat	U.S.A.	490–720	1.5	300
Wheat	U.S.A.	390–570	1.5	263
Tocopherol (mg / 100 g)				
Linseed oil	Worldwide	298–525	1.8	294
Rapeseed oil	Europe	381–574	1.5	301
Pumpkin oil	Worldwide	381–482	1.3	152
Sunflower oil	Worldwide	46–881	1.6	264
Sunflower oil	Worldwide	408–867	2.1	152
Soybean oil	Worldwide	866–1,239	1.4	302
Vitamin K_1(μg / 100 g)				
Cabbage (inner leaves)	U.S.A-Canada	72–228	3.2	303
Kale	U.S.A.-Canada	621–1657	2.7	303
Lettuce	U.S.A.-Canada	519–1,180	2.3	303
Spinach	U.S.A.-Canada	1,001–1,439	1.4	303

[a] In most cases comparison are made between the same plan varieties. In some cases, however, comparisons are made between different varieties grown in different locations, in which case part of the differences observed might be due to variety effect.
[b] Values are founded off.
[c] Dry weight basis.
[d] International units of vitamin A.
[e] β-carotene.
[f] Sum of α- and β-carotene.
[g] Different wheat classes grown in different countries.

TABLE 18
Ascorbic Acid Concentration in Sweet Oranges[a]
Grown in Different Locations in Iran[279]

Location	Ascorbic acid (mg / 100 ml Juice)
Noushahr	30[b]
Ramsar	34
Shahsavar	52
Bam	64

[a] It is not certain whether the same varieties were sampled in different locations.
[b] Values are rounded.

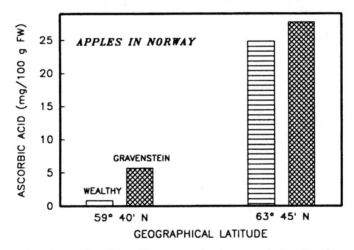

FIGURE 11. Ascorbic acid concentration in two varieties of apples grown at different geographical locations in Norway.[121]

(Figure 11). By comparing samples of two varieties of apples grown from Italy to Norway in two years, Nilsson[312] noted that the ascorbic acid content was higher in the apples grown in the more northern latitudes than in the more southern locations. Also, in the former Soviet Union, although the ascorbic acid content in the early cultivars of cabbage was found to be higher in the south (Krasnodor region) than in the north (in the area of the former Leningrad, present St. Petersburg), the content in the medium and late cultivars, however, was higher in the northern regions.[313] By considering the effect of light on ascorbic acid content in plants (page 90), it may be postulated that the relatively longer day lengths in the more northerly locations during the growing period in the northern hemisphere may be partly responsible for the higher ascorbic acid content found in the fruits and vegetables grown there. Also, cooler night temperature in the more northern latitudes has been implicated in favoring the accumulation of higher ascorbic acid in plants.[137, 287] In Southern California, oranges grown in the cooler coastal areas were noted to be higher in ascorbic acid than those grown in the warmer interior locations.[199]

In Finland, black currants grown in a more northerly place (Rovaniemi, average July day length of 20 h, 10 min) contained 30% higher ascorbic acid than those grown in a more southern location (Viik, July day length of 17 h, 35 min).[314]

Hårdh[137] noted that in the Scandinavian countries consumers have long considered the vegetables and berries grown in the northern parts to look and taste better than those grown in southern parts. This led Hårdh and co-workers to study the effect of geographical locations on the vitamin and other quality factors in several vegetables. The results of their very extensive studies in Finland, Sweden, and Norway[137, 138, 315, 316] are unfortunately hard to follow mainly because of the numerous factors that were investigated and the large number of tables containing results obtained with different rooting medium, growth conditions, varieties, or the year when the study was performed. Here some selected extracts of their data are reviewed.

As a whole, sweet pepper, tomato, swede, parsley, dill, leek, and cabbage had higher ascorbic acid when grown in the northern as compared with those grown in southern latitudes (Tables 19 and 20).[137, 138, 315] In potato,[137] spinach,[137] and lettuce,[316] however, the situation was reversed; i.e., the ascorbic acid content was higher in plants grown in the south than in the north. In numerous instances plants grown in plastic tunnels contained higher concentrations of ascorbic acid than those grown in greenhouses (Table 19), an observation that was interpreted to be due to the more predominant effect of light over that of temperature on the ascorbic acid content of some plants.

It is noteworthy that although plants grown in very different climatic conditions may product very different yields, their content of ascorbic acid is not proportionally different (Table 20) (For a discussion on the relationships between yield and plant vitamins see chapter 6.) For example, although the yield of leek grown in minerals soils in the southern parts of Norway (Landvik) was nearly seven times higher than when it was grown in the northern part (Holt) (300 vs. 40 g, respectively), the ascorbic acid content of southern- and northern-grown leek was very similar (41 vs. 44 mg/100 g DW)[138] (Table 20). Also, in

TABLE 19

Yield and Ascorbic Acid Concentration of Tomato and Pepper Grown under Similar Soil and Fertilizer Conditions in Greenhouse or in Plastic Tunnels in Different Geographical Locations in Finland[137]

Location	Yield (kg / m^2)		Ascorbic acid (mg / 100 g)	
	Green house	Plastic tunnel	Green house	Plastic tunnel
Tomato (Minerva)				
Muddusniemi (69°05′ N)	17.6	17.5	18.0	21.1
Rovaniemi (66°30′ N)	17.4	15.3	20.8	25.7
Viik (60°11′ N)	23.2	26.1	15.1	24.7
Tomato (Tummeliten)				
Muddusniemi	7.6	8.7	19.4	32.4
Rovaniemi	7.2	9.6	28.3	28.0
Viik	9.6	12.1	18.4	21.1
Pepper (Pedro)				
Muddusniemi	2.5	4.0	149.1	176.7
Rovaniemi	—	4.4	130.4	192.5
Viik	6.8	4.9	146.9	182.8

Finland, although the yield of parsley grown in the south (Viik) was 10 times higher than that in the north (Muddusniemi) (1.0 vs. 0.1 kg/m^2, respectively), the ascorbic acid content of plants grown in the south was only one half that of plants grown in the north (90 vs. 184 mg/100 g FW, respectively)[138] (Table 20). This is surprising if one considers the large differences in the temperature, amount of solar radiation, and proportion of blue and red light in these sites, which could all have played a role in the growth as well as the vitamin content of the plants, as mentioned by the authors.[137, 138, 315, 316]

Klein and Perry[275] measured the ascorbic acid in the vegetables purchased from local wholesale distributors in six different cities and in two different seasons in the eastern part of the United States and noted a significant difference in their ascorbic acid content. The following range of ascorbic acid concentrations (mg/100 g, rounded off) was found in vegetables from different locations: carrots, 5–11; celery, 5–8; tomatoes, 10–25; cabbage, 35–52; onion, 7–10; and corn, 5–8—differences that are in some cases greater than 100%. Strawber-

TABLE 20
**Yield and Ascorbic Acid (AA) Concentration in Parsley Grown in
Different Locations in Finland and in Leek Grown
in Norway**[138]

	Parsley	
Locations in Finland	Yield (kg / m^2)	Ascorbic acid (mg / 100 g FW)
Muddusniemi (69°05′ N)	0.1	184
Rovaniemi (66°33′ N)	2.7	163
Homantsi (62°40′ N)	1.8	131
Viik (60°11′ N)	1.0	90

	Leek	
Locations in Norway	Yield (g)	Ascorbic acid (mg / 100 g DW)
Holt (69°39′ N)	40	44
Kvithamar (63°28′ N)	180	16
As (59°40′ N)	280	14
Landvik (58°20′ N)	300	41

From tables 12 and 23 of Hårdh, J. E., Persson, A. R., and Ottosson, L., *Acta Agric. Scand.*, 27, 81, 1977.

ries grown in New Jersey were noted to have on the average 31 mg/ 100 g FW less ascorbic acid than those grown in Florida and California.[285]

The ascorbic acid content of kiwifruit may strongly vary depending on the location where it is grown. In Italy, for example, the ascorbic acid content of variety Hayward was noted to vary by almost 300% (67–199 mg/100 g) in the 20 farms sampled in Italy's main growing areas.[317] Comparisons between the reported ascorbic acid concentration in the kiwifruits (varieties Hayward and Bruno) grown in different parts of the world also show a range of more than 200%. In the variety Hayward, for example, it ranged from 70 mg/100 g in China[318] to 75, 96, and 190 mg/100 g in Italy, New Zealand, and Switzerland, respectively[319-321] (Figure 12). The variety Bruno had the following ascorbic acid concentrations (in mg/100 g): 146 in New Zealand,[320] 241 in Italy,[319] and 284 in Switzerland.[321]

Distances as little as several kilometers apart may also have an effect on the vitamin content of plants, provided there is a big enough difference in their climatic conditions. For example, tomatoes grown in

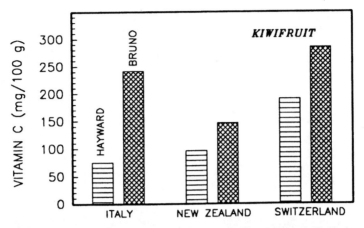

FIGURE 12. Vitamin C concentration in two varieties of kiwifruit grown in different countries.[319–321]

Geneva (New York) contained considerably less ascorbic acid than those grown in Phelps (New York) (20 vs. 31.4 g/100 g, respectively). This was attributed to the lower amount of sunshine in Geneva than in Phelps during the 18 days prior to harvest, although these two locations were only 13 km (8 miles) apart![322]

The relative effects of location of growth and chemical fertilizers were investigated by Reder,[323] who noted that the effect of location of growth on the ascorbic acid content of turnip greens was in some cases 13.75 times greater than the effect of chemical fertilizers.

B. Other Vitamins

The carotene content of plants may strongly vary depending on the location where they are grown in (Table 17). In general, plants grown in relatively warmer and southern latitudes seem to have higher carotene content, as observed in Finland[137] and in Norway.[324] At any given location, the carotene content was often higher in plants grown in the greenhouse than in plants grown in the open field (Table 21). Temperature in the greenhouse, however, may sometimes be too high for favorable β-carotene formation.[287]

Vitamin A activity in several vegetables collected from six cities across the United States was found to be significantly different. The following ranges of concentrations (in IU/100 g) were reported: carrots, 12,228–18,538; celery, 60–213; tomato, 530–1040; cabbage, 57–213; and corn, 99–351—which show in some cases a difference of more than 3 times![275]

TABLE 21
Total (Sum of α and β) Carotene Concentrations in Some Vegetables Grown in Peat in Greenhouse, Plastic Tunnel, or Open Field in Different Geographical Locations in Finland[137]

Location	Greenhouse	Plastic tunnel	Open field
		Spinach	
Muddusniemi (69°05′ N)	3.2[a]	2.4	2.0
Viik (60°11′ N)	3.6	2.2	3.2
		Dill	
Muddusniemi	4.6	1.4	3.4
Viik	3.9	3.8	2.4
		Lettuce	
Muddusniemi	1.7	0.6	0.8
Viik	2.3	0.7	1.1

[a] Rounded values

Location of growth also affects the thiamin, tocopherol, riboflavin, niacin, and vitamin K_1 content of some plants (Table 17). The thiamin content of "Kharkof" wheat grown in Manhattan (Kansas), for example, was nearly twice that of the same wheat grown in Ames (Iowa) (830 vs. 440 μg/100 g, respectively).[300] Putnam,[268] however, did not find any difference in the biotin or α-tocopherol content in wheat or barley grown in different geographical locations in England.

Eheart et al.[269] tested the ascorbic acid, thiamin, and riboflavin concentrations in one variety of lima beans grown for four years at four locations in the states of Georgia, Mississippi, South Carolina, and Virginia and noted that the thiamin and riboflavin content of beans differed significantly in different locations but no given locations produced beans highest in all three vitamins; beans grown in South Carolina had the highest thiamin but those grown in Georgia had the highest riboflavin.

Taro (*Colocasia esculenta*) roots from the Solomon Islands contained significantly lower riboflavin concentration than those from the Fiji Islands. Taros from the Solomon Islands were also lower in their content of nicotinic acid than those from Fiji, but the differences were not statistically significant.[291]

Ferland and Sadowski[303] compared the content of vitamin K_1 in several vegetables grown simultaneously in outdoor gardens in Boston

(U.S.A.) and in Montreal (Canada) during June to August. They noted that although seeds from the same lot was used in these experiments, lettuce, kale, and the inner leaves of cabbage grown in Montreal contained 2.3, 2.6, and 3.2 times more vitamin K_1 than those grown in Boston, respectively. They concluded that climate and soil properties could be responsible for the difference observed in the vitamin K_1 content of these vegetables in these two locations.

In some cases, plant vitamins seem to be independent of the geographical location the plants have been grown in. This has been observed to be the case for the thiamin content in snap beans[325] and oats[326] and the riboflavin and pyridoxine content in wheat[263,327] and pearl millet.[295]

IX. EFFECT OF ALTITUDE

A. Ascorbic Acid

The altitude above the sea level where the plants are grown may also affect their vitamin content. In general, plants grown at higher elevations seem to contain higher ascorbic acid than those grown at lower elevations. This has been observed in apples,[328,329] apricots,[329,330] pear,[329,331] plum,[329] sweet and sour cherry,[329,332] in vegetables such as tomato and cabbage,[333] rose hips,[334] in spruce needles,[335] and several other plants.[336] In the Azerbaijan (former U.S.S.R.), for example, apples grown at 1000 m above sea level contained 1.5–2 times more ascorbic acid than those grown at 50 m above sea level.[337] Vitamin C concentration in the fruits of sea buckthorn (*Hippophae rhamnoides* L.) was 5 times higher when plants were grown at higher elevations (508 vs. 110 mg/100 g at 3500 and 1900 m elevations, respectively).[341]

The ascorbic acid content of cultivated and wild (endogenous) plant genotypes may react differently when the plants are grown at different altitudes. For example, Shuruba[338] noted that in the Ukraine, the ascorbic acid content of wild apples was higher when plants were grown at higher altitudes than at lower altitudes. In the cultivated varieties of apples, however, the situation was reversed; i.e., apples grown at higher altitudes had lower ascorbic acid than those grown at lower altitudes.

In Colombia, citrus fruits grown at higher altitudes were found to be higher in vitamin C than those grown at lower altitudes.[339] In Armenia, the ascorbic acid content in potatoes grown at higher altitudes (at Sevan) was reported to increase throughout the tuber's growth and to reach 121–143 mg/100 g DW at the early stage of tuber maturity. These tubers, however, failed to reach full maturity. In the potatoes grown at lower altitudes (at Erevan), however, the ascorbic acid content of tubers reached 132–207 mg/100 g DW and decreased thereafter to 48–83 mg/100 g as the tubers reached maturity.[340]

B. Other Vitamins

Carotene concentration in spinach,[342] apricots,[330] carrots,[343] tomato,[333] and *Hypericum scabrum* and *H. elongatum*[336] was found to be changed by the latitude where plants were grown, which may be partly due to the pronounced effect of temperature on the carotene content in plants (see page 108). In Czechoslovakia, for example, carrots grown in lower (and warmer) elevations in the southern parts of the country were noted to have higher carotene content than those grown at higher elevations.[343] In Armenia, "Sateni" apricots grown at low elevation contained almost twice the carotene of those grown at high elevation (9.76 vs. 4.24 mg/100 g, at 870 and 1250 m, respectively).[330] The concentration of vitamin B_2 was, however, affected in the opposite direction so that "Yerevani" apricots, for example, contained less vitamin B_2 when grown at lower altitudes (59 vs. 79.8 mg/100 g at low and high altitudes, respectively).[330]

X. SUMMARY

Climatic factors such as light duration, light intensity and quality, and temperature were shown to strongly affect the concentration of vitamins, in particular that of ascorbic acid and carotene, in many different plants. Furthermore, vitamin content in plants may undergo strong diurnal, seasonal, and yearly fluctuations and may be different in plants grown at different geographical locations around the world and at different altitudes above sea level.

The nutritional implications of geographical variations in plant vitamins for the vitamin supply of people living in different parts of the globe is a matter that needs to be studied. It may be argued that with the increasing national and international movement of various plant foods, the foods from low- and high-vitamin areas would be mixed and thus no real influence on human nutrition would result in the long term. In the countries where locally grown products are mostly consumed, however, the situation might be different. If, for example, a given locality happens to be a "low-vitamin" area, then this could conceivably have some long-term consequences for the vitamin supply to the local population. This area of research, termed geomedicine by Låg[344] deserves much more attention, especially for the nutrition of people living in the developing countries.

REFERENCES

1. Franke, W. and Heber, U., Über die quantitative Verteilung der Ascorbinsäure innerhalb der Pflanzenzelle, *Z. Naturforsch.*, 19b, 1146, 1964.
2. Gerhardt, E., Untersuchungen Über beziehungen zwischen Ascorbinsäure and Photosynthese, *Planta*, 61, 101, 1964.
3. Law, M. Y., Charles, S. A., and Halliwell, B., Glutathione and ascorbic acid in spinach (*Spinacia oleracea*) chloroplasts, *Biochem. J.*, 210, 889, 1983.
4. Åberg, B., Ascorbic acid, in *Handbuch der Pflanzenphysiologie*, Vol. 6, Ruhland, W., Ed., Springer Verlag, Berlin, 1958, 479.
5. Loewus, F. A. and Loewus, M. W., Biosynthesis and metabolism of ascorbic acid in plants, *CRC Crit. Rev. Plant Sci.*, 5, 101, 1987.
6. Gillham, D. J. and Dodge, A. D., Chloroplast superoxide and hydrogen peroxide scavenging systems from pea leaves: seasonal variations, *Plant Sci.*, 50, 105, 1987.
7. Salunkhe, D. K. and Desai, B. B., Effect of agricultural practice, handling, processing, and storage on vegetables, in *Nutritional Evaluation of Food Processing*, Karmas, E. and Harris, R. S., Eds., Van Nostrand Reinhold, New York, 1988, chapter 3.
8. Loewus, F. A., Ascorbic acid and its metabolic products, in *The Biochemistry of Plants*, Vol. 14, Stumpf, P. K. and Conn, E. E., Eds., Academic Press, San Diego, 1988, 85.
9. Medawara, M. R., Notizen über Vitamin C in der Pflanze, *Phyton (Austria)*, 2, 193, 1950.
10. Mozafar, A. and Oertli, J. J., Vitamin C (Ascorbic acid): uptake and metabolism by soybean *J. Plant Physiol.*, 141, 316, 1993.
11. Nakasone, H. Y., Miyashita, R. K., and Yamane, G. M., Factors affecting ascorbic acid content of the acerola (*Malpighia glabra* L.), *Proc. Am. Soc. Hortic. Sci.*, 89, 161, 1966.
12. Matzner, F., Über den Gehalt und die Verteilung des Vitamin C in Äpfeln, *Erwerbsobstbau*, 4(2), 27, 1962.
13. Matzner, F., Vitamin-C-Gehalt in Früchten der "Schattenmorelle", *Erwerbobstbau*, 18(6), 83, 1976.
14. Eze, J. M. O., Growth of *Amaranthus hybridus* (African Spinach) under different daylight intensities in the dry season in southern Nigeria, *Exp. Agric.*, 23(2), 193, 1987; *Hortic. Abstr.*, 57, 7858, 1987.
15. Makus, D. J. and Gonzalez, A. R., Production of white asparagus using plastic row covers, in *Proc. Ann. Meetings—Arkansas St. Hortic. Soc.*, No. 110, Fayetteville, Arkansas, 1989, 136; *Hortic. Abstr.*, 61, 8001, 1991.
16. Matzner, F., Über den Vitamin-C-Gehalt der Kulturheidelbeeren, *Erwerbsobstbau*, 7(6), 105, 1965.
17. Finch, A. H., Jones, W. W. and van Horn, C. W., The influence of nitrogen upon the ascorbic acid content of several vegetable crops, *Proc. Am. Soc. Hortic. Sci.*, 46, 314, 1945.
18. Alekseev, R. V., Changes in ascorbic acid content during the formation of capsicum seeds, *Sbornik Nauchnykh Trudov VNII Oroshaemogo Ovoshchevodstva Bakhchevodstva*, 2, 98, 1974; *Hortic. Abstr.*, 45, 5008, 1975.
19. Henze, J., Einfluss unterscheidlicher NPK-Gaben auf Fruchinhaltsstoffe von Schattenmorellen auf F 12/1, *Erwerbsobstbau*, 18(12), 182, 1976.
20. Winston, J. R. and Miller, E. V., Vitamin C content and juice quality of exposed and shaded citrus fruits, *Food Res.*, 13, 456, 1948.
21. Hivon, K. J., Coty, D. M., and Quackenbush, F. W., Ascorbic acid and ascorbic acid oxidizing enzymes of green bean plants deficient in manganese, *Plant Physiol.*, 26, 832, 1951.

22. Åberg, B., Changes in the ascorbic acid content of darkened leaves as influenced by temperature, sucrose application, and severing from plants, *Physiol. Plant*, 2, 164, 1949.
23. Fernández, M. Del C. C., Variation in the β-carotene and ascorbic acid contents of lettuce and carrots as influenced by seasonal changes in Puerto Rico, *J. Agric. Univ. P. R.*, 48, 39, 1964; *Hortic. Abstr.* 34, 6839, 1964.
24. Grimstad, S. O., The effect of light source and irradiation on the content of L-ascorbic acid in lettuce, *Acta Hortic.*, 163, 213, 1984.
25. Mattei, F., Sebastiani, L. A., and Gibbon, D., The effect of radiant energy on growth of *Lactuca sativa* L., *J. Hortic. Sci.*, 48(4), 311, 1973.
26. Hulewicz, D. and Kalbarcyzk, M., Veränderlichkeit des Ertrages und einiger Nährkomponenten des Salats in Abhängigkeit vom Licht, *Arch. Gartenbau (Berlin)*, 24, 113, 1976.
27. Shinohara, Y. and Suzuki, Y., Effects of light and nutritional conditions on the ascorbic acid content of lettuce, *J. Jpn. Soc. Hortic. Sci.*, 50(2), 239, 1981; *Hortic. Abstr.*, 52, 2885, 1982.
28. Izumi, H., Ito, T., and Yoshida, Y., Effect of light intensity during growing period on ascorbic acid content and its histochemical distribution in the leaves and peel, and fruit quality of Satsuma mandarin, *J. Jpn. Soc. Hortic. Sci.*, 61(1), 7, 1992.
29. Hamner, K. C. and Nightingale, G. T., Ascorbic acid content of pineapples as correlated with environmental factors and plant composition, *Food Res.*, 11, 535, 1946.
30. Åberg, B., Effects of light and temperature on the ascorbic acid content of green plants, *Ann. Roy. Agric. Coll. Sweden*, 13, 239, 1946.
31. Crane, M. B. and Zilva, S. S., The influence of some genetic and environmental factors on the concentration of L-ascorbic acid in tomato fruits, *J. Hortic. Sci.*, 25, 36, 1949.
32. Hamner, K. C., Bernstein, L., and Maynard, L. A., Effect of light intensity, day length, temperature, and other environmental factors on the ascorbic acid content of tomatoes, *J. Nutr.*, 29, 85, 1945.
33. McCollum, J. P., Some factors affecting the ascorbic acid content of tomatoes, *Proc. Am. Soc. Hortic. Sci.*, 45, 382, 1944.
34. Kaski, I. J., Webster, G. L., and Kirch, E. R., Ascorbic acid content of tomatoes, *Food Res.*, 9, 386, 1944.
35. Fryer, H. C., Ascham, L., Cardwell, A. B., Frazier, J. C., and Willis, W. W., Effect of fruit cluster position on the ascorbic acid content of tomatoes, *Proc. Am. Soc. Hortic. Sci.*, 64, 360, 1954.
36. Fryer, H. C., Ascham, L., Cardwell, A. B., Frazier, J. C., and Willis, W. W., Relation between stages of maturity and ascorbic acid content of tomatoes, *Proc. Am. Soc. Hortic. Sci.* 64, 365, 1954.
37. Hobson, G. E. and Davies, J. N., The tomato, in *The Biochemistry of Fruits and Their Products*, Vol. 2, Hulme, A. C., Ed., Academic Press, London, 1971, 440.
38. McCollum, J. P., Effect of sunlight exposure on the quality constituents of tomato fruit, *Proc. Am. Soc. Hortic. Sci.*, 48, 413, 1946.
39. Ivanova, T. L., Tomato fruit quality in relation to different growing methods, *Nauchnye Tr., Omskii Sel'skokhozyaistvennyi Inst.*, 115, 41, 1973; *Hortic. Abstr.*, 44, 9725, 1974.
40. Venter, F., Solar radiation and vitamin C content of tomato fruits, *Acta Hortic.*, 58, 121, 1977.
41. Morimura, Y., Aoki, J., and Aimi, R., Effects of light intensity and temperature on L-ascorbic acid content in tender-green mustard (*Brassica campestris* L. var. *perviridis* L. H. Bailey), *Seibutsu Kankyo Chosetsu*, 20(2-3), 53, 1982; *Chem. Abstr.*, 98, 122837x, 1983.

42. Hamner, K. C. and Parks, R. Q., Effect of light intensity on ascorbic acid content of turnip greens, *Agron. J.*, 36(4), 269, 1944.
43. Somers, G. F., Kelly, W. C., and Hamner, K. C., Changes in ascorbic acid content of turnip-leaf discs as influenced by light, temperature, and carbon dioxide concentration, *Arch. Biochem.*, 18, 59, 1948.
44. Somers, G. F. and Kelly, W. C., Influence of shading upon changes in the ascorbic and carotene content of turnip greens as compared with changes in fresh weight, dry weight and nitrogen fractions, *J. Nutr.*, 62, 39, 1957.
45. Moldtmann, H. G., Untersuchungen über den Ascorbinsäuregehalt der Pflanzen in seiner Abhängigkeit von inneren und äusseren Faktoren, *Planta*, 30, 297, 1939.
46. Erner, Y., Goren, R., and Monselise, S. P., Influence of light of different spectral composition on growth and metabolism of citrus seedlings, *Physiol. Plant.*, 27, 327, 1972.
47. Murneek, A. E., Maharg, L., and Wittwer, S. H., Ascorbic acid (vitamin C) content of tomatoes and apples, *Univ. Missouri, Agric. Exp. Stn. Res. Bull.*, 568, 1954.
48. Somers, G. F., Hamner, K. C., and Kelly, W. C., Further studies on the relationship between illumination and the ascorbic acid content of tomato fruits, *J. Nutr.*, 40, 133, 1950.
49. Hassan, H. H. and McCollum, J. P., Factors affecting the content of ascorbic acid in tomatoes, *Univ. Illinois Agric. Exp. Stn. Bull.*, 573, 1954.
50. Brown, A. P. and Moser, F., Vitamin C content of tomatoes, *Food Res.*, 6, 45, 1941.
51. Krynska, W., Effect of irrigation on early cabbage and tomatoes grown on sloping ground, *Zeszyty Problemowe Postepow Nauk Rolniczych*, 181, 103, 1976; *Food Sci. Technol. Abstr.*, 9(10), J1504, 1977.
52. Kondo, S., Effect of environmental conditions on the contents of sugars and ascorbic acid in "Sendhu" apple fruit, *Nippon Shokuhin Kogyo Gakkaishi*, 39(12), 1112, 1992.
53. Ezell, B. D., Darrow, G. M., Wilcox, M. S., and Scott, D. H., Ascorbic acid content of strawberries, *Food Res.*, 12, 510, 1947.
54. Burkhart, L. and Lineberry, R. A., Determination of vitamin C and its sampling variation in strawberries, *Food Res.*, 7, 332, 1942.
55. Richardson, L. R., Effect of environment on the ascorbic acid content of plants, *South Coop. Series Bull.*, 36, 6, 1954.
56. Hansen, E. and Waldo, G. F., Ascorbic acid content of small fruits in relation to genetic and environmental factors, *Food Res.*, 9, 453, 1944.
57. Tafazoli, E. and Shaybany, B., Effects of short-day treatments on second crop summer-fruiting strawberries, *Exp. Agric.*, 14, 217, 1978.
58. Slate, G. L. and Robinson, W. B., Ascorbic acid content of strawberry varieties and selections at Geneva, New York in 1945, *Proc. Am. Soc. Hortic. Sci.*, 47, 219, 1946.
59. Takebe, M. and Yoneyama, T., Plant growth and ascorbic acid. 1. Changes of ascorbic acid concentrations in the leaves and tubers of sweet potato (*Ipomea batatas* Lam.) and potato (*Solanum tuberosum* L.), *Nippon Dojo Hiryogaku Zasshi*, 63(4), 447, 1992; *Chem. Abstr.*, 117, 190048v, 1992.
60. Singleton, V. L. and Gortner, W. A., Chemical and physical development of the pineapple fruit. II. Carbohydrate and acid constituents, *Food Sci.*, 30, 19, 1965.
61. Mironov, K. A., Effect of lowland fires on the biochemical composition of berries of some members of the Vacciniaceae family, *Izv. Vyssh. Uchebn. Zaved., Lesn. Zh.*, 6, 117, 1981; *Chem. Abstr.*, 96, 139748a, 1982.
62. Fernández, M. Del C. C., Effect of environment on the carotene content of plants, *South. Coop. Series Bull.*, 36, 24, 1954.
63. Kays, S. J., *Postharvest Physiology of Perishable Plant Products*, AVI Book, Van Nostrand Reinhold, New York, 1991.

64. Goodwin, T. W. and Goad, J. L., Carotenoids and triterpenoids, in *The Biochemistry of Fruits and Their Products*, Academic Press, London, 1970, 305.

65. McCollum, J. P., Effect of light on the formation of carotenoids in tomato fruits, *Food Res.*, 19(2), 182, 1954.

66. Thomas, R. L. and Jen, J. J., Red light intensity and carotenoid biosynthesis in ripening tomatoes, *J. Food Sci.*, 40, 566, 1975.

67. Raymundo, L. C., Chichester, C. O., and Simpson, K. L., Light-dependent carotenoid synthesis in the tomato fruit, *J. Agric. Food Chem.*, 24(1), 59, 1976.

68. Smith, L. L. W. and Morgan, A. F., The effect of light upon the vitamin A activity and the carotenoid content of fruits, *J. Biol. Chem.*, 101, 43, 1933.

69. Rouchaud, J., Moons, C., Meyer, J. A., Benoit, F., Ceustermans, N., and van Linden, F., Effects of soil fertilization and covering of the culture with plastic film on the provitamin A content of early lettuces, *Plant Soil*, 80(1), 139, 1984.

70. Eheart, J. F., Young, R. W., Massey, P. H., Jr., and Havis, J. R., Crop, light intensity, soil pH, and minor element effects on the yield and vitamin content of turnip greens, *Food Res.*, 20, 575, 1955.

71. Bondi, A. and Meyer, H., Carotene in Palestinian crops, *J. Agric. Sci.*, 36, 1, 1946.

72. Ben-Amotz, A. and Avron, M., On the factors which determine massive β-carotene accumulation in the halotolerant *Dunaliella bardawil*, *Plant Physiol.*, 72, 593, 1983.

73. Ben-Amotz, A. and Avron, M., The wavelength dependence of massive carotene synthesis in *Dunaliella bardawil* (Chlorophyceae), *J. Phycol.*, 25, 175, 1989.

74. Gustafson, F. G., Influence of light intensity upon the concentration of thiamin and riboflavin in plants, *Plant Physiol.*, 23, 373, 1948.

75. Crane, F. L., A light activated accumulation of niacin in tomato leaf disks, *Plant Physiol.*, 29, 188, 1954.

76. Crane, F. L., Precursor relationships in light activated accumulation of niacin in leaf disks from normal, light-starved, and zinc deficient tomato plants, *Plant Physiol.*, 29, 395, 1954.

77. Lyon, C. B., Beeson, K. C., and Ellis, G. H., Effects of micro-nutrient deficiencies on growth and vitamin content of the tomato, *Bot. Gaz. (Chicago)*, 104(4), 495, 1943.

78. Prudot, A. and Basiouny, F. M., Absorption and translocation of some growth regulators by tomato plants growing under UV-B radiation and their effects on fruit quality and yield, *Proc. Fla. St. Hortic. Soc.*, 95, 374, 1982 (Pub. 1983); *Chem. Abstr.*, 99, 66756b, 1983.

79. Grimstad, S. O., Light sources and plant irradiation. 3. Effect of light source and irradiation on the content of chlorophyll, L-ascorbic acid and glucose in lettuce (*Lactuca sativa* L.) grown in greenhouses under different natural light conditions, *Meld. Nor. Landbrukshoegsk.* 61(3), 1, 1982; *Chem. Abstr.*, 98, 31622g, 1983.

80. Sčerbakov, B. I., The effect of light in different parts of the spectrum on synthesis of vitamin C in plants, *Dokl. Akad. Nauk SSSR*, 66, 1149, 1949; *Field Crop Abstr.*, 3, 250, 1950.

81. Sid'ko, F. Ya., Tikhomirov, A. A., Zolotukhin, I. G., and Polonskii, V. I., Radish growth and development under the halogen arc lamp DRIF, *Fiziol. Biokhim. Kul't. Rast.*, 7(2), 181, 1975; *Hortic. Abstr.*, 46, 1121, 1976.

82. Zolotukhin, I. G., Lisovskii, I. G., and Bayanova, Y. I., Effect of light of different spectral composition and intensity on ascorbic acid biosynthesis in plants, *Fiziol Biokhim. Kul't. Rast.*, 11(2), 141, 1979; *Hortic. Abstr.*, 49, 7540, 1979.

83. Ellis, G. H. and Hamner, K. C., The carotene content of tomatoes as influenced by various factors, *J. Nutr.*, 25, 539, 1943.

84. Shewfelt, A. L. and Halpin, J. E., The effect of light quality on the rate of tomato color development, *Proc. Am. Soc. Hortic. Sci.*, 91, 561, 1967.

85. Thomas, R. L. and Jen, J. J., Phytochrome-mediated carotenoids-biosynthesis in ripening tomatoes, *Plant Physiol.*, 56, 452, 1975.
86. Jen, J. J., Influence of spectral quality of light on pigment systems of ripening tomatoes, *J. Food Sci.*, 39, 907, 1974.
87. Jen, J. J., Spectral quality of light and the ripening characteristics of tomato fruit, *HortScience*, 9(6), 548, 1974.
88. Mehta, P. M. and Bhavannarayana, K., Influence of red and far-red light on pigment systems in ripening tomato fruits, *Indian J. Plant Physiol.*, 24(3), 224, 1981.
89. Cohen, R. Z. and Goodwin, T. W., The effect of red and far-red light on carotenoid synthesis by etiolated maize seedlings, *Phytochemistry*, 1, 67, 1962.
90. Lopez, M., Candela, M. E., and Sabater, F., Carotenoids from *Capsicum annum* fruits: influence of spectral quality of radiation, *Biol. Plant.*, 28(2), 100, 1986.
91. Whitmore, R. A., Light and pigment development in the kidney bean, *Plant Physiol.*, 19(4), 569, 1944.
92. Lichtenthaler, H. K. and Becker, K., Kinetic of lipoquinone synthesis in etiolated *Raphanus* seedlings in continuous far-red and white light, *Z. Pflanzenphysiol.*, 75(4), 296, 1975.
93. Murneek, A. E. and Wittwer, S. H., Some factors affecting ascorbic acid content of apples, *Proc. Am. Soc. Hortic. Sci.*, 51, 97, 1948.
94. Horbowicz, M. and Bakowski, J., The effect of sprout position on the stalk on the chemical composition of Brussels sprouts. *Biuletyn Warzywniczy*, 31, 189, 1988; *Hortic. Abstr.*, 59, 2040, 1989.
95. Shinohara, Y., Suzuki, Y., Shibuya, M., Yamamoto, M., and Yamasaki, K., Effect of fertilization and foliar spray treatment on the ascorbic acid content of tomato and sweet pepper, *J. Jpn. Soc. Hortic. Sci.*, 49(1), 85, 1980; *Hortic. Abstr.*, 51, 8643, 1981.
96. Harding, P. L., Winston, J. R., and Fisher, D. F., Seasonal changes in the ascorbic acid content of juice of Florida oranges, *Proc. Am. Soc. Hortic. Sci.*, 36, 358, 1939.
97. Harding, P. L. and Thomas, E. E., Relation of ascorbic acid concentration in juice of Florida grapefruit to variety, rootstock, and position of fruit on the tree, *J. Agric. Res. (Washington, D.C.)*, 64(1), 57, 1942.
98. Cohen, A., The effect of different factors on the ascorbic acid content of citrus fruits. I. The dependence of the ascorbic acid content of the fruit on light intensity and on the area of assimilation, *Bull. Res. Council of Israel*, 3, 159, 1953.
99. Primo Yúfera, E., Royo Iranzo, J., and Sala Gomis, J. M., Quality of orange varieties. VI. Standardization of methods. Statistical significance of vitamin C values, *Rev. Agroquim. Tecnol. Aliment.*, 3, 341, 1963; *Hortic. Abstr.*, 34, 7570, 1964.
100. McCarty, C. D., Boswell, S. B., and Cole, D. A., Effect of tree density on maturity of Navels, *Citrograph*, 60(8), 291, 1975.
101. Izumi, H., Ito, T. and Yoshida, Y., Relationship between ascorbic acid and sugar content in citrus fruit peel during growth and development, *J. Jpn. Soc. Hortic. Sci.*, 57(2), 304, 1988.
102. Izumi, H., Ito, T., and Yoshida, Y., Fruit quality of mid- and late seasons citrus harvested from exterior and interior canopy of trees and its changes during storage (studies on vitamin C of fruits and vegetables part VI), *J. Jpn. Soc. for Cold Preservation of Food*, 16(2), 51, 1990.
103. Izumi, H., Ito, T., and Yoshida, Y., Sugar and ascorbic acid contents of Satsuma mandarin fruits harvested from exterior and interior canopy of trees during fruit development, *J. Jpn. Soc. Hortic. Sci.*, 58(4), 877, 1990.
104. Izumi, H., Ito, T., and Yoshida, Y., Changes in fruit quality of Satsuma mandarin during storage, after harvest from exterior and interior canopy of trees, *J. Jpn. Soc. Hortic. Sci.*, 58(4), 885, 1990; *Hortic. Abstr.*, 61, 1991.

105. Vidéki, L., Börzsei, J., and Báldy, B., The development of glasshouse tomato constituents by trusses, *Zöldségtermesztési Kutaó Intézet Bulletinje*, 7, 9, 1972; *Hortic. Abstr.*, 44, 3325, 1974.

106. Čirkova-Georgievska, M., Dzekova, M., Jankulovski, D., Peševska, V., and Mostafa, M., Changes in the chemical constituents of fruits at different tiers on tomatoes grown under plastic, *Godišen Zbornik Zemjodelsko-Šumarskiot Fakultet Univ. Skopje*, 29, 115, 1977/1978-1978/1979; *Hortic. Abstr.*, 51, 8666, 1981.

107. Paech, K., Über den Vitamin C-Gehalt deutscher Äpfel, Z., *Unters Lebensmitt.*, 76, 234, 1938.

108. Kessler, W., Über den Vitamin C-Gehalt deutscher Apfelsorten und seine Abhängigkeit von Herkunft, Lichtgenuss, Düngung, Dichte des Behanges und Lagerung, *Gartenbauwissenschaft*, 13, 619, 1939.

109. Murphy, E., Vitamin C and light, *Proc. Am. Soc. Hortic. Sci.*, 36, 498, 1939.

110. Simon, J., Ertrag und Vitamin C-Gehalt bei Paprika, *Bodenkultur*, 11, 208, 1960.

111. Bukatsch, F., Ascorbinsäure-gehalt und Atmungsintensität, *Phyton (Austria)*, 4, 35, 1952.

112. Keller, Th. and Schwager, H., Air pollution and ascorbic acid, *Eur. J. Forest Pathol.*, 7, 338, 1977.

113. Esterbauer, H., Grill, H., and Welt, R., Der jahreszeitliche Rhythmus des Ascorbinsäuresystems in Nadeln von *Picea abies*, *Z. Pflanzenphysiol.*, 98, 393, 1980.

114. Petcu, P., Phytochemische Untersuchungen an *Berberis virescens*, *Pharmazie*, 21, 54, 1966; *Hortic. Abstr.*, 36, 5229, 1966.

115. Izumi, H., Ito, T., and Yoshida, Y., Seasonal changes in ascorbic acid, sugar, and chlorophyll contents in sun and shade leaves of Satsuma mandarin and their interrelationships, *J. Jpn. Soc. Hortic. Sci.*, 59(2), 389, 1990.

116. Solov'eva, L. V., On the biological characteristics of upper and lower tiers of the crown in own-rooted and grafted apple trees, *Vestn. Mosk. Iniv. Biol. Počvoved.*, 6, 91, 1967; *Hortic. Abstr.*, 39, 2059, 1969.

117. Shattuck, V. I., Kakuda, Y., Shelp, B. J., and Kakuda, N., Chemical composition of turnip roots stored or intermittently grown at low temperature, *J. Am. Soc. Hortic. Sci.*, 116(5), 818, 1991.

118. Liptay, A. Papadopoulos, A. P., Bryan, H. H., and Gull, D., Ascorbic acid levels in tomato (*Lycopersicon esculentum* Mill.) at low temperatures, *Agric. Biol. Chem.*, 50(12), 3185, 1986.

119. Gutiev, O. G., The ascorbic acid content of tomato plants, *Tr. Prikl. Bot. Genet. Sel.*, 49(2), 295, 1973; *Hortic. Abstr.*, 44, 6781, 1974.

120. Rosenfeld, H. J., Ascorbic acid in vegetables grown at different temperatures, *Acta Hortic.*, 93, 425, 1979.

121. Skard, O. and Weydahl, E., Ascorbic acid—vitamin C—in apple varieties, *Medl. Nor. Landbrukshoegsk.*, 30, 477, 1950.

122. Efimova, K. E., Changes in the chemical composition of apples in relation to meteorological conditions, *Sadovodstvo Vinograd.*, 11, 16, 1988; *Hortic. Abstr.*, 59, 2697, 1989.

123. Shabalina, A. M., The relationship between certain indexes of apple fruit composition and weather conditions, *Byulleten' Glavnogo Botanicheskogo Sada*, 112, 34, 1979; *Hortic. Abstr.*, 49, 9239, 1979.

124. Kuusi, T., The most important quality criteria of some home-grown black-currant varieties. I. Ascorbic acid. II. Dry matter, pectin, acid content, colour and formol value, *Maataloust. Aikakausk.*, 37, 264, 1965; *Hortic. Abstr.*, 36, 4302, 1966.

125. Vestrheim, S., Ascorbic acid in black currants, *Meld. Nor. Landbrukshoegsk.*, 44(18), 1, 1965; *Hortic. Abstr.*, 36, 500, 1966.

126. Säkö, J. and Laurinen, E., The effect of fertilization on the black currant in two soils, *Ann. Agric. Fenniae*, 18, 96, 1979.

127. Rosenfeld, H. J., The effect of temperature on the ascorbic acid content of parsley (*Petroselinum crispum var. crispum* forma *crispum*), *Meld. Nor. Landbrukshoegsk.*, 54(20), 2, 1975; *Hortic. Abstr.*, 46, 10586, 1976.

128. Nilsson, F., Ascorbic acid in black currants, *Landbrukshoegsk. Ann.*, 35, 43, 1969; *Hortic. Abstr.*, 39, 6428, 1969.

129. Arndt, K., Der Einfluss der Temperatur auf den Vitamin-C-Gehalt in Rosenkohl, *Angew. Botanik*, 48(3/4), 125, 1974; *Hortic. Abstr.*, 45, 3997, 1975.

130. Kulikova, N. T., Content of ascorbic acid in green vegetables in polar region, *Sbornik Nauchnykh Trudov Poprikladnoi Bot. Genet. Sel.*, 107, 53, 1986; *Hortic. Abstr.*, 59, 1032, 1989.

131. Fritz, D. and Venter, F., Einfluss von Sorte, Standort und Anbaumassnahmen auf messbare Qualitätseigenschaften von Gemüse, *Landwirt. Forschung, Sonderh.*, 30(1), 95, 1974.

132. Hansen, E., Variations in the carotene content of carrots, *Proc. Am. Soc. Hortic. Sci.*, 46, 355, 1945.

133. Leclerc, J., Reuille, M. J., Miller, M. L., Lefebvre, J. M., Joliet, E., Autissier, N., Martinez, Y., and Perret, A., Effect of climatic conditions and soil fertilization on nutrient composition of salad vegetables in Burgundy, *Sci. Alimen.*, 10(6), 633, 1990.

134. Woolfe, J. A., *Sweet Potato*, Cambridge University Press, Cambridge, 1992.

135. Kimura, M., Rodriguez-Amaya, D. B., and Yokoyama, S. M., Cultivar differences and geographical effects on the carotenoid composition and vitamin A value of papaya, *Lebensmitt. Wiss. Technol.*, 24, 415, 1991.

136. Coggins, C. W. Jr., Hall, A. E., and Jones, W. W., The influence of temperature on regreening and carotenoid content of the "Valencia" orange rind, *J. Am. Soc. Hortic. Sci.*, 106(2), 251, 1981.

137. Hårdh, J. E., Der einfluss der Umwelt nördlicher Breitengrade auf die Qualität der Gemüse, *Qual. Plant.—Plant Foods Hum. Nutr.*, 25(1), 43, 1975.

138. Hårdh, J. E., Persson, A. R., and Ottosson, L., Quality of vegetables cultivated at different latitudes in Scandinavia, *Acta Agric. Scand.*, 27, 81, 1977.

139. Bradley, G. A., Smittle, D. A., Kattan, A. A., and Sistrunk, W. A., Planting date, irrigation, harvest sequence and varietel effects on carrot yields and quality, *Proc. Am. Soc. Hortic. Sci.*, 90, 223, 1967.

140. Bradley, G. A. and Dyck, R. L., Carrot color and carotenoids as affected by variety and growing conditions, *Proc. Am. Soc. Hortic. Sci.*, 93, 402, 1968.

141. Went, F. W., LeRosen, A. L., and Zechmeister, L., Effect of external factors on tomato pigments as studied by chromatographic methods, *Plant Physiol.*, 17, 91, 1942.

142. Tomes, M. L., Temperature inhibition of carotene synthesis in tomato, *Bot. Gaz. (Chicago)*, 124, 180, 1963.

143. Sayre, C. B., Robinson, W. B., and Wishnetsky, T., Effect of temperature on the color, lycopene, and carotene content of detached and vine-ripened tomatoes, *Proc. Am. Soc. Hortic. Sci.*, 61, 381, 1953.

144. Varga, A., and Bruinsma, J., Tomato, in *CRC Handbook of Fruit Set and Development*, Monselise, S., P., Ed., CRC Press, Boca Raton, Fla., 1986, 461.

145. Buescher, R. W., Influence of high temperature on physiological and compositional characteristics of tomato fruits, *Lebensmitt. Wiss. Technol.*, 12, 162, 1979.

146. Padula, M. and Rodriguez-Amaya, D. B., Characterization of carotenoids and assessment of the vitamin A value of Brazilian guavas (*Psidium guajava* L.), *Food Chem.*, 20, 11, 1986.

147. Mercadante, A. Z. and Rodriguez-Amaya, D. B., Carotenoid composition of a leafy vegetable in relation to some agricultural variables, *J. Agric. Food Chem.*, 39(6), 1094, 1991.

148. Gustafson, F. G., Influence of temperature on the vitamin content of green plants, *Plant Physiol.*, 25, 150, 1950.

149. O'Donnell, W. W., and Bayfield, E. G., Effect of weather, variety, and location upon thiamin content of some Kansas-grown wheats, *Food Res.*, 12, 212, 1947.

150. Mijil Dekker, L. P. van der and H. de Miranda, H., The vitamin B_1 content of dutch wheat and the factors which determine this content, *Neth. J. Agric. Sci.*, 2, 27, 1954.

151. Marquard, R., The influence of temperature and photoperiod on fat content, fatty acid composition, and tocopherols of rapeseed (*Brassica napus*) and mustard species (*Sinapis alba, Brassica juncea* and *Brassica nigra*), *Agrochimica*, 29, 145, 1985.

152. Marquard, R., Untersuchungen über den Einfluss von Sorte und Standort auf den Tocopherolgehalt verschiedener Pflanzenöle, *Fat Sci. Technol.*, 92(11), 452, 1990.

153. Beringer, H. and Saxena, N. P., Einfluss der Temperatur auf den Tocopherolgelaht von Samenfetten, *Z. Pflanzenerähr Bodenk.*, 120(2), 71, 1968.

154. Ruggeri, B. A., Gray, R. J. H., Watkins, T. R., and Tomlins, R. I., Effects of low-temperature acclimation and oxygen stress on tocopherol production in *Euglena gracilis, Z., Appl., Environ. Microbiol.*, 50, 1404, 1985.

155. Bonetti, D., Diurnal rhythm in ratio of reduced ascorbic acid to total ascorbic acid in leaves, *Boll. Soc. Ital. Biol. Sper.*, 25, 337, 1949; *Chem. Abstr.*, 44, 10050f, 1950.

156. Reid, M. E., Metabolism of ascorbic acid in cowpea plants, *Bull. Torrey Bot. Club.*, 68, 359, 1941.

157. Bregetova, L. G., Ascorbic acid content of cotton plant leaves, *Botan. Zhur.*, 36, 34, 1952; *Chem. Abstr.*, 45, 5772d, 1951.

158. Sugawara, T., Studies on the formation of ascorbic acid in plants. IV. Daily variation of ascorbic acid content and the concentration of carbohydrate in the leaves of plants, *Jpn. J. Botany*, 11, 344, 1941; *Chem. Abstr.*, 44, 10060i, 1950.

159. Hagen, U., Über die Tagesrythmik des Vitamin C-Gehaltes in Blättern, *Phyton* (*Austria*), 5, 1, 1953.

160. Ottosson, L., Changes in ascorbic acid in vegetables during the day and after harvest, *Acta Hortic.*, 93, 435, 1979.

161. Lee, E. H., Plant resistance mechanisms to air pollutants; rhythms in ascorbic acid production during growth under ozone stress, *Chronobiol. Int.*, 8(2), 93, 1991.

162. Smith, A. M. and Gillies, J., The distribution and concentration of ascorbic acid in the potato (*Solanum tuberosum*), *Biochem. J.*, 34, 1312, 1940.

163. Mekhanik, F. Ya., Factors that affect the content of ascorbic acid in leaves, *Botan. Zhur.*, 37(1), 71, 1952; *Chem. Abstr.*, 46, 7183f, 1952.

164. Tombesi, L. Variation of ascorbic acid, reduced glutathione, and of oxidase and catalase activity in plant tissues during the day, *Ann. Sper. Agrar.* (*Rome*), (N.S.), 5, 1021, 1951; *Chem. Abstr.*, 46, 2628b, 1952.

165. Zelenin, V. M., Biochemical characteristics of root-parsley in Perm region, *Tr. Perm. Sel'sk. Inst.*, 106, 92, 1974; *Hortic. Abstr.*, 46, 4922, 1976.

166. Kohman, E. F. and Porter, D. R., Physiological activity of ascorbic acid in plant life, *Science*, 95, 608, 1940.

167. Madsen, E., The effect of CO_2 concentration on the content of ascorbic acid in tomato leaves, *Ugeskrift for Agronomer*, 116(28), 592, 1971; *Hortic. Abstr.*, 42, 1483, 1972.

168. Bukatsch, F., Über den Askorbinsäuregehalt der Coniferennadeln, *Vitamine und Hormone*, 4, 192, 1943.

169. Matusis, I. I. and Jurova, G. G., Relationships between ascorbic acid accumulation in top and small fruits and some meteorogical factors, *Izv. Sib. Otd. Akad. Nauk SSSR*, No. 5, *Ser. Biol.-Med. Nauki*, No. 1, 76, 1968; *Hortic. Abstr.*, 40, 557, 1970.

170. Pirie, N. W., Optimal exploitation of leaf carotene, *Ecol. Food Nutr.*, 22, 1, 1988.

171. Speirs, M., Miller, J., Peterson, J. W. Wakeley, J. T., and Cochran, F. D., Variations in size of sample for the determination of thiamine and riboflavin content of turnip greens, *South Coop. Series Bull.*, 10, 69, 1951.

172. Reder, R., Richardson, L. R., Whitacre, J., Brittingham, W. H., and Kapp, L. C., Influence of time of day on the dry matter and vitamin content of turnip greens, *South. Coop. Series Bull.*, 10, 9, 1951.

173. Wills, R. B. H. and El-Ghetany, Y., Composition of Australian foods. 30. Apples and pears, *Food Technol. Aust.*, 38(2), 77, 1986.

174. Wills, R. B. H., Lim, J. S. K., and Greenfield, H., Variation in nutrient composition of Australian retail potatoes over a 12-month period, *J. Sci. Food Agric.*, 35, 1012, 1984.

175. Burrell, R. C., Brown, H. D., and Ebright, V. R., Ascorbic acid content of cabbage as influenced by variety, season, and soil fertility, *Food Res.*, 5, 247, 1940.

176. Robertson, G. L., and Nisperos, M. O., Changes in the chemical constituents of New Zealand grapefruit during maturation, *Food Chem.*, 11, 167, 1983.

177. Sachan, B. P., Pandey, D., and Shanker, G., Influence of weather on the chemical composition of guava fruit (*Psidium guajava* L.) var. Allahabad Safeda, *Punjab Hortic. J.*, 9, 119, 1969; *Hortic. Abstr.*, 41, 7788, 1971.

178. Tsytovich, K. I., Autumn-sown lettuce and spinach, *Tr. Prikl. Bot. Genet. Sel.*, 50(2), 60, 1973; *Hortic. Abstr.*, 44, 8498, 1974.

179. Narwadkar, P. R. and Pandey, R. M., Seasonal changes in ascorbic acid in developing mango buds, *Indian J. Plant Physiol.*, 26(4), 406, 1983.

180. Afzal, M., Saeed, M., and Roghani, M. S., Chemical composition of sweet orange at various stages of maturity, *J. Sci. Technol.*, 1(1), 84, 1977; *Hortic. Abstr.*, 49, 2907, 1979.

181. Ting, S. V., Nutrients and nutrition of citrus fruits, in *Citrus Nutrition and Quality*, Nagy, S., and Attaway, J. S., Eds., Am. Chem. Soc. Symp., 143, ACS, Washington, D.C., 1980, 3.

182. El-Sherbiny, G. A. and Rizk, S. S., Orange concentrates prepared by different techniques. 2. Storage stability, *Egypt J. Food Sci.*, 9(1-2), 93, 1981; *Chem. Abstr.*, 98, 52176e, 1983.

183. Zderkiewicz, T. and Dyduch, J., Changes in the accumulation of ascorbic acid and essential oil during different growth phases of parsley, *Acta Agrobotan.*, 25(2), 179, 1972; *Hortic. Abstr.*, 44, 648, 1974.

184. Franke, W., On the contents of vitamin C and thiamin during the vegetation period in leaves of three spice plants (*Allium schoenoprasum* L., *Melissa officinalis* L. and *Petroselinum crispum* (Mill.) Nym. ssp. *Crispum*) *Acta Hortic.*, 73, 205, 1978.

185. Hatton, T. T. Jr. and Reeder, W. F., Ascorbic acid concentrations in Florida-grown "Tahiti" (Persian) limes, *Proc. Tropical Region*, Am. Soc. Hortic. Sci., 15, 89, 1971; *Hortic. Abstr.*, 43, 8101, 1973.

186. Nakatoh, H., Ohta, H., and Nasiro, S., Quality of pineapple fruit and processed juice from different growing seasons in Okinawa, *J. Jpn. Soc. Food Sci. Technol.*, 32(12), 911, 1985; *Hortic. Abstr.*, 56, 10182, 1986.

187. Prihod'ko, S. N. and Musat, I. K., Studies on prickly pears in the Ukraine, Bjull. Glav. Bot. Sada, 56, 101, 1964; *Hortic. Abstr.*, 36, 3593, 1966.

189. Gadzhieva, G. G., Vitamin level in the fruit of roses introduced on the Apsheron peninsula (botanical garden), *Izv. Akad. Nauk. Az. SSR, Ser. Biol. Nauk*, 3, 23, 1978; *Chem. Abstr.*, 90, 118093u, 1979.

190. Kurosaki, T. and Kawakami, I. K., Histochemical and biochemical studies on Satsumas. I. Seasonal fluctuations in the distribution of concentration of ascorbic acid within pulp and peel tissues during the growth of Satsuma fruit, *J. Jpn. Soc. Hortic. Sci.*, 43(2), 189, 1974; *Hortic. Abstr.*, 45, 7832, 1975.

191. Franke, L. W., Vitamin C-Gehalt von frischem, tiefgekühltem und gekochtem Spinat (*Spinacia oleracea*), Z. Lebensm.-Unters. Forsch., 131, 11, 1966.

192. Stino, K. R., Abdelfattah, M. A. and Nassar, H., Studies on vitamin C and oxalic acid concentration in spinach, *Agric. Res. Rev.*, 51(5), 109, 1973; *Hortic. Abstr.*, 46, 304, 1976.

193. Holmes, A. D., Jones, C. P., and Ritchie, W. S., The ascorbic acid content of late-winter tomatoes, *N. Engl. J. Med.*, 229(12), 461, 1943.

194. Arora, S. K., Pandita, M. L., and Singh, K., Chemical composition of tomato (*Lycopersicon esculentum* Mill.) varieties as influenced by seasonal variation, *Haryana J. Hortic. Sci.*, 4(3/4), 230, 1975; *Hortic. Abstr.*, 47, 3724, 1977.

195. Shinohara, Y., Suzuki, Y., Shibuya, M., Effects of cultivation method, growing season and cultivar on the ascorbic acid content of tomato fruits, *J. Jpn. Soc. Hortic. Sci.*, 51(3), 338, 1982; *Hortic. Sci.*, 53, 4265, 1983.

196. Russell, L. F. and Mullin, W. J., Vitamin C content of fresh tomatoes, *Nutr. Report Int.*, 34, 575, 1986.

197. Grill, D., Esterbauer, H., and Welt, R., Über das Ascorbinsäuresystem in *Larix* und *Aesculus*, *Phyton (Austria)*, 20, 251, 1980.

198. Grill, D., Pfeifhofer, H., and Tschulik, A., Untersuchungen über die jahreszeitlichen Schwankungen von Nadelinhaltsstofen unter besonderer Berücksichtigung von Frosthärtefaktoren, *Phyton (Austria)*, 27(2), 221, 1987.

199. Nauer, E. M., Goodale, J. H., Summers, L. L., and Reuther, W., Climate effects on mandarins and Valencia oranges, *Calif. Agric.*, 28(4), 8, 1974.

200. Ting, S. V., Nutritional labeling of citrus products, in *Citrus Science and Technology*, Nagy, S., Shaw, P. E. and Veldhuis, M. K., Eds., AVI Publishing, Westport, Conn., 1977, 401.

201. Akimoto, N., Ochi, S., and Narita, H., Seasonal variations of sugar acid and vitamin contents in commercial Satsuma mandarin (*Citrus unshiu Marc.*), *Shizuoka-ken Eisei Kankyo Senta Hokoku*, 31, 15, 1988; *Chem. Abstr.*, 112, 175685k, 1990.

202. Heinze, P. H., Kanapauk, M. S., Wade, B. I., Grimball, P. C., and Foster, R. L., Ascorbic acid content of 39 varieties of snap beans, *Food Res.*, 9, 19, 1944.

203. Hayden, F. R., Heinze, P. H., and Wade, B. L., Vitamin content of snap beans grown in South Carolina, *Food Res.*, 13, 143, 1948.

204. Hibbard, A. D. and Flynn, L. M., Effect of maturity on the vitamin content of green snap beans, *Proc. Am. Soc. Hortic. Sci.*, 46, 350, 1945.

205. Hansen, E., Seasonal variation in the mineral and vitamin content of certain green vegetable crops, *Proc. Am. Soc. Hortic. Sci.*, 46, 229, 1945.

206. Poole, C. F., Heinze, P. H., Welch, J. E., and Grimball, P. C., Differences in stability of thiamin, riboflavin, and ascorbic acid in cabbage varieties, *Proc. Am. Soc. Hortic. Sci.*, 45, 396, 1944.

207. Tressler, D. K., Mack, G. L., and King, C. G., Factors influencing the vitamin C content of vegetables, *Am. J. Publ. Health.*, 26, 905, 1936.

208. Çevik, B., Kirda, C., and Dinç, G., Effect of some irrigation systems on yield and quality of tomato grown in a plastic covered greenhouse in the south of Turkey, *Acta Hortic.*, 119, 333, 1981.

209. El-Sherbiny, G. A., Rizk, S. S., and El-Shiaty, M. A., Quality aspects as related to seasonal variations in tomatoes, *Egypt. J. Food. Sci.*, 11(1-2), 73, 1983; *Chem. Abstr.*, 100, 155404s, 1984.

210. Ishii, G. and Saijo, R., Effect of various cultural conditions on total sugar content, vitamin C content and β-amylase activity of Daikon radish root (*Raphanus sativus*, L.), *J. Jpn. Soc. Hortic. Sci.*, 55(4), 468, 1987; *Hortic. Abstr.*, 59, 9059, 1989.

211. Rathore, D. S., Effect of season on the growth and chemical composition of guava (*Psidium guajava* L.) fruits, *J. Hortic. Sci.*, 51, 41, 1976.

212. Mitra, S. K., Ghosh, S. K., and Dhua, R. S. Ascorbic-acid content in different varieties of guava grown in West Bengal, *Sci. Cult.*, 50(7), 235, 1984.

213. Džikirba, V. V., Ascorbic acid and peroxidase in the leaves of some forms of citrus, *Tr. Suhum. Opyt. Stan. Subtrop. Kul't.*, 1, 267, 1967; *Hortic. Abstr.*, 39, 3612, 1969.

214. Khasanov, R. A. and Makhmadbekov, S., Studies on the seasonal changes in the activity of growth regulating substances and ascorbic acid in Meyer lemon leaves, *Subtrop. Kul't.*, 4, 73, 1990; *Hortic. Abstr.*, 62, 5224, 1992.

215. Seybold, A. and Mehner, H., *Über den Gehalt von Vitamin C in Pflanzen*, Springer Verlag, Heidelberg, 1948.

216. Kazakova, M. I., Vegetation and its effect on content of vitamin C in plant products, *Gigiena Sanit.*, 1, 37, 1951; *Chem. Abstr.*, 45, 5772g, 1951.

217. Repyakh, S. M., Dynamics of biologically active substances of pine (*Pinus sylvestris* L.) and Siberian spruce (*Picea obovata* Ldb.) needles, *Izv. Vyssh. Uchebn. Zaed., Lesn. Zh.*, 6, 91, 1983; *Chem. Abstr.*, 100, 32214x, 1984.

218. Liu, P., Yu, B., Xing, S., and Yu, B., Preliminary analysis of the contents of several organic substances in the leaves of five evergreen and deciduous species in Beijin [China] in different seasons, *Linye Kexue*, 18(1), 107, 1982; *Chem. Abstr.*, 97, 52501m, 1982.

219. Kaludin, K., Investigation on the content of vitamin C, carotene and mineral substances in the foliage of Scots pine (*Pinus sylvestris* L.), *Gorskostopanska Nauka*, 23(4), 51, 1986; *Forestry Abstr.*, 48, 3284, 1987.

220. Fujita, T., Kawai, N., Kamei, M., Itano, K., Okazaki, K., Nakayama, Y., Kanbe, T., and Sasaki, K., Composition of eucalyptus browses, *Ann. Rep. Osaka City Inst. Public Health Environ. Sci.*, 49, 147, 1986; *Chem. Abstr.*, 108, 149126b, 1988.

221. Bermadinger, E., Grill, D., and Guttenberger, H., Thiols, ascorbic acid, pigments and epicuticular waxes in needles from spruce in the altitude profile "Zillertal," *Phyton (Austria)*, 29(3), 163, 1989; *Chem. Abstr.*, 112, 115796n, 1990.

222. Madamanchi, N. R., Hausladen, A., Alscher, R. G., Amundson, R. G., and Fellows, S., Seasonal changes in antioxidants in red spruce (*Picea rubens* Sarg.) from three field sites in the northern United States, *New Phytol.*, 118, 331, 1991.

223. Rodahl, K., Content of vitamin C (L-ascorbic acid) in arctic plants, *Trans. Bot. Soc. Edinburgh*, 34, 205, 1944.

224. Vitez, L., Sluga, H., Golc-Wondra, A., and Mihelic, E., Contribution to the composition of dandelion. I. Characterization of water-soluble carbohydrates of dandelion by thin-layer chromatography, *Nova Proizvod.*, 37(5-6), 193, 1986; *Chem. Abstr.*, 108, 19243b, 1988.

225. Airapetova, S. A., Tarosova, E. O., and Stepanyan, T. G., Dynamics of accumulation of vitamin C in *Lycopersicon esculentum* M. var. *pimpinellifolium* x tomato hybrids, *Biol. Zh. Arm.*, 42(7), 690, 1989; *Chem. Abstr.*, 112, 74007k, 1990.

226. Feltwell, J. S. E. and Valadon, L. R. G., Carotenoid changes in *Brassica oleracea var. capitata* L. with age, in relation to the large white butterfly, *Pieris brassicae* L., *J. Agric. Sci. (Cambridge)*, 83, 19, 1974.

227. Booth, V. H. and Dark, S. O. S., The influence of environment and maturity on total carotenoids in carrots, *J. Agric. Sci.*, 39, 226, 1949.

228. Yamaguchi, M., Robinson, B., and MacGillivray, J. H., Some horticultural aspects of the food value of carrots, *Proc. Am. Soc. Hortic. Sci.*, 60, 351, 1952.

229. Kulikova, N. T., Characteristics of the chemical composition of carrot roots in the polar region, *Tr. Prikl. Bot. Genet. Sel.*, 45(1), 325, 1971; *Hortic. Abstr.*, 43, 5375, 1973.

230. Bernstein, L., Hamner, K. C., and Parks, R. Q., The influence of mineral nutrition, soil fertility, and climate on carotene content and ascorbic acid content of turnip greens, *Plant Physiol.*, 20, 540, 1945.

231. Krylov, S. V. and Baranova, N. D., The dynamics of carotene accumulation in carrot roots in relation to the date of sowing. *Dokl. Mosk. Sel'.-hoz. Akad. K. A. Timirjazeva*, 121, 85, 1966; *Hortic. Abstr.*, 37, 7088, 1967.

232. Kiatoko, M., McDowell, L. R., Bertrand, J. E., Chapman, H. L., Pate, F. M., Martin, F. G., and Conrad, J. H., Evaluating the nutritional status of beef cattle herds from four soil order regions of Florida. I. Macroelements, protein, carotene, vitamins A and E, hemoglobin and hematocrit, *J. Anim. Sci.*, 55(1), 28, 1982.

233. Bureau, J. L. and Bushway, R. J., HPLC determination of carotenoids in fruits and vegetables in the United States, *J. Food Sci.*, 51(1), 128, 1986.

234. Pirtskhalaishvili, E. S., Thiamine and riboflavin in citrus leaves in relation to fertilizer application, *Subtrop. Kul't.*, 1, 85, 1973; *Hortic. Abstr.*, 44, 741, 1974.

235. Booth, V. H., The α-tocopherol content of forage crops, *J. Sci. Food Agric.*, 15, 342, 1964.

236. Kawamur, S., Haraguchi, K., Kokura, H., Matsumura, Y., and Mori, T., Seasonal changes in α-tocopherol concentration in tea bushes, *Nippon Nogeikagaku Kaishi*, 62(9), 1355, 1988; *Hortic. Abstr.*, 59, 5399, 1989.

237. Huq, R. S., Abalaka, J. A., and Stafford, W. L., Folate content of various Nigerian foods, *J. Sci. Food Agric.*, 34, 404, 1983.

238. Pirtskhalaishvili, E. S., The effect of fertilizers on the folic acid content of leaves and fruit of Unshiu mandarin, *Subtrop. Kul't.*, 2, 161, 1971; *Hortic. Abstr.*, 42, 4879, 1972.

239. Koch, J., Über den Vitamin-C-Gehalt verschiedener Früchte während des Reifevorganges, *Die Industr. Obst-Gemüseverwertung*, 39, 231, 1954.

240. Marx, Th., Über den Oxydationsschutz der L-Ascorbinsäure in frischem Pflanzenmaterial, *Z. Lebensm. Unters. Forsch.*, 99(3), 180, 1954.

241. Pätzold, G. and Finke, M., Ergebnisse analytischer Untersuchungen beim Apfel, *Gartenbau*, 31(2), 52, 1984.

242. Trautner, K. and Somogyi, J. C., Änderungen der Zuker- und Vitamin-C-Gehalte in Früchten während der Reifung, *Mitt. Geb. Lebensmittelunters. Hyg.*, 69, 431, 1978.

243. Hagen, R. E., Elkin, E. R., and Farrow, R. P., Nutrient variations in canned fruits and vegetables. *HortScience*, 14(3), 251, 1979.

244. Massey, L. M., Jr., Nutritive quality of long-distance shipped green beans for processing, *J. Food Sci.*, 48, 1564, 1983.

245. Hanning, F., Comparison of the biological and chemical methods for the determination of vitamin C in canned strained vegetables and a study of its variation from year to year. *J. Nutr.*, 12(4), 405, 1936.

246. Spellerberg, B., and Ohms, J. P., Fruchtqualität von Beerenobstarten, *Obstbau* (*Bonn*), 15, 394, 1990.

247. Massey, P. H., Jr., Eheart, J. F., and Young, R. W., Variety, year, and pruning effects on the dry matter, vitamin content, and yield of broccoli, *Proc. Am. Soc. Hortic. Sci.*, 68, 377, 1956.

248. Branion, H. D., Roberts, J. S., Cameron, C. R., and McCready, A. M., The ascorbic acid content of cabbage, *J. Am. Diet. Assoc.*, 24, 101, 1948.

249. Koch, J., and Bretthauser, G., Über den Vitamin C Gehalt reifender Früchte, *Landwirt. Frorsch.*, 9, 51, 1956.

250. Guadagni, D. G., Nimmo, C. C., and Jansen, E. F., Time-temperature tolerance of frozen foods. VI. Retail packages of frozen strawberries, *Food Technol.*, 11, 389, 1957.

251. Speirs, M., Cochran, H. L., Peterson, W. J., Sherwood, F. W., and Weaver, J. G., The effects of fertilizer treatments, curing, storage, and cooking on the carotene and ascorbic acid content of sweetpotatoes, *South. Coop. Series Bull.*, 3, 5, 1945.

252. Müller-Haslach, W., Arold, G., and Kimmel, V., Einfluss der Düngungsintensität auf die Qualität von Tomaten, *Bayerisches Landwirt. Jahresb., Sonderh.*, 63(1), 81, 1986.

253. Eheart, M. S., Fertilization effects on the chlorophyll, carotene, pH, total acidity, and ascorbic acid in broccoli, *J. Agr. Food. Chem.*, 14, 18, 1966.

254. Evers, A. M., Effect of different fertilization practices on the carotene content of carrot, *J. Agric. Sci. Finland*, 61, 7, 1989.

255. Speirs, M., Dempsey, A. H., Miller, J., Peterson, W. J., Wakeley, J. T., Cochran, F. D., Reder, R., Cordner, H. B., Fieger, E. A., Hollinger, M., James, W. H., Lewis, H., Eheart, J. F., Eheart, M. S., Andrews, F. S., Young, R. W., Mitchell, J. H., Carrison, O. B., and McLean, F. T., The effect of variety, curing, storage, and time of planting and harvesting on the carotene, ascorbic acid, and moisture content of sweet potatoes, *South Coop. Series Bull.*, 30, 3, 1953.

256. Nik-Khah, A., Hoppner, K. H., Sosulski, F. W., Owen, B. D., and Wu, K. K., Variation in proximate fractions and B-vitamins in Saskatchewan feed grains, *Can. J. Animal Sci.*, 52, 407, 1972.

257. Gough, H. W. and Lantz, E. M., Relation of variety and locality to niacin, thiamine, and riboflavin content of dried beans grown in three locations, *Food Res.*, 15, 308, 1950.

258. Speirs, M., Effect of environment on the thiamine content of plants, *South. Coop. Series. Bull.*, 36, 48, 1954.

259. Hunt, C. H., Rodriguez, L. D., and Bethke, R. M., The environmental and agronomical factors influencing the thiamine, riboflavin, niacin, and pantothenic acid content of wheat, corn, and oats, *Cereal Chem.*, 27, 79, 1950.

260. Eheart, J. F., Young, R. W., and Allison, A. H., Variety, type, year and location effects on the chemical composition of peanuts, *Food Res.*, 20(5), 497, 1955.

261. Whiteside, A. G. O. and Jackson, S. H., The thiamin content of canadian hard red spring wheat varieties, *Cereal Chem.*, 20, 542, 1943.

262. Hunt, C. H., Ditzler, L., and Bethke, R. M., Niacin and pantothenic acid content of corn hybrids, *Cereal Chem.*, 24, 355, 1947.

263. Davis, K. R., Cain, R. F., Peters, L. J., Le Tourneau, D., and McGinnis, J., Evaluation of the nutrient composition of wheat. II. Proximate analysis, thiamin, riboflavin, niacin, and pyridoxin, *Cereal Chem.*, 58, 116, 1981.

264. Marquard, R., Schuster, W., und Seibel, K. K., Fettsäuremuster und Tokopherolgehalt im Öl verschiedener Sonnenblumensorten aus weltweitem Anbau, *Fette Seifen Anstrichmittel*, 79(1), 137, 1977.

265. Schuster, W., Zipse, W., and Marquard, R., Der Einfluss von Genotype und Anbauort auf verschiedene Inhaltsstoffe von Samen des Ölkürbis (*Cucurbita Pepo* L.), *Fette Seifen Anstrichmittel*, 85(2), 56, 1983.

266. Davis, K. R., Litteneker, N., Le Tourneau, D., Cain, R. F., Peters, L. J. and McGinnis, J., Evaluation of the nutrient composition of wheat. I. Lipid composition, *Cereal Chem.*, 57, 178, 1980.

267. Szkilladziowa, W. et al., Results of studies on nutrient content in selected varieties of edible potatoes, *Chem. Abstr.*, 87, 182878u, 1977.

268. Putnam, M. E., Thiamin, riboflavin, biotin and α-tocopherol contents of wheat and barley grown in the UK in 1975, 1976, 1977 and 1978. *Animal Nutrition Events*, 1, 25pp, 1978; *Food Sci. Technol. Abstr.*, 12(1), M35, 1980.

269. Eheart, J. F. Wakeley, J. T., Speirs, M., Cowart, F. F., Miller, J., Heinze, P. H., Kanapaux, M. S., Sheets, O. A., McWhirter, L., Geiger, M., Moore, R. C., Effect of different planting dates, bean maturity, and location on the vitamin content of lima beans, *South. Coop. Series Bull.*, 12, 5, 1951.

270. Teich, A. H. and Menzies, J. A., The effect of nitrogen, phosphorus and potassium on the specific gravity, ascorbic acid content and chipping quality of potato tubers, *Am. Potato J.*, 41, 169, 1964.

271. Fernández, M. C., Llanos, E., Martin, O., and Ancín, M. C., Vitamins B_1, B_2, B_6, and C content variations in asparagus in function of production area, harvest period, canning process and storage time of canned products, (in Spanish), *Rev. Agroqúm Tecnol. Aliment*, 31(4), 532, 1991.

272. Janes, B. E., The relative effect of variety and environment in determining the variations of per cent dry weight, ascorbic acid, and carotene content of cabbage and beans, *Pro. Am. Soc. Hortic. Sci.*, 45, 387, 1944.

273. Eheart, J. F., Moore, R. C., Speirs, M., Cowart, F. F., Cochran, H. L., Sheets, O. A., McWhirter, L., Geiger, M., Bowers, J. L., Heinze, P. H., Hayden, F. R., Mitchell, J. H., Carolus, R. L., Vitamin studies on lima beans, *South. Coop. Series Bull.*, 5, 1, 1946.

274. Janes, B. E., Variations in the dry weight, ascorbic acid and carotene content of collards, broccoli and carrots as influenced by geographical location and fertilizer level, *Proc. Am. Soc. Hortic. Sci.*, 48, 407, 1946.

275. Klein, B. P. and Perry, A. K., Ascorbic acid and vitamin A activity in selected vegetables from different geographical areas of the United States, *J. Food Sci.*, 47, 941, 1982.

276. Leveille, G. A., Bedford, C. L., Kraut, C. W., and Lee, Y. C., Nutrient composition of carrots, tomatoes and red tart cherries, *Fed. Proc.*, 33(11), 2264, 1974.

277. Godfrey-Sam-Aggrey, W., Haque, I., and Garber, M. J., Relation of citrus leaf nutrients to internal fruit quality in Sierra Leone, *J. Plant Nutr.*, 1, 185, 1979.

278. Malena, T. V., Lykova, R. V., and Chigireva, E. A., Biologically active substance content of *Hippophae rhamnoides* L. fruits in natural populations of Issyk-Kul' region, *Khimiko Farmatsevticheskii Zhurnal*, 18(10), 1226, 1984; *Hortic. Abstr.*, 55, 2925, 1985.

279. Edrissi, M. and Kooshkabadi, H., Determination of vitamin C in Iranian citrus fruits, *Iran J. Agric. Res.*, 3(2), 81, 1975.

280. Mullin, W. J., Jui, P. Y. Nadeau, L. and Smyrl, T. G., The vitamin C content of seven cultivars of potatoes grown across Canada, *Can. Inst. Food Sci. Technol. J.*, 24(3/4), 169, 1991.

281. Kolbe, H. and Müller, K., Einfluss differenzierter Nährstoffgaben auf einige wertgebende Inhaltstoffe in Speisekartoffeln, *Veröff. Arbeitsgem. Kartoffelforsch.*, 6, 12, 1984.

282. Karikka, K. J., Dudgeon, L. T., and Hauck, H. M., Influence of variety, location, fertilizer, and storage on the ascorbic acid content of potatoes grown in New York State, *J. Agric. Res. (Washington, D. C.)*, 68, 49, 1944.

283. Augustin, J., Johnson, S. R., Teitzel, C., Toma, R. B., Shaw, R. L., True, R. H., Hogan, J. M. and Deutsch, R. M., Vitamin composition of freshly harvested and stored potatoes, *J. Food Sci.*, 43(5), 1566, 1978.

284. Allison, R. M. and Driver, C. M., The effect of variety, storage and locality on the ascorbic acid content of the potato tuber, *J. Sci. Food Agric.*, 4, 386, 1953.

285. Hudson, D. E., Mazur, M. M. and Lachance, P. A., Ascorbic acid, riboflavin, and thiamin content of strawberries during postharvest handling, *HortScience*, 20(1), 71, 1985.

286. Reder, R., Speirs, M., Chochran, H. L., Hollinger, M. E., Farish, L. R., Geiger, M., McWhirter, L., Sheets, O. A., Eheart, J. F., Moore, R. C., and Carolus, R. L., The effects of maturity, nitrogen fertilization, storage and cooking, on the ascorbic acid content of two varieties of turnip greens. *South. Coop. Series Bull.*, 1, 1, 1943.

287. Hårdh, K. and Hårdh, J. E., Studies on quality of vegetables and strawberries at different latitudes in Finland, *Ann. Agric. Fenniae*, 16, 19, 1977.

288. Boukin, V. N., Notes on the study of vitamins in plants, *Qual. Plant.*, 3-4, 374, 1958.

289. Udayasekhara, Rao, P. and Belavady, B., Chemical composition of high yielding varieties of pulses: varietal, locational and year to year differences, *Indian J. Nutr. Dietet.*, 16, 440, 1979.

290. Speirs, M., Effect of environment on the riboflavin content of plants, *South. Coop. Series. Bull.*, 36, 42, 1954.

291. Bradbury, J. H. and Singh, U., Thiamin, riboflavin, and nicotinic acid contents of tropical root crops from the South Pacific, *J. Food Sci.*, 51(6), 1563, 1986.

292. Dressel, J. and Jung, J., Gehaltsniveau an Vitaminen des B-Komplexes in Abhängigkeit von Stickstoffzufuhr und Standort, *Landwirt. Forsch. Sonderh.*, 35, 261, 1978.

293. Pomeranz, Y., Chemical composition of kernel structure, in *Wheat: Chemistry and Technology*, Vol. 1, Pomeranz, Y., Ed., Am. Association of Cereal Chemists. St. Paul, Minn., 1988, 97.

294. Marquard, R., Schuster, W., and Iran-Nejad, H., Untersuchungen über Tokopherol- und Thiamingehalt in Leinsaat aus weltweitem Anbau und aus dem Phytotron unter definierten Klimabedingungen, *Fette Seifen Anstrichmittel*, 79(7), 265, 1977.

295. Simwemba, C. G., Hoseney, R. C., Varriano-Martson, E., and Zeleznak, K., Certain B vitamins and phytic acid contents of pearl millet [*Pennisetum americanum* (L.) Leeke], *J. Agric. Food Chem.*, 32, 31, 1984.

296. Aitkin, Y., Influence of environment and variety on nitrogen and thiamin in field peas (*Pisum sativum*), *Proc. Roy. Soc. Victoria*, 67, 257, 1955.

297. Lee, J. W. and Underwood, E. J., The influence of variety on the thiamin and nitrogen contents of wheat, *Aust. J. Exp. Biol. Med. Sci.*, 28, 543, 1950.

298. Hoffer, A., Alcock, A. W., and Geddes, W. F., The effect of variations in Canadian spring wheat on the thiamine and ash of long extraction flours, *Cereal Chem.*, 21, 210, 1944.

299. Schuphan, W., Kling, M., and Overbeck, G., Einfluss geneticher und umweltbedingter Faktoren auf die Gehalte an Vitamin B_1, Vitamin B_2 und Niacin von Winter- und Sommerweizen, *Qual. Plant.*, 15, 177, 1968.

300. Somers, G. F. and Beeson, K. C., The influence of climate and fertilizer practices upon the vitamin and mineral content of vegetables, *Adv. Food Tech.*, 1, 291, 1948.

301. Marquard, R., Der Einfluss von Sorte und Standort sowie einzelner definierter Klimafaktoren auf den Tokopherolgehalt im Rapsöl, *Fette Seifen Anstrichmittel*, 78(9), 341, 1976.

302. Marquard, R. and Schuster, W., Protein- and Fettgehalte des Kornes sowie Fettsäuremuster und Tokopherolgehalte des Öles bei Sojabohnensorten von stark differenzierten Standorten, *Fette Seifen Anstrichmittel*, 82(4), 137, 1980.

303. Ferland, G. and Sadowski, J. A., Vitamin K_1 (Phylloquinone) content of green vegetables: effect of plant maturation and geographical growth location, *J. Agric. Food Chem.*, 40, 1874, 1992.

304. Esteves, M. T. da C., Carvalho, V. D. de, Chitarra, M. I. F., Chitarra, A. B., and Paula, M. B. de, Characteristics of fruits of six guava (*Psidium guajava* L.) cultivars during ripening. II. Vitamin C and tannin contents, in *Anais do VII Congresso*

Brasileiro de Fruitcultura, Vol. 2, Florianópolis, Brazil, 1984, 490; *Hortic. Abstr.*, 55, 7323, 1985.

305. Baqar, M. R., Technical note: vitamin C content of some Papua New Guinean fruits, *J. Food Technol.*, 15, 459, 1980.

306. Samorodova-Bianki, G. B., Zhmurko, L. A., and Stepanova, F. P., Vitamins C and P in apples, *Sadovodstvo*, 4, 30, 1973; *Hortic. Abstr.*, 43, 7417, 1973.

307. Nesterov, Ya. S., Phenotypic variability of apple fruits in relation to growing conditions, *Vestnik Sel'skokhozyaistvennoi Nauki, Moscow, USSR*, 6, 54, 1975; *Hortic. Abstr.*, 46, 2924, 1976.

308. Kochanov, E. M., Stankevich, K. V., Chemical composition of fruit of new apple cultivars in relation to geographical location, *Tr. Tsentral'noi Geneticheskoi LaboratoriiI.V. Michurina*, 16, 134, 1975; *Hortic. Abstr.*, 47, 190, 1977.

309. Udachina, E. G., Sokolova, S. M., Samokhina, T. V., and Budarina, T. D., Chemical composition of the fruit of apple varieties introduced into the Moscow area, *Byull. Gl. Bot. Sada*, 127, 55, 1983; *Chem. Abstr.*, 101, 87499v, 1984.

310. Sazonova, L. V., Variability in the characters and properties of chinese radish (*Raphanus sativus sp. Sinensis*), *Tr. Prikl. Bot. Genet. Sel.*, 51(3), 89, 1974; *Hortic. Abstr.*, 45, 8387, 1975.

311. Streighthoff, F., Munsell, H. E., Ben-dor, B. A., Orr, M. L., Cailleau, R., Leonard, M. H., Ezekiel, S. R., Kornblum, R., and Koch, F. G., Effect of large-scale methods of preparation and the vitamin content of food. I. Potatoes, *J. Am. Diet. Assoc.*, 22, 117, 1946.

312. Nilsson, F., Fruit quality and climate, *Nord. JordbrForskn.*, 49, 274, 1967; *Hortic. Abstr.*, 38, 2606, 1968.

313. Lukovnikova, G. A., Stepanova, V. M. and Amerikantseva, G. S., The chemical composition of white cabbage in relation to meteorological conditions in different geographical regions of the USSR, *Byulleten'Vsesoyuznogo Ordena Lenina Instituta Rastenievodstva Imeni N.I. Vavilova*, 66, 67, 1976; *Hortic. Abstr.*, 48, 407, 1978.

314. Hårdh, J. E., Factors affecting the vitamin C content of black currants, *Maataloust. Aikakausk*, 36, 14, 1964; *Hortic. Abstr.*, 34, 6469, 1964.

315. Hårdh, J. E. and Hårdh, K., Quality tests on greenhouse vegetables, *Ann. Agric. Fenniae*, 11, 342, 1972.

316. Hårdh, J. E. and Hårdh, K. Effect of radiation, day-length and temperature on plant growth and quality: a preliminary report, *Hortic. Res. (Edinburgh)*, 12, 25, 1972.

317. Castaldo, D., Lo Voi, A., Trifiro, A., and Gherardi, S., Composition of Italian kiwi (*Actinidia chinensis*), puree, *J. Agric. Food Chem.*, 40(4), 594, 1992.

318. Zhang, Y. P. and Long, Z. X., Seedling selections of *Actinidia deliciosa*, *J. Fruit Sci.*, 8(2), 124, 1991; *Hortic. Abstr.*, 62, 5610, 1992.

319. Lintas, C., Adorisio, S., and Cappeloni, M., Composition and nutritional evaluation of kiwifruit grown in Italy, *New Zealand J. Crop Hortic. Sci.*, 19, 341, 1991.

320. Cotter, R. L., Macrae, E. A., Ferguson, A. R., McMath, K. L., and Brennan, C. J., A comparison of the ripening, storage and sensory qualities of seven cultivars of Kiwifruit, *J. Hortic. Sci.*, 66(3), 291, 1991.

321. Gersbach, K., Vitamin-C in Früchten (vor allem in Äpfeln), *Züri-Obst*, 45(10), 2, 1984.

322. Somers, G. F., Hamner, K. C., and Nelson, W. L., Field illumination and commercial handling as factors in determining the ascorbic acid content of tomatoes received at the cannery, *J. Nutr.*, 30, 424, 1945.

323. Reder, R., Ascham, L., and Eheart, M. S., Effect of fertilizer and environment on the ascorbic acid content of turnip greens, *J. Agric. Res. (Washington, D.C.)*, 66(10), 375, 1943.

324. Balvoll, G., Carrot quality in south and north Norway, *Gartneryrket*, 66(36/37), 598, 1976; *Hortic. Abstr.*, 47, 7590, 1977.

325. Kelly, E., Dietrich, K. S., and Porter, T., Vitamin B₁ content of eight varieties of beans grown in two locations in Michigan, *Food Res.*, 5, 253, 1940.

326. Holman, W. I. M. and Godden, W., The aneurin (vitamin B₁) content of oats. I. The influence of variety and locality. II. Possible losses in milling, *J. Agric. Sci.*, 37, 51, 1947.

327. Andrews, J. S., Boyd, H. M., and Terry, D. E., The riboflavin content of cereal grains and bread and its distribution in products of wheat milling, *Cereal Chem.*, 19, 55, 1942.

328. Dragavcev, A. P., A new region of fruit growing in the mountains, *Vestn. S-Kh. Nauki*, 6(3), 95, 1961; *Hortic. Abstr.*, 31, 5786, 1961.

329. Olisaev, A. A., The effect of altitude on fruit tree phenology and fruit quality in the Central Caucasus foothills, *Tr. Sev.-Kavkaz. NII Gornogo i Predgorn. Sel'skogo Khozyaistva*, 1, 122, 1974; *Hortic. Abstr.*, 46, 4165, 1976.

330. Maroutian, S. A., Michaelian, V. M., and Petrossian, J. A., Biochemical changes in apricot trees with vertical zonality, *Acta Hortic.*, 192, 29, 1985.

331. Stamboliev, M., Lazarov, K., and Ivanov, A., The effect of certain ecological factors on the size and chemical composition of fruits of the pear cultivars Curé and Beurré d'Hardenpont under the conditions of northwest Bulgaria, *Gradinar. Lozar. Nauka*, 9(1), 15, 1972; *Hortic. Abstr.*, 44, 1390, 1974.

332. Stamboliev, M. and Ivanov, A., The picking maturity, size and chemical composition of certain sweet and sour cherry varieties grown in north-west Bulgaria, *Gradinar. Lozar. Nauka*, 8(6), 11, 1971; *Hortic. Abstr.*, 42, 5452, 1972.

333. Kryńska, W. and Martyniak, B., Nutritional value of early cabbage and tomatoes grown on sloping grounds, *Roczniki Nauk Rolniczych, A*, 103(4), 79, 1978; *Hortic. Abstr.*, 50, 2613, 1980.

334. Alekseev, B. D., Chemical study of rosa L. of Dagestan, *Rastit. Resur.*, 17(4), 557, 1981; *Chem. Abstr.*, 96, 3666q, 1982.

335. Polle, A., Chakrabarti, K., and Pennenberg, H., Detoxification of peroxides in spruce tree needles (*Picea abies* L.) at the mountain site Wank, Calcareous Alps, *Chem. Abstr.*, 116, 261465c, 1992.

336. Lavygina, I. E., The content of biologically active substances in the aerial part of *Hypericum scabrum* L. and *H. Elongatum* Ledeb. (Uzbek SSR), *Rastitel'nye Resursy*, 24(4), 561, 1988; *Hortic. Abstr.*, 59, 4301, 1989.

337. Feteliyev, E. T., The effect of altitude on ascorbic acid content, *Sbornik Trudov Azer. NII Sadovodstva, Vinogradarstva Subtrop. Kul't.*, 8, 105, 1975; *Hortic. Abstr.*, 47, 1131, 1977.

338. Shuruba, G. A., The effect of ecological conditions on the content of biologically substances in apples in western Ukraine, *Nauk. Pratsi. Lviv. Sil's'kogospod. Institut*, 34, 102, 1972; *Hortic. Abstr.*, 43, 6600, 1973.

339. Moncada, B. J., Ríos-Castaño, D., and Torres, M. R., Citrus fruit quality in Colombia, in *Proc. Trop. Reg. Am. Soc. Hortic. Sci. Vol.* 12, 1969, 126; *Hortic. Sci.*, 40, 7124, 1970.

340. Tsovyan, Zh. V., Accumulation of ascorbic acid and the activity of oxidative enzymes in potato tubers grown under different conditions, *Biol. Zh. Arm.*, 20(1), 90, 1967; *Field Crop Abstr.*, 22, 2991, 1969.

341. Glazunova, E. M., Gachechiladze, N. D., Bondar, V. V., Korzinnikov, Yu. S., Potapova, I. M., and Gur'yanov, A. F., Biochemical fruit characteristics of *Hippophae rhamnoides* L. growing in the Western Pamirs, *Rast. Resursy*, 20(2), 232, 1984; *Hortic. Abstr.*, 54, 6135, 1984.

342. Čirkova-Georgievska, M., Peševska, V., Petrovska, V., and Vesova, N., The carotene content of some spinach (*Spinacia oleracia*) population in Macedonia, *Godišen Zbornik Zemjodelsko-Šumarskiot Fakultet Univ. Skopje, Zemjodelstvo*, 24, 65, 1970/1971; *Hortic. Abstr.*, 44, 1019, 1974.

343. Toul, V., The influence of site of the β-carotene content of carrots, *Bull Vyzk. Ust. Zelin., Olomouc*, 18, 49, 1974; *Hortic. Abstr.*, 46, 4736, 1976.

344. Låg, J., *Geomedicine*, CRC Press, Boca Raton, Fla., 1990.

Chapter 5

PLANTS' NUTRITIONAL STATUS AND VITAMIN CONTENT

I. USE OF MINERAL FERTILIZERS

The extensive use of mineral fertilizers in the industrialized nations during the last several decades has increased the yield per hectare of many food and forage crops in these countries.[1-3] The use of fertilizers has long been known to also change the chemical composition of plants and in some cases alter their resistance to various pests.[4-23] Effect of fertilizers on plant vitamins, however, has received relatively less attention, in particular from scientists in the Western countries. A major portion of reports on the effect of fertilizers on plant vitamins have appeared in the non-English journals and thus they seem to have gone mostly unnoticed by the English-speaking scientists. Here I will present the collected information available on the effect of mineral nutrients on plant vitamins.

A number of studies indicate that application of fertilizers to plants can change their vitamin content, in some cases by a considerable amount.[24-70] The majority of these experiments, however, have been conducted with a simultaneous addition of more than one macro- and/or micronutrients to the plants. Thus, based on the composite effect of several nutrients on some plant vitamins, one cannot gain insight into the single effect of each mineral nutrient on any given vitamin. Moreover, in some experiments, despite the very wide range of fertilizers used, no specific information is given (in the published reports or in the abstracts available) as to whether at some application rates plants showed signs of nutrient deficiency or toxicity. It is thus often hard to establish the relative changes in the plant's growth and its vitamin content as a result of soil fertilization.

Before we continue the discussion on the effect of nutrient supply and vitamins, one point needs to be clarified with respect to the summary of literature presented in Table 1. In some instances, plants subjected to nutrient deficiency were noted to have higher vitamin content than those sufficiently supplied with that nutrient.[54,57] For the sake of simplicity, these cases have been reported as if an increased supply of that nutrient would decrease the content of that vitamin in that plant.

In spite of some inconsistencies, the majority of reports are surprisingly consistent with respect to the effect of a given nutrient on a given vitamin and in a given plant. This can be appreciated if one remembers

TABLE 1

**Effect of Mineral Fertilization on the Concentration of Vitamins
in Fruits and Vegetables[a]**

Fertilizer	Effect[b]	Plant	Ref.
		Ascorbic acid	
N	Increase	Apples	71, 72
N	Increase	Cabbage	73–78
N	Increase	Cauliflower	75, 79
N	Increase	Cherry (sour)	80
N	Increase	Chili	81
N	Increase	*Chrysanthemum coronarium*	82
N	Increase	*Corchorus olitorius*	83
N	Increase	*C. olitorius*	84
N	Increase	Cucumber	85
N	Increase	Eggplant	86
N	Increase	Guava	87
N	Increase	Horse radish	88
N	Increase	Lettuce	89, 90
N	Increase	Mango	91, 92
N	Increase	Oats (leaf)	93, 94
N	Increase	Peaches	95
N	Increase	Pineapple	96
N	Increase	Potato	97–99
N	Increase	Spinach	100, 101
N	Increase	Sweet potato	102
N	Increase	Tomato	75, 103–108
N	Increase	Kohlrabi (leaf)	108
N	No effect	Cabbage	109, 110
N	No effect	Carrots	111
N	No effect	Cauliflower	112
N	No effect	Cherry (sour)	113
N	No effect	Currant (black)	114, 127
N	No effect	Grapefruit	115
N	No effect	Oranges	116, 117
N	No effect	Pineapple	118
N	No effect	Potato	119, 120
N	No effect	Radish	121
N	No effect	Satsuma	122
N	No effect	Sweet potato	123
N	No effect	Tomato	124
N	No effect	Turnip greens	125, 126
N	Decrease	*Amararanthus tristis*	128
N	Decrease	Apples	129, 130
N	Decrease	Brussels sprouts	110, 131
N	Decrease	Cabbage	131, 132, 134, 135
N	Decrease	Cabbage (Chinese)	133
N	Decrease	Cantaloupe	136
N	Decrease	Carrots	137

TABLE 1 (continued)

Fertilizer	Effect[b]	Plant	Ref.
		Ascorbic acid	
N	Decrease	Cauliflower	90, 138
N	Decrease	Chard	110
N	Decrease	Chard (Swiss)	139, 140
N	Decrease	Chili (green)	141
N	Decrease	Cucumber	142
N	Decrease	Endive	143
N	Decrease	Kale	110, 131, 139
N	Decrease	Grapefruit	144–148
N	Decrease	Leeks	149
N	Decrease	Lemon	150–152
N	Decrease	Lime (Kagzi)	153
N	Decrease	Lime (Persian)	154, 155
N	Decrease	Mandarin	156–158
N	Decrease	Onion	53, 56
N	Decrease	Oranges	146, 159–163
N	Decrease	Peaches	164
N	Decrease	Pepper (green)	136
N	Decrease	Pomegranate	165
N	Decrease	Potato	136, 166–175
N	Decrease	Radish	176, 177
N	Decrease	Raspberry	178
N	Decrease	Spinach	140, 179, 181–183
N	Decrease	Spinach (New Zealand)	180
N	Decrease	Stock beet	131
N	Decrease	Strawberry	184, 185
N	Decrease	Tomato	90, 130, 186–199
N	Decrease	Turnip greens	126
P	Increase	Cabbage	76, 200
P	Increase	Cauliflower	75
P	Increase	Chili (green)	141
P	Increase	Cotton (leaf)	201
P	Increase	Cucumber	142
P	Increase	Peanut	202
P	Increase	Pepper	203
P	Increase	Potato	97, 173, 204
P	Increase	Radish	121, 177
P	Increase	Spinach	205
P	Increase	Sweet potato	206
P	Increase	Tomato	104, 106, 107, 195
P	No effect	Cabbage	134
P	No effect	Currant (black)	114, 127
P	No effect	Lettuce	98
P	No effect	Mandarin	207
P	No effect	Potato	119, 166

TABLE 1   (continued)

Fertilizer	Effect[b]	Plant	Ref.
		Ascorbic acid	
P	No effect	Several vegetables	131
P	No effect	Tomato	190
P	Decrease	Citrus	209
P	Decrease	Grape (leaf)	210
P	Decrease	Horseradish	88
P	Decrease	Lemon	151
P	Decrease	Onion	53, 54, 57
P	Decrease	Oranges	146, 160, 212
P	Decrease	Mandarin	158, 211
P	Decrease	Pepper	203
P	Decrease	Spinach	100
P	Decrease	Stock beet	131
P	Decrease	Swiss chard	208
K	Increase	Bananas	213, 214
K	Increase	Brussels sprouts	131
K	Increase	Cabbage	75, 76, 78, 101, 131, 134, 215–219
K	Increase	*Capsicum*	220, 221
K	Increase	Carrots	216, 222
K	Increase	Cauliflower	75, 90
K	Increase	Celeriac	222
K	Increase	Citrus	147, 209
K	Increase	*Cucurbita pepo*	223
K	Increase	Cucumber	224
K	Increase	Grapefruit	115, 226
K	Increase	Guava	225
K	Increase	Kohlrabi	216
K	Increase	Leeks	149
K	Increase	Lemon	151, 227, 228
K	Increase	Mandarin	158, 229
K	Increase	Onion	53, 55, 230
K	Increase	Oranges	146, 212, 228, 231–233
K	Increase	Peaches	95
K	Increase	Potato	216, 234
K	Increase	Radish (horse)	88
K	Increase	Radish	121, 177
K	Increase	Rice (leaf)	235
K	Increase	Spinach	90, 216, 236
K	Increase	Stock beet	131
K	Increase	Tomato	35, 103, 195, 203, 216, 237–240
K	No effect	Currant (black)	114, 127
K	No effect	Grape (leaf)	210
K	No effect	Lettuce	89, 241, 242
K	No effect	Pepper	221
K	No effect	Sweet potato	123
K	Decrease	Currant (black)	243
K	Decrease	Oranges	161

TABLE 1 (continued)

Fertilizer	Effect[b]	Plant	Ref.
		Ascorbic acid	
K	Decrease	Potato	97, 166, 170, 173
K	Decrease	Sweet potato	102
K	Decrease	Swiss chard	208
K	Decrease	Turnip greens	244
Ca	Increase	Apples	245, 246
Ca	Increase	Oro (*Antiaris africana*)	247
Ca	Increase	Potato	248
Mg	Increase	Cauliflower	75
Mg	Increase	Currant (black)	249
Mg	Increase	Peaches	95
Mg	Increase	Pineapple	250
Mg	Increase	Potato	90, 251
Mg	Increase	Spinach	108, 205, 216
Mg	Increase	Tomato	35
Mg	No effect	Acerola	252
Mg	No effect	Oranges	161
Mg	No effect	Potato	253
Mg	Decrease	Beans (leaf)	254
Mg	Decrease	Lettuce	255
Mg	Decrease	Mandarin	256
Mg	Decrease	Oats (leaf)	94
Mg	Decrease	Spinach	131
Mn	Increase	Apples	257, 258
Mn	Increase	Cabbage	259, 260
Mn	Increase	Carrots	261
Mn	Increase	Cucumber	262
Mn	Increase	Currant (black)	249, 263–265
Mn	Increase	Lettuce	266
Mn	Increase	*Rumex tianschanicus*	267
Mn	Increase	Spinach	263
Mn	Increase	Strawberry	268
Mn	Increase	Tomato	191, 240, 260 263, 269–272
Mn	Increase	Tomato (leaf)	273
Mn	No effect	Beans (leaf)	274
Mn	No effect	Mandarin	275
Mn	No effect	Soybean (leaf)	274, 276
Mn	No effect	Tomato (leaf)	277
Mn	No effect	Tomato	278, 279
Mn	Decrease	Mandarin	280
Mn	Decrease	Potato	281
B	Increase	Apples	258, 282, 283
B	Increase	Beet (leaf)	278
B	Increase	Cabbage	260
B	Increase	Carrots	284, 285
B	Increase	Cauliflower	286

TABLE 1   (continued)

Fertilizer	Effect[b]	Plant	Ref.
		Ascorbic acid	
B	Increase	Cucumber	262, 287
B	Increase	Currant (black)	288, 289
B	Increase	Eggplants	290
B	Increase	*Hippophae rhamnoides*	291
B	Increase	Jerusalem artichoke tuber	292
B	Increase	Potato	293
B	Increase	Pumpkin	294
B	Increase	Radish	295
B	Increase	Strawberry	296
B	Increase	Tomato	296
B	Increase	Tomato	191, 260, 297–299
B	No effect	*Capsicum*	300
B	No effect	Potato	281
B	No effect	Tomato	279, 301
B	Decrease	Mandarin	280
B	Decrease	Mango	302
B	Decrease	Potato	285
B	Decrease	Spinach	131
Fe	Increase	Mandarin	275
Fe	Increase	Oranges	303
Fe	Increase	Potato	304
Fe	Increase	Tomato	305
Fe	No effect	*Capsicum*	300
Fe	No effect	Rice (leaf)	306
Fe	No effect	Mandarin	275
Fe	Decrease	Oranges	212
Fe	Decrease	Pineapple	307
Fe	Decrease	Tomato	308
Mo	Increase	Beans (leaf)	309, 310
Mo	Increase	Cabbage	311, 312
Mo	Increase	Cauliflower	138
Mo	Increase	Clover (leaf)	313
Mo	Increase	Currant (black)	288
Mo	Increase	Lettuce (*romana*)	314
Mo	Increase	Mandarin oranges	315
Mo	Increase	Potato	281, 316
Mo	Increase	Spinach	309
Mo	Increase	Sunflower	317
Mo	Increase	Vegetables	318
Mo	No effect	Lettuce (*capitata*)	314
Cu	Increase	Aubergine	319
Cu	Increase	Barley (leaf)	320
Cu	Increase	Cabbage	259, 321
Cu	Increase	Carrots	284
Cu	Increase	Cauliflower	322
Cu	Increase	Currant (black)	264
Cu	Increase	Lettuce	321

TABLE 1   (continued)

Fertilizer	Effect[b]	Plant	Ref.
		Ascorbic acid	
Cu	Increase	Oats (leaf)	320
Cu	Increase	Oranges	323
Cu	Increase	Spinach	321
Cu	Increase	Tomato	305
Cu	No effect	Mandarin	275
Cu	No effect	Tomato fruit	320
Cu	Decrease	*Celosia argentea* (leaf)	324
Cu	Decrease	*Amaranthus dubius* (leaf)	324
Zn	Increase	Alfalfa	325
Zn + Fe	Increase	Apples	326
Zn	Increase	Cabbage	259
Zn	Increase	*Capsicum*	327
Zn	Increase	Carrots	261
Zn	Increase	Cauliflower	286
Zn	Increase	Chickpea	328
Zn	Increase	Corn	329
Zn	Increase	Cucumber	262
Zn	Increase	Cucumber (leaf)	330
Zn	Increase	Currant (black)	249, 289
Zn	Increase	*Hippophae rhamnoides*	291
Zn	Increase	Mandarin	331
Zn	Increase	Guava	332
Zn + Cu	Increase	Oranges	323, 333
Zn + B	Increase	Peas	334
Zn	Increase	Potato	281
Zn	Increase	Rice (leaf)	306
Zn	Increase	Rowan (*Sorbus*)	335
Zn	Increase	Sunflower	329
Zn	Increase	Tomato	305, 336–339
Zn	No effect	*Capsicum*	300
Zn	No effect	Mandarin	275
Zn	No effect	Oranges	117
Zn	Decrease	Aubergine	319
Zn	Decrease	Grape (leaf)	210
Zn	Decrease	Tomato	272
Co	Increase	Apples	257
Co	Increase	Cabbage	311
Co	Increase	*H. rhamnoides*	291
Co	Increase	Maize (leaf)	340
Co	Increase	Potato	281, 341
I	Increase	*H. rhamnoides*	291
I	Increase	Potato	281
I	Decrease	Tomato	342
Ni	Increase	Alfalfa	343
S	Increase	Cabbage	75
S	Increase	Cauliflower	75
S	Increase	Chili	344

TABLE 1   (continued)

Fertilizer	Effect[b]	Plant	Ref.
		Ascorbic acid	
S	Increase	Potato	345
S	Increase	Tomato	75
S	No effect	Peanut	202
V	Increase	Beans (leaf)	310
Clinoptilolite (zeolite)	Increase	Carrots	346
Clinoptilolite	Increase	Potato	347
Clinoptilolite	Increase	Red pepper	348
Zeolite + manure	Increase	Cabbage	349
Zeolite + manure	Increase	Carrot	349
Zeolite + manure	Increase	Tomato	349
		Biotin	
N	Increase	Oats kernel	393
P	Increase	Oats (leaf)	390
		Carotene[c]	
N	Increase	*Amaranthus tristis*	128
N	Increase	Bermuda grass	350
N	Increase	Cabbage	179
N	Increase	*Capsicum*	351
N	Increase	Carrots	101, 111, 352–356
N	Increase	Celeriac	222
N	Increase	Collards	357
N	Increase	Grass (*lolium* sp.)	358
N	Increase	Lettuce	358, 359
N	Increase	Maize (leaf)	358
N	Increase	Parsley	108
N	Increase	Rye (leaf)	358
N	Increase	Spinach	108, 216, 360
N	Increase	Stock beet	179
N	Increase	Sweet potato	206
N	Increase	Swiss chard	208
N	Increase	Spinach	100, 182, 216, 361
N	Increase	Tomato	362
N	Increase	Turnip greens	126
N	No effect	Sweet potato	123
N	No effect	Carrots	363
N	No effect	Lettuce	364
P	Increase	Carrots	365
P	No effect	Carrots	352

TABLE 1   (continued)

Fertilizer	Effect[b]	Plant	Ref.
		Carotene[c]	
P	No effect	Sweet potato	366
P	Decrease	Carrots	363
P	Decrease	Spinach	100, 182
K	Increase	*Capsicum*	221
K	Increase	Carrots	365
K	Increase	Lettuce	359
K	Increase	Tomato	238
K	No effect	Carrots	352, 354, 363
K	No effect	Sweet potato	123, 366
K	Decrease	Spinach	182
Mg	Increase	Carrots	358, 363, 367
Mg	Increase	Spinach	108, 216
Mg	Increase	Swiss chard	208
Ca	Decrease	Carrots	363
Mn	Increase	Carrots	261
Mn	Increase	Chokeberry (black)	368
Mn	Increase	Corn (leaf)	369
Mn	Increase	*H. rhamnoides*	291
Mn	Increase	Oats (leaf)	369
Mn	Increase	Soybean (leaf)	369
Mn	Increase	Wheat (leaf)	369
Mn	No effect	Lettuce	266
Cu	Increase	Barley (leaf)	320
Cu	Increase	Carrots	320
Cu	Increase	Cauliflower	322
Cu	Increase	*H. rhamnoides*	291
Cu	Increase	Oats (leaf)	320
Cu	Increase	Spinach	320
Cu	Increase	Wheat (leaf)	320
Zn	Increase	Alfalfa	325
Zn	Increase	Carrots	261, 367
Zn	Increase	Chokeberry (black)	368
Zn	Increase	Corn (leaf)	329
Zn	Increase	Cotton (leaf)	329
Zn	Increase	Peas	370
B	Increase	Carrots	367, 371
B	Increase	Jerusalem artichoke leaf	292
B	Increase	Pumpkin	294
B	Increase	Tomato	372
Mo	Increase	Beans (leaf)	310
Mo	Increase	Lettuce (*Romana*)	314
Mo	No effect	Lettuce (*Capitata*)	314
Ni	Increase	Alfalfa	343
Co	Increase	Timothy grass	341
		Folic acid	
N	Increase	Tea (leaf)	381

TABLE 1   (continued)

Fertilizer	Effect[b]	Plant	Ref.
		Niacin	
N	Increase	Brown rice	389
N	Increase	Potato	168
N	Increase	Spinach	216, 378
N	Increase	Wheat	376
P	Increase	Oats (leaf)	390
		Pantothenic acid	
N	Increase	Oats kernel	393
P	Increase	Oats (leaf)	390
		Riboflavin	
N	No effect	Maize kernel	374
N	No effect	Sweet potato	391
N	No effect	Wheat kernel	375
N	Increase	Collards	357
N	Increase	Oats (leaf)	94
N	Increase	Spinach	182, 216, 361, 378–380
N	Increase	Tea (leaf)	381
N	Increase	Turnip greens	126
N	Decrease	Brown rice	389
B	Increase	Beet	278
B	Increase	Cucumber	262, 387
B	Increase	Tomato	278
Mn	Increase	Beet	278
Mn	Increase	Cucumber	262
Mn	Increase	Tomato	278
Cu	Decrease	Barley	386
Mo	Decrease	Barley	386
Zn	Increase	Cucumber	262
		Thiamin	
N	Increase	Barley kernel	373
N	Increase	Cabbage	78
N	Increase	Collards	357
N	Increase	Maize kernel	374, 375
N	Increase	Oats kernel	216, 375, 376
N	Increase	Peas	377
N	Increase	Potato	167
N	Increase	Spinach	182, 216, 361, 378–380
N	Increase	Tea (leaf)	381
N	Increase	Turnip greens	126
N	Increase	Wheat kernel	374, 375, 382–385

TABLE 1 (continued)

Fertilizer	Effect[b]	Plant	Ref.
		Thiamin	
P	Increase	Oats	375
P	Increase	Peas	375
P	Increase	Spinach	216
P	Increase	Wheat kernel	375, 376
K	Decrease	Millet kernel	375
K	Decrease	Wheat kernel	376
K	Decrease	Maize kernel	376
K	Decrease	Lettuce	374
K	Decrease	Peas	375
B	Increase	Barley	386
B	Increase	Cucumber	262, 387
Mn	Increase	Cucumber	262
Zn	Increase	Cucumber	262
		Tocopherol	
N	Decrease	Grape (leaf)	210
N	Decrease	Sunflower oil	392
N	Decrease	Spinach	182
P	Decrease	Grape (leaf)	210
K	No effect	Grape (leaf)	210
K	Decrease	Spinach	182
Mn	Increase	Soybean (leaf)	369
Zn	Increase	Grape (leaf)	210

[a] Edible parts of plants unless otherwise indicated.

[b] In designating the effect of nutrients, the range of nutrient applications considered here is mostly from sufficiency to luxury consumption. In the cases, however, where deficiency of a nutrient, for example, was reported to decrease the concentration of a given vitamin, for the sake of this table, the reverse was used; i.e., increasing the supply of that nutrient was set to increase the concentration of that particular vitamin, and vice versa.

[c] In some cases the original article is about vitamin A activity.

that the information in Table 1 encompasses a very diverse number of plants, plant varieties, experimental conditions, and the laboratories where the investigations were carried out. With this in mind, in the following discussion a selected number of those reports for which relatively more detailed information was available are reviewed.

A. Ascorbic Acid

1. Effect of N Fertilizers

Increased application of nitrogen fertilizers has been shown by numerous investigators to decrease the content of ascorbic acid in many different fruits and vegetables. There are also reports indicating that nitrogen fertilizer may have no effect on or may even increase the content of vitamin C in some plants (Table 1). Part of the controversy may be due to (a) the amount of nitrogen fertilizer used by different authors and (b) the fact that the response of ascorbic acid to increased application of fertilizers may go through a maxima and then decrease if the application rate is increased beyond a given value. This has been noted for the effects of nitrogen,[86, 99, 100, 108] phosphorus,[100] and potassium[100] fertilizers on the ascorbic acid content and the effects of potassium,[100, 358] calcium,[358] and sodium[358] on the carotene content in plants.

Another source of complications may be the fact that fertilizers may interact with climatic conditions, and thus the effect of a given fertilizer on plant vitamins may differ from year to year. For example, in New Zealand, nitrogen fertilization decreased the ascorbic acid content of strawberries in one year but did not have any effect on this vitamin in another year.[185] For further discussions on the yearly fluctuations in plant vitamins see chapter 4.

Åberg and Ekdahl,[394] based on a very extensive study with numerous vegetables, concluded that (a) in the suboptimal range, i.e., the range where plant yield increases as the nitrogen supply increases, nitrogen fertilization may slightly decrease the ascorbic acid content of plant leaves, and (b) in the supraoptimum range, i.e., the range where plant yield decreases as a result of increased nitrogen supply, the ascorbic acid may show an increase with increased nitrogen application rates. This pioneering observation seem to indicate that the ascorbic acid synthesis and plant growth are affected by different extent by the nitrogen fertilization, so that a relative dilution or accumulation of ascorbic acid may take place in the plant tissues and thus the decrease or increase observed by different authors. Another reason for the decreased ascorbic acid by nitrogen fertilization may be the well-known stimulating effect of this fertilizer on the vegetative growth of plants. Increased foliage could then reduce the intensity and composition of light reaching other plant tissues (leaves or fruits), and as a result, a reduction in the ascorbic acid content in the shaded plant parts could be expected.[126] One may postulate that this mechanism of reduction in ascorbic acid may be more relevant under greenhouse than under field conditions.

Numerous reports indicate that increased use of nitrogen fertilizers decreases the ascorbic acid content in two important vegetables, namely potatoes and tomatoes (Table 1), the extent of which appears to be relatively large (Figure 1). Since potatoes and tomatoes are among the major sources of vitamin C in human nutrition, the long-term effects of heavy use of nitrogen fertilizers in producing these crops on the vitamin C supply of humans need to be investigated.

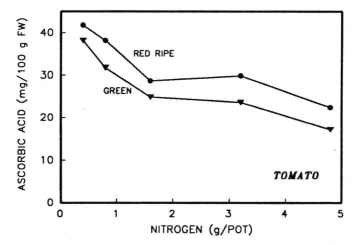

FIGURE 1. Adverse effect of increased nitrogen fertilizer on the ascorbic acid concentration in immature and red ripe tomatoes.[187]

Nitrogen fertilizers may also reduce the ascorbic acid concentration in leafy vegetables such as different *Brassica* species (Brussels sprouts, cauliflower, and cabbage), chard, and spinach. The total amount of ascorbic acid in the plant, however, may increase since nitrogen fertilization usually increases plant yield (Figure 2).[140] Sengewald[100] noted that too high or too low a nitrogen supply to spinach may decrease the total ascorbic acid produced per plant (Table 2), which was interpreted as the effect of nitrogen on the allocation of available carbohydrate for vitamin synthesis. In cabbage, another important vegetable in many countries, Sørensen[134] noted that applying 600 kg N/ha increased the weight of each cabbage head by almost threefold (535.6 vs. 1836.5 g/head) but reduced the ascorbic acid content by about 34% (71.0 vs. 46.8 mg/100 g in the 0 and 600 kg N/ha, respectively).

Nitrogen fertilization and the kind of soil in which plants are grown may also affect the amount of ascorbic acid in the fruits and vegetables long after they have been harvested, i.e., during their storage or processing. For example, Paschold and Scheunemann[395] reported that although very high applications of N did not always reduce the ascorbic acid concentration in white cabbage, depending on the kind of soil on which the cabbage was grown, however, the concentration of ascorbic acid in the sauerkraut produced from these cabbages was differently affected by the N fertilization. In oranges, not only did high nitrogen fertilization reduce the content of ascorbic acid in the fresh fruits, fruits from the fertilized trees also lost their ascorbic acid much faster during the subsequent storage![162] In apples, foliar application of urea did not affect the vitamin C content in the fresh fruits. After harvest, however,

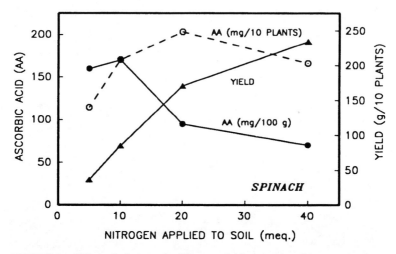

FIGURE 2.   Effect of nitrogen fertilizer on yield, ascorbic acid concentration, and total ascorbic acid in spinach plants. Data are from tables 1 and 2 of Wittwer, S. H., Schroeder, R. A., and Albrecht, W. A., *Soil Sci.*, 59, 329, 1945.

TABLE 2
Effect of N, P, and K Fertilization on the Growth, Carotene, and Ascorbic Acid (AA) Concentration in Spinach[a]

Fertilizer (g / pot)	Growth (g DW / plant)	Vitamin content			
		Carotene	AA	Carotene	AA
		(µg / plant)		(mg / 100 g DW)	
NH$_4$NO$_3$					
0.2	72[b]	56	281	77	390
1	117	110	462	93	395
2	114	118	340	103	298
P$_2$O$_5$					
0.2	57	47	263	82	461
1	95	67	441	70	464
3	110	74	485	67	441
K$_2$O					
0.2	58	23	244	39	421
1	92	42	474	45	515
3	109	38	428	34	393

[a] Data of first harvest of Sengewald, E., *Nahrung*, 3, 428, 1959.
[b] Values rounded.

fruits from urea-treated trees lost their vitamin C more rapidly than those from untreated trees.[396]

a. Opposite Effects on NO₃ and Vitamin C

Heavy use of nitrogen fertilizers on plants, along with decreasing their vitamin C content, may also increase the concentration of NO_3 in their edible parts, sometimes to potentially dangerous levels.[397, 398] For example, in iceberg lettuce a reverse relationship was found between the NO_3 and ascorbic acid content of the leaves.[399] The outer leaves, which are normally discarded, had the highest ascorbic acid and lowest concentration of NO_3, but the innermost leaves, normally consumed, were found to be highest in NO_3 and lowest in ascorbic acid (see Figure 9).

What are the possibilities for reducing the NO_3 content without much reduction in the vitamin C content in plants? One possible method might be by subjecting the plants to nitrogen fertilizer depravation, especially toward the end of their growing period, a method that could reduce the nitrate content and simultaneously increase the ascorbic acid in plants. This method, which can only be used when plants are grown in media where their nutrient supply can be controlled (NFT, hydroponics, etc.), has been tested by some investigators. For example, in Japan, Shinohara and Suzuki[400] noted a negative effect of high leaf NO_3 concentration on the ascorbic acid content of lettuce and Japanese honewort and reported that under hydroponic conditions, transferring the plants two to seven days before their harvest to solutions low in nitrate reduced the concentration of NO_3 in their leaves and simultaneously increased their content of ascorbic acid.

Watanabe et al.[401] tested the effect of intermittent supply of nutrients in order to reduce the NO_3 and increase the ascorbic acid in spinach. They noted that subjecting the plants for three days to cycles of 15 minutes of nutrient supply (on cycle) followed by 120 minutes of nutrient deprivations (off cycle) one can reduce the NO_3 content and at the same time increase the ascorbic acid in the spinach leaves without affecting their yield. Other investigators have noted that subjecting the lettuce grown with the NFT method to NPK deficiency during the final 16 days of its growth during the summer months can increase the ascorbic acid in the leaves but it may also reduce the yield.[402] These are important findings with possible immediate application possibilities for greenhouse-grown plants raised with various nutrient solution techniques and deserve serious attention on the part of scientists and commercial growers to optimize the technique so that vegetables with low NO_3 and high vitamin-C could be produced without much decrease in yield.

TABLE 3

Effects of NO_3 or NH_4 Forms of Nitrogen Fertilizer on the Ascorbic Acid Concentration in the Leaves of Plant Seedlings[309]

	Ascorbic acid (mg / 100 g DW)		
Treatment	Beans	Sunflower	Maize
NO_3	1004	648	703
NH_4	550	471	629

A further incentive for reducing the NO_3 and increasing the ascorbic acid content in fruits and vegetables should be the ever-increasing number of reports that ascorbic acid may reduce the conversion of NO_3, to NO_2 and thus reduce the formation of cancer-forming *N*-nitroso compounds in humans.[403, 404] It is of interest that heavy use of nitrogen fertilization appears to have a double negative effect on the nutritional quality of fruits and vegetables since it increases their nitrate and at the same time lowers their ascorbic acid content. Thus any attempt, such as the recommendation made above, is strongly recommended for improving the quality of foods we consume. For a discussion on the relative effects of organic and inorganic fertilizers on the NO_3 and vitamin C content of plants see page 196.

b. Kind of N Fertilizers

Numerous reports show that plants supplied with the NH_4 form of nitrogen fertilizers contain less ascorbic acid than the same plants supplied with the NO_3 form.[4, 131, 405-408] Even mixture of $NH_4 + NO_3$ fertilizers resulted in lower ascorbic acid in tomatoes than when NO_3 was given alone.[409] The negative effect of NH_4 on ascorbic acid content, however, strongly depends on the kind of plant under study (Table 3).

The mechanism by which NH_4 salts, as compared with NO_3 salts, lower the ascorbic acid content of plants is controversial, and the nature of counter-ions may play be important in this respect. Mapson and Cruickshank,[406] for example, noted that in cress seedlings the ascorbic acid content was reduced much greater when the counteranions in the NH_4 salts used were SO_4, Cl, or HPO_4. Ammonium salts of HCO_3, NO_3, or CH_3COO, however, did not influence the ascorbic acid content of the seedlings as compared with the control plants. Somers and Kelly,[407] by suspending leaf discs of turnip and broccoli in solutions containing various concentrations of several salts, showed that although NH_4 salts in general depressed the accumulation of ascorbic acid, the chloride and sulfate salts of NH_4 reduced the ascorbic acid accumulation more than the nitrate salt (NH_4NO_3) did. This may suggest that the rate of uptake of the anion part of the salts may affect the ascorbic acid similar to the effect on other organic acids in plants. According to some authors, the adverse effect of $(NH4)_2SO_4$ on ascorbic acid content in plants is due to the effect of this salt on vitamin C synthesis and not to an accelerated catabolism of this vitamin in the plant tissues.[407]

The relative effect of NO_3 and NH_4 on the ascorbic acid may also depend on the amount of fertilizer used. For example, fertilization of a Nigerian vegetable (*Corchorus olitorius*) with nitrogen at the rate of 60 kg/ha in the form of NO_3 or NH_4 was noted to increase the ascorbic acid concentration in the leaves. The NO_3 fertilizer, however, increased the ascorbic acid somewhat more than the NH_4 fertilizer did. Further increase in the N application rate to 120 kg/ha reduced the ascorbic acid content if the nitrogen was applied in the form of NO_3 but continued to increase the ascorbic acid if the form of nitrogen was NH_4. As a result of this, at the rate of 120 kg N/ha, the NO_3-fertilized plants had less ascorbic acid than the NH_4-fertilized plants, a situation that is the reverse of the situation at lower (60 kg/ha) N application rate.[83]

2. Effect of P Fertilizers

Reports on the effect of phosphorus fertilizer on the ascorbic acid content of plants seem to strongly vary with the kind of plants under consideration and appear to be inconsistent. There are about the same number of reports indicating that phosphorus fertilizer may have no effect, decrease, or increase the plant's ascorbic acid content (Table 1). Somers and Kelly[407] reported that increasing the concentration of H_3PO_4 as a single anion source and in the presence of a balanced mixture of Ca, Mg, K, and NH_4 caused a marked decrease in the rate of ascorbic acid accumulation in leaves of turnip and broccoli. They suggested that the cation/anion balance in plant cells may be the determining factor for the amount of ascorbic acid accumulated in them. Pankov and Pavlova[57] noted that a slight P deficiency might increase while a severe deficiency decreases the ascorbic acid content in onions. Also, Kanesiro et al.[411] reported that tomatoes grown in soil contained the highest ascorbic acid when plants were grown with no added nitrogen and low phosphorus supply. Phosphorus deficiency at the beginning of growth, however, was noted to decrease the ascorbic acid content of tomatoes.[106]

3. Effect of K Fertilizers

Potassium fertilization has been found to increase the ascorbic acid content in many different plants grown under very different conditions

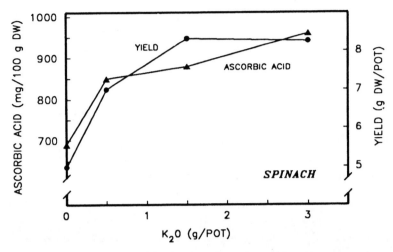

FIGURE 3. Effect of K fertilization on yield and ascorbic acid concentration in spinach. Data are from the 2 g N treatment of Pfützer, G., Pfaff, C., and Roth, H., *Landw. Forsch.*, 4(2), 105, 1952.

(Table 1). The positive effect of K on the ascorbic acid content of important vitamin C sources such as *Brassica* crops, citrus fruits, and tomatoes needs to be emphasized. Some reports indicate that the response of plants' ascorbic acid to potassium fertilizers may go through a maximum, and thus plants subjected to too little or too much K fertilizer may contain less ascorbic acid than those supplied with an optimum amount (Table 2). Other observations, however, show that potassium fertilizer may increase the ascorbic acid concentration in plants even at those high applications rates where it does not affect the yields any longer (Figure 3).[216]

Scharrer and Werner[131] studied the effect of NPK fertilizer on the ascorbic acid content of spinach, kale, winter endive, carrots, Brussels sprouts, stock beet, and mustard and noted that among the three nutrients tested, only K gave a consistent positive effect on the ascorbic acid content of these plants. According to these authors, although the highest level of potassium used was higher than what the plants needed for optimum growth, it still increased the ascorbic acid content of the plants.

The positive effect of K fertilizer on the ascorbic acid content of plants may be much smaller than the variations that might occur in the ascorbic acid content of plants, depending on the geographical locations where they are grown (chapter 4). For example, Reder et al.,[244] based on their comprehensive experiments, reported that although K fertilization resulted in some significant changes in the ascorbic acid content of turnip greens, the overall effect of location where plants were grown was 13.75 times higher than the most important effect of K.

4. Effect of Other Nutrients

Data summarized in Table 1 show that compared with N, P, and K, far fewer research reports are available on the effect of other macro- and micronutrients on the vitamin C content of plants. With some exceptions, increasing the supply of most other nutrients seems to increase the concentration of ascorbic acid in a wide range of crops and under a wide range of experimental conditions. Here we will review some selected cases.

An increased supply of Ca seems to increase the ascorbic acid content in several plants in all cases known to us (Table 1). This is in agreement with the view that the internal concentration of calcium in some plants seems to be related to their ascorbic acid content. In *Capsicum*, for example, a significant and positive correlation was found between the Ca and ascorbic acid content.[412] Pre- or postharvest application of Ca was noted to increase the ascorbic acid concentration in tomatoes,[246] apples, and pears.[413] Furthermore, treatment of apples and pears with an 8% solution of $CaCl_2$ was noted to increase the Ca content of the fruits and at the same time increased the resistance of these fruits to fungal rots caused by *Penicillium expansum*, *Botrytis cinerea*, and *Cladosporium macrocarpum* when compared with fruits and low Ca content (usually the untreated ones). Fruits treated with Ca solution also showed a lower rate of ascorbic acid loss during the postharvest period.[413,414] For a discussion on the possible relationships between the vitamin content of plants and their resistance or suscepti- bility to various pathogenic organisms, see chapter 8.

Deficiencies of certain micronutrients are known to reduce the ascorbic acid content in some plants. In the Florida citrus fruits, for example, application of Zn, Mg, Mn, or Cu to sandy soils to correct their deficiency was noted to increase the ascorbic acid content in the fruits. Supplying the trees with these elements above the levels required for normal growth, however, did not increase the ascorbic acid content of the fruits any further.[156] Also, in Egypt, spraying Valencia oranges with zinc sulfate was noted to increase the ascorbic acid content of the fruit juice.[415]

Watson and Noggle[94] reported that when oats previously grown for three weeks under sufficient supply of nutrients were subjected to a three-week deficiency of various nutrients, the leaves of the plants subjected to K and Mg deficiency showed a strikingly higher concentra- tion of ascorbic acid. Mg deficiency, however, reduced the concentra- tion of riboflavin in the leaves (Table 4). In contrast, addition of Mg to

TABLE 4
Effect of Single Deficiency of Various Mineral Nutrients on the Growth and Concentration of Vitamins in the Leaves of Oats[94]

Treatment	Growth (mg DW / plant)	Ascorbic acid (mg / 100 g DW)	Riboflavin (μg / g DW)
Control[a]	267	440	21.5
−K[b]	243	635	20.7
−Ca	248	356	17.5
−Mg	229	814	14.8
−N	119	369	10.6
−S	193	533	19.3
−P	205	488	18.7

[a] Control plants were grown in complete nutrient.
[b] Plants were grown for 21 days in an otherwise complete nutrient solution that lacked one mineral nutrient.

the germinating soybean seeds was noted to increase the ascorbic acid concentration of cotyledons significantly.[416] Dube and Misra[75] noted that deficiencies of N, P, K, Ca, Mg, and S all decreased the concentration of ascorbic acid in cabbage, cauliflower, and tomato; the effects were, however, different in different plants.

Although several authors have noted an inconsistent effect of boron on the ascorbic acid content of plants, the majority of reports, however, indicate that increased boron supply can increase the ascorbic acid content in many different plants (Table 1). It is not certain whether plants in which boron fertilization increased their ascorbic acid content originally had boron deficiency. Based on some reports, it appears that there is no clear relationship between the boron content in the plant tissues and their ascorbic acid content. In tomatoes, for example, boron deficiency was noted to strongly reduce the concentration of boron in the leaves (from 92 to 8 μg/g DW) and to reduce the number of fruits (from 9.7 to 0.2 fruits per vine). It did not, however, affect the ascorbic acid concentration in the few fruits that still grew on each vine (27 and 26 mg/100 g FW of well-nourished and boron-deficient fruits, respectively).[301]

Response of a plant's vitamin C to some micronutrient fertilizers is noteworthy. For example, Krupyshev[289] noted that spraying black currants twice with boron not only increased the fruit's sugar content, but also increased its ascorbic acid content by more than twofold (82.8 vs. 170.4 mg/100 g in the control and treated fruits, respectively). Also, spray application of Zn was noted to increase the fruit's ascorbic acid to

TABLE 5
Effect of Injecting Molybdenum-Deficient Plants with
Molybdenum on the Ascorbic Acid Content of Their Leaves[417]

Plant	Ascorbic acid (mg / 100 g)	
	– Mo	+ Mo
Sugar beet	26[a]	75
Barley	57	139
Tomato	56	121
Brussels sprouts	49	155
Alaska clover	92	209
Cauliflower	66	264

[a] Values rounded.

165.5 mg/100 g. Whether the untreated plants had shown any B or Zn deficiency symptoms was unfortunately not specified.

Lyon et al.[308] grew tomatoes in nutrient solutions deficient in Cu, Mn, Zn, Fe, or Mo and reported that although withholding of any one nutrient (with the exception of Mo) decreased growth and fruitfulness, iron was the only nutrient whose deficiency actually increased the ascorbic acid content of tomato fruits by a small but significant degree. Iron deficiency was also noted to increase the ascorbic acid in pineapple.[307]

In Australia, addition of Zn to the soils where plants had previously shown signs of Zn deficiency increased the concentration of ascorbic acid in herbage legumes (barrel medic, subterranean clover, and strawberry clover) but only at certain harvests and under winter conditions of comparatively low light intensity.[241]

Application of Cu to Cu-deficient muck soils increased growth of barley, oats, and fall-grown spinach and increased the ascorbic acid content in the leaves of both plants. Cu application, however, did not affect the ascorbic acid content in the spring-grown spinach and in tomato fruit.[320]

Although addition of Mo to soil did not affect the ascorbic acid content of lettuce and peas,[241] injection (through the petiole) of Mo-deficient plants with Mo was found to markedly increase the ascorbic acid concentration in several vegetables within a 24-h period (Table 5). Also, increasing the Mo concentration in the nutrient solution was noted to increase the ascorbic acid in the leaves and stem of cauliflower.[138] Treating the seeds of leek, spinach, and lettuce prior to planting with solutions (0.1–0.1%) of $KMnO_4$ or $CuSO_4$, or H_3BO_3 was noted to increase their ascorbic acid content by up to 50%.[418]

Mineral elements may also affect the ascorbic acid content of plants without having any effect on the plant growth as such. For example, a low concentration of iodine in the nutrient solution (4 ppm) was found to considerably decrease the ascorbic acid concentration in the tomato leaves without affecting the growth of the plants to any considerable degree.[342]

Mineral nutrients are known to interact with each other with respect to their uptake and, in some cases, their function within the plants. Although little is known about the interactions between various mineral nutrients as far as plant vitamins are concerned, available data indicate that some interactions may take place. For example, Ca and K are noted to interact for their effect on the ascorbic acid content of tomatoes grown with the hydroponic method.[419] Also, a high N : K ratio was found to favor the accumulation of ascorbic acid in the roots of Hamburg parsley.[420]

Finally, application of micronutrients may, in some cases, affect the rate of vitamin loss after the fruits have been harvested. For example, apples from trees sprayed with Zn during their flowering and again in August were found to lose only 27–38% of their original ascorbic acid during four months' storage, but the loss in the untreated apples was much higher, i.e., 52–56%.[421]

B. Carotene

1. Effect of Macronutrients

Numerous reports indicate that an increased supply of N, Mg, Mn, Cu, Zn, and B increases the concentration of carotene in a wide range of fruits and vegetables (Table 1). In the following, we will discuss some of these reports in more detail.

Bernstein et al.,[422] based on the results of an extraordinarily large experiment performed in both sand and soil, reported that although deficiencies of S, N, and K lowered the carotene content in turnip greens, the deficiency of P was without any effect on the carotene content in this plant. They concluded that only those treatments that resulted in a visible leaf chlorosis also caused a considerable decrease in the carotene content of plant leaves. This is in agreement with the report that in young oats, reduction in the supply of nitrate, calcium, or potassium to the plants (under water culture conditions) reduced the concentration of carotene in the leaves.[423]

Total carotenes and the relative amounts of α- and β-carotenes may all be affected differently by mineral fertilizers so that the ratio of β/α carotenes in the plant tissues may change when plants are supplied with

TABLE 6
**Effect of Soil Application of Fertilizers on the Carotene Content
and the Ratio of β / α Carotene in Carrots[424]**

| Treatment | Carotene (mg / kg FW) | | β / α |
	β-carotene	α-carotene	
Control[a]	106[d]	44	2.4
NPKMg[b]	134	52	2.6
−N[c]	144	48	3.0
−P	123	55	2.2
−K	150	59	2.5
−Mg	137	59	2.3

[a] Control treatment did not receive any fertilizer.
[b] Plants received complete NPKMg fertilizers.
[c] Plants received complete without fertilizers without N, P, K, or Mg, respectively.
[d] Values rounded.

From Vereecke, M., Van Maercke, D., *Acta Hortic.*, 93, 197, 1979. With permission.

increasing amounts of some fertilizers. For example, carrots grown in unfertilized soil were found to contain markedly less α- and β-carotene than those grown in soil amended with a complete fertilizer. Single omission of P, K, or Mg from the otherwise complete fertilizer treatment, however, altered the α- and β-carotene differently, so that the ratio of β/α carotene in the carrots was somewhat changed (Table 6). In tomatoes, although an increased supply of K decreased the concentration of β-carotene, it increased the lycopene and total carotene of the fruits (Figure 4).[353]

Sacharrer and Bürke[358] studied the effect of nutrient supply on the carotene content in the leaves of lettuce, maize, carrots, *Lolium westerwoldicum*, *Secale cereale* L., and *Beta vulgaris* var. saccharif and concluded that the deficiency of any one of N, P, K, Mg, Ca, and Na decreased the carotene content in the leaves and that an optimal supply of N, P, and K was necessary for high levels of carotene in the leaves of these plants. Excess supply of K, however, was noted to decrease the carotene concentration of plants, an observation also made by Ijdo[236] on the carotene content of spinach.

The comparative effect of NO_3 and NH_4 fertilizers on the carotene content of plants is not clear. For example, the leaves of several forage and vegetable plants supplied with NO_3 fertilizers were noted to contain higher concentrations of carotene than those supplied with NH_4 fertilizers.[358,425] In contrast, Miller et al.[126] reported that turnip

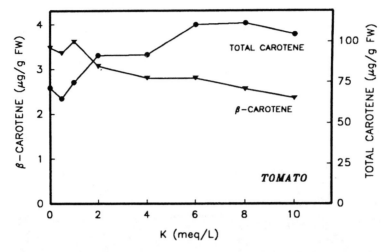

FIGURE 4. Effect of potassium fertilization on the total carotene and β-carotene concentrations in tomato. Data are from tables 2 and 4 of Trudel, M. J. and Ozbun, J. L., *J. Am. Soc. Hortic. Sci.*, 96(6), 763, 1971.

greens contained more carotene when fertilized with NH_4 as compared with NO_3. Eheart,[426] however, did not find any different in the carotene content of broccoli fertilized with NO_3 or NH_4.

2. Effect of Micronutrients

Among the micronutrients, Mn, Cu, Zn, and B are often noted to increase the carotene content in different plants (Table 1). This is in contrast to the earlier report by Lyon et al.,[308] who, after subjecting tomato plants to various micronutrient deficiencies, did not detect any difference in the carotene content of tomato (expressed on the fresh weight basis) subjected to deficiencies of Fe, Mn, Cu, Zn, or Mo.

Boron has been frequently noted to increase the carotene content of many plants. Kelly et al.,[371] for example, noted that boron-deficient carrots had lower yield and lower carotene content than plants receiving adequate B. No apparent relationship, however, could be found between the boron and carotene content of the roots: that is, once boron deficiency was alleviated, carotene concentration stayed constant regardless of the level of boron in the tissue (Table 7). In contrast, Gum et al.[278] noted that leaves of beet showing symptoms of boron deficiency contained much less carotene than the normal-appearing leaves taken from the same plant.

Application of Ni and Zn to a native Chinese variety of peas (*Pisum sativum* L.) was noted to increase the concentration of carotene and citrin (a compound sometimes referred to as vitamin P) in the shoots

TABLE 7
Effect of Boron Supply[a] on Yield, Boron, and Carotene
Concentrations in the Roots of Carrots[371]

Boron (ppm)	Yield (g DW / plant)	Root B (ppm DW)	Root carotene (mg / 100 g DW)
0.0	2.8	13	60
0.1	6.5	22	74
0.5	7.8	22	76
2.0	9.8	36	79
5.0	8.6	50	78
Significance	**b	**	**

[a] In the irrigation water used in the sand-culture experiment.
[b] ** indicates significance at the 1% level.

and in the green peas. Addition of Mo salts, however, increased the content of citrin only.[370]

Application of Cu to organic (muck) soils known to be deficient in Cu increased the growth of barley, oats, spinach, and carrots and also increased the carotene concentration in the leaves of barley, oats, and spinach and in the roots of carrots under greenhouse conditions.[320] Also, application of Mn to Mn-deficient soils of Indiana was reported to increase the carotene concentration in the leaves of soybean, maize, wheat, and oats.[369]

C. Thiamin

Thiamin content in plants has consistently been found to be increased by increased application of N, P, B, and Zn and to be decreased by K fertilization (Table 1). Here some of the reports are reviewed in more detail.

Thiamin concentrations in spinach and oats show positive correlations with the nitrogen supply to the plants (Figure 5).[216] Also, in wheat, thiamin concentration shows a close and positive correlation with nitrogen fertilization and the protein content of the kernels (Figure 6).[216, 427–430] Since kernel protein in cereals is also increased by nitrogen fertilization,[431] it thus appears that nitrogen fertilization (Figure 5), kernel protein, and kernel thiamin are closely interrelated components of cereal grains. Also, cereal varieties naturally rich in kernel nitrogen and protein tend to be also higher in thiamin (Figure 6). The effect of N fertilizer on the thiamin content of plants is so pronounced that nitrogen fertilizer rates above that required for optimum growth of plants were noted to still increase the thiamin as well as riboflavin content in spinach leaves (Figure 5).

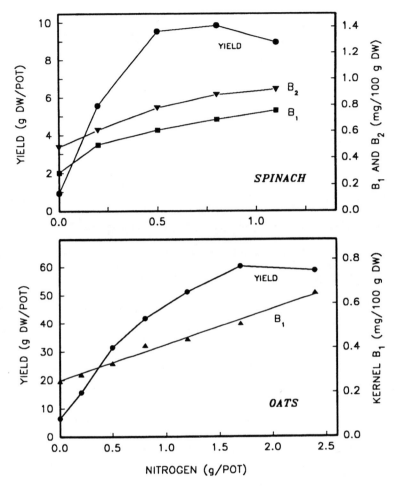

FIGURE 5. Effect of nitrogen fertilization on the yield and vitamin concentrations in spinach leaf and oats Kernel. Data from tables 5 and 6 of Pfützer, G., Pfaff, C., and Roth, H., *Landw. Forsch.*, 4(2), 105, 1952.

The response of a plant's thiamin to nitrogen fertilizers may vary in different soils. For example, application of $(NH_4)_2SO_4$ was found to consistently increase the thiamin content of rice kernels in three different soils. In one of the soils, however, although $(NH_4)_2SO_4$ increased the thiamin content, NH_4NO_3 and urea had little effect.[432]

Late application of N fertilizers at the time of booting was noted to be most effective in raising the concentration of both protein and thiamin in the kernels of four out of eight winter wheats and in four out of six summer wheat varieties.[385] The positive effect of nitrogen on the

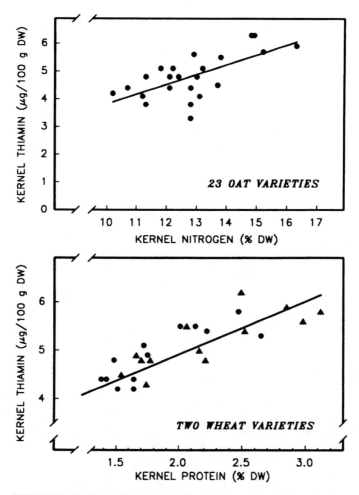

FIGURE 6. Relationships between the kernel concentrations of nitrogen and thiamin in 24 varieties of oats[429] and between the kernel concentrations of protein and thiamin in two varieties of wheat.[382]

thiamin content of plants has also been confirmed under sand culture conditions. In contrast to nitrogen, increasing the P concentration in the nutrient solution, however, consistently decreased the thiamin concentration in the leaves of two maize varieties.[433]

Hunt et al.,[376] based on a very comprehensive study, came to the conclusion that the effect of fertilizer (other than lime) on the vitamin content of plants was largely dependent on the kind of vitamin and the plant under study. In wheat, for example, application of P and K fertilizers alone increased and decreased the thiamin content, respec-

TABLE 8
**Effect of Phosphorous Concentration in the Rooting Media on the Growth
and Vitamin Concentrations (μg / g DW) in the
Leaves of Oats[390]**

Phosphorus (mmol l^{-1})	Growth (g DW)	Folic acid	Biotin	Niacin	Pantothenic acid
0.10	8.95	1.02	0.102	48.5	21.5
0.25	13.90	1.11	0.148	56.2	26.1
1	14.38	1.05	0.164	57.6	30.2
2	13.68	1.16	0.168	60.1	28.5
5	13.02	1.56	0.183	61.1	33.7

tively; NO_3 alone increased the niacin, and NO_3 and K fertilizers together increased the thiamin content in the kernels. The effect of liming and season of growth was also highly significant for the content of several vitamins. For example, liming increased the thiamin content of wheat, oats, and corn and the content of niacin and pantothenic acid in the wheat kernel.

The response of a plant's vitamin to fertilization may vary from one variety to another. For example, Syltie and Dahnke[430] noted that P and K fertilization of the variety Waldon of hard red spring wheat grown in North Dakota increased the vitamin B_1 and B_2 in the grains but decreased the concentration of the same vitamins in the grains of the variety Era. Also, application of K and/or P fertilizers to the Era wheat decreased the concentration of B_1 and B_2 in the grains. The concentrations of vitamin B_6 and B_{12} were not affected by fertilizer treatments.

D. Other Vitamins

With the exception of riboflavin, studies on the effect of mineral nutrients on the concentration of other plant vitamins are rather limited (Table 1). Based on the limited information gathered, it appears that the concentrations of most other vitamins studied are positively affected when increased amounts of various mineral nutrients are supplied to the plants.

McCoy et al.[390] reported that under nutrient-culture condition, oats grown with 5 mmol l^{-1} of P contained significantly higher folic acid, biotin, niacin, and calcium pantothenate than plants supplied with 50 times less (i.e., 0.1 mmol l^{-1}) phosphorus (Table 8). Increasing nitrogen supply was also noted to significantly increase the niacin concentration in both leaves and kernels of oats.[423]

Boron and manganese deficiencies were found to slightly reduce the concentration of riboflavin in the foliage and roots of beet and in the foliage and fruits of tomato.[278] In Mn-deficient soils, soil application of Mn was noted to increase the tocopherol concentration in soybean leaves.[369]

In a marine microalga (*Dunaliella tertiolecta*), maximum yield of β-carotene and vitamin C occurred when algae was supplied with urea and maximum vitamin E occurred when it was supplied with nitrate.[434]

E. Summary

The data summarized in (Table 1) along with reports cited in more detail in the above discussions could be summarized as follows:

1.   Most of the studies on plant vitamins have been done with main plant nutrients, namely, N, P and K. Relatively less is known about the effect of other mineral nutrients on plant vitamins.
2.   On the whole, the effect of nutrients on plant vitamins strongly depends on the kind of nutrient, vitamin, and plant under consideration. Thus in a given plant, different vitamins may be affected differently by a given nutrient and also different nutrients may affect the concentration of a given vitamin differently.
3.   The most consistent effect of mineral fertilizers on plant vitamins are:
      -Decrease in ascorbic acid by N fertilizers
      -Increase in ascorbic acid by P and K, Mn, B, Mo, Cu, Zn, Co
      -Increase in carotene by N, Mg, Mn, Cu, Zn, B
      -Increase in thiamin by N, P, and B
      -Increase in riboflavin by N

The above effects may not hold true in all cases for all plants but they are found to be true in the majority of cases. Some 40 years ago, Richardson[435] reported that the effect of mineral nutrients on plant vitamins is of little importance from the practical nutritional point of view. By considering the data presented above, I conclude that Richardson's conclusion may no longer be valid, especially with respect to the negative effect of nitrogen fertilizers on the ascorbic acid content of fruits and vegetables (Table 1). Due to the ever-increasing evidence on the possible protecting role of ascorbic acid for some forms of cancer,[436-451] the negative effect of nitrogen fertilizer on the content of vitamin C in fruits and vegetables and its nutritional significance needs to be investigated.

As noted above, one possible means of reducing the negative effect of high nitrogen fertilization is nitrogen withdrawal prior to plant harvest, a technique that can be used only if plants are grown in nutrient solution (or any of its variations). When plants are grown in soil, however, the only method to improve the ascorbic acid in plants is by the optimization of the amount of nitrogen fertilizer used and a concerted attempt and willingness to reduce nitrogen fertilizers to a practicable and economical feasible minimum. An alternative solution may be a shift to the use of slow-release fertilizers, especially for greenhouse-grown crops such as tomatoes, since, based on the only report known to this author on this subject, their use appears to increase the vitamin C content of tomatoes when compared with the standard (fast-release) fertilizers.[452] This is certainly an area that deserves much more research before definite recommendations can be made to farmers.

II. MINERAL VERSUS ORGANIC FERTILIZERS

Environmental and health consequences of conventional farming as compared with the so called organic-, alternative-, biological-, sustainable-, or biodynamic-farming methods have recently been the subject of much debate among the general public and scientists in various disciplines.[2, 453, 454] Although an exact definition of each method of food production is not available, heavy or exclusive use of inorganic fertilizers, along with the use of various plant protection chemicals, are the hallmarks of conventional farming. The "alternative" methods, however, rely solely, or mainly, on the manure, composts, and compounds such as bone meals as sources of plant fertilizers, use less or no plant protection chemicals such as pesticides, herbicides, and fungicides, and the products are distributed without the use of artificial preservatives or dyes.[455]

Often, claims have been made (directly or indirectly) that the organically grown foods are of better quality than those grown by other methods. This, along with their lower yield and/or their mystical appeal to certain people, is the reason why the organically produced fruits and vegetables are often marketed at a much higher price than those produced by the conventional method. The differences in the content of heavy metal and pesticide residues in plants grown with different methods of farming and the relative risks attributed to natural versus man-made toxins in our foods are covered in some recent books.[438, 439] Here we concentrate on the effect of conventional versus organic methods of farming on the content of plant vitamins. In the discussion that follows, for the sake of simplicity, organic versus conventional farming methods are treated somewhat synonymous with the use of organic versus inorganic fertilizers, respectively, although in some cases this simplification might be a very loose approximation of the respective methods, and with the awareness that the use of organic fertilizer by itself may not justify a product to be called organically grown.

Available information summarized in Table 9 indicates that in the majority of cases, use of organic fertilizers does not affect the concentration of most plant vitamins with the exception of thiamin, which seems to be higher in the plants grown with organic instead of inorganic fertilizers. The number of studies conducted up to now in this area is, however, relatively small and thus an unequivocal conclusion cannot be drawn at this time. Another major source of difficulty is the shortcomings in the inclusion of proper control treatment(s) when different soil treatments with very different concentrations of easily available (inorganic) and slowly available (organic) fertilizers are compared for their

TABLE 9
Relative Concentration of Vitamins in Plants Grown with Organic or Mineral Fertilizers[a]

Plant	Relative difference	Ref.
	Ascorbic acid	
Cabbage	Organic > mineral	456
Celeriac	Organic > mineral	457
Lemon	Organic > mineral	460
Lettuce	Organic > mineral	457, 459
Spinach	Organic > mineral	456, 459
Tomato	Organic > mineral	461
Cabbage	Organic = mineral	462
Carrots	Organic = mineral	456, 457, 463–465
Celeriac	Organic = mineral	456
Leek	Organic = mineral	462, 466
Pepper	Organic = mineral	465
Potato	Organic = mineral	4, 456, 463, 465, 467
Rye (leaf)	Organic = mineral	463
Snap beans	Organic = mineral	463, 464
Tomato	Organic = mineral	199, 465
Turnip	Organic = mineral	466
Capsicum anuum	Organic < mineral	458
Lettuce	Organic < mineral	399
Raspberry	Organic < mineral	468
Spinach	Organic < mineral	469, 240

TABLE 9   (continued)

Plant	Relative difference	Ref.
	Carotene	
Carrots	Organic > mineral	457, 461, 476
Carrots	Organic = mineral	456, 464
Ryegrass	Organic = mineral	4
Spinach	Organic = mineral	469
	Niacin	
Carrots	Organic = mineral	457
Celeriac roots	Organic = mineral	457
	Pantothenic acid	
Carrots	Organic = mineral	457
Celeriac roots	Organic = mineral	457
	Pyridoxine	
Carrots	Organic = mineral	457
Celeriac roots	Organic = mineral	457
Wheat kernel	Organic = mineral	457
	Riboflavin	
Carrots	Organic = mineral	457
Celeriac roots	Organic = mineral	457
Wheat kernel	Organic = mineral	475, 477
	Thiamin	
Cabbage	Organic > mineral	471
Carrots	Organic > mineral	457
Cauliflower	Organic > mineral	472
Spinach	Organic > mineral	469
Wheat kernel	Organic > mineral	474
Celeriac roots	Organic = mineral	457
Wheat kernel	Organic = mineral	475
Capsicum	Organic < mineral	458
	Tocopherol	
Wheat kernel	Organic < mineral	458

[a] Or with organic vs. conventional methods of farming.

TABLE 10
Effect of Mineral (NPK) or Organic Fertilizers on the Ascorbic Acid Concentration (mg / 100 g) in Several Vegetables Grown in Two Different Soils[456]

Treatment	Spinach	Lettuce	Savoy	Celeriac	Carrots	Potato
			FEN Soil			
NPK	29.9	9.7	41.8	11.8	8.0	25.3
Stable manure	53.1	15.4	73.5	14.0	8.0	27.7
NPK + stable manure	30.2	8.8	44.0	14.8	8.4	23.6
Composted manure	49.1	15.4	80.1	14.1	8.1	26.4
			Sand			
NPK	37.4	15.4	46.2	—	6.3	28.3
Stable manure	57.8	16.3	85.5	—	4.3	33.1
NPK + stable manure	34.3	15.4	41.8	—	6.8	32.2
Composted manure	48.8	14.1	75.7	—	5.3	33.0

Compiled from tables 4 and 5 of Schuphan, W., *Qual. Plant.—Plant Foods Hum. Nutr.*, 23(4), 333, 1974.

effects on plant growth and plant vitamins. Here we will discuss some selected reports in more detail.

A. Ascorbic Acid

The concentration of ascorbic acid in plants may increase, decrease, or not change by the use of organic fertilizers as compared with inorganic fertilizers; the number of reports showing no advantageous effect of organic fertilizers on the ascorbic acid content of plants, however, outweighs the other reports (Table 9).

Schuphan,[456] based on very extensive experiments lasting for 12 years on the effect of organic and inorganic (NPK) fertilizer on the ascorbic acid and carotene content of several vegetables, concluded that fertilizing the plants with stable manure or composted manure (when compared with the same plants fertilized with inorganic fertilizer) increased the ascorbic acid concentration in spinach, lettuce, and savoy cabbage but did not affect its concentration in carrots and celeriac roots and in potato tubers; this increase in spinach, for example, was as much as 78% when grown in FEN soil (53.1 vs. 29.9 mg/100 g, respectively) (Table 10). Celeriac tubers grown by organic farmers were noted to contain higher ascorbic acid than those grown by conventional farmers

TABLE 11

Effect of Organic or Inorganic (Conventional) Methods of Production on the Vitamin Concentrations in Carrots and Celeriac[457]

Farming method	β-Carotene[a]	Ascorbic acid[a]	Thiamin[b]	Riboflavin[b]	Niacin[b]	Pantothenic acid[b]	Nitrate[c]
Carrots							
Organic	8.3x[d]	4.5	43	17	440	180	413
Conventional	7.2y	3.8	36	16	405	178	433
Celeriac roots							
Organic	—	8.1x	33	49	660	280	250x
Conventional	—	7.3y	36	47	629	269	572y

[a] Concentration in mg/100 g FW.
[b] Concentration in μg/100 g FW.
[c] Concentration in ppm.
[d] For each plant and vitamin combination, only those values followed by different letters are significant at 0.05 level; the rest are not statistically different.

(Table 11). Also, Pettersson[478] noted that potatoes grown in Sweden by the "biodynamic farming system" contained significantly higher ascorbic acid than those grown with the conventional method (18.1 vs. 15.5 mg/100 g FW, respectively. In contrast, Beeson[4] cited a 25-year fertilizer experiment according to which the ascorbic acid content of potato tubers was not affected whether plants were grown with manure or inorganic fertilizer. Also, Svec et al.[465] did not find any difference in the ascorbic acid contents in potato, tomato, or pepper grown with organic or conventional fertilizers (Table 12).

TABLE 12

Ascorbic Acid Concentration in Potato, Tomato, and Peppers Grown with Organic or Conventional (Mineral) Fertilizers[465]

Treatment	Ascorbic acid (mg / 100 g FW)[a,b]		
	Potato	Tomato	Pepper
Organic	13	23	113
Conventional	13	22	112

[a] Values rounded.
[b] In no cases did organic and conventional treatments affect the ascorbic acid concentration significantly.

Montagu and Goh[199] studied different rates of compost, blood and bone meals, and two mineral fertilizers on the yield, color, visual qualities, and ascorbic acid content of tomatoes and noted that if one considers the effect of rates of fertilizer applied, then compost was not as effective as the mineral fertilizers in increasing the fruit yield. The concentration of ascorbic acid in the fruits, however, decreased with increasing application of fertilizers irrespective of the kind used. The compost-fertilized fruits, however, contained much less nitrogen than those fertilized with inorganic fertilizers.

B. Carotene

Leclerc et al.[457] compared the vitamin contents in carrots and celeriac roots produced by organic and conventional farmers in France and noted that carrots produced by organic farmers had significantly higher β-carotene (Table 11). In contrast, Scuhphan (1974) noted that organic versus inorganic fertilizers either slightly reduced or did not affect the carotene content in spinach and carrots.

Evers[476] compared the carotene content of organically grown carrots in two locations in Finland with those grown in another trial at the experimental farm encompassing 10 different inorganic fertilizer and irrigation treatment combinations and noted that on the average, organically grown carrots contained higher carotene than the plants grown with inorganic fertilizers (65.6 vs. 57.7 mg/100 g DW, respectively).

Plants grown by the "biological farmers" are reported to be more resistant against infection by fungi, and thus vegetables grown with this method keep better during storage.[479, 480] Also, organically grown carrots were noted to kept better in storage. Thus although after four to six months of storage only 47.7% of the inorganically grown carrots were considered to be marketable, as much as 86% of the organically grown carrots were still in a marketable condition.[480] Organically grown carrots, however, despite similar amount of sugars to those grown with inorganic fertilizers, did not taste good.[481]

C. Thiamin and Other Vitamins

After vitamin C, most studies on the effect of organic fertilizers on plant vitamins have concentrated on B vitamins and specially on thiamin (Table 1). The limited information available indicates that organic fertilizers in comparison with inorganic fertilizers do not change the concentration of riboflavin, niacin, pantothenic acid, or pyridoxine in most cases studied. Here some of the reports on the effect of organic fertilizers on thiamin are reviewed.

The first known work on the effect of organic fertilizers on B vitamins in plants dates back to 1926 when McCarrison and Viswanath,[482] by feeding vitamin-deficient pigeons with grains of wheat and millet grown on plots fertilized with either animal manure or chemical fertilizer, reported that "...millet grown on soil treated with cattle manure contained more B vitamins than millet grown on soil treated with complete chemical manure." Later Rowlands and Wilkinson,[483] using vitamin-deficient rats, reported that animals fed with grass seeds obtained from plants fertilized with pig manure gained weight at twice the rate as rats fed with seeds from plots fertilized with "artificial" or inorganic fertilizers. They suggested that seeds of plants grown on plots fertilized with manure contained more B vitamins. They suspected that plants had absorbed the B vitamins from the soils fertilized with the pig manure.

Organically grown spinach purchased from local health-food stores was noted to contain more than twice the thiamin present in the plants grown with the conventional methods (0.4 vs. 0.18 mg/100 g DW).[469] Grains of green gram and wheat grown with cow dung or cane-bagasse contained significantly more thiamin than control plants.[473] Although these authors unfortunately did not indicate the nature of their control treatment, from their data it appears that combined organic and inorganic fertilizers resulted in the highest thiamin concentration in most of the grains tested.

De and Laloraya[458] compared the effects of ammonium sulfate, urea, green manure, poultry, and cowdung manure on the contents of thiamin, ascorbic acid, and α-tocopherol in the leaves of 20-day-old *Capsicum annuum* (chili) and noted that ammonium sulfate was the best source of nitrogen for the production of thiamin and tocopherol. Urea decreased the concentration of these vitamins but resulted in the highest ascorbic acid level in the leaves (Table 13).

Leclerc et al.[457] noted that organically grown carrots tended to have higher thiamin and niacin than carrots grown by conventional methods; the differences, however, were not statistically significant. The concentrations of other vitamins were not different in the plants grown by organic or conventional method (Table 11).

Do plants grown with organic fertilizers synthesize more thiamin or do the plant roots absorb this vitamin from manure, as suspected by Rowlands and Wilkinson[483] more than 60 years ago? Based on the above-cited circumstantial evidence and the direct uptake experiments discussed below, the answer to this question is believed to be the latter.

TABLE 13
Vitamin Concentrations in Chili Seedlings as Affected by the Use of Organic or Inorganic Fertilizers[458]

Treatment[a]	Thiamin (μg / 100 g)	Ascorbic acid (mg / 100 g)	α-Tocopherol (μg / 100 g)
Control	275	166	2,120
Ammonium sulfate	284	205	5,010
Urea	190	287	2,950
Green manure	220	134	4,080
Poultry excreta	160	145	4,180
Cow dung manure	206	125	2,550

[a] The amount of fertilizer was adjusted so that plants in all treatments received an equivalent of 0.27 g of nitrogen per 5 kg of soil.

From the unsaturated soil treatment of De, S. K. and Laloraya, D., *Plant Biochem. J.*, 7(2), 116, 1980.

III. ORGANIC WASTES AND SOIL AS SOURCES OF VITAMINS FOR PLANTS

Human and animal excrements are known to contain various vitamins,[484-488] which are returned directly to land or end up in the sewer waters,[489-490] so that sewage water may have as much as 0.15 μg/l of thiamin![494] It is thus conceivable that addition of animal wastes or sewer water to the soil can raise the natural vitamin concentration in the soil.[491,492] This could in turn result in higher content of vitamins in the plants fertilized with organic wastes if plants roots can absorb vitamins by the roots.

Irrigation of plants with waste water from industrial complexes is shown to increase the concentrations of vitamins B_1, B_2, and niacin in corn and that of vitamin B_1 in cabbage.[493] Also, irrigation with "uncleared" sewage water from the city of Milan, Italy, was found to double the vitamin B_1 concentration of fodder plants as compared to irrigation of forage with "clean" water (363.7 vs. 114.2 μg/100 g DW).[494] Unfortunately, it was not clearly stated whether the clean water used was sewer water after being treated (cleaned) by the sewage treatment process or was from a completely different source.

Numerous soil microorganisms are able to synthesize several B vitamins.[497] For example, some 70% of the soil bacteria are shown to be able to produce vitamin B_{12}.[499] Addition of organic fertilizers to soil could also increase the activity of soil microorganisms and increase the

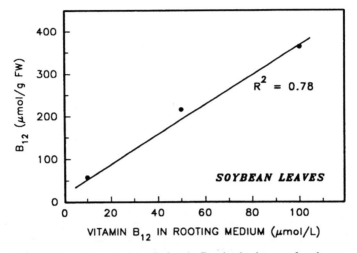

FIGURE 7. Concentration of vitamin B_{12} in the leaves of soybeans whose roots were placed in nutrient solutions containing increasing concentrations of vitamin B_{12} and allowed to absorb this vitamin for a period of five days. (From Mozafar, A. and Oertli, J. J., *Plant Soil*, 139, 23, 1992. With permission.)

soil's vitamin content.[500] Soil concentration of thiamin is estimated to be ca. 19.3 μg/kg and that of vitamin B_{12} ca. 15 μg/kg (see Mozafar and Oertli[495, 496]).

A. Uptake of Vitamins by the Plant Roots

Although the above reports on the increase in the vitamin content of plants fertilized with organic wastes point to a possible intake of vitamins by the plant roots, direct evidence for the uptake of vitamins by the plant roots became available only recently[495, 496] (Figures 7 and 8). It was shown that plant roots can easily absorb not only the relatively small vitamins such as thiamin, but can also absorb the relatively large and complex molecules such as vitamin B_{12} (molecular weight of 1355 Daltons and diameter of 8 angstrom) with ease.[496] Experiments conducted with the addition of cow dung to soil have also shown that plant roots can absorb the vitamin B_{12} present in the cow dung, so much so that fertilizing spinach with cow dung increased the vitamin B_{12} in their leaves.[496] These results show that observations made by McCarrison and Viswanath,[482] Rowland and Wilkinson,[483] Antonioni and Monzini,[494] Selivanov and Smolenskaya,[493] and others on the increased content of vitamins B_1 and B_{12} in plants fertilized with animal wastes were in fact due to the uptake of these vitamins by the plant roots. Also, the observations on the vitamin B_{12} nutrition of

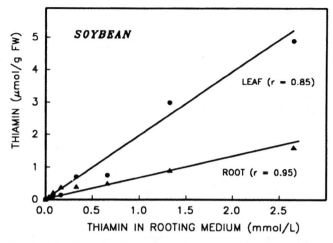

FIGURE 8. Concentration of thiamin in the roots and leaves of soybean plants whose roots were places in nutrients containing increasing concentration of thiamin and were allowed to absorb this vitamin for a period of 2 weeks. (From Mozafar, A. and Oertli, J. J., *J. Plant Physiol.*, 139, 436, 1992. With permission.)

villagers consuming vegetables fertilized with human wastes, which was interpreted as the surface contamination of vegetables with vitamin B_{12} from manure (see chapter 1, page 9), may have been a misinterpretation of the data and plants could have in fact contained more vitamin B_{12} through absorbing this vitamin by their roots from the soil, as shown to be true in spinach.[496]

a. Benefits to Plants and Consumers

The ability of plants to absorb the soil vitamin raises questions as to the possible benefit plants could drive from the vitamins they absorb by their roots and the nutritional consequences for the consumers of organically fertilized plants. Although vitamin B_{12} apparently does not have a function to play in the plants, thiamin is considered to be a growth factor for the plant roots (see Mozafar and Oertli[495]), and thus under certain soil conditions, increased content of soil thiamin (when organic wastes are added to the soil, for example) may benefit the plant growth by providing the roots with extra amounts of this growth factor. Experimental evidence is, however, needed to establish the rate of thiamin turnover in the soil and the threshold of soil thiamin that could effectively increase the root content of this vitamin.

Whether differences in the vitamin content of plants grown with organic or inorganic fertilizers would have any long-term nutritional consequences for humans cannot be ascertained at this time since apparently no study has been done on this subject. One promising candidate for nutritional studies might be vitamin B_{12}, which apparently cannot be synthesized by the plants. Thus the vegetarian people who, by choice or necessity, consume only plant products might be in danger of vitamin B_{12} deficiency.[501] Since plants grown on soils rich in this vitamin or fertilized with animal manure could, as shown above, may have more of this vitamin than those grown with inorganic fertilizer, they may prove to be beneficial for the vegetarian people. This is an area deserving closer attention by the nutritionists and medical scientists.

In summary, there is strong evidence that organic fertilizers may affect the concentration of some (but not all) vitamins in plants. Whether the magnitude of increase or decrease in the content of different vitamins is nutritionally relevant remains to be seen. Furthermore, since the culinary qualities of plants do not seem to change in parallel manners as their vitamin contents when plants are fertilized with organic or inorganic fertilizers, more research is needed to clarify the relative importance of nutritional, economic, and ecological merits of so-called organic farming under different socioeconomic conditions.

IV. ORGANIC VERSUS INORGANIC FERTILIZER AND THE NO₃ CONTENT OF PLANTS

One of the most consistent differences between plants grown with organic versus inorganic fertilizers is the lower nitrate content of the former, which is much more clear than the difference in their vitamin content.[8, 199, 466, 470, 502–504] There are, however, exceptions. As an example, Nilsson[462] did not find any significant difference between the organic and inorganic fertilizers in the ascorbic acid concentration in cabbage and leek or in the carotene content in carrots. Leek and cabbage grown with organic fertilizer, however, contained significantly higher nitrate in their tissues than those grown with inorganic fertilizer (Table 14). Schudel et al.[459] reported that spinach and stock beet (*Spinacia oleracea* L. var. cicla) grown with organic fertilizers produced less yield but contained lower nitrate and higher ascorbic acid than those fertilized with inorganic fertilizers (Table 15). Lower nitrate content in the plants grown with organic fertilizers may have nutritional significance because high nitrate in the foods is suspected to have an adverse effect on human health under special conditions.[438, 439]

Plant varieties may react differently to nitrogen fertilizers and thus may have different nitrate content even when grown under similar conditions. For example, Vogtmann et al.[399] compared the NO_3 and vitamin C content in 12 varieties of iceberg lettuce grown with organic and inorganic fertilizers and noted that plants grown with inorganic fertilizer contained significantly more NO_3 than those grown with

TABLE 14
Nitrate and Ascorbic Acid Concentrations[a] in Three Vegetables as Affected by the Use of Inorganic (NPK) or Organic Fertilizers[462]

Treatment	Nitrate (mg / kg FW)			(mg / 100 g FW)		
	Cabbage	Leek	Carrots	Cabbage[b]	Leek[b]	Carrots[c]
NPK	84	239	141	54	13	18
Organic	171	324	112	50	13	19
Organic + PK	132	417	243	51	14	18

[a] Average of 1975 and 1976 experiments and the 1, 1 level treatments from Nilsson, T., *Acta Hortic.*, 93, 209, 1979.
[b] Ascorbic acid.
[c] Carotene.

TABLE 15
Effects of Farm Yard Manure (FYM) and Mineral Fertilizers (MIN) on Yield, Nitrate, and Ascorbic Acid Concentrations in Spinach (Variety Nores) Leaf[459]

Treatment[a]	Yield[b] (ton / ha)	NO$_3$ (mg / 100 g DW)	Ascorbic acid (mg / 100 g FW)
Control	12.1a	1229a	67.6b
FYM 100	11.4a	1247a	69.4b
FYM 300	13.2a	939a	70.5b
MIN 100	21.3b	2672b	54.9a
MIN 300	28.4c	3968c	49.9a

[a] Based on the assumption that 40% of the total nitrogen in farm yard manure would be available to plants, the amount of manure application was adjusted to give 100 and 300 kg N/ha of nitrogen.
[b] Values inside each column followed by the different letters are significantly different at the 5% level.

organic fertilizer. The highest concentration measured, however, was 1700 mg NO$_3$/kg FW, which was still lower than the limit set in Germany, which is 3000 mg NO$_3$/kg. These authors also noted that the ascorbic acid content was in some (but not all) varieties considerably higher in the lettuce fertilized with inorganic as compared with organic fertilizer. Thus, in producing lettuce with organic fertilizers one needs to select those lettuce varieties that would produce a reasonable yield, contain the highest vitamin C, and the lowest possible NO$_3$ content. One interesting finding of these authors was that the core regions of iceberg lettuce, a part that is preferably eaten, have the highest concentration of NO$_3$ and the lowest concentration of ascorbic acid as com-

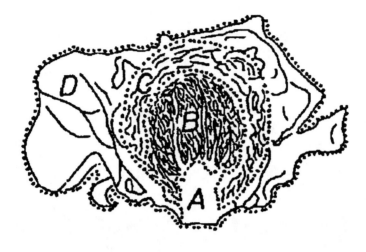

Section	NO₃ (mg/kg)	Vitamin C (mg/kg)
A	> 1000	58
B	> 500	160
C	> 400	147
D	> 100	361

FIGURE 9. Concentrations of vitamin C and nitrate in different leaves from outside to the inner parts of lettuce. (From Vogtmann, H., Kaeppel N., and Fragstein, P. V., Nitrat- und Vitamin C-Gehalt bei verschiedenen Sorten von Kopfsalat und unterschiedlicher Düngung, *Ernährungs-Umschau*, 34, 12, 1987. With permission.)

pared with the outer leaves, which have the lowest NO₃ and highest vitamin C but are discarded by most people prior to consumption (Figure 9).

Decrease in the plant vitamins due to organic fertilizer has also been observed by others. For example, Cheng[468] noted that farmyard manure increased the yield but reduced the ascorbic acid content of raspberry (*Rubus idaens*). Also, spinach grown under organic gardening methods was noted to be lower in ascorbic acid than that grown under conventional farming (65 vs. 103 ppm DW, respectively) but the difference was not statistically significant.[469] *Capsicum* grown with cow-dung or poultry manure was noted to have less ascorbic acid than that grown with ammonium sulfate or urea fertilizers.[458]

Organic fertilizer may also affect the ascorbic acid content of some plants during their storage. For example, fresh lettuce, eggplant, cabbage, and tomato did not show any consistent difference in their ascorbic acid content based on the organic or inorganic fertilizer they received during growth. Vegetables grown with organic fertilizer, however, seemed to retain more of their original ascorbic acid after a period of storage.[505] Also, spinach retained more (lost less) of its original vitamin C and accumulated less nitrate during storage when grown with organic versus inorganic fertilizers.[506]

Finally, the effect of organic fertilizer on plant vitamins may be strongly altered depending on the prevailing weather conditions, presumably through the effect of climate on the rate of mineralization of native and added soil organic matter and thereby the amount of nutrients available to the plants. For example, Yoshida et al.[507] based on a three-year experiment, noted that depending on the year, organically fertilized tomatoes had lower or higher ascorbic acid than those grown with mineral fertilizer. Thus the ascorbic acid contents in tomatoes were 14–15 versus 9–12 in 1980 and 14–25 versus 13–20 in 1981 in the plants grown with organic and inorganic fertilizers, respectively. In 1982 the two treatments contained very similar ascorbic acid content.

V. SOIL TYPE

Soil properties appear to affect the concentration of some vitamin in plants although the exact reason (such as soil physical and water-holding properties, soil chemical or biological characteristics, etc.) that may be responsible for the effects observed is hard to ascertain in most cases. For example, fruits and vegetables grown on sandy, loam, or clay soils may differ in their contents of various vitamins. This has been noted for the ascorbic acid content in black currant,[127] citrus fruits,[508,509] cabbage,[395,510,511] radish,[512] potato,[167,168] and tomato[513] and the concentration of carotene in carrots[355,511,514–516] and in rye leaves.[517] Here we will review some selected reports in more detail.

Oranges grown in Israel on sandy loam soil were found to be larger, have a thicker peel, and contain more total acids and vitamin C in their juice than those grown on poorly aerated loamy soil.[509] In Egypt, citrus fruits grown on sandy soils were also found to be larger and had higher juice but they contained less ascorbic acid than those grown on loamy soil.[508] Tomatoes grown in Egypt on sandy soil were also noted to have less ascorbic acid than those grown on sandy loam or clay loam soils.[513]

Black currants grown in Finland on coarse fine sandy soils were found to be significantly higher in ascorbic acid than fruits grown on sandy clay soils.[127] Radish roots grown on alluvial soil tended to have higher vitamin C than those grown on andosol soil.[512]

Although some reports indicate that potatoes grown on sandy soils may be higher in ascorbic acid, riboflavin, and niacin and lower in thiamin than those grown on loamy soil,[167,168] Lampitt et al.[518] did not find any effect of soil types on the ascorbic acid content of potatoes.

TABLE 16
**Vitamin Concentrations (mg / kg DW) in the Kernels of Cereals
Grown in Different Soils[524]**

Soil	Thiamin	Riboflavin	Nicotinic acid	Pantothenic acid	Pyridoxine
			Wheat		
Loam	4.8a[a, b]	1.6	65.6	11.4a	5.9a
Clay	4.7b	1.6	65.3	10.7b	5.7b
			Barley		
Loam	5.1a	1.9a	96.6a	9.2	7.8a
Clay	4.9b	1.8b	92.8b	9.3	7.5b
			Oats		
Loam	7.2	1.8a	15.4	7.9	3.0
Clay	7.2	1.7b	14.8	7.7	2.8

[a] Data are the average of a five-year experiment in 16 locations in Canada.
[b] For each crop and vitamin combination, values followed by different letters are significantly different at the 5% level.

Thiamin content of wheat kernels was found to be higher in the plants grown on brown soils than on gray soils[520, 521] and was higher in plants grown on sea-clay soil than on sandy soil.[522] Niacin content of wheat, however, was not affected by the type of soil where the plants were grown.[523] Nik-khah et al.[524] compared the concentration of several vitamins in the kernels of wheat, barley, and oats and found a significant effect of soil type on the concentration of some vitamins (Tables 16 and 17). For example, wheat grown on brown soil was higher in pyridoxine but lower in riboflavin than wheat grown on gray soil. Both barley and oats grown on brown soil contained significantly higher thiamin but lower riboflavin than those grown on gray soils. The concentration of pantothenic acid, however, was not affected by the soil type in any of the cereals tested. The concentration of nicotinic acid in barley was significantly lower when grown on brown soil than on black or gray soils (Table 17).

The effect of soil type on the carotene content of tomato was studied by Ellis and Hamner.[525] They grew tomato plants in a greenhouse at Ithaca, New York, on soils collected from four different locations in the United States and noted that soil origin had no effect on the carotene

TABLE 17
Vitamin Concentrations (mg / kg DW) in the Kernels of Cereals Grown in Different Soils[524]

Soil	Thiamin	Riboflavin	Nicotinic acid	Pantothenic acid	Pyridoxine
			Wheat		
Brown	4.8[a, b]	1.6b	65.3	11.5	6.1a
Dark brown	4.7	1.6b	67.2	10.6	5.8b
Black	4.8	1.6b	65.4	10.6	5.8b
Gray	4.8	1.7a	64.0	11.6	5.6b
			Barley		
Brown	5.2a	1.7b	90.6b	9.5	7.3b
Dark brown	4.9b	1.8b	91.8b	9.0	7.3b
Black	5.0b	1.8b	97.2a	9.4	7.6b
Gray	4.9b	1.9a	99.3a	9.3	8.2a
			Oats		
Brown	7.5a	1.7b	14.6	8.4	3.0
Dark brown	7.3b	1.7b	15.6	7.7	2.8
Black	7.2b	1.7b	14.8	7.4	2.8
Gray	7.0c	1.8a	15.4	7.8	2.9

[a] Data are the average of a five-year experiment in 16 locations in Canada.
[b] For each crop and vitamin combination, values followed by different letters are significantly different at the 5% level.

content in the tomato fruits. They also noted that although under sand culture experiments, variations in the supply of several micro- and macronutrients greatly affected the growth and fruitfulness of the plants, variation in nutrient supply did not have any consistent effect on the carotene content of tomato fruits.

The carotene content of carrots is markedly affected by the kind of soil they are grown in. Geissler,[526] for example, reported that in Germany carrots grown on sandy soil contained more than twice the carotene content of those grown on loamy soil. Gormley et al.[527] noted that carrots grown on mineral soil contained significantly more carotene than those grown on peat soil (Table 18). Under Swedish conditions, however, when plants were grown at different latitudes, carrots grown on mineral soil produced more yield but were only slightly lower in carotene than those grown on peat soil (Table 19). By considering the

TABLE 18
Carotene Concentration[a,b] in Carrots
(mg / 100 g FW) Grown in Two
Different Soils[527]

Soil	Harvest 1	Harvest 2
Peat	7	10
Mineral	15	21

[a] Values rounded.
[b] Differences between the peat and mineral soils
are significant at the 0.001 level.

TABLE 19
Effects of Rooting Substrate on the Yield and the Total Carotene
Concentration[a,b] in Carrots Grown in Two Years
in Three Locations in Sweden[528]

Location	Yield (g)		Carotene (mg / 100 g)	
	Peat	Mineral soil	Peat	Mineral soil
1973				
Ojebyn (65° 03′ N)	76	103	5	7
Ultuna (59° 49′ N)	35	44	9	8
Alnarp (55° 05′ N)	68	70	16	15
1974				
Ojebyn	62	73	13	13
Ultuna	58	55	18	12
Alnarp	51	59	19	16

[a] Sum of α- and β-carotenes.
[b] Values rounded.

strong positive effect of temperature on the carotene content of carrots (chapter 4), it may be postulated that part of the above differences in the carotene content of plants grown on sandy versus other soils or media might be due to difference in the soil temperature brought about by the difference in the water-holding capacities of different soils. For a discussion of the possible relationships between soil temperature and moisture and the carotene content in carrots, see chapters 4 and 7, respectively.

TABLE 20
Effect of Nitrogen Fertilizer[a] and Soil Type on the Yield and Ascorbic Acid Concentration in White Cabbage and in the Sauerkraut Produced from Them[395]

Nitrogen (kg / ha)	Alluvial	Diluvial	Loam
Cabbage Yield (ton / ha)			
250	119	111	120
550	131	123	121
Ascorbic Acid in Cabbage (mg / 100 g FW)			
250	49	57	44
550	48	56	45
Ascorbic Acid in Sauerkraut (mg / 100 g FW)			
250	23	11	12
550	6	9	11

[a] Given in five dressings.

Compiled and rounded from the tables 4, 5, and 7 of Paschold, P. J. and Scheunemann, C., *Arch. Gartenbau* (*Berlin*), 32(6), 229, 1984.

Vitamin loss during the postharvest period may also be affected by the kind of soil in which plants were originally grown. For example, Paschold and Scheunemann[395] reported that sauerkraut made from the cabbage grown on alluvial soil was about twice as rich in vitamin C as that made from cabbage grown on diluvial soil, despite the fact that the freshly harvested cabbage grown on alluvial soil had somewhat lower vitamin C than that grown on diluvial soil (Table 20). Also, Stepanishina[529] noted that cabbages grown on peaty-bog soil lost more of their ascorbic acid during storage than those grown on sod podzolic soil.

VI. SOIL pH

Although pH of rooting media (soil or artificial media) could strongly alter the availability of various mineral nutrients to the plant roots, effect of soil pH on plant vitamins has rarely been directly investigated. Based on the very limited information available, however, it appears

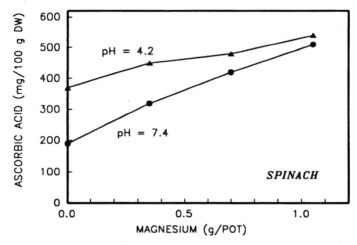

FIGURE 10.   Effect of Mg fertilization on the ascorbic acid concentration in spinach grown in soils with different pH values.[108]

that the effect of pH on plant vitamins may be of indirect nature, perhaps through its effect on plant growth. Thus, despite the two reports showing no effect of pH on the ascorbic acid content of potato[530] and the carotenoid content of sweet potato roots,[531] other available reports point to a possible interaction between the pH of the rooting media and other growth factors on the plant vitamins, which warrants more research to clarify the reasons for it.

Pfaff and Pfützer[108] noted that Mg fertilization of spinach was more effective in increasing the ascorbic acid in the leaves if the rooting medium had a lower pH (Figure 10). Virtanen et al.[405] reported that in sand culture, in the pH range of 4.0–7.0, concentrations of both ascorbic acid and carotene in peas and wheat foliage, determined at the flowering stage, were highest at that pH value at which plant growth was at its highest. They noted that in general, plants grown at lower pH ranges had lower carotene (Table 21) and ascorbic acid than those grown at higher pH.[405]

The thiamin content of turnip greens was also notably affected by the interactions between the soil pH and the available light intensity so that plants grown at high light intensity and low soil pH had the highest thiamin content. The content of ascorbic acid and riboflavin in the leaves, however, was not affected by soil pH.[532]

VII. ARTIFICIAL ROOTING SUBSTRATE VERSUS SOIL

A. Ascorbic Acid

There is an increasing sense of apprehension, mostly among people following the "back to nature" line of thinking, that vegetables such as

TABLE 21
Effect of the pH of the Rooting Medium (Sand Culture) on Growth and Carotene Concentration in the Leaves of Peas and Wheat at the Beginning of the Flowering Stage[405]

pH	Dry weight (g / 10 plants)		Carotene (µg / g DW)	
	Wheat	Peas	Wheat	Peas
7	7.389	2.605	93.7	193.1
6.5	7.996	—	112.8	—
6	6.861	4.167	108.1	264.3
5	5.666	2.565	60.1	111.9
4	4.250	1.104	37.6	106.0

From the KNO_3 treatment of Virtanen, A. I., Hausen, S. V., and Saastamoinen, S., *Biochem. Z.*, 267, 179, 1933.

tomatoes and cucumbers grown on "unnatural" synthetic media (solution culture, peat, rock wool, nutrient film technique [NFT], etc.) may be of inferior quality to plants grown in the natural soil. Scientific proof for or against this concern, however, is very limited and often controversial. One reason may be the fact that strict scientific comparison between the degree of nutrient supply to the roots of a plant grown in a complex system such as soil with that of a plant grown in a relatively well-defined medium such as nutrient solution, NFT, and so forth is very difficult. In other words, roots of plants grown in a solid medium (soil, sand, rock wool, etc.) and irrigated with a nutrient solution may not be exposed to exactly the same concentration of a given nutrient at their surface when compared with the roots of plants freely floating in a similar nutrient solution (hydroponics, NFT, etc.). With these difficulties in mind and by considering the reports cited in previous parts of this chapter that a given mineral nutrient may affect the concentration of different vitamins differently, it should not come as a surprise that the available information (as reviewed in the following) does not provide a more clear answer as to whether the vegetables grown on synthetic media have the same, more, or less of certain vitamins than their counterparts grown in soil. This might appear very astonishing for some if one considers the impact that some newer methods of vegetable production have had in making some vegetables such as tomatoes available all year round in some countries. Here are some of the available reports in more detail.

Tomatoes grown on synthetic media were noted to have a slightly higher[533-536] or the same[537,538] concentration of ascorbic acid as tomatoes grown in soil (Table 22). Cucumbers grown by hydroponic method

TABLE 22

Ascorbic Acid Concentration in Tomatoes Grown by the Nutrient Film Technique (NFT) or in Soil[538]

	1978		1979	
Substrate	**Sept. 5**	**Sept. 23**	**June 19**	**Sept. 3**
Soil	14.8	20.0	15.5	17.0
NFT	15.8	21.5	17.5	22.0
Significance[a]				
Growth substrate:		NS		NS
Date of sampling:		*		NS

[a] NS = not significant;
* significant at 5% level.

TABLE 23

Ascorbic Acid Concentration in Tomatoes Grown in Peat or in Nutrient Solution[541]

	Ascorbic acid (mg / 100 m*l* juice)	
Growth medium	**At harvest**	**One week after harvest**
Nutrient solution	14.4	14.7
Peat	17.8	17.9
SE	0.9	0.4

contained higher ascorbic acid content than those grown in soil.[533] In contrast to these reports, ascorbic acid content in the leaf[539] or fruits[540] of the tomato plants grown by solution-culture methods was found to be lower than that of plants grown in soil. Also, tomatoes grown on nutrient solution were noted to be significantly lower in ascorbic acid than those grown on peat (Table 23).[541] Janse and Winden,[542] however, noted that the tomatoes grown on rock wool had a slightly higher ascorbic acid than those grown in soil. This is in contrast to the findings of Nakayama et al.[543] according to which minitomatoes grown in rock wool were significantly lower in vitamin C than those grown in soil. Mokrzecka,[544] who compared the effect of various nitrogen sources and growth medium on the ascorbic acid content of savoy cabbage reported that in most cases plants grown in sand contained more ascorbic acid than those grown in peat.

Part of the discrepancies in the above reports may be due to the difference in the climatic conditions (light intensity, etc.) under which plants on different rooting substrates were grown and also to differences in the planting densities common in different methods of production. For example, Cronin and Walsh[545] compared the ascorbic acid content of tomatoes grown with high-density planting (35,000 plants/acre) with an arch frame system in the NFT method with those grown in peat at conventional density (exact value not specified by the authors) and noted that plants grown in peat had much higher ascorbic acid than those grown with NFT method (23.1 vs. 13.3, respectively, in the variety Pipo). These authors noted that the higher ascorbic acid content of peat-grown tomatoes was undoubtedly due to the more favorable light conditions experienced by these plants. Although the fruits produced by the NFT system were lower in ascorbic acid than those grown in peat, they, however, scored much better in panel evaluation for their flavor.[545]

B. Carotene

Reports on the effect of rooting media on the carotene content of plants are also inconsistent. For example, although tomatoes grown in soil were found to contain 30% more total carotenoids[540] or 70% more β-carotene[537] than those grown in sand (and supplied with complete nutrient solution) or in soil, no such differences were obtained by others.[533, 546]

C. Thiamin and Other Vitamins

Cucumbers grown in soil, sand, or solution did not show any considerable difference in their contents of several vitamins (Table 24). There is, however, one report indicating that rose geranium grown in hydroponics contained more tocopherol than those grown in soil,[549] and there are several reports showing the effect of rooting media on the thiamin content of plants.

For example, Hurni[550] noted that *Melandrium album* grown in soil had higher thiamin content than those grown with sand-culture method. This difference, which was in some cases as much as 55%, led to the conclusion that the location of plant growth may significantly affect the thiamin content in plant tissues. Also, wheat plants grown in organic soil (Moorland) were found to have considerably higher thiamin content in their kernels than those grown in quartz sand and irrigated with the same nutrient solution. This was also true irrespective of changes in the amount of N, P, and K fertilizers used.[383] By considering the numerous reports on the positive effect of nitrogen fertilizer (Table 1) and soil type (page 200) on the thiamin content of plants, and the ability of plants to absorb soil vitamin (page 194), one may postulate that plants grown in soil (a) might have received more nitrogen than those grown on quartz sand or (b) soils might have supplied the plants with thiamin synthesized by the soil microorganisms or that originally present in some organic wastes added to the soil at some earlier time (see page 193).

TABLE 24

Effect of Type of Rooting Medium on the Concentration (μg / 100 g FW) of Several Vitamins in Cucumber[457,458]

Rooting medium	Ascorbic acid	Choline	Pantothenic acid	Nicotinic acid	Pyridoxine	β-carotene	Riboflavin	Folic acid	Biotin	B$_{12}$
Sand	12,300	4,800	190	120	30	21.6	8	1.2	1	0.012
Solution	—	4,500	210	120	40	22.9	9	0.9	1	0.01
Soil	14,100	4,500	180	100	30	17.9	9	1.1	0.9	0.01

REFERENCES

1. Hauck, R. D., Ed., *Nitrogen in Crop Production*, ASA-CSSA-SSSA, Madison, Wis., 1984.
2. National Research Council, *Alternative Agriculture*, National Academy Press, Washington, D.C., 1989.
3. Mays, D. A., Ed., *Forage Fertilization*, American Society of Agronomy, Madison, Wis., 1974.
4. Beeson, K. C., The effect of fertilizers on the nutritive quality of crops and health of animals and men, *Plant Food J. (Washington)*, 5, 6, 1951.
5. Beeson, K. C. and Matrone, G., *The Soil Factor in Nutrition*, Marcel Dekker, New York, 1976.
6. White, P. L. and Selvey, N., Eds., *Nutritional Quality of Fresh Fruits and Vegetables*, Futura Publ., Mount Kisco, N.Y., 1974.
7. International Potash Institute, *Fertilizer Use and Plant Health*, Int. Potash Inst., Bern, Switzerland, 1976.
8. Schuphan, W., *Mensch und Nahrungspflanze*, Dr. Junk Verlag, Den Haag, 1976.
9. Deckard, E. L., Tsai, C. Y., and Tucker, T. C., Effect of nitrogen nutrition on quality of agronomic crops, in *Nitrogen in Crop Production*, Hauck, R. D., Ed., American Society of Agronomy, Madison, Wis., 1984, chapter 41.
10. Hill, W. A., Effect of nitrogen nutrition on quality of three important root/tuber crops, in *Nitrogen in Crop Production*, Hauck, R. D., Ed., American Society of Agronomy, Madison, Wis., 1984, chapter 43.
11. Locascio, S. L., Wiltbank, D. D., and Gull, D. D., Fruit and vegetable quality as affected by nitrogen nutrition, in *Nitrogen in Crop Production*, Hauck, R. D., Ed., American Society of Agronomy, Madison, Wis., 1984, chapter 42.
12. Gormley, T. R. and Maher, M. J., Quality of intensively produced tomatoes, in *The Effects of Modern Production Methods on the Quality of Tomatoes and Apples*, Gormley, T. R., Sharples, R. O., and Dehandtschtter, J., Eds., Commission of the European Communities, Luxembourg, 1985, 63.
13. Usherwood, N. R., The role of potassium in crop quality, in *Potassium in Agriculture*, Munson, R. D., Ed., ASA-CSSA-SSSA, Madison, Wis., 1985, 489.
14. Pattee, H. E., Ed., *Evaluation of Quality of Fruits and Vegetables*, AVI Publishing, Westport, Conn., 1985.
15. Marschner, H., *Mineral Nutrition of Higher Plants*, Academic Press, London, 1986.
16. Olson, R. A., and Frey, K. J., *Nutritional Quality of Cereal Grains: Genetic and Agronomic Improvement*, Agronomy 28, Am. Soc. Agron. Madison, Wis., 1987.
17. Karmas, E. and Harris, R. S., Eds., *Nutritional Evaluation of Food Processing*, Van Nostrand Reinhold, New York, 1988.
18. Engelhard, A., Ed., *Managment of Diseases with Macro- and Micronutrients*, Am. Phytopathol. Press, St. Paul, Minn., 1989.
19. Eskin, N. A. M., Ed., *Quality and Preservation of Fruits*, CRC Press, Boca Raton, Fla., 1991.
20. Eskin, N. A. M., Ed., *Quality and Preservation of Vegetables*, CRC Press, Boca Raton, Fla., 1989.
21. Kays, S. J., *Postharvest Physiology of Perishable Plant Products*, AVI Book, Van Nostrand Reinhold, New York, 1991.
22. Gould, W. A., *Tomato Production, Processing and Quality Evaluation*, 3rd Ed., CTI Publications Inc., Baltimore, Md., 1992, 439.
23. Harris, P. M., *The Potato Crop*, Chapman & Hall, London, 1992.
24. Pankratova, E. M., Increasing the physiological activity and yields of fruit trees by foliar nutrition, *Fiziol. Rast.*, 7, 584, 1960; *Hortic. Abstr.*, 31, 3936, 1961.

25. Kapelev, I. G., The biological aspects of raising winter cabbages in the south, *Vestn. Sel'sk. Nauki*, 9(7), 34, 1964; *Hortic. Abstr.*, 35, 3309, 1965.

26. Kraut, H. and Wirths, W., Die Bedeutung der Düngung für die menschliche Ernährung, in *Handbuch der Pflanzenernährung und Düngung*, Linser, H., Ed., III/2 Springer Verlag, Vienna, 1965, 1355.

27. Nehring, K., Düngung, Qualität und Futterwert, in *Handbuch der Pflanzenernährung und Düngung*, III/2, Linser, H., Ed., Springer Verlag, Vienna, 1965, 1260.

28. Petkov, M., The effects of manuring with mineral fertilizers on the quantity and quality of paprika yields, *Grad. Lozar. Nauka*, 1(5), 61, 1964; *Hortic. Abstr.*, 35, 5854, 1965.

29. Böttcher, H., and Ziegler, G., Experiments on the influence of manuring on the yield per unit area, quality and storage loses of green bush beans, *Arch. Gartenbau*, 15, 75, 1967; *Hortic. Abstr.*, 38, 3230, 1968.

30. Gnanakumari and Satyanarayana, G., Effect of N, P and K fertilizers at different rates on flowering, yield and composition of brinjal (*Solanum melongena* L.). *Indian. J. Agric. Sci.*, 41(6), 554, 1971. *Hortic. Abstr.*, 42, 7876, 1972.

31. Nechaeva, E. I., The effect of fertilizers on the chemical composition of tomatoes, *Tr. Dal'nevostochnogo Nauchno Issledovatel'skogo Instituta Sel'skogo Khozyaistva*, 13(2), 207, 1973; *Hortic. Abstr.*, 44, 5782, 1974.

32. Stanilova, D., Boboshevska, D., and Vitanova, G., The effect of nutrition on the yield and chemical composition of spinach, *Nauchni Trudove, Vissh Selskostopanski Institut "Georgi Dimitrov,"* *Obshcho Zemedelie*, 24, 35, 1972; *Hortic. Abstr.*, 45, 6489, 1975.

33. Goryachova, L. O., The effect of pre-sowing seed treatment with minor elements on watermelon yields and quality in southern Ukraine, *Visnik Sils'kogospod. Nauki*, 7, 76, 1974; *Hortic. Abstr.*, 45, 5835, 1975.

34. Paduchikh, L. V., The effect of long-term fertilizer application on changes in the biochemical characteristics of apples, *Nauchn. Tr. Ukr. Sel'sk. Akad.*, 113, 95, 1974; *Hortic. Abstr.*, 45, 9137, 1975.

35. Pankov, V. V., Leaf diagnosis of tomato potassium nutrition, *Agrokhimiya*, 7, 139, 1974; *Hortic. Abstr.*, 45, 5914, 1975.

36. Pankov, V. V., Effect of magnesium on chemical composition and yield of tomatoes, *Khimiya v Sel'skom Khozyaistve*, 12(9), 582, 1974; *Food Sci. Technol. Abstr.*, 7(3), J420, 1975.

37. Ananyan, A. M., Garibyan, B. A., Avakyan, A. G., and Avundzhyan, E. S., The effect of major and minor elements on the accumulation of dry matter and vitamin C in tomato fruits, *Izvestiya Sel'Skokhozyaistvennoi Nauki*, 6, 32, 1975; *Hortic. Abstr.*, 46, 11399, 1976.

38. Henze, J., Einfluss unterschiedlicher NPK-Gaben auf Fruchtinhaltsstoffe von Schattenmorellen auf F 12/1, *Erwerbsobstbau*, 18(12), 182, 1976.

39. Luchnik, N. A., The effect of fertilizers on nutrient utilization during plant development and on productivity of head cabbage grown on dark chestnut soil, *Sbornik Nauchnykh Statei Karagand. Gos. S.-Kh. Opytnoi Stantsii*, 5, 104, 1975; *Hortic. Abstr.*, 46, 7597, 1976.

40. Pirovski, M. and Dyankova, D., The effect of mineral fertilization on the vegetative development, yield and quality of cauliflower, *Gradinar. Lozar. Nauka*, 12(1), 80, 1975; *Hortic. Abstr.*, 46, 1107, 1976.

41. Khachidze, V. S., Mosashvili, L. A., and Abesadze, G. E., The effect of mineral fertilizers on the vitamin C, B_1, B_2 contents of vine leaves and grapes, *Sadovod. Vinograd. Vinodel. Mold.*, 8, 61, 1976; *Hortic. Abstr.*, 47, 10357, 1977.

42. Kryńska, W., Kawecki, Z., and Piotrowski, L., The effect of fertilization, irrigation and cultivar on the quality of fresh, sour and pickled cucumbers, *Zeszyty Naukowe*

Akademii Rolniczo-Technicznej Olsztynie, *Rolnictwo*, 15, 109, 1976; *Hortic. Abstr.*, 47, 4552, 1977.
43. Lenartowicz, W., Plocharski, W., and Wlodek, L., The influence of fertilization on the quality of small fruits. I. The influence of mineral fertilization of black currant on the chemical composition of extracted juice, *Fruit Sci. Rep.*, 3(3), 43, 1976; *Hortic. Abstr.*, 47, 6381, 1977.
44. Kulesza, W. and Szafranek, R. C., The effect of differential fertilization on the growth and cropping of apple trees, *Roczniki Nauk Rolniczych*, A, 103(3), 79, 1978; *Hortic. Abstr.*, 49, 9207, 1979.
45. Emura, K. and Hosoya, T., Effect of fertilizer on the quality and yield of spring-sown carrots, *Bull. Saitama Hortic. Exp. Stn.*, 8, 13, 1979; *Hortic. Abstr.*, 50, 9153, 1980.
46. Sagare, B. N. and Badhe, N. N., Effect of molybdenum alone and in combination with nitrogen and phosphorus on yield, uptake, protein and ascorbic acid content of cauliflower (*Brassica oleracea* Linn.), *Punabrao Krishi Vidyapeeth Res. J.*, 4(1), 28, 1980; *Hortic. Abstr.*, 51, 9364, 1981.
47. Mallick, M. F. R. and Muthukrishnan, C. R., Effect of micronutrients on the quality of tomato (*Lycopersicon esculentum* Mill.), *Vegetable Sci.*, 7, 6, 1980; *Hortic. Abstr.*, 52, 871, 1982.
48. Nosal, K., Effect of five foliar-applied chelates on the chemical composition and storability of McIntosh apples. Part II., *Zeszyty Naukowe Akademii Rolniczej im. H. Kollataja w Krakowie*, *Ogrodnictwo* 7, 158, 149, 1980; *Hortic. Abstr.*, 52, 4520, 1982.
49. Singh, H. K., Singh, B. P., and Chauhan, K. S., Effect of foliar feeding of various chemicals on physico-chemical quality of guava fruits, *Haryana Agric. Univ. J. Res.*, 11(3), 411, 1981; *Hortic. Abstr.*, 52, 5833, 1982.
50. Sorokina, G. I., Change in growth rate and metabolic processes of potatoes in relation to mineral nutrition level, in *Fitogorm. Ikh Deistvie Rast.*, Yakushkina, N. I., Ed., 1982, 125; *Chem. Abstr.*, 98, 178100u, 1983.
51. Takhirov, M. T., Pulatov, B. A., and Khasanov, Yu. A., Data for the hygienic substantiation of permissible residual amounts of nitrate nitrogen in melon and vegetable crops, *Gig. Sanit.*, 10, 10, 1982; *Chem. Abstr.*, 98, 3646z, 1983.
52. Mazur, K., Mazur, T., and Mazgaj, M., Effect of nitrogen dose, nitrogen:phosphorus:potassium ratio, and microelement addition on the content of amino acids and vitamin C in potatoes, *Rocz. Glebozn.*, 34(4), 69, 1983; *Chem. Abstr.*, 101, 209877q, 1984.
53. Pankov, V. V., Leaf analysis in relation to onion nutrition, in *Proc. VIth Int. Colloquium for the Optimization of Plant Nutrition*, Vol. 2, Montpellier, France, AIONP/GERDAT, 1984, 449.
54. Pankov, V. V., Chemical composition and productivity of onions in relation to nitrogen and phosphorus nutrition, in *Agrotekhnika Ovoshchnykh Kul't. Gor'kii*, USSR, 1983, 9; *Hortic. Abstr.*, 55, 4306, 1985.
55. Pankov, V. V., Effect of potassium on the chemical composition and productivity of onions, *Agrotekhnika Ovoshchnykh Kul't. Gor'kii*, USSR, 1983, 20; *Hortic. Abstr.*, 55, 4307, 1985.
56. Pankov, V. V., Leaf diagnosis of nitrogen nutrition of onions, *Agrokhimiya*, 11, 65, 1984; *Chem. Abstr.*, 102, 61427r, 1985.
57. Pankov, V. V. and Pavlova, V. P., Effect of phosphorus on the yield and chemical composition of onions, in *Desistvie Udobrenii i Otkhodov Promyshlennosti na Produktivnosi Sel'skokhozyaistvennykh Kul'tur, Kachestov Urozhaya i Svoistva Pochvy. Gorkii*, USSR, 1984, 16; *Hortic. Abstr.*, 56, 4147, 1986.
58. Bubarova, M., Atanasova, E., Vulev, P., and Ruskova, M., Effect of cultivar and soil acidity on the yield-fertilizer relation and related biochemical characteristics in lettuce, *Rast. Nauki*, 24(10), 78, 1987; *Chem. Abstr.*, 108, 149502w, 1988.

59. Singh, R. R., Chauhan, K. S., and Singh, H. K., Effect of various doses of N, P, K, on physicochemical composition of ber fruit cultivar—Gola, *Prog. Hortic.*, 18(1/2), 35, 1986; *Hortic. Abstr.*, 58, 5353, 1988.

60. Kukh, I. A. and Protsyuk, G. E., Effect of fertilizer rate on the yield and quality of potatoes, *Agrokhimiya*, 4, 51, 1988; *Chem. Abstr.*, 109, 5698n, 1988.

61. Levchenko, L. A., Antraptseva, N. M., and Shchegrov, L. N., Regulatory action of binary magnesium-zinc dihydrophosphate in the optimization of tomato nutrition under greenhouse conditions, *Fiziol. Biokhim. Kul't. Rast.*, 21(3), 273, 1989; *Chem. Abstr.*, 111, 77023c, 1989.

62. Almazov, B. N. and Kholuyako, L. T., Effect of organic and mineral fertilizers on yields in a vegetable crop rotation and on the fertility of a leached chernozem. Part 1. Effect of peat and mineral fertilizers on the yield and quality of potatoes and other vegetables. *Agrokhimiya*, 1, 53, 1990; *Chem. Abstr.*, 112, 197097x, 1990.

63. Almazov, B. N. and Kholuyako, L. T., Effect of organic and mineral fertilizers on yields in a vegetable crop rotation and on the fertility of a leached chernozem. Part 4, After effect of peat and mineral fertilizers on soil nutrients, crop yields and quality in the rotation, *Agrokhimiya*, 5, 52, 1990; *Chem. Abstr.*, 113, 57954v, 1990.

64. Ponomareva, A. N., Krotkikh, T. A., and Racheva, G. A., Effect of micronutrients on the yield and quality of cabbage grown in the Cis-Urals, *Agrokhimiya*, 2, 98, 1990; *Chem. Abstr.*, 112, 177479v, 1990.

65. Rupasova, Zh. A. and Ignatenko, V. A., Effect of mineral nutrition and varietal characteristics on the accumulation of secondary metabolites by cranberry fruits during ripening, *Vestsi Akad. Navuk BSSR, Ser. Biyal. Navuk*, 4, 24, 1990; *Chem. Abstr.*, 113, 190327j, 1990.

66. Barinova, E. V., Koverzanova, E. V., Zhuravlev, V. I., Gumargalieva, K. Z. and Moiseev, Yu. V., Changes in the content of ascorbic acid and potassium ions in potatoes, *Pishch. Prom-st. (Moscow)*, 5, 90, 1991; *Chem. Abstr.*, 115, 278460z, 1991.

67. Ghosh, B. and Mitra, S. K., Effect of varying levels of nitrogen, phosphorus and potassium on yield and quality of litchi (*Litchi chinensis* Sonn.) cv. Bombai, *Haryana J. Hortic. Sci.*, 19(1-2), 7, 1990; *Hortic. Abstr.*, 61, 4368, 1991.

68. Snapyan, G. G., Arutyunyan, A. Ts., Astabatsyan, G. A., and Bakryan, Ch. A., Effect of mineral fertilizers on the quality of Beurre Ardanpont pears, *Agrokhimiya*, 10, 70, 1990; *Chem. Abstr.*, 115, 91212c, 1991.

69. Bagal, S. D., Shaikh, G. A., and Adsule, R. N., Influence of different levels of N, P and K fertilizers on the protein, ascorbic acid, sugars and mineral contents of tomato, *J. Maharashtra Agric. Univ.*, 14(2), 153, 1989; *Hortic. Abstr.*, 62, 4046, 1992.

70. Voican, V., Davidescu, V., and Atanasiu, N., Determining the optimum fertilizer regime for glasshouse tomatoes on reddish-brown soil, *Lucrări Stiințifice, Institutul Agronomic'Nicolae Bălcescu', Bucuresti, Seria B, Horticultură*, 32(1), 15, 1989; *Hortic. Abstr.*, 62, 6623, 1992.

71. Ul'yano, A. M. and Ul'yanova, D. A., Significance of agricultural technological measures in the accumulation of several vitamins by fruits and berries, in *Tr. Vses. Semin. Biol. Aktiv. (Lech.) Veshchestvam Plodov Yagod*, 4., Franchuk, E. P., Ed., Michurinsk, USSR, 1982, 275; *Chem. Abstr.*, 81, 76927a, 1974.

72. Hamad, M. S., Abbas, A. S., and Akil, A. M., The effect of urea foliar sprays on tree growth and physical and chemical properties of apple fruits cv. Golden Delicious, *Zanco, Ser. A*, 6(4), 1, 1980; *Chem. Abstr.*, 97, 22687j, 1982.

73. Burrell, R. C., Brown, H. D., and Ebright, V. R., Ascorbic acid content of cabbage as influenced by variety, season, and soil fertility, *Food Res.*, 5, 247, 1940.

74. Largskij, J. N., The effect of fertilizers on vegetable quality, *Agrohimija*, 5, 77, 1969; *Hortic. Abstr.*, 40, 6022, 1970.

75. Dube, S. D. and Misra, P. H., The effect of macronutrient deficiencies on the ascorbic acid content of some vegetables, *Sci. Cult.*, 41, 485, 1975.

76. Umarov, Kh. Z. and Kaziev, G. M., Rates and proportions of mineral fertilizers for late head cabbage on typical grey soils in Uzbekistan, *Agrokhimiya*, 8, 80, 1977; *Hortic. Abstr.*, 48, 4523, 1978.

77. Gurgul, E., Effect of irrigation and nitrogen and magnesium fertilization on the activity of some enzymes and the yield and chemical composition of Savoy cabbage and red cabbage, *Rozpr., Akad. Roln. Szczecinie*, 84, 105 pp, 1982; *Chem. Abstr.*, 98, 88210a, 1983.

78. Shanker, J., Saka, S. K., and De, S. K., Effect of various concentrations of nitrogen, phosphorus, and potassium, applied to the soil or leaves, on the levels of vitamins C and B_1 in cabbage. *C. R. Seances Soc. Biol. Ses Fil.*, 182(3), 244, 1988; *Chem. Abstr.*, 111, 22611n, 1989.

79. Käppel, R., Einfluss von Düngung, Erntetermin, Aufbereitung und Lageratmosphäre auf die Qualität von Lagerblumenkohl (*Brassica oleracea L. convar. botrytis* (L.) Alef. var. botrytis 1.), Ph.D. Thesis, Technischen Universität München, Freising-Weihen-stephan, Germany, 1977.

80. Ul'yanova, D. A., The quality of top fruit and berries at different nitrogen and potassium rates, in *Sbornik Nauchnykh Rabot, Plodovodstvo Yagodovodstvo Nechernozemnoi Polosy, Vol. III, Moscow, USSR, Ministerstvo Sel'skogo Khozyaistva*, 1971, 173; *Hortic. Abstr.*, 44, 202, 1974.

81. Dass, R. C. and Mishra, S. N., Effect of nitrogen, phosphorus, and potassium on growth and quality of chili (*Capsicum annuum*, L.), *Plant Sci.*, 4, 78, 1972.

82. Yang, S. B., Park, K. W., and Chiang, M. H., The effect of fertilizer application, spacing and sowing date on the growth and quality of *Chrysanthemum coronarium* L., *Abstracts of Communicated Papers* (*Horticulture Abstracts*), *Korean Soc. for Hortic. Sci.*, 7(1), 72, 1989; *Hortic. Abstr.*, 59, 8284, 1989.

83. Fawusi, M. O. A. and Fafunso, M., Effects of the source and level of nitrogen on the yield and nutritive value of *Corchorus olitorius*, *J. Plant Foods*, 3, 225, 1981.

84. Fawusi, M. O. A., Quality and compositional changes in *Corchorus olitorius* as influenced by nitrogen fertilization and post-harvest handling, *Sci. Hortic.* (*Amsterdam*), 21(1), 1, 1983.

85. Maurya, K. R., Effect of nitrogen and boron on sex ratio, yield, protein and ascorbic acid content of cucumber (*Cucumis sativus* Linn.), *Indian J. Hortic.*, 44, 239, 1987.

86. Addae-Kagya, K. A. and Norman, J. C., The influence of nitrogen levels on local cultivars of eggplant (*Solanum integrifolium* L.), *Acta Hortic.*, 53, 397, 1977.

87. Tiwari, M. D., Upadhiyaya, J. S., and Singh, M. P., Effect of varying levels of nitrogen with and without micronutrients on guava production, *Plant Food Rev.* (*Bombay*), 8(10), 1, 1968; *Hortic. Abstr.*, 39, 3784, 1969.

88. Haraszthy, J., Changes in the vitamin C content of horse-radish as affected by different rates of mineral fertilizers, *Debreceni Agrártud. Föisk. Tud. Közlem.*, 8, 371, 1963; *Hort. Abstr.*, 34, 5372, 1964.

89. Harding, C. F. and David, J. J., The effect of certain mineral nutrients on the ascorbic acid content of leaf lettuce, *Food Res.*, 19, 138, 1954.

90. Müller, K. and Hippe, J., Influence of differences in nutrition on important quality characteristics of some agricultural crops, *Plant Soil*, 100, 35, 1987.

91. Singh, B. P., Singh, S. B., Singh, D. C., and Singh, T. B., Effect of soil and foliar application of urea on the physico-chemical composition of mango fruit (*Mangifera indica* L.) cv. Langra, *Bangladesh Hortic.*, 5(1), 29, 1977; *Hortic. Abstr.*, 51, 3115, 1981.

92. Singh, R. L., Singh, B., and Singh, R., Effects of foliar application of urea on the chemical composition of mango fruit (*Mangifera indica* L.) Langra, *Plant Sci.* (*India*), 11, 94, 1979; *Hortic. Abstr.*, 52, 3390, 1982.

93. Wynd, F. L. and Noggle, G. R., Influence of chemical characteristics of soil on production of vitamin C in leaves of oats, *Food Res.*, 10, 537, 1945.

94. Watson, S. A. and Noggle, G. R., Effect of mineral deficiencies upon the synthesis of riboflavin and ascorbic acid by the oat plant, *Plant Physiol.*, 22, 228, 1947.

95. Radajewska, B., Effect of cultural practices on growth, cropping and processing value of the peach cultivars Siewka Jerzykowska and Siewka Rakoniewicka, *Roczniki Akademii Rolniczej Poznaniu, Rozprawy Naukowe*, 166, 1, 1987; *Hortic. Abstr.*, 58, 5492, 1988.

96. Mustaffa, M. M., Influence of plant population and nitrogen on fruit yield, quality and leaf nutrient content of Kew pineapple, *Fruits*, 43(7-8), 455, 1988; *Hortic. Abstr.*, 59, 8769, 1989.

97. Ott, M., Über den Gehalt von Feld- und Gartenfrüchten und Vitamin C und Carotin bei verschiedener Düngung, *Angewand. Chem.*, 50(2), 75, 1937.

98. Mondy, N. I., Koch, R. L., and Chandra, S., Influence of nitrogen fertilization on potato discoloration in relation to chemical composition. 2. Phenols and ascorbic acid, *J. Agric. Food Chem.*, 27(2), 418, 1979.

99. Takebe, M. and Yoneyama, T., Plant growth and ascorbic acid. 1. Changes of ascorbic acid concentrations in the leaves and tubers of sweet potato (*Ipomea batatas* Lam.) and potato (*Solanum tuberosum* L.). *Nippon Dojo Hiryogaku Zasshi*, 63(4), 447, 1992; *Chem. Abstr.*, 117, 190048v, 1992.

100. Sengewald, E., Untersuchungen über den Einfluss der Düngung auf den Carotin- und Vitamin-C-Gehalt von Spinat (*Spinacia oleracea* L.) unter berücksichtigung der Entwicklung, *Nahrung*, 3, 428, 1959.

101. Mengel, K., Influence of exogenous factors on the quality and chemical composition of vegetables, *Acta Hortic.*, 93, 133, 1979.

102. Sharfuddin, A. F. M. and Voican, V., Effect of plant density and NPK dose on the chemical composition of fresh and stored tubers of sweet-potato, *Indian J. Agric. Sci.*, 54(12), 1094, 1984.

103. Barooah, S. and Ahmed, A. Z., NPK trial on tomato. Response to NPK fertilizers, at differenrt levels, on growth and ascorbic acid content of tomato, *Indian J. Agron.*, 9, 268, 1964; *Hortic. Abstr.*, 35, 8038, 1965.

104. Sharma, C. B. and Mann, H. S., Effect of phosphatic fertilizers at varying levels of nitrogen and phosphate on the quality of tomato fruits, *Indian J. Hortic.*, 28, 228, 1971.

105. Varma, A. N., Srivastava, D. C., and Sharma, R. K., Effect of urea spray on growth, yield and quality of tomato (*Lycopersicon esculentum* Mill.), *Maysore J. Agric. Sci.*, 4, 107, 1970.

106. Ledovskii, S. Ya. and Korzun, G. P., The effect of fertilizers and soil moisture on the chemical composition of tomatoes, *Vestn. S-kh. Nauki, Moscow, USSR*, 2, 97, 1976; *Hortic. Abstr.*, 46, 11437, 1976.

107. Patil, A. A. and Bojappa, K. M., Effects of cultivars and graded levels of nitrogen and phosphorus on certain quality attributes of tomato (*Lycopersicon esculentum* Mill.). I. TSS, acidity, ascorbic acid and puffiness, *Mysore J. Agric. Sci.*, 18(1), 35, 1984.

108. Pfaff, C. and Pfützer, G., Über den Einfluss der Ernährung auf den Carotin- und Ascorbinsäuregehalt verschiedener Gemüse- und Futterpflanzen, *Angewandte Chemie*, 50(9), 179, 1937.

109. Janes, B. E., The relative effect of variety and environment in determining the variations of per cent dry weight, ascorbic acid, and carotene content of cabbage and beans, *Pro. Am. Soc. Hortic. Sci.*, 45, 387, 1944.

110. Bomme, U., Eid, K., and Kraus, A., Nitrogen fertilization of white cabbage for sauerkraut production, *Gemüse*, 23(2), 62, 1987; *Soils and Fertilizers*, 50, 9492, 1987.

111. Venter, F., Nitrate content in carrots (*Daucus carota* L.) as influenced by fertilization, *Acta Hortic.*, 93, 163, 1979.

112. Fritz, P. D., Käppel, R., and Weichmann, J., Vitamin C in gelagertem Blumenkohl, *Landwirt. Forsch.*, 32(3), 275, 1979.

113. Matzner, F., Vitamin-C-Gehalt in Früchten der "Schattenmorelle," *Erwerbsobstbau*, 18(6), 83, 1976.

114. Hårdh, J. E., Factors affecting the vitamin C content of black currants, *Maataloust. Aikakausk.*, 36, 14, 1964; *Hortic. Abstr.*, 34, 6469, 1964.

115. Smith, P. F., Quality measurements on selected size of Marsh grapefruit from trees differentially fertilized with nitrogen and potash, *Proc. Am. Soc. Hortic. Soc.*, 83, 316, 1963.

116. Keleg, F. M., El Gazzar, A. M., and Mansour, A. M., Effect of different forms, rates, and time of application of nitrogen fertilizers on the nutational status, growth and yield of Washington navel orange and Balady mandarin trees, *Alexandria J. Agric. Res.*, 22, 281, 1974.

117. Sahota, G. S. and Arora, J. S., Effect of nitrogen and zinc on "Hamlin" sweet orange (*Citrus sinensis*), *Engei Gakkai Zasshi*, 50(3), 281, 1981; *Chem. Abstr.*, 96, 141717h, 1982.

118. Hamner, K. C. and Nightingale, G. T., Ascorbic acid content of pineapples as correlated with environmental factors and plant composition, *Food Res.*, 11, 535, 1946.

119. Baker, L. C., Parkinson, T. L., and Lampitt, L. H., The vitamin C content of potatoes grown on reclaimed land, *J. Soc. Chem. Ind.*, 65, 428, 1946.

120. Baker, L. C., Lampitt, L. H., Money, R. W., and Parkinson, T. L., The composition and cooking quality of potatoes from fertilizer trials in the East Riding of Yorkshire, *J. Sci. Food Agric.*, 1, 109, 1950.

121. Roy, R. N. and Seth, J., Nutrient uptake and quality of radish (*Raphanus sativus* L.) as influenced by levels of nitrogen, phosphorus and potassium and methods of their application, *Indian J. Hortic.*, 28(2), 144, 1971.

122. Kodama, M., Akamatsu, S., Bessho, Y., Owada, A., and Kubo, S., Effect of nitrogen fertilizing on the composition of satsuma juice, *J. Jpn. Soc. Food Sci. Technol.*, 24(8), 398, 1977; *Hortic. Abstr.*, 48, 3934, 1978.

123. Hammett, L. K. and Miller, C. H., Influence of mineral nutrition and storage on quality factors of "Jewel" sweet potatoes, *J. Am. Soc. Hortic. Sci.*, 107(6), 972, 1982.

124. Dastane, N. G., Kulkarni, G. N., and Cherian, E. C., Effects of different moisture regimes and nitrogen levels on quality of tomato, *Indian J. Agron.*, 8, 405, 1963.

125. Reder, R., Speirs, M., Chochran, H. L., Hollinger, M. E., Farish, L. R., Geiger, M., McWhirter, L., Sheets, O.l A., Eheart, J. F., Moore, R. C., and Carolus, R. L., The effects of maturity, nitrogen fertilization, storage and cooking, on the ascorbic acid content of two varieties of turnip greens. *South. Coop. Series Bull.*, 1, 1, 1943.

126. Miller, E. V., Army, T. J., and Krackenberger, H. F., Ascorbic acid, carotene, riboflavin, and thiamine content of turnip greens in relation to nitrogen fertilization, *Proc. Soil. Sci. Soc. Am.*, 20, 379, 1956.

127. Säkö, J. and Laurinen, E., The effect of fertilization on the black currant in two soils, *Ann. Agric. Fenniae*, 18, 96, 1979.

128. Singh, U. C., Sundararajan, S., and Veeraragavathatham, D., Effect of split application of nitrogen on crude protein, carotene and ascorbic acid contents of clipping amaranthus (*Amaranthus tristis* L.) cv. Co.3., *S. Indian Hortic.*, 34(3), 150, 1986; *Hortic. Abstr.*, 58, 2937, 1988.

129. Kessler, W., Über den Vitamin C-Gehalt deutscher Apfelsorten und seine Abhängigkeit von Herkunft, Lichtgenuss, Düngung, Dichte des Behanges und Lagerung, *Gartenbauwissenschaft*, 13, 619, 1939.
130. Murneek, A. E., Maharg, L., and Wittwer, S. H., Ascorbic acid (vitamin C) content of tomatoes and apples, *Univ. Missouri, Agric. Exp. Stn. Res. Bull.*, 568, 1954.
131. Scharrer, K. and Werner, W., Über die Abhängigkeit des Ascorbinsäure-Gehaltes der Pflanze von ihrer Ernährung, *Z. Pflanzenernhähr. Düng. Bodenk.*, 77(2), 97, 1957.
132. Kryńska, W. and Martyniak, B., Nutritional value of early cabbage and tomatoes grown on sloping grounds, *Roczniki Nauk Rolniczych, A*, 103(4), 79, 1978; *Hortic. Abstr.*, 50, 2613, 1980.
133. Venter, F., Der Nitratgehalt in Chinakohl (*Brassica pekinensis* (Lour.) Rupr.), *Gartenbauwissenschaft*, 48(1), 9, 1983; *Hortic. Abstr.*, 53, 4118, 1983.
134. Sørensen, J. N., Dietary fibers and ascorbic acid in white cabbage as affected by fertilization, *Acta Hortic.*, 163, 221, 1984.
135. Amirov, B. V., Manankov, M. E., and Saparov, A. S., Effectiveness of combined application of herbicide and fertilizer in cabbage cultivation, *Vestn. S-Kh. Nauki Kaz.* 11, 34, 1990; *Chem. Abstr.*, 115, 87419h, 1991.
136. Finch, A. H., Jones, W. W., and van Horn, C. W., The influence of nitrogen upon the ascorbic acid content of several vegetable crops, *Proc. Am. Soc. Hortic. Sci.*, 46, 314, 1945.
137. Böttcher, H., Ziegler, G., and Diwisch, F., Einfluss überhöhter Stickstoffdüngung auf Haltbarkeit und Qualitätserhaltung bei der Lagerung von Möhren, *Arch. Gartenbau*, 17, 43, 1969.
138. Agarwala, S. C. and Hewitt, E. J., Molybdenum as a plant nutrient. IV. The interrelationships of molybdenum and nitrate supply in chlorophyll and ascorbic acid fractions in cauliflower plants grown in sand culture, *J. Hortic. Sci.*, 29, 291, 1954.
139. Comis, D., Nitrogen overload may shrivel vitamin content, *Agric. Res.*, 37, 10, 1989.
140. Wittwer, S. H., Schroeder, R. A., and Albrecht, W. A., Vegetable crops in relation to soil fertility. II. Vitamin C and nitrogen fertilizers, *Soil Sci.*, 59, 329, 1945.
141. Uddin, M. M. and Begum. S., Effect of fertilizers on vitamin C content of green chili (*Capsicum sp.*), *Bangladesh J. Sci. Ind. Res.*, 25(1-4), 118, 1990; *Chem. Abstr.*, 115, 231197d, 1991.
142. Largskii, Yu. N., The effect of fertilizers on changes in mobile forms of soil nutrients and on cucumber quality, *Agrokhimiya*, 2, 138, 1971; *Hortic. Abstr.*, 42, 1140, 1972.
143. Venter, F., Nitrate content of endive (*Cichorium endiva* L.), *Gartenbauwissenschaft*, 48(5), 230, 1983.
144. Jones, W. W., van Horn, C. W., Finch, A. H., Smith, M. C., and Caldwell, E., A note on ascorbic acid: nitrogen relationships in grapefruit, *Science*, 99, 103, 1944.
145. Jones, W. W., van Horn, C. W., and Finch, A. H., The influence of nitrogen nutrition of the tree upon the ascorbic acid content and other chemical and physical characteristics of grapefruit, *Univ. of Arizona, Ariz. College of Agric., Agric. Exp. Stn. Tech. Bull.*, 106, 456, 1945.
146. Kefford, J. F., The chemical constituents of citrus fruits, *Adv. Food Res.*, 9, 285, 1959.
147. Embleton, T. W., Reitz, H. J., and Jones, W. W., Citrus fertilization, in *The Citrus Industry*, Vol. III, Reuther, W., Ed., Univ. California, Division of Agricultural Sciences, Riverside, Calif., 1973, 122.
148. Erdman, J. W., Jr. and Klein, B. P., Harvesting, processing, and cooking influences on vitamin C in foods, in *Ascorbic Acid: Chemistry, Metabolism, and Uses*, Seib, P. A. and Tolbert, B. M., Eds., Adv. Chem. Series, No. 200, Am. Chem. Soc., Washington, D.C., 1982, 499.
149. Kolota, E., The effect of increasing NPK rates and of the number of top dressings with N on the yield and nutritive value of leeks. Part I. The effects on yield, dry

matter, vitamin C and sugar content, *Roczniki Nauk Rolniczych*, *A*, 99(4), 95, 1973; *Hortic. Abstr.*, 45, 3930, 1975.

150. Bzhalava, U. Sh., Effectiveness of nitrogen fertilizer forms and rates in a Meyer lemon orchard, *Subtrop. Kul't.*, 6, 87, 1983; *Hortic. Abstr.*, 54, 7641, 1984.

151. Babu, J. D., Lavania, M. L., and Misra, K. K., Qualitative changes in the fruit of pant lemon-1 *Citrus limon* Burm.) as influenced by N, P and K treatments, *Prog. Hortic.*, 16(3/4), 188, 1984; *Hortic. Abstr.*, 56, 3748, 1986.

152. Balraj Singh and Ranvir Singh, Effect of nitrogen fertilization on quality of lemon (*Citrus limon* Burm.), *Prog. Hortic.*, 16(3/4), 308, 1984; *Hortic. Abstr.*, 56, 3749, 1986.

153. Bhattacharya, A., Singh, R. P., and Singh, A. R., Studies in the effect of nitrogen on the growth, yield and quality of Kagzi lime (*Citrus aurantifolia* Swingle), *Prog. Hortic.*, 5(1), 41, 1973; *Hortic. Abstr.*, 45, 5354, 1975.

154. Hernández, J., Effect of nitrogen, phosphorus and potassium on yield, fruit quality and foliar contents of Persian lime, *Cult. Trop.*, 1(3), 49, 1979; *Hortic. Abstr.*, 51, 9755, 1981.

155. Hernández, J., Effect of NPK fertilization on the yield, fruit quality and leaf composition of Persian limes, *Cult. Trop.*, 5(1), 137, 1983; *Food Sci. Technol. Abstr.*, 17(3), J49, 1985.

156. Nagy, S. and Wardowski, W. F., Effect of agricultural practice, handling, processing, and storage of fruits, in *Nutritional Evaluation of Food Processing*, Karmas, E. and Harris, R. S., Eds., Van Nostrand Reinhold, New York, 1988, chapter 4.

157. Singh, M. P. and Agrawal, K. C., Studies in the nitrogen requirements of citrus. II. The effect of four schedules of N-levels on vigor, yield and fruit quality of young mandarins, *A. R. Hortic. Res. Inst.*, *Saharanpur*, 57, 1960; *Hortic. Abstr.*, 32, 3695, 1962.

158. Mann, M. S. and Sandhu, A. S., Effect of NPK fertilization of fruit quality and maturity of Kinnow mandarin, *Punjab Hortic. J.*, 28(1-2), 14, 1988; *Hortic. Abstr.*, 61, 6463, 1991.

159. Jones, W. W. and Parker, E. R., Ascorbic acid-nitrogen relations in Navel orange juice, as affected by fertilizer application, *Proc. Am. Soc. Hortic. Sci.*, 50, 195, 1947.

160. Smith, P. F., Reuther, W., and Gardner, F. E., Phosphate fertilizer trials with oranges in Florida. II. Effect on some fruit qualities, *Proc. Am. Soc. Hortic. Sci.*, 53, 85, 1949.

161. Reuther, W. and Smith, P. F., Relation of nitrogen, potassium, and magnesium fertilization to some fruit qualities of Valencia orange, *Proc. Am. Soc. Hortic. Sci.*, 59, 1, 1952.

162. Reig Feliú, A. and Albert Bernal, A., Cold storage of citrus. III. A second year of experiments on cold storage of Washington Navel oranges in relation to different levels of nitrogen fertilizer application, *An. Inst. Nac. Invest. Agron.* (*Madrid*), 15, 93, 1966; *Hortic. Abstr.*, 37, 1740, 1967.

163. Aso, P. J. and Stein, E., The timing and rate of nitrogen fertilizing for Valencia oranges, *Rev. Industr. Agríc. Tucumán*, 45, 107, 1967; *Hortic. Abstr.*, 39, 1422, 1969.

164. Wittwer, S. H., and Hibbard, A. D., Vitamin C–nitrogen relations in peaches as influenced by fertilizer treatment, *Proc. Am. Soc. Hortic. Sci.*, 49, 116, 1947.

165. Sen, N. L. and Chauhan, K. S., Effect of differential NPK fertilization on physico-chemical characters of pomegranate, *Punjab Hortic. J.*, 23(1/2), 59, 1983; *Hortic. Abstr.*, 54, 2959, 1984.

166. Teich, A. H. and Menzies, J. A., The effect of nitrogen, phosphorus and potassium on the specific gravity, ascorbic acid content and chipping quality of potato tubers, *Am. Potato J.*, 41, 169, 1964.

167. Augustin, J., Variation in the nutritional composition of fresh potatoes, *J. Food Sci.*, 40, 1295, 1975.

218   *Plant Vitamins*

218   *Plant Vitamins*

168. Augustin, J., McDole, R. E., McMaster, G. M., Painter, C. C., and Sparks, W. C., Ascorbic acid content in Russet Burbank potatoes, *J. Food Sci.*, 40, 415, 1975.
169. Shekhar, V. C., Iritani, W. M., and Arteca, R., Changes in ascorbic acid content during growth and short-term storage of potato tubers (*Solanum tuberosum* L.), *Am. Potato J.*, 55, 663, 1978.
170. Goldstein, W., Potato quality, *Bio-Dynamics*, 140, 3, 1981.
171. Barannikova, Z. D. and Mel'nikova, I. E., Qualitative composition of the protein and productivity in relation to increased rates of nitrogen fertilizers, *Sel'sk. Biol.*, 9, 53, 1984; *Chem. Abstr.*, 101, 209868n, 1984.
172. Rojek, S., Effect of spray irrigation and nitrogen fertilization on the quantity and quality of late potato crops, *Zesz. Nauk. Akad. Roln. Wroclawiu, Melior*, 142, 119, 1983; *Chem. Abstr.*, 102, 23584g, 1985.
173. Asenov, R. and Matakov, N., Effect of fertilizer and combined manure and fertilizer application on the yield and quality of potatoes for seed production, *Rast. Nauki*, 22(7), 48, 1985; *Biol. Abstr.*, 81(2), 13871, 1986.
174. Kaczorek, S., The effects of different proportions of N and P and different dates of application on some quality characteristics of young tubers of early potatoes, *Biuletyn Instytutu Ziemniaka*, 29, 81, 1983; *Field Crop Abstr.*, 39, 2947, 1986.
175. Kozlowski, M., Rydzik, W., and Ciecko, Z., Effect of nitrogen fertilization on potato yield, composition, and nutritional value, and the quality of potato silage, *Acta Acad. Agric. Tech. Olstenensis: Zootech.*, 32(361), 119, 1989; *Chem. Abstr.*, 114, 80475u, 1991.
176. Park, K. W. and Fritz, D., Influence of fertilization on quality components of radish grown in greenhouse, *Gartenbauwissenschaft*, 48(5), 227, 1983.
177. Patil, H. B. and Patil, A. A., Effect of nitrogen, phosphorus, potassium and their method of application on nutrients and ascorbic acid contents of radish cv. Japanese White, *S. Indian J. Hortic.*, 34(4), 266, 1986; *Hortic. Abstr.*, 58, 7569, 1988.
178. Ljones, B. and Sakshaug, K., Nitrogen effects on composition and yield components of raspberry cultivars, *Meld. Nor. Landbrukshoegsk.*, 46(12), 1, 1967.
179. Pfützer, G. and Pfaff, C., Untersuchungen auf Gehalt an Carotin und Vitamin C bei Gemüsen und Futterstoffen, *Angewandte Chemie*, 48(36), 581, 1935.
180. Wittwer, S. H., Vegetable crops in relation to soil fertility. IV. Nutritional quality of New Zealand spinach, *J. Nutr.*, 31, 59, 1946.
181. Schuphan, W., Depression physiologisch aktiver Kationen in Nahrunspflanzen als Folge moderner landwirtschaftlicher Kulturmassnahmen, *Ernährungs- Umschau*, 4, 148, 1971; *Hortic. Abstr.*, 42, 1008, 1972.
182. Avdonin, N. S. and Kochubei, I. V., The effect of a derno-podzolic soil and fertilizers on spinach chemical composition, *Vliyanie Svoistva Pochv Udobr. Kachestvo Rast.*, 4, 43, 1978; *Hortic. Abstr.*, 49, 4970, 1979.
183. Zhang, C. L., Watanabe, Y., and Shimada, N., The effect of nitrogen concentration in the nutrient solution on the growth and nutrient contents of hydroponically grown spinach, *Techn. Bull. Fac. Hortic.*, *Chiba University*, 43, 1, 1990; *Hortic. Abstr.*, 62, 2950, 1992.
184. Panova, Z. M., Kondakov, A. K., and Krainova, V. V., Fertilization and the chemical composition of strawberries, *Tr. Tsentr. In-ta Agrokhim. Obsluzh, S.-kh.*, 4(2), 10, 1976; *Hortic. Abstr.*, 47, 7300, 1977.
185. Haynes, R. J. and Goh, K. M., Effect of nitrogen and potassium applications on strawberry growth, yield and quality, *Commun. Soil Sci. Plant Anal.*, 18(4), 457, 1987.
186. Somers, G. F., Kelly, W. C., and Hamner, K. C., Influence of nitrate supply upon the ascorbic acid content of tomatoes, *Am. J. Bot.*, 38, 472, 1951.
187. Neubert, P., Untersuchungen über den Einfluss der Stickstoffdüngung auf Reifung, Ertrag und Qualität der Tomatenfrucht, *Arch. Gartenbau.*, 7, 29, 1959.

188. Hashad, M. N., Moursi, M. A., and Gomma, A., Some factors affecting ascorbic acid content of tomatoes, *Ann. Agric. Sci.*, (*Cairo*), 3(2), 81, 1958; *Hortic. Abstr.*, 33, 1080, 1963.
189. Anonymous, Trace elements for tomatoes, *Indian Fmg.*, 17(1), 42, 1967; *Hortic. Abstr.*, 38, 1249, 1968.
190. Chaudhuri, B. B. and De, R., Effect of soil and foliar application of nitrogen and phosphorus on ascorbic acid (vitamin C) in tomato, *Sci. Cult.*, 35(7), 317, 1969.
191. Aliev, D. A., The effect of a combination of microelements and N and P on the yield and quality of tomatoes, *Tr. Azerb. Nauč. Issled. Inst. Ovošč.*, 2, 175, 1970; *Hortic. Abstr.*, 41, 4220, 1971.
192. Chaudhuri, B. B. and De, R., Effect of soil and foliar application of nitrogen and phosphorus on the quality of tomato fruits, *J. Food Sci. Technol.*, 9, 16, 1972.
193. Burge, J., Mickelsen, O., Nicklow, C., and Marsh, G. L., Vitamin C in tomatoes: comparison of tomatoes developed for mechanical or hand harvesting, *Ecology of Food and Nutrition*, 4, 27, 1975.
194. Nedranko, L. V., Quality of the tomato cultivar Moldavskii Ranii in relation to changes in the nitrogen and phosphorus level of the nutrient solution, *Tr. Kishinev. S-Kh. Institut*, 161, 17, 1976; *Hortic. Abstr.*, 47, 7559, 1977.
195. Cheng, B. T., A greenhouse manuring experiment on *Cucumis sativas* and *Lycopersicon esculentum* in Québec, *J. Chinese Agric. Chem. Soc.*, 20(1/2), 9, 1982; *Food Sci. Technol. Abstr.*, 15(9), J1339, 1983.
196. Müller-Haslach, W., Arold, G., and Kimmel, V., Influence of the intensity of fertilization on the fruit quality of tomatoes, in *The Effects of Modern Production Methods on the Quality of Tomatoes and Apples*, Gormley, T. R., Sharples, R. O., and Dehandtschtter, J., Eds., Commission of the European Communities, Luxembourg, 1985, 47.
197. Müller-Haslach, W., Arold, G., and Kimmel, V., Einfluss der Düngungsintensität auf die Qualität von Tomaten, *Bayerisches Landwirt. Jahresb.*, Sonderh., 63(1), 81, 1986.
198. Kaniszewski, S., Elkner, K., and Rumpel, J., Effect of nitrogen fertilization and irrigation on yield, nitrogen status in plants and quality of fruits of direct seeded tomatoes, *Acta Hortic.*, 200, 195, 1987.
199. Montagu, K. D. and Goh, K. M., Effects of forms and rates of organic and inorganic nitrogen fertilizers on the yield and some quality indices of tomatoes (*Lycopersicon esculentum* Miller), *N. Z. J. Crop Hortic. Sci.*, 18(1), 31, 1990.
200. Zamanova, M. N., The effect of mineral fertilizers on the yield and quality of head cabbage, in *Puti Povysheniya Produktivn. Ob'-Irtysh Poimy*. Novosibirsk, USSR, 1976, 155; *Hortic. Abstr.*, 47, 11374, 1977.
201. Asamov, D. K. and Beknazarov, B. O., Effect of phosphorus nutrition on the content of ascorbic acid in cotton leaves, *Uzb. Biol. Zh.*, 5, 38, 1989; *Chem. Abstr.*, 112, 75862x, 1990.
202. Mishra, S. N. and Singh, A. P., Changes in carbohydrate fraction and ascorbic acid content of groundnut under the influence of sulfur and phosphorus fertilizers, *Indian J. Plant Physiol.*, 30(1), 134, 1987.
203. Shinohara, Y., Suzuki, Y., Shibuya, M., Yamamoto, M., and Yamasaki, K., Effect of fertilization and foliar spray treatment on the ascorbic acid content of tomato and sweet papper, *J. Jpn. Soc. Hortic. Sci.*, 49(1), 85, 1980; *Hortic. Abstr.*, 51, 8643, 1981.
204. Klein, L. B., Chandra, S., and Mondy, N. I., The effect of phosphorus fertilization on the chemical quality of Katahdin potatoes, *Am. Potato J.*, 57, 259, 1980.
205. Chauhan, D. S., Effect of manuring with phosphate, magnesium and manganese alone and in combination on growth, mineral matter and vitamin contents of spinach (*Beta vulgaris* L.) leaves, *Agra Univ. J. Res. (Science)*, 19(3), 87, 1970.

206. Maurya, K. R. and Dhar, N. R., Influence of phosphatic fertilizers alone and in combination with water hyacinth on the yield and composition of sweet potato (*Ipomoea batatas* Poir), *Mysore J. Agric. Sci.*, 10(3), 387, 1976; *Chem. Abstr.*, 88, 135545u, 1978.

207. Dumbadze, N. M., The effect of different forms of phosphorus fertilizers on some qualitative indices of satsumas, *Subtrop. Kul't.*, 5/6, 143, 1976; *Hortic. Abstr.*, 48, 1773, 1978.

208. Brown, H. D., Patton, M. B., and Blythe, A., Influence of mineral levels upon carotene and ascorbic acid contents of Swiss chard grown in the greenhouse, *Food Res.*, 12, 4, 1947.

209. Cohen, A., Citrus Fertilization, *Bull. Int. Potash Inst.* (*Bern, Switzerland*), 4, 1976.

210. Krieg, F., Einfluss der Nährstoffversorgung auf die Ozonanfälligkeit der Weinrebe (*Vitis vinifera* L. cv. Pinot noir), Thesis, Institute of Plant Sciences, Swiss Federal Institute of Technology, Zürich, Switzerland, 1992.

211. Lekvindaze, P. A., The effect of different superphosphates rates on some quality indices in satsumas, *Subtrop. Kul't.*, 5, 83, 1972; *Hortic. Abstr.*, 43, 8072, 1973.

212. Sites, J. W. and Camp, A. F., Producing Florida citrus for frozen concentrate, *Food Technol.*, 9, 361, 1955.

213. Mustaffa, M. M., Growth and yield of Robusta banana in relation to potassium nutrition, *J. Potassium Res.*, 3(3), 129, 1987; *Hortic. Abstr.*, 59, 1649, 1989.

214. Mustaffa, M. M., Studies on growth, yield and quality of hill banana as a result of potassic fertilizer use, *J. Potassium Res.*, 4(2), 75, 1988; *Hortic. Abstr.*, 59, 7902, 1989.

215. Ott, M., Pflanzenqualität, Volksernährung und Düngung, *Forschungsdienst*, 5, 546, 1938.

216. Pfützer, G., Pfaff, C., and Roth, H., Die Vitaminbildung der höheren Pflanze in Abhängigkeit von ihrer Ernährung, *Landw. Forsch.*, 4(2), 105, 1952.

217. Umarova, M. Z., Effect of K fertilization on yield and nutrient utilization by cabbages, *Agrokhimiya*, 11, 56, 1973; *Food Sci. Technol. Abstr.*, 6(5), J615, 1974.

218. Luchnik, N. A., The effect of potassium fertilizers on the yield and quality of head cabbage grown on chestnut brown soil, *Sbornik Nauchnykh Statei Karagand. Gos. S.-Kh. Opytnoi Stantsii*, 5, 110, 1975; *Hortic. Abstr.*, 46, 9290, 1976.

219. Zhukova, P. S. and Belova, V. I., Effectiveness of herbicides on white cabbage plantings with the application of various doses of potassium fertilizers, *Khim. Sel'sk. Khoz.*, 10, 36, 1983; *Chem. Abstr.*, 99, 208058b, 1983.

220. Richter, R., Svoboda, J., and Chmela, V., The effect of graduated rates of K_2O on the yield and quality of sweet peppers, *Bull. Vyzk. Ust. Zelin., Olomouc*, 12/13, 93, 1968/1969; *Hortic. Abstr.*, 41, 6829, 1971.

221. Bubicz, M., Korzen, A., and Perucka, I., Effect of potassium fertilization on the content of L-ascorbic acid, β-carotene, and capsaicin in fruits of pepper (*Capsicum annuum* L.), *Rocz. Nauk Roln., Ser. A.*, 104(4), 43, 1981; *Chem. Abstr.*, 96, 103068q, 1982.

222. Wolf, E., Einfluss der Wachstumsdauer und steigender Nährstoffgaben auf den Karotin- und Vitamin-C-Gehalt von Möhren und Sellerie, *Landwirt. Forschung*, 7(2), 139, 1955.

223. Sanchez Conde, M. P., The response of courgette plants (*Cucurbita pepo*) to different nitrogen and potassium doses, *Agrochimica*, 32(2-3), 108, 1988; *Hortic. Abstr.*, 59, 3866, 1989.

224. Veveris, Ya. Ya., The optimum potassium rate for cucumbers grown in greenhouses on sphagnum peat, in *Nauch. Osnovy Progressiv. Tekhnol. Ovoshchevodstve, Riga, Latvian SSR, "Zinatne"*, 1974, 71; *Hortic. Abstr.*, 45, 2389, 1975.

225. Mitra, S. K., Studies on guava nutrition with special reference to potassium and nitrogen, *J. Potassium Res.*, 3(4), 160, 1987; *Soils and Fertilizers*, 52, 8470, 1989.

226. Smith, P. F. and Rasmussen, G. K., Relationship of fruit size, yield and quality of Marsh grapefruit to potash fertilization, *Proc. Fla. St. Hortic. Soc. 1960*, 73, 42, 1961; *Hortic. Abstr.*, 31, 6871, 1961.

227. Embleton, T. W. and Jones, W. W., Potassium builds lemon quality, *Bett. Crops*, 52(1), 18, 1968; *Hortic. Abstr.*, 39, 3565, 1969.

228. Embleton, T. W., Jones, W. W., Platt, R. G., and Burns, R. M., Potassium nutrition and deficiency in citrus, *Calif. Agric.*, 28(8), 6, 1974.

229. Lekvinadze, F. A., The effect of various potassium rates on the quality of satsumas, *Subtrop. Kul't.*, 6, 95, 1971; *Hortic. Abstr.*, 42, 8393, 1972.

230. Mustafa, A. M. and Chzhao, A. E., Effect of various fertilizer combinations on the resistance of onion to neck rot and bulb storability, *Izv. Timiryazevsk. S-kh. Akad.*, 5, 181, 1982; *Chem. Abstr.*, 97, 181075z, 1982.

231. Smith, P. F., Effect of potassium level and substrate lime on growth, fruit quality, and nutritional status of Valencia orange trees, *Plant Analy. Fert. Problems*, 4, 332, 1964.

232. Embleton, T. W., Jones, W. W., Pallares, C., and Platt, R. G., Effects of fertilization of citrus on fruit quality and ground water nitrate-pollution potential, in *Proc. Int. Soc. Citriculture, 1978*, Griffith, Australia, 1980, 280; *Hortic. Abstr.*, 51, 4033, 1981.

233. Desai, U. T., Choudhari, K. G., and Chaudhari, S. M., Studies on the nutritional requirements of sweet orange, *J. Maharashtra Agric. Univ.*, 11(2), 145, 1986; *Chem. Abstr.*, 105, 190081j, 1986.

234. Müller, K., Zur Frage der Kalidüngung zu Kartoffeln, *Kartoffelbau*, 39(3), 102, 1988.

235. Velazhahan, R., and Ramabadran, R., Changes in ascorbic acid content of rice following potassium fertilization and infection by *Sarocladium oryzae*, *Sci. Cult.*, 56, 460, 1990.

236. Ijdo, J. B. H., The influence of fertilizers on the carotene and vitamin C content of plants, *Biochem J.*, 30, 2307, 1936.

237. Kaziev, M. Z. and Tursumetov, A. A., Tomato yields and quality in relation to potassium rates, *Nauch. Tr., Tashkent. S. Kh. Inst.*, 32, 54, 1972; *Hortic. Abstr.*, 44, 3297, 1974.

238. Anand, N. and Muthukrishnan, C. R., Effect of potassium on growth, yield and quality of tomato, *Potash Rev.*, 8/9, 1, 1974; *Hortic. Abstr.*, 46, 4686, 1976.

239. Sobulo, R. A. and Olorunda, A. O., The effects of nitrogen, phosphorus and potassium on the canning quality of tomatoes (*Lycopersicon esculentum*) in southwestern Nigeria, *Acta Hortic.*, 53, 171, 1977.

240. Gould, W. A., *Tomato Production, Processing and Quality Evaluation*, 2nd Ed., AVI Publishing, Westport, Conn., 1983.

241. Ferres, H. M. and Brown, W. D., The effect of mineral nutrients on the concentration of ascorbic acid in legumes and two leaf vegetables, *Aust. Exp. Biol. Med. Sci.*, 24, 111, 1946.

242. Sanchez Conde, M. P., Evaluation de la influencia de diferentes niveles de potasio y de la luz sobre el contenido de vitamina C en lechuga (*Lactuca sativa*), *An. Edafol. Agrobiol.*, 41, 1047, 1982.

243. Ul'yanova, D. A., Plant nutrition and fruit quality, *Sadovodstvo*, 6, 22, 1978; *Hortic. Abstr.*, 48, 8938, 1978.

244. Reder, R., Ascham, L., and Eheart, M. S., Effect of fertilizer and environment on the ascorbic acid content of turnip greens, *J. Agric. Res. (Washington, D.C.)*, 66(10), 375, 1943.

245. Bangerth, F., Cooper, T., and Filsouf, F., Increasing the ascorbic acid content of apples with calcium, *Naturwissenschaften*, 61(9), 404, 1974; *Hortic. Abstr.*, 45, 3733, 1975.

246. Bangerth, F., Beziehungen zwischen dem Ca-Gehalt Bzw. der Ca versorgung von Apfel-, Birnen- und Tomatenfrüchten und ihren Ascorbinsäuregehalt, *Qual. Plant. —Plant Foods Hum. Nutr.*, 26(4), 341, 1976.

247. Lasekan, O. O., Effect of calcium on the storage life of oro (*Antiaris africana*), *J. Sci. Food Agric.*, 51(2), 281, 1990.

248. Al'shevskii, N. G., Effect of magnesium and boron on metabolism and productivity of potato on sodpodzolic sandy loam soils of Polessie, *Agrokhimiya*, 7, 93, 1986; *Chem. Abstr.*, 105, 114274j, 1986.

249. Sapatyi, S. E., The effect of minor elements on the improvement of fruit quality in black currants, *Nauchn. Tr. Ukr. Sel'sk. Akad.*, 57, 117, 1971; *Hortic. Abstr.*, 42, 7493, 1972.

250. Torres Chinea, E., Dominguez Martin, Q., and Crispin, E. C., Effects of zinc, magnesium, and filter cake on the yield and quality of the pineapple variety "Red Spanish," *Cent. Agric.*, 13(2), 74, 1986; *Chem. Abstr.*, 106, 101307t, 1987.

251. Al'shevskii, N. G. and Shulyarenko, P. I., Comparative effectiveness of applying magnesium ammonium phosphate and epsomite to potatoes, *Agrokhimiya*, 5, 65, 1983; *Chem. Abstr.*, 99, 37554d, 1983.

252. Hernández-Medina, E. and Vélez-Santiago, J., Response of acerola (*Malpighia punicifolia* L.) to lthe application of lime and foliar sprays of magnesium and minor elements, *Proc. Carib. Reg. Am. Soc. Hortic. Sci.*, 4, 20, 1960; *Hortic. Abstr.*, 31, 5547, 1961.

253. Mondy, N. I. and Ponnampalam, R., Potato quality as affected by source of magnesium fertilizer nitrogen, minerals, and ascorbic acid, *J. Food Sci.*, 51(2), 352, 1986.

254. Cakmak, I. and Marschner, H., Magnesium deficiency and high light intensity enhances activities of superoxide dismutase, ascorbate peroxidase, and glutathione reductase in bean leaves, *Plant Physiol.*, 98, 1222, 1992.

255. Sanchez Conde, M. P., Response of lettuce (*Lactuca sativa*) to different levels of magnesium and light in controlled cultivation, *Agrochimica*, 30(6), 465, 1986; *Hortic. Abstr.*, 58, 876, 1988.

256. Kuznetsov, A. V. and Treshchov, A. G., Effect of magnesium on the quality and amino acid composition of mandarins, *Tr. Univ. Druzhby Nar. im. Patrisa Lumumby*, 90(14), 114, 1980; *Chem. Abstr.*, 97, 22708s, 1982.

257. Chikalova, E. A., Dynamics of vitamin C accumulation in apples during foliar spraying of the trees with manganese and cobalt salts, *Tr. Vses. Semin. Biol. Aktiv. (Lech.) Veshchestvam Plodov Yagod*, 4th, 257, 1970; *Chem. Abstr.*, 81, 76922v, 1974.

258. Jovanović, M., The effect of boron and manganese nutrition on the properties of Golden Delicious apple fruits, *Zbornik Radova Poljoprivrednog Fac.*, 20(540), 1, 1972; *Hortic. Abstr.*, 44, 8237, 1974.

259. Zaporozhan, Z. E., Changes in yield quality of early cabbage under the effect of minor elements applied on degraded chernozems, *Nauchn. Tr. Ukr. Sel'sk. Akad.*, 57, 158, 1971; *Hortic. Abstr.*, 42, 7687, 1972.

260. Dzhavakhishvili, D. L. and Egorashvili, N. V., Use of manganese and boron fertilizers for tomatoes and cabbage on forest cinnamonic soils of Eastern Georgia [USSR], *Khim. S-Kh. Khoz.*, 8, 38, 1985; *Chem. Abstr.* 103, 177496n, 1985.

261. Nikolaevskaya, A. A., Gromovaya, N. D., and Dumenko, M. F., Preplanting treatment of seeds and its effect on growth, development, yield and quality of carrots, *Nauk. Pr.-Ukr. Sil's'kogospod. Akad.*, 96, 1981; *Chem. Abstr.*, 98, 124832r, 1983.

262. Pidoplichko, V. M., Kaznachei, R. Ya., and Kononko, L. N., Effect of trace elements on cucumber immunity against root rot on the content of some vitamins during hydroponic culture, *Visn. Sil's'kogospod. Nauki*, 11(6), 77, 1968; *Chem. Abstr.*, 70, 2853z, 1969.

263. Bronsart, H. V., Erhöhung des Vitamin-C-Gehaltes durch Mangandüngung, *Z. Pflanzenernähr. Düng. Bodenk.*, 51(2), 153, 1950.

264. Naichenko, V. M. and Shamotienko, G. D., Effect of some trace elements on the accumulation and preservation of ascorbic acid in black currants, *Konservn. Ovoshchesush. Prom-st.*, 9, 26, 1982; *Chem. Abstr.*, 98, 52530r, 1983.

265. Tolpeikina, G. I., The effect of manganese on the productivity of red currants, in *Voprosy Khimii Biokhimii Sistem, Soderzh. Marganets Polifenoly, Chelyabinsk, USSR*, 3, 47, 1975; *Hortic. Abstr.*, 47, 5353, 1977.

266. Lešina, A. V., The effect of manganese on the pigment and ascorbic acid contents of lettuce, *Dokl. Akad. Nauk Beloruss. SSR*, 10(4), 279, 1966; *Hortic. Abstr.*, 37, 7007, 1967.

267. Klyshev, L. K., Pershukova, A. M., and Lysenko, A. I., The role of manganese in tannin-bearing plants, *Izv. Akad. Nauk Kaz. SSR, Ser. Biol.*, 5, 32, 1986; *Chem. Abstr.*, 106, 99471w, 1987.

268. Stanchev, L. B., Effect of manganese and zinc on the quality of strawberries, *Pochvoznanie Agrokhimiya*, 9(5), 62, 1974; *Hortic. Abstr.*, 45, 5737, 1975.

269. Hester, J. B., Manganese and vitamin C, *Science*, 93, 401, 1941.

270. Ruszkowska, M., Some experiments on the physiological role of manganese in tomato plants, *Acta Soc. Bot. Polon.*, 29, 553, 1960; *Hortic. Abstr.*, 32, 3139, 1962.

271. Stancu, E., The influence of the microelements manganese, boron, iron and zinc on the chemical composition of tomatoes, *Lucr. Šti. Inst. Agron. N. Balcescu, Ser. B*, 7, 119, 1964; *Hortic. Abstr.*, 35, 8041, 1965.

272. Mehrotra, O. N., Saxena, H. K., and Dube, S. D., Effect of trace elements on ascorbic acid content of tomatoes (*Lycopersicon esculentum*, Mill), *Labdev J. Sci. Technol., B*, 8(1), 38, 1970; *Hortic. Abstr.*, 45, 394, 1975.

273. Nezhnev, Yu. N. and Zubanova, L. S., The effect of manganese on tomato yield and fruit quality, *Agrokhimiya*, 4, 104, 1978; *Hortic. Abstr.*, 49, 1254, 1979.

274. El-Fouly, M. M., Über den Einfluss der Manganernährung auf den Vitamin C-Gehalt von Pflanzenblättern, *Plant Soil*, 24(3), 473, 1966.

275. Aiyappa, K. M., Srivasrava, K. C., Bojappa, K. M., and Sulladmath, U. V., Studies on the effect of micronutrient sprays singly and in various combinations on *Citrus reticulata* Blanco (Coorg mandarin seedling trees), *Indian J. Hortic.*, 25, 104, 1968.

276. Hivon, K. J., Doty, D. M., and Quackenbush, F. W., Ascorbic acid and ascorbic-acid-oxidizing enzymes of manganese-deficient soybean plants growing in the field, *Soil. Sci.*, 71, 353, 1951.

277. Lyon, C. B. and Beeson, K. C., Manganese and ascorbic acid formation, *J. Am. Soc. Agron.*, 35, 166, 1943.

278. Gum, O. B., Brown, H. D., and Burrell, R. C., Some effects of boron and manganese on the quality of beets and tomatoes, *Plant Physiol.*, 20, 267, 1945.

279. Marx, Th., and Sahm, U., Über den Einfluss von Mangan- und Bordüngungen auf den L-Ascorbinsäuregehalt der Tomaten, *Z. Pflanzenernähr. Düng. Bodenk.*, 59, 157, 1952.

280. Kodua, M. A., Effect of boron and manganese on mandarin yield and quality, in *Effectivn. Primeneniya Mikroudobr. v Respublikakh Zakavkaz'ya Materialy Nauch. Soveshch., Tiblisi, Georgian SSR*, 1980, 165; *Hortic. Abstr.*, 51, 7384, 1981.

281. Klimiene, I., Effect of trace element doses on vitamin C content in potato tubers, *Agron., Melior., Gidrotekh.*, 22, 1980; *Chem. Abstr.*, 96, 51303r, 1982.

282. Ljubkin, Ju. I., The effect of boron on bearing apple orchards, *Tr. Krasnojarsk. sel'.-hoz. Inst.*, 19, 396, 1968; *Hortic. Abstr.*, 40, 249, 1970.

283. Dube, S. D., Tewari, J. D., and Ram, C. B., Effect of carrier of boron on the quality of apple variety Pymer, *Prog. Hortic.*, 5(3), 67, 1973; *Hortic. Abstr.*, 45, 7081, 1975.

284. Kononovich, A. L., The effect of microelements on the quality of carrot yield, *Uchenye Zapiski Blagoveshchenskogo Gosudarstvennogo Pedagogicheskogo Instituta*, 17, 17, 1971; *Hortic. Abstr.*, 42, 4130, 1972.

285. Novikova, O. S., Titova, O. I., Blinova, M. B., Safonov, A. P., and Mukhitdinova, D. R., Efficiency of boron in combination with ammonium nitrate, *Khimizatsiya Sel'skogo Khozyaistva*, 5, 10, 1991; *Hortic. Abstr.*, 62, 4919, 1992.

286. Singh, J., Singh, S. B., Addy, S. K., Singh, A., Raghav, M., Pathak, S., and Srivastava, S., Effect of micronutrients on the synthesis of major biochemical constituents of cauliflower (*Brassica oleracea* Botrytis group), *J. Recent Adv. Appl. Sci.*, 2(2), 324, 1987; *Chem. Abstr.*, 109, 91807b, 1988.

287. Gluntsov, N. M., Zabolotnova, L. A., Skvortsova, N. K., Utkina, T. N., and Puzanova, O. A., Effect of boron on the yield and quality of cucumber and tomato fruits, *Agrokhimiya*, 8, 75, 1989; *Chem. Abstr.*, 111, 213834x, 1989.

288. Baikov, G. K. and Muryseva, N. M., The effect of minor elements on the yield and vitamin content of black currants, in *Dikorastushchie Introdutsiruemye Poleznye Rasteniya Bashkirii*. Ufa, USSR, No. 4, 1974, 126; *Hortic. Abstr.*, 46, 953, 1976.

289. Krupyshev, P. V., The effect of boron and zinc on the chemical composition of black currant fruit, *Puti Adaptatsii Rast. pri Introduktsii Severe. Petrozavodsk, USSR*, 1977, 154; *Hortic. Abstr.*, 48, 5339, 1978.

290. Chaturvedi, G. S., Gautam, N. C., Padmakar, Prasad, R., and Awasthi, C. P., Effect of boron on biochemical constituents and yield of brinjal plants (*Solanum melongea* L.), *Proc. Natl. Acad. Sci., India, Sect. B*, 57(4), 513, 1987.

291. Mishulina, I. A., The effect of foliar nutrition with minor elements of *Hippophaë rhamnoides* on the fruit content of biologically active substances, *Biol. Aktiv. Veshchestva Plodov Yagod.*, Moscow, USSR, 97, 1976; *Hortic. Abstr.*, 48, 320, 1978.

292. Tabin, S., The effect of boron nutrition on the growth, development, yield and chemical composition of Jerusalem artichokes, *Roczn. Nauk Rol., Ser. A*, 91, 721, 1966; *Hortic. Abstr.*, 37, 4777, 1967.

293. Lim, S. U. and Kim, J. H., The effects of boron application on the yield and quality of potatoes (*Solanum tuberosum* L.), *Han'guk Nonghwa Hakhoechi*, 26(3), 191, 1983; *Chem. Abstr.* 100, 155745d, 1984.

294. Verma, V. K., Singh, N., and Choudhury, B., Studies on the effect of chemicals on the nutritional content of pumpkin (*Cucurbita moschata* Duch. ex Poir.), *S. Indian Hortic.*, 33(4), 261, 1985; *Hortic. Abstr.*, 57, 1141, 1987.

295. Maurya, K. R. and Singh, B. K., Effect of boron on growth, yield, protein and ascorbic acid content of radish, *Indian J. Hortic.*, 42(3/4), 281, 1985.

296. Freeman, J. A., The control of strawberry fruit rot in coastal British Columbia, *Can. Plant Dis. Survey*, 44, 96, 1964; *Hortic. Abstr.*, 35, 774, 1965.

297. Oza, A. M. and Rangnekar, Y. B., Effect of soil and foliar application of boron on ascorbic acid content of tomato fruits, *J. Indian Soc. Soil. Sci.*, 16, 423, 1968.

298. Cheng, B. T., Interaction among B, Cu and Zn on the growth of *Avena sativa* L. and *Lycopersicon esculentum* Mill, *J. Chinese Agric. Chem. Soc.*, 25(2), 198, 1987; *Food Sci. Technol. Abstr.*, 20(4), J63, 1988.

299. Ahmed, A. J. and Hargitai, L., Effect of boron on soil nitrogen dynamics and on tomato yield in model experiments, *Agrokem. Talajtan*, 38, 270, 1989; *Chem. Abstr.*, 111, 152694c, 1989.

300. Husain, S. A., Shaik Mohammad, and Rao, B. V. R., Response of chili (*Capsicum annuum*) to micronutrients, *Indian J. Agron.*, 34(1), 117, 1989.

301. Lyon, C. B. and Parks, R. Q., Boron deficiency and the ascorbic-acid content of tomatoes, *Bot. Gaz. (Chicago)*, 105, 392, 1944.

302. Singh, Z. and Dhillon, B. S., Effect of foliar application of boron on vegetative and panicle growth, sex expression, fruit retention and physicochemical characters of

fruits of mango (*Mangifera indica* L.) cv Dusehri, *Trop. Agric.* (*Guildford, U. K.*), 64(4), 305, 1987; *Hort. Abstr.*, 59, 1687, 1989.

303. Chitkara, S. D. and Bhambota, J. R., Effect of different concentrations of iron sprays on the incidence of chlorosis in sweet orange (*Citrus sinensis*), *Indian J. Hortic.*, 28(1), 16, 1971.

304. De, S. K., Singha, A., and Verma, S. N., Ascorbic acid content in potato (*Solanum tuberosum*) as affected by pyrite and ferrous sulfate application (both as soil application and foliar spray) with normal dose of manure under different soil moisture conditions, *Indian J. Agric. Chem.*, 16(2), 253, 1983; *Chem. Abstr.*, 102, 45043g, 1985.

305. Fazalur Rahman Mallick, M. and Muthukrishnan, C. R., Effect of micronutrients on the quality of tomato (*Lycopersicon esculentum* Mill.), *Vegetable Sci.*, 7(1), 6, 1980.

306. Tandon, P. K. and Saxena, H. K., Influence of zinc, iron and phosphorus supply on the rate of photosynthesis and contents of sugars and ascorbic acid in rice, *Indian J. Agric. Chem.*, 19(2), 75, 1986; *Chem. Abstr.*, 108, 5484r, 1988.

307. Sideris, C. P. and Young, H. Y., Effects of iron on chlorophyllous pigments, ascorbic acid, acidity and carbohydrates of *Ananas comosus* (L.) Merr., supplied with nitrate or ammonium salts, *Plant Physiol.*, 19, 52, 1944.

308. Lyon, C. B., Beeson, K. C., and Ellis, G. H., Effects of micro-nutrient deficiencies on growth and vitamin content of the tomato, *Bot. Gaz.* (*Chicago*), 104(4), 495, 1943.

309. Amberger, A. and El-Fouly, M. M., Ascorbinsäureoxydaseaktivität und Ascorbinsäuregehalt in Pflanzen bei verschiedener Stickstoff- und Spurenelementeernährung, *Z. Pflanzenernähr. Düng. Bodenk.*, 105, 37, 1964.

310. Nikolov, B. A. and Peterburgskij, A. V., The effect of micronutrients on the yield and quality of legumes, *Izv. Timiryazevsk. S-kh. Akad.*, 4, 141, 1967; *Hortic. Abstr.*, 38, 989, 1968.

311. Kosinova, V. P. and Rudin, V. D., The effect of the micro-elements cobalt and molybdenum on white cabbage yield and quality, *Nauchn. Tr., Stavrop. Sel'sk. Inst.*, 3(36), 89, 1973; *Hortic. Abstr.*, 45, 2359, 1975.

312. Domska, D., Benedycka, Z., and Krauze, A., Effect of molybdenum fertilization on protein yield, vitamin C and nitrogen compounds content in white head cabbage, *Acta Acad. Agric. Tech. Olstenensis: Agric.*, 46, 85, 1988; *Chem. Abstr.*, 112, 6642p, 1990.

313. Avdonin, N. S. and Arens, I., Effect of molybdenum on clover biochemical processes and quality, *Vliyanie Svoistv Pochv Udobr. Kach. Radt.*, 114, 1966; *Chem. Abstr.*, 67, 63375e, 1967.

314. Petrov, H. and Gorbanov, S., The effect of foliar nutrition with molybdenum on lettuce yield and quality, *Nauchni Trudove, Vissh Selskostopanski Inst. "Vasil Kolarov,"* 21(2), 71, 1972; *Hortic. Abstr.*, 43, 7630, 1973.

315. Tavadze, A. M. and Dzhibladze, A. D., Effect of molybdenum on the yield and quality of mandarin oranges in the western Georgian SSR, *Khim. Sel'sk. Khoz.*, 11, 27, 1983; *Chem. Abstr.*, 100, 50571r, 1984.

316. Munshi, C. B. and Mondy, N. I., Effect of soil application of sodium molybdate on the quality of potato: polyphenol oxidase activity, enzymatic discoloration, phenols, and ascorbic acid, *J. Agric. Food Chem.*, 36, 919, 1988.

317. Stoimenov, S. and Stoyanova, I., Effect of molybdenum and glucose in reversing the unfavorable consequences of high-rate ammonium nitrate application in relation to nitrogen and carbohydrate metabolism, *Pochvozn., Agrokhim. Rastit. Zasht.*, 20(4), 17, 1985; *Chem. Abstr.*, 104, 5092g, 1986.

318. Luo, X., Hu, G., Shang, A., Wei, H., Feng, Y., and Qin, Q., Nitrogen [nitrate and nitrite] decontamination [of cereals and vegetables] by molybdenum in high incidence area of esophageal cancer, *J. Environ. Sci.* (*China*), 1(1), 85, 1989; *Chem. Abstr.*, 112, 54280r, 1990.

319. Dhakshinamoorthy, M. and Krishnamoorthy, K. K., Stuies on the effect of zinc and copper on the yield and quality of brinjal, *Madras Agric. J.*, 76(6), 345, 1989; *Hortic. Abstr.*, 61, 3785, 1991.

320. Lucas, R. E., Effect of copper fertilization on carotene, ascorbic acid, protein, and copper contents of plants grown on organic soils, *Soil Sci.*, 65, 461, 1948.

321. Livdane, B. and Ozolina, G., Activity of ascorbate oxidase and content of ascorbic acid in plant leaves differently supplied with copper, *Regul. Rosta Metab. Rast.*, 221, 1983; *Chem. Abstr.*, 100, 208540s, 1984.

322. Lukovnikova, G. A. and Kuliev, K. A., The effect of copper and boron on certain biochemical processes in carrots and cauliflower, *Byulleten' Vsesoyuznogo Ordena Lenina Instituta Rastenievodstva Imeni N.I. Vavilova*, 59, 66, 1976; *Hortic. Abstr.*, 47, 5695, 1977.

323. Bacha, M. A. A., Responses of "Succary" and "Balady" orange trees to foliar sprays of zinc and copper, *Indian J. Agric. Sci.*, 45(5), 189, 1975; *Hortic. Abstr.*, 47, 11883, 1977.

324. Adedeji, F. O. and Fanimokun, V. O., Copper deficiency and toxicity in two tropical leaf vegetables (*Celosia argentea* L. and *Amaranthus dubius* Mart. ex Thell.), *Environ. Exper. Bot.*, 24(1), 105, 1984.

325. Gyul'akhmedov, A. N., Agaev, N. A., Azimov, A. M., and Agaeva, T. M., Effect of the trace element zinc applied together with mineral fertilizers on alfalfa, *Dokl. Akad. Nauk Az. SSR*, 41(4), 58, 1985; *Chem. Abstr.*, 104, 5091f, 1986.

326. Stojkovska, A., Trpevski, V. Vojnovski, B., and Čepujonovska, V., The effect of Fe, Zn, Cu, and Mn chelates on fruit quality in the apple cultivar Golden Delicious, *Godišen Zbornik Zemjodelsko-Sumarskiot Fakultet Univ.—Skopje, Ovoštarstvo*, 25, 145, 1972/1973; *Hortic. Abstr.*, 45, 9147, 1975.

327. Fekete, L., The effect of foliar nutrition on eating capsicum yields and vitamin C content, in *Agrártudományi Egyetem Közleményei*, Gödöllo (Hungary), 1974, 201; *Hortic. Abstr.*, 46, 8474, 1976.

328. Reddy, K. J., Effect of zinc on ascorbic acid and ascorbic acid oxidase activity in chickpea, *Indian J., Bot.*, 8(2), 203, 1985.

329. Stoyanov, D. and Gikov, G. N., Effect of zinc nutrition on crops grown in a leached chernozemsmonitza with high levels of mobile phosphorus, *Pochvozn. Agrokhim.*, 25(2), 34, 1990; *Chem. Abstr.*, 113, 151378g, 1990.

330. Vekirchik, K. N., Effect of zinc on sexual development of cucumber plants, *Rost. Rast. Puti Ego Regul.*, Yakushkina, N. I., Ed., Moscow, 1981, 119; *Chem. Abstr.*, 97, 22713q, 1982.

331. Mdinaradze, T. D. and Kechakmadze, M. S., Trace elements and shelf life of mandarin oranges, *Subtrop. Kul't.*, 2, 101, 1982; *Chem. Abstr.*, 97, 90664z, 1982.

332. Arora, J. S. and Singh, J. R., Some effects of foliar spray of zinc sulphate on growth, yield and fruit quality of guava (*Psidium guajava*), *J. Jpn. Soc. Hortic. Sci.*, 39, 207, 1970; *Hortic. Abstr.*, 41, 9944, 1971.

333. Manchanda, H. R., Effect of different micronutrient sprays on chlorosis, granulation, fruit fall, yield and fruit quality of sweet oranges var. Blood-red, *J. Res. (Ludhiana)*, 4, 508, 1967; *Hortic. Abstr.*, 39, 1433, 1969.

334. Nguyen Quang La, The effect of boron and zinc on the ascorbic acid content of peas, *Kertészeti Egyetem Közleményei*, 38(6), 43, 1974; *Hortic. Abstr.*, 45, 9646, 1975.

335. Sil'janova, Ju. I., The effect of boron and zinc on enzyme activity and accumulation of ascorbic acid in the fruit of cultivated rowan, Tr. *Kazansk. Sel'sk. Inst.*, 44, 31, 1964; *Hortic. Abstr.*, 36, 342, 1966.

336. Somers, G. F. and Beeson, K. C., The influence of climate and fertilizer practices upon the vitamin and mineral content of vegetables, *Adv. Food Tech.*, 1, 291, 1948.

337. Khomchak, M. E., Shiyan, O. I., and Zaporozhan, Z. E., The effect of minor elements on tomato yield and fruit quality, *Nauch. Tr. Ukr. Sel'sk. Akad.*, 57, 165, 1971; *Hortic. Abstr.*, 42, 7921, 1972.

338. Mohapatra, A. R. and Kibe, M. M., Response of tomato to zinc fertilization in a zinc-deficient soil of Maharashtra, *Indian J. Agric. Sci.*, 41(8), 650, 1971; *Hortic. Abstr.*, 42, 6210, 1972.

339. Seitkuliev, Ya. S., Tailakov, N., and Rozyeva, M., Effect of fertilizer on the quality of tomato fruits grown on irrigated light Sierozems of Turkmenistan, *Izv. Akad. Nauk Turkm. SSR, Ser. Biol. Nauk*, 5, 35, 1990; *Chem. Abstr.*, 115, 231199f, 1991.

340. Zamfirescu, N., and Tacu, F., The effect of some trace elements on the synthesis of ascorbic acid in maize, *Lucrări Stiintifice, Institutul Agronomic'N. Bălcescu', Bucuresti, A (Agronomie)*, 13, 215, 1970; *Field Crop Abstr.*, 26, 149, 1973.

341. Lipskaya, G. A. and Ivanov, N. P., Role of cobalt in the enhancement of yields on peat bog soils of the transitional type, *Agrokhimiya*, 12, 86, 1988; *Chem. Abstr.*, 110, 93987e, 1989.

342. Hageman, R. H., Hodge, E. S., and McHargue, J. S., Effect of potassium iodide on the ascorbic acid content and growth of tomato plants, *Plant Physiol.*, 17, 465, 1942.

343. Ogoleva, V. P. and Cherdakova, L. N., Effect of nickel on biochemical processes in alfalfa, *Khim. Sel'sk. Khoz.*, 3, 58, 1986; *Chem. Abstr.*, 104, 185458n, 1986.

344. Niranjana, K. V. and Devi, L. S., Influence of P and S on yield and quality of chillies, *Curr. Res.—Univ. Agric. Sci. (Bangalore)*, 19(6), 93, 1990; *Hortic. Abstr.*, 62, 5838, 1992.

345. Ramamurthy, N. and Devi, L. S., Effect of different sources of sulfur on the yield and quality of potato, *J. Indian Soc. Soil Sci.*, 30(3), 405, 1982; *Chem. Abstr.*, 99, 52442j, 1983.

346. Tsitsishvili, G. V., Zardalishvili, O. G., Kikodze, K. O., Shatirishvili, I. Sh., and Andronikashvili, T. G., Some biochemical indexes of carrots grown on clinoptilolite-containing soil, *Soobshch. Akad. Nauk Gruz. SSR*, 128(1), 125, 1987; *Chem. Abstr.*, 108, 111299z, 1988.

347. Shevchenko, L. A., Sidorenko, V. P., and Balyabo, S. A., Effect of clinoptilolite on agrochemical characteristics of sod-podzolic friable sandy soil and on potato yield, *Agrokhimiya*, 3, 63, 1986; *Chem. Abstr.*, 104, 206190m, 1986.

348. Tsitsishvili, G. V., Zardalishvili, O. Yu., Andronikashvili, T. G., Kikodze, K. O., and Shatirishvili, I. Sh., Biochemical indexes of pepper grown on clinoptilolite-containing soil, *Soobshch. Akad. Nauk Gruz. SSR*, 127(3), 641, 1987; *Chem. Abstr.*, 108, 130556v, 1988.

349. Kardava, M. A., Mikhailova, N. N., Tsitsishvili, G. V., Andronikashvili, T. G., Maisuradze, G. V., and Koval'chuk, N. I., After effect of organo-zeolitic fertilizers, *Izv. Akad. Nauk Gruz. SSR, Ser. Khim.*, 16(1), 53, 1990; *Chem. Abstr.*, 113, 77159t, 1990.

350. Monson, W. G., Burton, G. W., and Wilkinson, W. S., Effet of N fertilization and simazine on yield, protein, amino-acid content, and carotenoid pigments of coastal Bermudagrass, *Agron. J.*, 63, 928, 1971.

351. Margóczi, K., Takács, E., Técsi, L., and Maróti, I., Photosynthesis and production of two *Capsicum annuum* L. cultivars at different nitrate supplies, *Photosynthetica*, 23(4), 441, 1989; *Hortic. Abstr.*, 61, 3777, 1991.

352. Freeman, J. A. and Harris, G. H., The Effect of nitrogen, phosphorus, potassium and chlorine on the carotene content of the carrot, *Sci. Agric.*, 31, 207, 1951.

353. Trudel, M. J. and Ozbun, J. L., Influence of potassium on carotenoid content of tomato fruit, *J. Am. Soc. Hortic. Sci.*, 96(6), 763, 1971.

354. Habben, J., Quality constituents of carrots *Daucus carota* L., as influenced by nitrogen and potassium fertilization, *Acta Hortic.*, 29, 295, 1973; *Hortic. Abstr.*, 44, 5815, 1974.

355. Moussa, A. G., Geissler, T., and Markgraf, G., Ertrag, Qualität und Nitratgehalt von Speisemöhre und Spinat bei steigender Stickstoffdüngung im Modellversuch. *Arch. Gartenbau*, 34(1), 45, 1986.

356. Cserni, I., Prohászka, K., and Patócs, I., The effect of different N-doses on changes in the nitrate, sugar and carotene contents of carrots, *Acta Agron. Hung.*, 38(3-4), 341, 1989; *Hortic. Abstr.*, 61, 3839, 1991.

357. Sheets, O. A., Prementer, L., Wade, M., Gieger, M., Anderson, W. S., Peterson, W. J., Rigney, J. A., Wakeley, J. T., Cochran, F. D., Eheart, J. F., Young, R. W., and Massey, P. H., Jr., The nutritive value of collards, *South. Coop. Series Bull.*, 39, 5, 1954.

358. Scharrer, K. and Bürke, R., Der Einfluss der Ernährung auf Provitamin-A-(Carotin)-Bildung in Landwirtschaftlichen Nutzpflanzen, *Z. Pflanzenernähr Düng. Bodenk.*, 62(3), 244, 1953.

359. Hulewicz, D. and Kalbarczyk, M., Veränderlichkeit des Ertrages und einiger Nährkomponenten des Salats in Abhängigkeit vom Licht, *Arch. Gartenbau (Berlin)*, 24, 113, 1976.

360. Salunkhe, D. K. and Desai, B. B., Effect of agricultural practice, handling, processing, and storage on vegetables in *Nutritional Evaluation of Food Processing*, Karmas, E. and Harris, R. S., Eds., Van Nostrand Reinhold, New York, 1988, chap. 3.

361. Bünemann, G., Fritz, D., Schwerdtfeger, E., and Venter, F., Düngen wir richtig im Blick auf die Qualität von Obst und Gemüse? *Landwirt. Forsch. Sonderh.*, 35, 72, 1978.

362. Škrbić, K., Tomato yield and quality in relation to nitrogen nutrition, *Agrohemija*, 4, 251, 1987; *Hortic. Abstr.*, 59, 2190, 1989.

363. Southards, C. J. and Miller, C. H., A greenhouse study on the macroelement nutrition of the carrot, *Proc. Am. Soc. Hortic. Sci.*, 81, 335, 1962.

364. Leclerc, J., Reuille, M. J., Miller, M. L., Lefebvre, J. M., Joliet, E., Autissier, N., Martinez, Y., and Perret, A., Effect of climatic conditions and soil fertilization on nutrient composition of salad vegetables in Burgundy, *Sci. Alimen.*, 10(6), 633, 1990.

365. Michalik, H., Effect of fertilization with macro and micro elements on the dry matter, sugar and β-carotene contents of carrots, *Biuletyn Warzywniczy*, 28, 141, 1985; *Hortic. Abstr.*, 57, 5649, 1987.

366. Constantin, R. J., Jones, L. G., and Hernandez, T. P., Effect of potassium and phosphorus fertilization on quality of sweet potatoes, *J. Am. Soc. Hortic. Sci.*, 102(6), 779, 1977.

367. Florescu, M. and Cernea, S., A study on the variation in the contents of carotenoids and sugars in carrots as a function of the microelements magnesium, boron, copper, zinc and molybdenum applied to differently fertilized plots, *Lucr. Şti. Inst. Agron. Cluj*, 17, 75, 1961; *Hortic. Abstr.*, 34, 6885, 1964.

368. Yuditskaite, S., The effect of micro-elements on the carotene content of black chokeberry fruit, in *Introduktsiya Rastenii Botan. Sadakh Pribaltiki. Riga*, USSR, 1974, 156; *Hortic. Abstr.*, 45, 7177, 1975.

369. Burger, O. J. and Hauge, S. M., Relation of manganese to the carotene and vitamin contents of growing crop plants, *Soil Sci.*, 72, 303, 1951.

370. Lo, T. Y., Carotene and citrin content of peas as influenced by chemical treatment, *Food Res.*, 10, 308, 1945.

371. Kelly, W. C., Somers, G. F., and Ellis, G. H., The effect of boron on the growth and carotene content of carrots, *Proc. Am. Soc. Hortic. Sci.*, 59, 352, 1952.

372. Ankush, J. A., Hargitai, L., Biacs, P. A., and Daood, H. G., Influence of boron on quality attributes of tomato fruit, *Acta Aliment.*, 19(1), 63, 1990.
373. Pfützer, G., Pfaff, C., and Roth, H., Der Einfluss des Stickstoffgehalts von Gerstenkörnern auf die Entwicklung und den Vitamingehalt der jungen Pflanzen, *Biochem. Z.*, 297, 137, 1938.
374. Dressel, J. and Jung, J., Gehaltsniveau an Vitaminen des B-Komplexes in Abhängigkeit von Stickstoffzufuhr und Standort, *Landwirt. Forsch. Sonderh.*, 35, 261, 1978.
375. Scharrer, K. and Preissner, P., Der Vitamin-B$_1$-Gehalt der Pflanze in Abhängigkeit von ihrer Ernährung, *Z. Pflanzenernähr. Düng. Bodenk.*, 67(2), 166, 1954.
376. Hunt, C. H., Rodriguez, L. D., and Bethke, R. M., The environmental and agronomical factors influencing the thiamine, riboflavin, niacin, and pantothenic acid content of wheat, corn, and oats, *Cereal Chem.*, 27, 79, 1950.
377. Aitkin, Y., Influence of environment and variety on nitrogen and thiamin in field peas (*Pisum sativum*), *Proc. Royal Soc. Victoria*, 67, 257, 1955.
378. Döring, H., Einfluss der Düngung auf die Inhaltsstoffe der Kulturepflanzen. I. Mitt. Einfluss auf die Vitaminbildung, *Nahrung*, 4, 1159, 1960.
379. Fritz, D. and Venter, F., Einfluss von Sorte, Standort und Anbaumassnahmen auf messbare Qualitätseigenschaften von Gemüse, *Landwirt. Forschung*, *Sonderh.*, 30(1), 95, 1974.
380. Kochubei, I. V., Effect of the properties of sod-podzolic soil and fertilizers on the yield and levels of vitamins in spinach leaves, *Vestn. Mosk. Univ., Biol., Pochvoved.*, 30(6), 88, 1975; *Chem. Abstr.*, 84, 178851a, 1976.
381. Adeishvili, N. I. and Tokhadze, Z. B., Dynamics of the content of some group B vitamins of tea leaves in relation to varying background of mineral nutrition, *Subtrop. Kult.*, 1-2, 66, 1977; *Chem. Abstr.*, 88, 103842x, 1978.
382. Lee, J. W. and Underwood, E. J., The influence of variety on the thiamin and nitrogen contents of wheat, *Aust. J. Exp. Biol. Med. Sci.*, 28, 543, 1950.
383. Witsch, H. V. and Flügel, A., Untersuchungen über den Aneurinhaushalt höherer Pflanzen, *Ber. Dtsch. Bot. Ges.*, 64, 107, 1951.
384. Finch, L. R. and Underwood, E. J., The influence of clover leys on the thiamin and nitrogen contents of wheat, *Aust. J. Exp. Biol. Med. Sci.*, 29, 131, 1951.
385. Jahn-Deesbach, W. and May, H., The effect of variety and additional late spring nitrogen application on the thiamin (vitamin B$_1$) content of the total wheat grain, various flour types, and secondary milling products, *Z. Acker-Pflanzenbau*, 135, 1, 1972.
386. Pavel, J. and Žaková, J., Thiamine and riboflavin contents in hydroponically cultivated barley at increased doses of certain microelements, *Acta Univ. Agric. Brno, Fac. Agron.* (*A*), 16(4), 567, 1968; *Field Crops Abstr.*, 23, 2045, 1970.
387. Kononko, L. N., Comparative characteristics of the chemical composition of cucumbers grown in nutrient solution containing trace nutrient additives, *Vop. Ratsion. Pitan.*, 6, 47, 1970; *Chem. Abstr.*, 76, 152554r, 1972.
389. Taira, H., Taira, H., Matsuzaki, A., and Matsushima, S., Effect of nitrogen topdressing on B-vitamin content of rice grain, *Nihon Sakumostsu Gakkai Kiji*, 45(1), 69, 1976; *Food Sci. Technol. Abstr.*, 11(2), M139, 1979.
390. McCoy, T. A., Bostwick, D. G., and Devich, A. C., Some effects of phosphorus on the development, the B vitamin content, and the inorganic composition of oats, *Plant Physiol.*, 26, 784, 1951.
391. Villareal, R. L., Tsou, S. C. S., Lin, S. K., and Chiu, S. C., Use of sweet potato (*Ipomea batatas*) leaf tips as vegetables. II. Evaluation of yield and nutritive quality, *Exp. Agric.*, 15, 117, 1979.

392. Panchenko, T. A., Relation of the quality of sunflower oil to the mineral nutrition conditions of plants, *Maslo-Zhir. Prom-st.*, 4, 11, 1978; *Chem. Abstr.*, 89, 41312j, 1978.

393. Langston, R., Effects of different concentrations of nitrate on the growth and vitamin content of oat plants, *Plant Physiol*, 26, 115, 1951.

394. Åberg, B. and Ekdahl, I., Effect of nitrogen fertilization on the ascorbic acid content of green plants, *Physiol. Plant.* 1, 290, 1948.

395. Paschold, P. J. and Scheunemann, C., Effect of cultural measures on the yield and selected quality characters of white cabbage (*Brassica oleracea* L. *convar. capitata* (L) Alef. *f. capitata*), *Arch. Gartenbau (Berlin)*, 32(6), 229, 1984.

396. Kropp, K. and Ben, J., Effect of foliar application of urea on the yield, content of some chemical components and storability of Cox's Orange Pippin apples, *Zeszyty Naukowe Akademii Rolniczej Hugona Kollataja Krakowie, Ogrodnictwo*, 8, 119, 1981; *Hortic. Abstr.*, 53, 8353, 1983.

397. Lorenz, O. A. and Weir, B. L., Nitrate accumulation in vegetables, in *Environmental Quality and Food Supply*, White, P. L., and Robbins, D., Eds., Futura Publ. Co. Mount Kisco, N.Y., 1974, 93.

398. Maynard, D. N., Barker, A. V., Minotti, P. L., and Peck, N. H., Nitrate accumulation in vegetables, *Adv. Agron.*, 28, 71, 1976.

399. Vogtmann, H., Kaeppel N., and Fragstein, P. V., Nitrat- und Vitamin C-Gehalt bei verschiedenen Sorten von Kopfsalat und unterschiedlicher Düngung, *Ernährungs-Umschau*, 34, 12, 1987.

400. Shinohara, Y. and Suzuki, Y., Quality improvement of hydroponically grown leaf vegetables, *Acta Hortic.*, 230, 279, 1988.

401. Watanabe, Y., Shiwa, S., and Shimada, N., Effects of intermittent solution supply on contents of ascorbic acid sugars, nitrate, and soluble oxalate of spinach plants, *Nippon Dojo Hiryogaku Zasshi*, 59(6), 563, 1988; *Chem. Abstr.*, 111, 4507d, 1989.

402. Sady, W., Wojtaszek, T., Rozek, S., and Myczkowski, J., Greenhouse lettuce production by the nutrient film technique (NFT) with limited NPK fertilization. I. Cropping and the content of some compounds in the leaves, *Zeszyty Naukowe Akademii Rolniczej im. Hugona Kollataja Krakowie, Ogrodnictwo*, 210(15), 197, 1987; *Hortic. Abstr.*, 58, 218, 1988.

403. Mirvish, S., Wallcave, L., Eagen, M., and Shubik, P., Ascorbate-nitrite reaction: possible means of blocking the formation of carcinogenic N-nitroso compounds, *Science*, 177, 65, 1972.

404. Kyrtopoulos, S. A., Ascorbic acid and the formation of N-nitroso compounds: possible role of ascorbic acid in cancer prevention, *Am. J. Clin. Nutr.*, 45, 1344, 1987.

405. Virtanen, A. I., Hausen, S. V., and Saastamoinen, S., Untersuchungen über die Vitaminbildung in Pflanzen. I, *Biochem. Z.*, 267, 179, 1933.

406. Mapson, L. W. and Cruickshank, E. M., Effect of various salts on the synthesis of ascorbic acid and carotene in cress seedlings, *Biochem. J.*, 41, 197, 1947.

407. Somers, G. F. and Kelly, W. C., Ascorbic acid and dry matter accumulation in turnip and broccoli leaf discs after infiltration with inorganic salts, organic acids, and some enzyme inhibitors, *Plant Physiol.*, 26, 90, 1951.

408. De, S. K. and Shanker, J., Effect of temik 10g and other nitrogen compounds on the contents of vitamins C and E in cabbage, *Pesticides*, 21, 33, 1987.

409. Cheng, B. T., Suitable substrate for soilless culture, *J. Chinese Agric. Chem. Soc.*, 16(1/2), 103, 1978; *Food Sci. Technol. Abstr.*, 12(10), J1433, 1980.

410. Sastry, K. S. and Sarma, P. S., The nature of the effect of ammonium sulphate on the biosynthesis of ascorbic acid in plants, *Biochem J.*, 62, 451, 1956.

411. Kanesiro, M. A. B., Faleiros, R. R. S., and Nascimento, V. A., Effect of fertilizer treatments on the vitamin C content of tomatoes, *Científica*, 6(2), 225, 1978; *Hortic. Abstr.*, 49, 5105, 1979.

412. Walker, R. R., Hawker, J. S., and Törökfalvy, E., Effect of NaCl on growth, ion composition and ascorbic acid concentration of *capsicum* fruit, *Scientia Hortic.*, 12, 211, 1980.

413. Novobranova, T. I., Gudkovskii, V. A., and Uryupina, T. L., Effect of calcium on resistance of apples and pears to fungal rot during preservation, *Vestn. S-kh. Nauki Kaz.*, 4, 46, 1982; *Chem. Abstr.*, 97, 22367e, 1982.

414. Lu, C. W. and Ouyang, S. R., The effect of preharvest calcium sprays on the storage of table grapes, *Acta Hortic. Sinica*, 17(2), 103, 1990; *Hortic. Abstr.*, 62, 8124, 1992.

415. El-Naggar, S., Gaafar, A. A., El-Hammady, A. M., and Badr, A., Effect of fruit thinning and zinc sulphate spray on alternate bearing habit and fruit quality in "Valencia" oranges, *Gartenbauwissenschaft*, 38, 343, 1973.

416. Shen, T., Hsieh, K. M., and Chen, T. M., Effects of magnesium chloride and manganous nitrate upon the content of ascorbic acid in soybean during germination, with observations on the activity of ascorbic acid oxidase, *Biochem. J.*, 39, 107, 1945.

417. Hewitt, E. J., Agrawala, S. C., and Jones, E. W., Effect of molybdenum status on the ascorbic acid content of plants in sand culture, *Nature (Lond.)*, 166, 1119, 1950.

418. Rubzow, M. I., Die Spurenelementdüngung erhöht den Gehalt an Vitamin C im Gemüse, *Gartenbauwissenschaft*, 25, 61, 1960.

419. Matev, I. and Stanchev, L., Effect of Na^+, K^+, Ca^{2+} and Mg^{2+} imbalance on the development and biological value of glasshouse tomatoes, *Gradinar. Lozar. Nauka*, 16(1), 76, 1979; *Hortic. Abstr.*, 50, 7832, 1980.

420. Hulewicz, D., Nurzyński, J., and Mokrzecka, E., Dependence of some quality indices in parsley on mineral fertilization, *Roczniki Nauk Rolniczych, A*, 94(4), 95, 1970; *Hortic. Abstr.*, 42, 1971, 1972.

421. Kolesnik, A. A. and Cerevitinov, O. B., A trial on the foliar nutrition of apple trees with zinc salts, *Sadovodstvo*, 4, 48, 1964; *Hortic. Abstr.*, 34, 6331, 1964.

422. Bernstein, L., Hamner, K. C., and Parks, R. Q., The influence of mineral nutrition, soil fertility, and climate on carotene content and ascorbic acid content of turnip greens, *Plant Physiol.*, 20, 540, 1945.

423. McCoy, T. A., Free, S. M. Jr., Langston, R. G., and Snyder, J. Q., Effect of major elements on the niacin, carotene, and inorganic content of young oats, *Soil Sci.*, 68, 375, 1949.

424. Vereecke, M., Van Maercke, D., Substrative fertilization experiment on carrots (*Daucus carota* L.) in relation to soil- and leaf analysis, yield and quality, *Acta Hortic.*, 93, 197, 1979.

425. Fernández, M. Del C. C., Effect of environment on the carotene content of plants, *South. Coop. Series. Bull.*, 36, 24, 1954.

426. Eheart, M. S., Fertilization effects on the chlorophyll, carotene, pH, total acidity, and ascorbic acid in broccoli, *J. Agric. Food Chem.*, 14, 18, 1966.

427. Greer, E. N. and Kent, N. L., The effect of nitrogen top dressing on the quality of winter wheat, *Agriculture*, 57, 59, 1950.

428. Greer, E. N., Ridyard, H. N., and Kent, N. L., The composition of British-grown winter wheat. I. Vitamin-B_1 content, *J. Sci. Food Agric.*, 3, 12, 1952.

429. Holman, W. I. M. and Godden, W., The aneurin (vitamin B_1) content of oats. I. The influence of variety and locality. II. Possible losses in milling, *J. Agric. Sci.*, 37, 51, 1947.

430. Syltie, P. W. and Dahnke, W. C., The vitamin B_1, B_2, B_6, B_{12}, and E contents of hard red spring wheat as influenced by fertilization and cultivar, *Qual. Plant.—Plant Foods Hum. Nutr.*, 32, 51, 1983.

431. Wokes, F., Proteins, *Plant Foods Hum. Nutr.*, 1, 23, 1968.

432. Sathe, V., Venkitasubramanian, T. A., and De, S. S., Effect of fertilizers on the thiamine content of rice, *Sci. Cult.*, 18, 33, 1952; *Chem. Abstr.*, 47, 2918i, 1953.

433. Burkholder, P. R. and McVeigh, I., Studies on thiamin in green plants with the *Phycomyces* assay method, *Am. J. Bot.*, 27, 853, 1940.

434. Abalde, J. and Fabregas, J., β-carotene, vitamin C and E content of the microalga *Dunaliella tertiolecta* cultured with different nitrogen sources, *Bioresource Technol.*, 38, 121, 1991.

435. Richardson, L. R., Effect of environment on the ascorbic acid content of plants, *South. Coop. Series Bull.*, 36, 6, 1954.

436. Doll, R., Nutrition and cancer: a review, *Nutr. Cancer*, 1, 35, 1979.

437. Weisburger, J. H., Mechanism of action of diet as a carcinogen, *Nutri. Cancer*, 1, 74, 1979.

438. Calabrese, E. J., *Nutrition and Enviornmental Health. The Influence of Nutritional Status on Pollutant Toxicity and Carcinogenicity*, Vol. 1, *The Vitamins*, Wiley-Interscience Publ., New York, 1980.

439. Calabrese, E. J., *Nutrition and Environmental Health. The Influence of Nutritional Status on Pollutant Toxicity and Carcinogenicty*, Vol. 2, *Minerals and Macronutrients*, Wiley-Interscience Publ., New York, 1981.

440. Calabrese, E. J., Does exposure to environmental pollutants increase the need for vitamin C? *J. Environ. Path. Toxicol. Oncol.*, 5(6), 81, 1985.

441. Rensberger, B., Cancer, the new synthesis, *Science 84*, 5(7), 28, 1984.

442. Gey, K. F., Brubacher, G. B., and Stähelin, H. B., Plasma levels of antioxidant vitamins in relation to ischemic heart disease and cancer, *Am. J. Clin. Nutr.*, 45, 1368, 1987.

443. Marchand, L., Yoshizawa, C. N., Kolonel, L. N., Hankin, J. H., and Goodman, M. T., Vegetable consumption and the lung cancer risk: a population-based case-control study in Hawaii, *J. Natl. Cancer Inst.*, 81, 1158, 1989.

444. Ziegler, R. G., A review of epidemiologic evidence that carotenoids reduce the risk of cancer, *J. Nutr.*, 119, 116, 1989.

445. Thurnham, D. I., Anti-oxidant vitamins and cancer prevention, *J. Mironutr. Anal.*, 7, 279, 1990.

446. Narbonne, J. F., Cassand, P., Daubeze, M., Colin, C., and Leveque, F., Chemical mutagenicity and cellular status in vitamin A, E and C, *Food Addit. Contam.*, 7, S48, 1990.

447. Butterworth, C. E., Jr, and Bendich, A., Introduction, in *Micronutrients in Health and in Disease Prevention*, Bendich, A. and Butterworth, C. E., Jr., Eds., Marcel Dekker, New York, 1991, 1.

448. Gaby, S. K., Bendich, A., Singh, V. N., and Machlin, L. J., *Vitamin Intake and Health*, Marcel Dekker, New York, 1991, 1.

449. Byers, T. and Perry, G., Dietary carotenes, vitamin C, and vitamin E as protective antioxidants in human cancers, *Annu. Rev. Nutr.*, 12, 139, 1992.

450. Fraga, C. G., Motchnik, P. A., Shigenaga, M. K., Hellbock, H. J., Jacob, R. A., and Ames, B. N., Ascorbic acid protects against endogenous oxidative DNA damage in human sperm, *Proc. Natl. Acad. Sci. (USA)*, 88, 11003, 1991.

451. Niki, E., Vitamin C as an antioxidant, in Selected Vitamins, Minerals, and Functional Consequences of Maternal Malnutrition, Simopoulos, A. P., Ed., *World Rev. Nutr. Diet.*, 64, 1, 1991.

452. Gull, D. D., Locascio, S. J., and Kostewicz, S. R., Composition of greenhouse tomatoes as affected by cultivar, production media and fertilizer, *Proc. Fla. St. Hortic. Soc.*, 90, 395, 1977; *Food Sci. Technol. Abstr.*, 11(5), J767, 1979.
453. Kral, D. M., *Organic Farming: Current Technology and its Role in a Sustainable Agriculture*, ASA-ASSA-SSSA, Madison, Wis., 1984.
454. Poincelot, R. P., *Toward a More Sustainable Agriculture*, AVI Publishing, Westport, Conn., 1986.
455. Senauer, B., Asp, E., and Kinsey, J., *Foods Trends and the Changing Consumer*, Eagan Press, St. Paul, Minn., 1991.
456. Schuphan, W., Nutritional value of crops as influenced by organic and inorganic fertilizer treatments. Results of 12 years' experiments with vegetables, *Qual. Plant—Plant Foods Hum. Nutr.*, 23(4), 333, 1974.
457. Leclearc, J., Miller, M. L., Joliet, E., and Rocquelin, G., Vitamin and mineral contents of carrot and celeriac grown under mineral or organic fertilization, *Biol. Agric. Hortic.*, 7, 339, 1991.
458. De, S. K. and Laloraya, D., Effect of inorganic and organic fertilizers on the contents of vitamin B_1, C and E in chili (*Capsicum annuum*), *Plant Biochem. J.*, 7(2), 116, 1980.
459. Schudel, P., Eichenberger, M., Augstburger, F., Kläy, R., and Vogtmann, H., Über den Einfluss von Kompost- und NPK-Düngung auf Ertrag, Vitamin-C- und Nitratgehalt von Spinat und Schnittmangold, *Schweiz. Landwirt. Forsch.*, 18, 337, 1979.
460. Bagaturiya, N. Sh., Bziava, R. M., Tsanava, N. G. and Lominadze, Sh. D., Effect of mineral nutrition on the yield, quality, and composition of essential oils in lemon and orange fruits, *Subtrop. Kul't.*, 2, 68, 1990; *Chem. Abstr.*, 114, 60967c, 1991.
461. Eggert, F. P. and Kahrmann, C. L., Response of three vegetable crops to organic and inorganic nutrient sources, in *Organic Farming: Current Technology and Its Role in a Sustainable Agriculture*, Kral, D. M., Ed., ASA-CSSA-SSSA, Madison, Wis., 1984, Chap. 8.
462. Nilsson, T., Yield, storage ability, quality and chemical composition of carrot, cabbage, and leek at conventional and organic fertilizing, *Acta Hortic.*, 93, 209, 1979.
463. Brandt, C. S. and Beeson, K. C., Influence of organic fertilization on certain nutritive constituents of crops, *Soil Sci.*, 71, 449, 1951.
464. Basel, Karger, 1991, Salomon, M., Natural foods—myth or magic, *Q. Bull. Assoc. Food Drug Officials U.S.*, 36, 131, 1972.
465. Svec, L. V., Thoroughgood, C. A., and Mok, H. C. S., Chemical evaluation of vegetables grown with conventional or organic soil amendments, *Commun. Soil Sci. Plant Analy.*, 7(2), 213, 1976.
466. Termine, L., Lairon, D., Taupier-Letage, B., Gautier, S., Lafont, R., and Lafont, H., Yield and content in nitrates, minerals and ascorbic acid of leek and turnip grown under mineral or organic nitrogen fertilization, *Plant Foods Hum. Nutr.*, 37, 321, 1987.
467. Scheunert, A., Reschke, J., and Kohlemann, E., Über den Vitamin C-Gehalt der Kartoffeln. IV. Mitteilung: Über den Einfluss verschiedener Düngung, *Biochem. Z.*, 305, 1, 1940.
468. Cheng, B. T., Farmyard manure and chemical fertilizers as a source of nutrients for raspberry, *Commun. Soil Sci. Plant Anal.*, 13(8), 633, 1982.
469. Wilberg, E., Über die Qualität von Spinat aus "biologischem Anbau," *Landwirt. Forschung*, 25, 167, 1972.
470. Yoshida, R., Quality and nitrate content of vegetable crops, *Kenkyu Hokoku—Toyama-kenritsu Gijutsu Tanki Daigaku*, 23, 106, 1989; *Chem. Abstr.*, 112, 6402k, 1990.
471. Shanker, J., Saha, S. K., and De, S. K., Effect of certain nitrogen compounds on the biosynthesis of thiamine in cabbage plant, *Indian J. Plant Physiol.*, 25(3), 306, 1982.

472. De, S. K., Prasad, R., and Shanker, J., Effect of nitrogenous fertilizers on the content of vitamin B_1 in cauliflower (*Brassica oleracea*), *Biovigyanam*, 8, 43, 1982.

473. Singh, R. D. and Dhar, N. R., Effect of organic matter without and with rockphosphate on crop yield, quality and soil characteristics, *Indian J. Agric. Sci.*, 56(7), 539, 1986.

474. Harris, L. J., Note on the vitamin B_1 potency of wheat as influenced by soil treatment, *J. Agric. Sci.*, 24, 410, 1934.

475. Belderok, B., Effects of alternative farming procedures on the nutritional value and processing characteristics of wheat, *Voeding*, 39(12), 352, 1978; *Food Sci. Technol. Abstr.*, 11(4), M385, 1979.

476. Evers, A. M., Effect of different fertilization practices on the carotene content of carrot, *J. Agric. Sci. Finland.*, 61, 7, 1989.

477. Davis, K. R., Peters, L. J., and Le Tourneau, D., Variability of vitamin content in wheat, *Cereal Foods World*, 29(6), 364, 1984.

478. Pettersson, B. D., A comparison between the conventional and biodynamic farming systems as indicated by yield and quality, in *Towards a Sustainable Agriculture*, Beeson, J. M. and Vogtmann, H., Eds., Proc. IFOAM Conf., Sissach, 1977, Switzerland, Wirz AG, Aarau, Switzerland, 1978, 87.

479. Vogtmann, H., Organic farming practices and research in Europe, in *Organic Farming: Current Technology and Its Role in a Sustainable Agriculture*, Kral, D. M., Ed., ASA-CSSA-SSSA, Madison, Wis., 1984, 19.

480. Evers, A. M., Effect of different fertilization practices on the quality of stored carrot, *J. Agric. Sci. Finland*, 61, 123, 1989.

481. Evers, A. M., Effect of different fertilization practices on the glucose, fructose, sucrose, taste and texture of carrot, *J. Agric. Sci. Finland*, 61, 113, 1989.

482. McCarrison, R. and Viswanath, B., The effect of manurial conditions on the nutritive and vitamin values of millet and wheat, *Indian J. Med. Res.*, 14, 351, 1926.

483. Rowlands, M. J. and Wilkinson, B., The vitamin B content of grass seeds in relationship to manures, *Biochem. J.*, 24, 199, 1930.

484. Maynard, L. A. and Loosli, J. K., *Animal Nutrition*, 6th, ed., McGraw-Hill, New York, 1969, chapter 9.

485. Pratt, J. M., *Inorganic Chemistry of Vitamin B_{12}*, Academic Press, London, 1972.

486. Friedrich, W., *Handbuch der Vitamine*, Urban & Schwarzenberg, Munich, Germany, 1987.

487. Machlin, L. J., Ed., *Handbook of Vitamins*, 2nd Ed., Marcel Dekker, New York, 1991.

488. Bender, D. A., *Nutritional Biochemistry of the Vitamins*, Cambridge University Press, Cambridge, 1992.

489. Kocher, V. and Corti, U. A., Beitraege zur Kenntnis des Vitamingehaltes von Belebtschlamm aus Abwasserreinigungsanlagen, *Schweiz. Z. Hydrol.*, 14, 333, 1952.

490. Firth, J. A. and Johnson, B. C., Sewage sludge as a feed ingredient for swine and poultry, *Agric. Food Chem.*, 3(9), 795, 1955.

491. Kononova, M. M., *Soil Organic Matter. Its Nature, Its Role in Soil Formation and in Soil Fertility*, 2nd Ed., Pergamon Press, London, 1966, 212.

492. Vyblov, N. F., Changes in the biological activity of grey forest soils in Upper Altai resulting from the application of manure, *Izvestiya Sibirskogo Otdeleniya Akademii Nauk SSSR, Biologicheskikh Nauk*, 6(1), 44, 1988; *Soils and Fertilizers*, 53, 1526, 1990.

493. Selivanov, A. P. and Smolenskaya, I. Ya., Effects of irrigation with industrial waste waters on biochemical indexes and vitamin content of agricultural crops, *Gig. Sanit.*, 1, 11, 1983; *Chem. Abstr.*, 98, 88205c, 1983.

494. Antoniani, C. and Monzini, A., The vitamin content of produce and products on the farm. III. Vitamin B_1 content of fodder from meadows irrigated by sewage as

compared with that irrigated by clean water, *Ann. Sper. Agrar.*, (*Rome*), 4, 625, 1950; *Chem. Abstr.*, 44, 10957g, 1950.

495. Mozafar, A. and Oertli, J. J., Uptake and transport of thiamin (vitamin B₁) by barley and soybean, *J. Plant Physiol.*, 139, 436, 1992.

496. Mozafar, A. and Oertli, J. J., Uptake of a microbially-produced vitamin (B₁₂) by soybean roots, *Plant Soil*, 139, 23, 1992.

497. Mansurova, M. L. and Noskova, L. N., Ability of some bacteria isolated from Uzbek soils to produce B group vitamins, *Nauchn. Tr.—Tashk. Gos. Univ.*, 514, 1976; *Chem. Abstr.*, 88, 148667k, 1978.

498. Lochhead, A. G., Soil bacteria and growth-promoting substances, *Bact. Rev.*, 22, 145, 1958.

499. Lochhead, A. G. and Thexton, R. H., Vitamin B₁₂ as a growth factor for soil bacteria, *Nature* (*Lond.*), 167, 1034, 1951.

500. Manorik, A. V., Vasil'chenko, V. F., Derevyanko, S. I., Nichik, M. M., Malichenko, S. M., and Zhelyuk, V. M., Significance of B vitamins in the interrelations of plants and microorganisms in root nutrition of plants, *Mater. Vses. Simp. Fiziol.-Biokhim. Osn. Form. Rast. Soobshchestv.* (*Fitotsenozov*), *1st*, Moscow 1965, 242, 1966; *Chem. Abstr.*, 67, 90091r, 1967.

501. Lewis, N. M., Kies, C., and Fox, H. M., Vitamin B₁₂ status of lacto-ovo-vegetarian and omnivore subjects fed controlled lacto-vegetarian, vegan and omnivore diets, *Nutr. Rep. Int.*, 34, 197, 1986.

502. Lairon, D., Termine, E., and Lafont, H., Valeur nutritionnelle comparée des légumes obtenus par les méthodes de l'agriculture biologique ou de l'agricultrue conventionnelle, *Cahiers Nutr. Diet.*, 19(6), 331, 1984.

503. Lairon, D., Spitz, N., Termine, E., Ribaud, P., Lafont, H., and Hauton. J., Effect of organic and mineral nitrogen fertilization on yield and nutritive value of butterhead lettuce, *Qual. Plant.—Plant Foods Hum. Nutr.*, 34, 97, 1984.

504. Linder, M. C., Ed., *Nutritional Biochemistry and Metabolism*, Elsevier, New York, 1985.

505. Asano, J., Effect of oganic manures on quality of vegetables, *Jpn. J. Res. Q.*, 18, 31, 1984.

506. Wistinghausen, E. V., and Richter, M., Fertilizer type and the quality of vegetables, *Revista Facultad Agronomia*, *Univ. Buenos Aires*, 4(2), 123, 1983; *Hort. Abstr.*, 54, 2334, 1984.

507. Yoshida, K., Mori, S., Hasegawa, K., Nishizawa, N., and Kumazawa, K., Reducing sugar, organic acid, and vitamin C contents of tomato fruits cultured with organic fertilizers in comparison with inorganic fertilizers, *Nippon Eiyo, Shokuryo Gakkaishi*, 37(2), 123, 1984; *Chem. Abstr.*, 101, 169903m, 1984.

508. El-Zorkani, A. S., Seasonal changes in the physical characters and juice composition of some citrus varieties as affected by soil type, *Agric. Res. Rev.*, (*Cairo*), 46(3), 72, 1968; *Hortic. Abstr.*, 40, 2234, 1970.

509. Bar-Akiva, A. and Hamou, M., Soil factors influencing fruit quality and mineral composition of leaves of Valencia orange trees, *Commun. Soil Sci. Plant Analy.*, 5(3), 203, 1974.

510. Jankovskaja, N. M., The effect of fertilizers on the chemical composition of cabbages and carrots, *Himija Sel'. Hoz.*, 5(5), 11, 1967; *Hortic. Abstr.*, 38, 5483, 1968.

511. Chipman, E. W. and Forsyth, F. R., Characteristics of the epidermal layer of carrot roots grown on peat and mineral soil, *Can. J. Plant Sci.*, 51, 513, 1971.

512. Ishii, G. and Saijo, R., Effect of various cultural conditions on total sugar content, vitamin C content and β-amylase activity of Daikon radish root (*Raphanus sativus, L.*), *J. Jpn. Soc. Hortic. Sci.*, 55(4), 468, 1987; *Hortic. Abstr.*, 59, 9059, 1989.

513. El Mahmoodi, L. T., El Shiati, M. A., and Atwa, A. A., Effect of soil type on the keeping quality and storage life of tomatoes, *Agric, Res. Rev.* (*Cairo*), 42(4), 82, 1964; *Hortic. Abstr.*, 36, 3102, 1966.

514. Habben, J., Einfluss der Stickstoff- und Kaliumdüngung auf Ertrag und Qualität der Möhre (*Daucus carota L.*), *Landwirt. Forsch.*, 26(2), 156, 1973.

515. Kulikova, N. T., Characteristics of the chemical composition of carrot roots in the polar region, *Tr. Prikl. Bot. Genet. Sel.*, 45(1), 325, 1971; *Hortic. Abstr.*, 43, 5375, 1973.

516. Elkner, K. and Michalik, H., The effect of soil type and irrigation on nutrient content in 3 carrot cultivars, *Biuletyn Warzywniczy*, 15, 175, 1973; *Hortic. Abstr.*, 45, 7528, 1975.

517. Wynd, F. L. and Noggle., Influence of chemical properties of soil on production of vitamin C in leaves of rye. *Food Res.*, 11, 169, 1946.

518. Lampitt, L. H., Baker, L. C., and Parkinson, T. L., Vitamin-C content of potatoes. II. The effect of variety, soil, and storage, *J. Soc. Chem. Industry* (*Lond.*), 64, 22, 1945.

520. Spencer, E. Y. and Galgan, M. W., Relations of thiamine content in Saskatchewan wheat to protein content, variety, and soil zone, *Can. J. Res. Sec. F*, 27, 450, 1949.

521. McElroy, L. W., Kastelic, J., and McCalla, A. G., Thiamine and riboflavin content of wheat, barley, and oats grown in different soil zones in Alberta, *Can. J. Res. Sec. F*, 26, 191, 1948.

522. Mijll Dekker, L. P. van der and H. de Miranda, H., The vitamin B_1 content of dutch wheat and the factors which determine this content, *Neth. J. Agric. Sci.*, 2, 27, 1954.

523. McElroy, L. W. and Simonson, H., The niacin content of wheat, barley, and oats grown in different soil zones in Alberta, *Can. J. Res. Sec. F*, 26, 201, 1948.

524. Nik-Khah, A., Hoppner, K. H., Sosulski, F. W., Owen, B. D., and Wu, K. K., Variation in proximate fractions and B-vitamins in Saskatchewan feed grains, *Can. J. Anim. Sci.*, 52, 407, 1972.

525. Ellis, G. H. and Hamner, K. C., The carotene content of tomatoes as influenced by various factors, *J. Nutr.*, 25, 539, 1943.

526. Geissler, T. and Rüdiger, K., Zwekmässige Mineraldüngung-Voraussetzung für hohe Erträge und guten Gebrauchswert bei Speisemähren, *Gartenbau*, 27, 168, 1980.

527. Gormley, T. R., O'Ríordáin, F., and Prendiville, M. D., Some aspects of the quality of carrots on different soil types, *J. Food Technol.*, 6, 393, 1971.

528. Hårdh, J. E., Persson, A. R., and Ottosson, L., Quality of vegetables cultivated at different latitudes in Scandinavia, *Acta Agric. Scand.*, 27, 81, 1977.

529. Stepanishina, T. V., The effect of soil type on the chemical composition of head cabbage and on changes occurring during storage, *Puti Povysheniya Urozhainosti Ovoshchnykh Kul'tur*, *Mezhved. Temat. Sbornik*, 2, 128, 1972; *Hortic. Abstr.*, 44, 7597, 1974.

530. Karikka, K. J., Dudgeon, L. T., and Hauck, H. M., Influence of variety, location, fertilizer, and storage on the ascorbic acid content of potatoes grown in New York State, *J. Agric. Res.* (*Washington, D.C.*), 68, 49, 1944.

531. Constantin, R. J., Jones, L. G., and Hernandez, T. P., Sweet potato quality as affected by soil reaction (pH) and fertilizer, *J. Am. Soc. Hortic. Sci.*, 100(6), 604, 1975.

532. Eheart, J. F., Young, R. W., Massey, P. H., Jr., and Havis, J. R., Crop, light intensity, soil pH, and minor element effects on the yield and vitamin content of turnip greens, *Food Res.*, 20, 575, 1955.

533. Gzirishvili, O. Sh., Changes in the contents of ascorbic acid and carotene in plants grown in the soil and hydroponically, *Soobshcheniya Akad. Nauk Gruzinskoi SSR*, 76(3), 689, 1974; *Hortic. Abstr.*, 45, 8520, 1975.

534. Nanko, Y., Namiki, T. and Takashima, S., Ascorbic acid and reducing sugar contents and titratable acidity of tomato fruit grown in nutrient solution or in soil, *Studies from the Institute of Horticulture, Kyoto University*, 9, 66, 1979; *Hortic. Abstr.*, 51, 5527, 1981.

535. Benoit, F. and Ceustermans, N., Qualitative aspects of tomatoes grown by NFT, *Boer en de Tuinder*, 90, 21, 1984; *Hortic. Abstr.*, 54, 9207, 1984.

536. Benoit, F. and Ceustermans, N., Some qualitative aspects of tomatoes grown by NFT, *Soilless Cult.*, 3(2), 3, 1987; *Hortic. Abstr.*, 58, 4331, 1988.

537. Granges, A., The chemical composition of tomatoes grown in nutrient film culture, *Acta Hortic.*, 98, 219, 1980.

538. Baevre, O. A., A comparison of the fruit quality of tomatoes grown in soil and in a nutrient solution (NFT), *Meld. Nor. Landbrukshoegsk.*, 64(12), 1, 1985.

539. Ledovskij, S. Ja., Characteristics of tomato plants grown by the hydroponic methods, *Fiziol. Biohim. Kul't. Rast.*, 2(1), 30, 1972; *Hortic. Abstr.*, 41, 1496, 1971.

540. López-andréu, F. J., Esteban, R. M., Molla, E., and Carpena, O., Effect of nutrition on tomato fruit quality. II. Carotenoids, ascorbic acid, pectic substances and flavonoids, *An. Edafol. Agrobiol.*, 47(7–8), 1191, 1988; *Hortic. Abstr.*, 61, 8048, 1991.

541. Gormley, T. R. and Egan, J. P., Studies on the quality of tomato fruit grown in peat and nutrient solution media, *Acta Hortic.*, 82, 213, 1978.

542. Janse, J. and Van Winden, C. M. M., Quality and chemical composition of tomatoes grown on rockwool, in *The Effects of Modern Production Methods on the Quality of Tomatoes and Apples*, Gormley, T. R., Sharples, R. O., and Dehandtschtter, J., Eds., Commission of the European Communities, Luxembourg, 1985, 93.

543. Nakayama, K., Kuwahara, T., and Nakayama, M., Distribution and behavior of the contents of components in foods. VII. Mini-tomatoes cultured by using rock-wool, *Kochi Joshi Daigaku Kiyo, Shizen Kagakuhen*, 36, 21, 1988; *Chem. Abstr.*, 109, 229471w, 1988.

544. Mokrzecka, E., The effect of different forms of nitrogen fertilization on the crop, vitamin C content and carbohydrates content of savoy (*Brassica oleracea* var. sabauda L.), *Acta Agrobot.*, 33(1), 51, 1980; *Hortic, Abstr.*, 51, 5458, 1981.

545. Cronin, D. A. and Walsh, P. C., A comparison of chemical composition, firmness and flavor in tomatoes from peat and nutrient film growing systems, *Irish J. Food Sci. Technol.*, 7, 111, 1983.

546. Gormley, T. R., Maher, M. J., and Walshe, P. E., Quality and performance of eight tomato cultivars in a nutrient film technique system, *Crop Res. (Hortic. Res.)*, 23, 83, 1983.

547. Geissler, Th., Nickisch, K., and Haenel, H., Ein Beitrag zur Frage der Hydrokultur von Treibgurken, *Arch. Gartenbau*, 2, 84, 1954.

548. Haenel, H., Mikrobiologische Untersuchungen über den Gahalt von Faktoren des B-Komplexes in Lebens- und Futtermitteln, *Ernährungsforschung*, 1, 533, 1956.

549. Akopyan, G. O., Mairpetyan, S. Kh., and Stepanyan, B. T., Tocopherols of rose geranium in outdoor hydroponics, *Biol. Zh. Arm.*, 37(7), 594, 1984; *Hortic. Abstr.*, 55, 5591, 1985.

550. Hurni, H., Die Biosynthese von Aneurin in der höher Pflanze. Der B$_1$-Gehalt von *Melandrium album* unter verschiedenen Bedingungen, *Z. Vitaminforschung*, 15, 198, 1945.

Chapter 6

EFFECTS OF AGRONOMIC PRACTICES AND CONDITIONS ON PLANT VITAMINS

I. EFFECT OF RIPENING AND MATURITY ON PLANT VITAMINS

In plants, the terms aging, senescence, maturing and ripening are used interchangeably. Ripening process in fruits or vegetables may or may not be concomitant with continuous changes in their size after a given size has been reached. Thus any alterations in the vitamin concentration as plants ripen may be, in cases, partly associated with simultaneous changes occurring in their size and weight. Starting vitamin concentration in the immature plants could be quite different in different years, and thus the vitamin concentration may increase or decrease as the plants approach maturity in different years (Figure 1). This point needs to be kept in mind when consulting the reports on the effect of maturity on plant vitamins as summarized in Table 1.

Another point to be emphasized is the fact that the physiological maturity of plants may not necessarily coincide with the stage they are considered best for consumption. For example, bitter melon (*Momordica charantia*, L.), cucumber, squash, okra (*Abelmoschus esculentus*, L.), peas, sweet corn, snap beans, and many leafy vegetables are usually harvested and consumed at immature or mature-unripe conditions. Bananas and potatoes are preferred when fully mature, and broccoli and cauliflower when overmature or postmature.[194, 195] The preference for a mature or immature fruit or vegetable could have nutritional consequences especially in cases where maturity reduces some plant vitamins. In Southeast Asian countries, for example, bitter melon is always consumed at the mature green stage, when its β-carotene content is relatively low, and not at the fully ripe stage, when its carotene content is high.[137]

Finally, maturity may also change the bioavailability of some vitamins. A case in point is the niacin in corn: feeding assays with rats have shown that kernels harvested at the milky stage have relatively more bioavailable niacin than those harvested in the fully mature stage.[196] Thus the effect of maturity on the changes in the vitamin concentrations in plants as discussed below, and their nutritional implications, must be considered with the above points in mind.

A. Ascorbic Acid

Ascorbic acid concentration may increase or decrease as fruits or vegetables mature (Table 1). The decrease in the ascorbic acid content

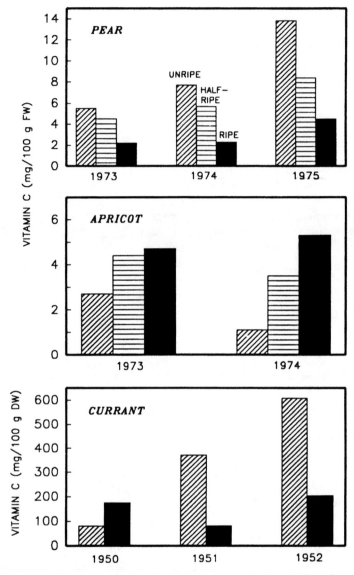

FIGURE 1. Changes in the vitamin C concentration during the ripening of pear (variety Williams),[3] apricots (variety Luizet),[3] and currant (variety Rote Hollandische)[4] in different years.

TABLE 1
Changes in the Vitamin Concentration in Plants[a] as They Mature, Ripen, or Age

Plant	Change with maturity	Ref.
	Ascorbic acid	
Apples	Increase	4–8
Apricots	Increase	1–3
Bananas	Increase	9
Beans (snap)	Increase	10–12
Ber	Increase	13, 14
Capsicum	Increase	15
Cashew apple	Increase	16, 17
Cherry (sweet)	Increase	4
Cherry	Increase	3
Cherry (tart)	Increase	4
Currant (red)	Increase	4
Currant (black)	Increase	4
Gourd (Bottle)	Increase	18
Grape (Veeport)	Increase	19
Guava	Increase	20–24
Jujube	Increase	25
Kiwifruit	Increase	26, 27
Lemon	Increase	28
Litchi (lychee)	Increase	29
Mandarin (Satsuma)	Increase	30
Mango	Increase	31
Papaya	Increase	27, 32, 33
Passion fruit	Increase	34
Peas	Increase	35
Peaches	Increase	1, 3, 36
Pear	Increase	4
Pepper	Increase	37–43
Pimento	Increase	44
Plum	Increase	3, 45
Potato	Increase	46
Rowan (*Sorbus*)	Increase	47
Spinach	Increase	48
Strawberry	Increase	4, 49
Tomato	Increase	50–63
Turnip greens	Increase	64
Watermelon	Increase	65
Ziziphus	Increase	66
Apples	No change	67
Blueberry	No change	68
Cabbage	No change	69
Citrus sudachi	No change	70
Grape	No change	3
Kiwifruit	No change	71
Pepper fruit	No change	72
Strawberry	No change	3
Tomato	No change	73

TABLE 1   (continued)

Plant	Change with maturity	Ref.
	Ascorbic acid	
Acerola (*Malpighia punicifolia* L.)	Decrease	24, 74–76
Bananas	Decrease	27, 77
Beans (lima)	Decrease	78, 79
Beans (broad)	Decrease	80
Ber	Decrease	81
Blackthorn (*Frangula alnus*)	Decrease	82
Carrots	Decrease	83
Cherry (tart)	Decrease	84
Citrus fruits	Decrease	85
Collards	Decrease	86
Cowpea	Decrease	87
Currant (black)	Decrease	88–91
Date	Decrease	92
Fig	Decrease	93
Gooseberry	Decrease	4
Grape	Decrease	94
Grapefruit	Decrease	95
Jujube	Decrease	96
Kale (fodder)	Decrease	97
Kiwifruit	Decrease	98
Lemon	Decrease	99
Lettuce	Decrease	100
Lime (Persian)	Decrease	101
Lychee, pericarp	Decrease	102
Mango	Decrease	103–112
Okra	Decrease	113
Oranges	Decrease	95, 114–117
Passion fruit	Decrease	27
Peas	Decrease	118–120
Peaches	Decrease	121–122
Pear	Decrease	3
Persimmon (Japanese)	Decrease	123
Petunia (leaf)	Decrease	124
Phyllanthus niruri	Decrease	125
Pineapple	Decrease	126
Plantain	Decrease	127
Potato	Decrease	128–130
Radish root	Decrease	131
Raspberry	Decrease	132
Rose hips	Decrease	133
Sapodilla (*Manilkara zapota*)	Decrease	134
Sapota (*Achras sapota*)	Decrease	135
Spinach	Decrease	136
Tomato (leaf)	Decrease	57
Wheat kernel	Decrease	137

TABLE 1   (continued)

Plant	Change with maturity	Ref.
	Biotin	
Gourd	Decrease	176
Grape	Decrease	169
	Carotene	
Acerola	Increase	27
Barbados cherry	Increase	75
Bitter melon (Charantia)	Increase	138
Cactus (prickly pear)	Increase	139
Cantaloupe	Increase	27
Capsicum	Increase	15
Carrots	Increase	38, 83, 140–147
Cucurbita	Increase	148
Endive	Increase	149
Jujube	Increase	25
Lettuce	Increase	149
Mango	Increase	103, 108, 112, 150
Mango (African)	Increase	151
Muskmelon	Increase	152
Oranges	Increase	153
Palm fruit (*Elaeis guineensis*)	Increase	154
Papaya	Increase	27, 32, 33, 155
Passion fruit	Increase	156
Peas	Increase	157
Peaches	Increase	121
Plantain	Increase	27
Squash	Increase	158
Sweet potato	No change	161, 162
Tomato	Increase	62, 159, 160
Tomato	No change	73
Bananas	Decrease	77
Beans (snap)	Decrease	12, 163
Cabbage	Decrease	164
Clover (leaf)	Decrease	165
Date	Decrease	92
Grass	Decrease	166
Guava	Decrease	167
Jujube	Decrease	96
Kale (fodder)	Decrease	97
Pepper	Decrease	38
Sapodilla (*Manilkara zapota*)	Decrease	134
Spinach	Decrease	143
Sweet potato (leaf)	Decrease	168

TABLE 1   (continued)

Plant	Change with maturity	Ref.
	Niacin	
Asparagus	Increase	12
Beans (snap)	Increase	11, 12
Capsicum	Increase	179
Grape	Increase	169
Spinach	Increase	181
Potato	No change	170
Bananas	Decrease	77
Beans (lima)	Decrease	161
Carrots	Decrease	83
Gourd	Decrease	176
Tomato	Decrease	182
	Pantothenic acid	
Carrots	Increase	83
Grape	Increase	169
Tomato	Increase	176
Tomato	Decrease	182
	Pyridoxine	
Bananas	Increase	192
Capsicum	Increase	179
Carrots	Increase	83
Grape	Increase	169
Tomato	Decrease	182
	Riboflavin	
Grape	Increase	169
Pepper (sweet)	Increase	41
Potato	Increase	170
Tomato	Increase	171
Bananas	No change	77
Beans (snap)	No change	10, 172
Alfalfa	Decrease	173, 174
Beans (lima)	Decrease	78, 79, 174
Beans (snap)	Decrease	11, 12
Carrots	Decrease	83
Cowpea	Decrease	175
Gourd	Decrease	176
Grape	Decrease	177
Grass (timothy)	Decrease	173
Peas	Decrease	35
Soybean	Decrease	178
Wheat	Decrease	136

TABLE 1 (continued)

Plant	Change with maturity	Ref.
	Thiamin	
Asparagus	Increase	12
Beans (snap)	Increase	10, 172
Capsicum	Increase	179
Citrus	Increase	27
Grape	Increase	27
Potato	Increase	170, 180
Spinach	Increase	181
Tomato	Increase	171
Wheat	Increase	137
Peas	No change	35, 118
Beans	Decrease	11
Beans (lima)	Decrease	38, 79, 161
Beans (snap)	Decrease	12
Collards	Decrease	86
Cowpea	Decrease	175
Gourd	Decrease	176
Pepper (sweet)	Decrease	41
Tomato	Decrease	182
	Tocopherol	
Beans (leaf)	Increase	183
Clover (leaf)	Increase	165
Kale (fodder)	Increase	97
Maize	Increase	184
Paprika	Increase	185–188
Xanthium strumarium (Leaf)	Increase	189
Maize	Decrease	190
Dactylis glomerata	Decrease	191
Festuca pratensis	Decrease	191
Lolium prenne	Decrease	191
Phleum pratense	Decrease	191
	Vitamin K$_1$	
Spinach	Increase	193
Swiss chard	Increase	193
Cabbage (inner leaves)	Decrease	193

[a] The edible part unless otherwise is indicated.

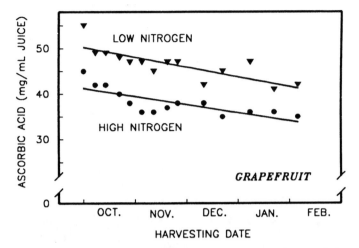

FIGURE 2. Decrease in the ascorbic acid concentration in grapefruit
(fertilized with low or high amounts of nitrogen) as the harvesting
season progresses.[197]

of mature citrus fruits (Figure 2)[197] is, for example, considered to be
partly due to dilution effect because the fruit size increases as it
matures.[113, 114, 198] When calculated per fruit basis, however, the total
content of ascorbic acid increases as citrus fruits ripen.[161]

In apples, the sum of L-ascorbic acid and dehydrascorbic acid per
unit of fruit weight remains constant throughout the growth of apples;
the proportion of L-ascorbic acid to that of dehydroascorbic acid,
however, decreases as the fruit matures.[67]

In peaches, the ascorbic acid concentration of most varieties decreases as the fruit ripens; in some varieties, however, ascorbic acid
content may not change, or it may even increase as fruit goes from
firm-ripe to ripe stage of maturity.[120] Thus the reports by Schroder
et al.[36] and Trautner and Somogyi[3], according to which ascorbic acid
content of peaches increases as the fruit matures, may be due to the
difference in genotypes used and the stage of maturity used by different
authors.

In black currants and gooseberries, the ascorbic acid content per
berry increases very rapidly during the early maturing stage but stays
constant thereafter. Thus as the fruit ripens and its weight increases, a
decrease in its ascorbic acid concentration (per unit fruit weight) is
observed[49, 199]. In strawberries, however, total ascorbic acid per berry
remains low up to the color development stage, after which there is a
rapid increase followed by a decrease at the end of the season.[49]

Ripening of *Capsicum* is accompanied by an increase in its ascorbic acid concentration,[15, 37, 38, 200] which may range from 136.1 mg/100 g in the green-mature peppers to 202.5 mg/100 g in the red (ripe) peppers.[38] According to Beckley and Notley,[37] *Capsicum* fruits taken from a single plant at different stages of their ripening showed the following ascorbic acid concentrations (mg/100 g): green, 231; purplish-black, 253; red with black patches, 272; fully red ripe, 363. Even different parts of a singly partly-ripe *Capsicum* showed the following ascorbic acid concentration (mg/100 g): green parts, 332; pink parts, 300; and deep-red parts, 515.

In some plants ascorbic acid concentration seems to undergo a biphasic change; i.e., it shows an initial increase and later decrease as they ripen or overripen. Such a decrease in the ascorbic acid concentration at the later stages of ripening has been observed in acerola,[209] apples,[6, 52, 210] peas,[38] edible-podded peas,[211] bananas,[9] gooseberry,[49] grape,[212] guava[213] tangerine,[214] jamun fruit (*Syzgium cumunii*),[215] passion fruit,[216] okra,[217] soybean seed,[218] and *Zizyphus jujuba*.[219]

In potatoes, the concentration of ascorbic acid in the newly formed tubers tends to increase and reach its maximum value shortly prior to the normal harvest time.[128, 170, 201, 202] In the nine potato varieties studied by Baird and Howatt,[203] for example, the highest amount of ascorbic acid occurred in the month of August, followed by a steady decrease at various stages of maturity until potatoes were harvested. Shekhar et al.[129] showed that the ascorbic acid content of potato tubers has two distinct phases: the first phase occurs as the tubers are growing in size when the ascorbic acid concentration increases steadily and reaches its maximum in August-September, and in the second phase, it decreases continuously as the tubers reach maturity.

A biphasic type of change in the ascorbic acid concentration has also been observed in tomato fruit.[204-207] Thus, as the tomatoes mature, their ascorbic acid concentration increases steadily and reaches its maximum when the fruits are at the yellow-red stage. As the fruits mature further and turn red (the stage at which most consumers prefer to have their tomatoes) or become overmature, they may lose some of their ascorbic acid. This type of change in the ascorbic acid of tomatoes takes place in varieties naturally producing small, medium, or large-size fruits (Figure 3). This decrease in ascorbic acid in tomatoes going from the turning stage to the pink and red stage seems to also take place in the fruits ripened off the vine (storage ripened).[208] For a more detailed account of differences in the vitamin content of fruits ripened on or off the vine, refer to page 220.

Ascorbic acid in snap beans generally increases as beans mature[12] or their harvest is delayed, indicating a possible relationship between the

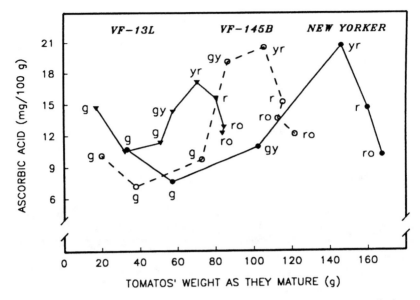

FIGURE 3.   Ascorbic acid concentrations in three varieties of tomatoes producing small, medium, and large fruits during fruit ripening. Fruits were allowed to ripen on the plants. g = green; gy = green yellow; yr = yellow red; r = red; ro = red over ripe.[205]

contents of ascorbic acid and total solids in this vegetable.[172, 220] For the best marketing qualities, snap beans are usually harvested at an immature stage, i.e., at a stage when their ascorbic acid is not at its the maximum.[172]

In green peas, total ascorbic acid increases as the peas grow in size and reach their maturity. At the same time, the relative distribution of this vitamin between different parts of the seed undergoes changes in a way that the concentration of ascorbic acid increases in the cotyledons and decreases in the testa (Figure 4).[120] Furthermore, changes in the ascorbic acid content of peas may be somewhat variety-specific. For example, Fan Yung and Pavlova[221] noted that although the ascorbic acid in some pea varieties changed very little as the peas ripened, in the early varieties of peas, which are very rich in this vitamin, however, it decreases sharply as they mature.

B. Other Vitamins

Depending on the plant and vitamin under consideration, the concentrations of several other vitamins may increase or decrease or undergo no major changes as plants mature (Table 1). In general, the

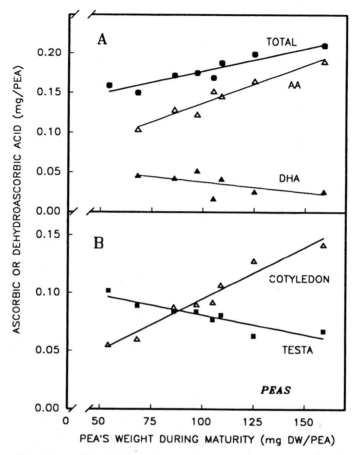

FIGURE 4. (A) Distribution of ascorbic acid (AA) and dehydroascorbic acid (DHA) in different parts of pea seeds as they increase in size during their maturity. (B) Changes in the total vitamin C content in cotyledon and testa as the peas mature.[120]

carotene content of the majority of fruits and vegetables increases as they mature. In carrots, for example, carotene content increases as the roots mature and increase in size. This is especially noticeable in the phloem tissues of the roots (Figure 5). Overmature carrots, however, may have relatively less carotene than the mature ones.[38] Decrease in carotene content at the later stages of maturation has also been noted in fenugreek leaves, in radish, turnip, and spinach,[222] and in leaves of several forage plants.[223]

In forage plants the β-carotene and vitamin A activity may decrease as the plants mature and pass the blooming stage,[116,224] so mature plants may have 50% or less of the maximum carotenoids they con-

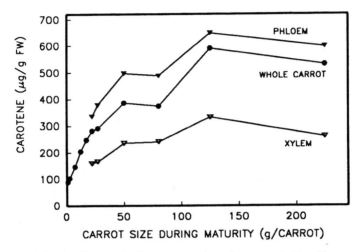

FIGURE 5.   Changes in the concentration of carotene in the phloem and xylem tissues (outer and inner tissues, respectively) of carrots as they mature and increase in size.[144]

tained in their immature stage.[224] The β-carotene content of grasses, for example, was noted to decrease with maturity and as a result of drying or fermentation and silage.[116] Since mature leaves contain relatively more carotene than mature stems, hays made of legumes are considered to be richer in provitamin A activity than hays made of grasses. The β-carotene in the green fodder is the main source of vitamin A for grazing animals and the vitamin A and β-carotene content in the cow's milk was noted to increase when cows were fed with forage high in carotene content.[116] The degree of greenness in a fodder is considered to be a good index for its carotene content.[224]

The concentrations of different vitamins may not necessarily change in the same direction as plants mature. For example, in tomato fruit, ascorbic acid, carotene, and riboflavin all increase as the fruit matures (Table 1). In peas, however, the concentration of ascorbic acid increases, that of thiamin remains constant, and that of riboflavin decreases as the pods mature.[35] In potato the concentration of vitamin B_1 increases progressively, that of ascorbic acid reaches its maximum in mid-July, and that of vitamin B_2 does not change as the tubers grow in size and reach maturity (Figure 6).

In snap beans, pods harvested in midsummer contain the most ascorbic acid and niacin, whereas beans harvested in the fall are highest in carotene and riboflavin.[12] It is thus recommended that snap beans should be harvested as soon as the pods reach their full length so that they contain the best balance of various vitamins.[11]

In wheat, the concentration of thiamin in the kernels increases continuously and that in the leaves, stems, and glumes decreases as

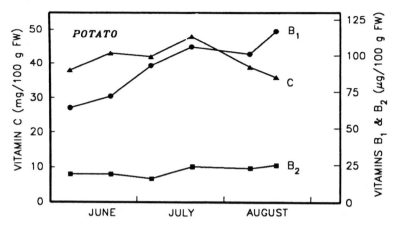

FIGURE 6. Changes in the concentration of three vitamins in the potato tuber during the growing season.[170]

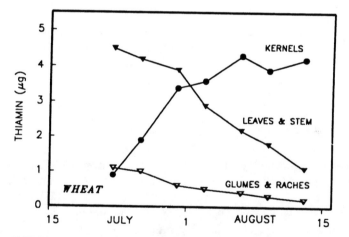

FIGURE 7. Changes in the thiamin concentration in the kernels and in the other parts of wheat plant during the kernel filling process.[225]

the plant matures (Figure 7). This has been taken as evidence for the transport of this vitamin from the vegetative parts of the plant to the kernels as they mature.[226, 227] In the maize kernel, the concentrations of pantothenic acid and niacin decrease progressively during the kernel ripening (Figure 8), which may be due to a dilution of this vitamin in the kernel as more storage substances accumulate in the kernels.[227]

Tocopherol content in plant tissues may also change as they mature and approach senescence. In the leaves of white clover (*Trifolium*

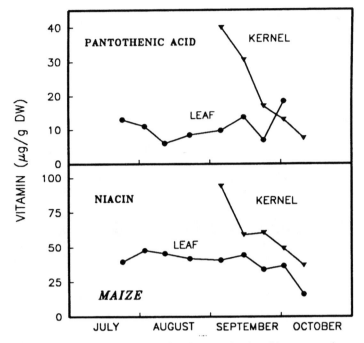

FIGURE 8.   Changes in the niacin and pantothenic acid concentrations in the leaf and kernels of maize (variety Ohio Gold) during the course of kernel development.[227]

repens), for example, the concentration of α-tocopherol increases continuously from March to November. In the leaves of cocksfoot grass (*Dactylis glomerata*)[228] and *Abies numidica*,[229] however, the tocopherol content shows a biphasic type of change: reaches a maximum level and then decreases. In broccoli and several other plants, the concentration of α-tocopherol in yellow leaves were much higher than in the green leaves, in some cases by a factor of more than 10 times.[229] Whether the increases observed in the tocopherol concentration of some plants as they approach senescence has any direct physiological significance in the aging physiology of plants is not clear at this time. We note that in animals, many age-related alterations in tissues are observed in vitamin E–deficient individuals, which indicates that increased production of free radicals in the aging tissues may play a role in the aging process in animals.[230]

II. NATURAL VERSUS ARTIFICALLY RIPENED FRUITS

Ripe tomatoes cannot be transported to distant markets without being damaged. Thus in the industrial production systems, tomatoes are

TABLE 2
Relative Vitamin Concentrations in Tomato and Papaya Ripened on the Plant as Compared with the Fruits Harvested at an Unripe Stage and Artificially Ripened at a Later Time

Plant	Difference	Ref.
	Ascorbic acid	
Tomato	Vine ripened > artifically ripened	59, 160, 232–241
Tomato	Artifically ripened > vine ripened	242, 243
Papaya	Tree ripened = artifically ripened	244
	Carotene	
Papaya	Tree ripened > artificially ripened	244
Tomato	Vine ripened > artificially ripened	235
Tomato	Vine ripened = artificially ripened	160, 232, 234, 235, 239, 240, 243, 244
	Thiamin	
Tomato	Vine ripened > artifically ripened	243
	Riboflavin	
Tomato	Vine ripened > artifically ripened	243

often harvested at mature-green or breaker stage (incipient red color) and are then "artificially" ripened in transit or at the destination market with the aid of ethylene. Whether such artificially ripened fruits have the same sensory qualities and same concentration of ascorbic acid as their counterparts left to ripen on the vine is a matter of much discussion.[231] Majority of reports, however, indicate that the vine-ripened fruits are higher in ascorbic acid than those that are artificially ripened (Table 2). In contrast, Lee[243] reported that artificially ripened tomatoes had a slightly (but statistically significant) higher ascorbic acid than those ripened on the vine. It was also noted the artificially ripened tomatoes were higher in thiamin one year, but lower in another year, than those ripened on the vine (Table 3).

The reason for the relatively higher ascorbic acid in the vine versus artifically ripened tomatoes may be due to difference in the rate of ascorbic acid synthesis (or accumulation) in the fruits ripened differently. Thus, although the concentration of ascorbic acid increases in tomatoes left on the vine to ripen as well as in those detached from the vine and ripened artificially, but (a) tomatoes picked at the breaker stage may have, depending on variety, only 43.6 to 69.2% of the

TABLE 3
Concentration of Thiamin, Riboflavin, Carotene, and Ascorbic Acid in Tomato Fruits Ripened on Vine or Harvested Green-Mature and Ripened Artifically at 19°C for One Week[243]

Ripening Method	Thiamin (μg / 100 g FW) 1978	1979	Riboflavin (μg / 100 g FW) 1978	1979	Carotene (μg / 100 g FW) 1978	1979	Ascorbic acid (mg / 100 g FW) 1978	1979
Vine ripened	60	46	49	34	690	1060	18.0	17.0
Artifically ripened	47	55	32	39	680	1050	19.3	19.0
Significance	1%	1%	0.05%	N.S.	N.S.	N.S.	5%	1%

Values are average of five cultivars compiled from tables 1, 3, 5, and 7 of Lee, Y. C., *Korean J. Food Sci. Technol.*, 16(1), 59, 1984.

potential ascorbic acid if they had ripened on the plant, and (b) there may be differences in the rate of ascorbic acid accumulation in the detached and attached fruits; artifically ripened tomatoes may contain 68–73% (variety Rick High Sugar) or 80–84% (variety Ace 55) of the ascorbic acid of vine-ripened tomatoes.[237] Also Pantos and Markakis[236] noted that vine-ripened tomatoes contained one-fourth to one-third more ascorbic acid than those ripened artificially.

Flavor and taste of tomatoes may be influenced by the way they are ripened, so tomatoes picked when fully or partly green or at the breaker stage and artificially ripened at 20°C may be less sweet, more sour, have more off flavor, and less tomato-like flavor than those left on the vine to ripen.[59] In other words, the sweeter the tomatoes, the higher was their ascorbic acid content.

III. FIELD VERSUS GREENHOUSE-GROWN PLANTS

A. Ascorbic Acid

Plants grown in greenhouses (mostly during the winter or colder parts of the year) may contain less ascorbic acid than those grown in the open field (mostly during the warmer parts of the year) (Table 4). This may be due to the general differences between greenhouse environment and open field with regard to light intensity and quality and day and night temperatures (see chapter 4).

Greenhouse-grown tomato plants, depending on variety, may produce much more yield than the field-grown plants (Table 5). The tomatoes produced in the greenhouse, however, may have close to half the ascorbic acid of those grown in the field (Figure 9; Table 5).[225, 259] Also, leafy vegetables grown in greenhouses may have lower ascorbic acid than those grown in the fields.[245] For example, the ascorbic acid contents (mg/100 g) of greenhouse- versus field-grown vegetables were: in head lettuce, 17 versus 20; in romaine lettuce, 16 versus 21; in endive, 20 versus 25; in spinach, 54 versus 60,[245] and in *Capsicum*, 60 versus 160,[246] i.e., a factor of more than 2.7 times.

B. Other Vitamins

The concentration of carotene in the greenhouse-grown tomatoes may be lower than in the field-grown ones (Table 4). In contrast, Hårdh[274] noted that the carotene contents of broccoli, dill, lettuce, and spinach were all higher in the greenhouse-grown plants than in the field-grown plants in Finland. It was argued that higher temperature in the greenhouse (as compared with light intensities in the field) was instrumental for the effects observed.

Thiamin contents of four wheat varieties grown under field conditions were found to be considerably lower (5.1–5.9 μg/g) than in the

TABLE 4
Relative Concentration of Vitamins in Plants Grown in Greenhouse[a] Compared with Those Grown in the Open Field

Plant	Difference	Ref.
	Ascorbic acid	
Beet (leaf)	Field > greenhouse	245
Endive	Field > greenhouse	245
Capsicum	Field > greenhouse	246
Cauliflower	Field > greenhouse	247
Cress	Field > greenhouse	248
Cucumber	Field > greenhouse	247, 249
Kohlrabi	Field > greenhouse	247
Lettuce	Field > greenhouse	245, 247, 250, 251
Prickly pear	Field > greenhouse	252
Radish	Field > greenhouse	249
Satsuma mandarin	Field > greenhouse	253
Spinach	Field > greenhouse	245
Strawberry	Field > greenhouse	254
Tomato	Field > greenhouse	52, 171, 247, 249, 255–267
Vegetables (several)	Field > greenhouse	268, 269
Tomato	Field = greenhouse	270, 271
Turnip root	Field = greenhouse	272
Cabbage	Greenhouse > field	249
Cauliflower	Greenhouse > field	249
Lettuce	Greenhouse > field	249
	Carotene	
Broccoli	Greenhouse > field	273
Carrots	Greenhouse > field	274
Dill	Greenhouse > field	273
Lettuce	Greenhouse > field	273
Papaw	Greenhouse > field	275
Spinach	Greenhouse > field	273
Lettuce	Field > greenhouse	274
Satsuma mandarin	Field > greenhouse	253
Tomato	Field > greenhouse	260, 261, 276, 277
	Riboflavin	
Tomato	Field = greenhouse	171
	Thiamin	
Wheat	Greenhouse > field	278

[a] Or under some kind of cover such as plastic covers.

TABLE 5

Yield and Ascorbic Acid Concentration in Tomatoes Grown Outdoors or in Greenhouse[259]

	Tomato variety			
	No. 10 X Bisson	No. 10	Kome	Sarja
Yield (kg / 1000 m^2)				
Outdoors[a]	1,638	923	544	395
Greenhouse[b]	5,678	3,710	2,198	1,958
Ascorbic acid (mg / 100 g)				
Outdoors	33	33	22	22
Greenhouse	18	20	16	14

[a] During the spring and summer months.
[b] During the winter and spring months.

Compiled and rounded data from tables 4 and 5 of Daskaloff, Chr. and Ognjanowa, A., *Arch. Gartenbau*, 10, 193, 1962.

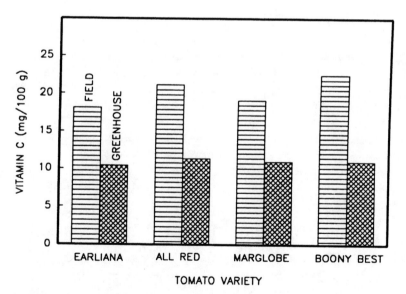

FIGURE 9. Relative ascorbic acid concentration in four tomatoe varieties grown in the field or in the greenhouse.[255]

wheat grown during winter in greenhouses from the same seeds (8.5–8.9 $\mu g/g$).[278] The environmental factors that might have been responsible for such a difference were not, however, specified.

Qualities other than vitamin contents may also be different in the plants grown in the greenhouse (or other types of protected environments such as plastic tunnels) and those grown in open fields. For example, Satsuma mandarins grown in vinyl houses were found to be substantially lower in ascorbic acid and carotenoids than those grown in the field. The fruits grown in vinyl houses, however, were preferred by the consumers to those grown in the field because they had better sensory qualities, such as peel color, sweetness, tartness, and odor at peeling.[253]

IV. GREENHOUSE VENTILATION AND CO$_2$ ENRICHMENT

The effect of greenhouse atmosphere and the CO_2 enrichment of the greenhouse air on plant growth has been the subject of numerous studies. Information as to the effect of chemical composition of greenhouse air on the vitamin content of greenhouse-grown plants, however, is extremely limited, despite some limited indications that changes in the concentration of CO_2 in the greenhouse may in fact alter the content of some vitamins in vegetables. For example, Madsen[279] grew tomatoes for up to 32 days in atmosphere enriched with CO_2 (from 0.035 to 0.50 vol %) and noted that increasing the CO_2 concentration to 0.10% and higher increased the ascorbic acid content in the tomato leaves by 50%. Kimball and Mitchell[280] noted that lack of proper greenhouse ventilation reduced the consumer acceptability of tomatoes. Both a lack of good ventilation and extra enrichment of the greenhouse air with CO_2, however, increased the vitamin A content but did not affect the vitamin C content. In contrast, other workers have reported that tomatoes grown on plants located in poorly ventilated parts of the greenhouse were lower in ascorbic acid than those grown in well-ventilated areas.[281] More research, however, is needed to clarify the nutritional significance of any changes that might occur in plant vitamins due to poor greenhouse ventilation and/or the enrichment of its air with CO_2.

V. EFFECT OF INCREASED PLANT YIELD AND SIZE ON PLANT VITAMINS

A. Planting Density

Economics of industrial crop production require that plants be planted at an optimum density (number per unit field or greenhouse

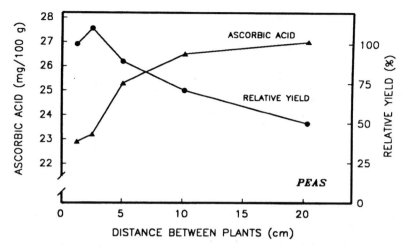

FIGURE 10.   Effect of increasing the distances between the pea plants on yield and the ascorbic acid concentration in the peas.[38]

area) to maximize their yield. Whether any "crowding" of the plants, with its inevitable effect on the amount of light reaching each individual plant (or its parts such as its fruits), would change the vitamin content of the plants or their edible parts is the subject of this chapter.

Based on the strong effect of light on the concentration of some vitamins, especially ascorbic acid (chapter 4), one may expect that higher planting densities, because of mutual shading, may decrease the ascorbic acid in plants. In agreement with this view are the reports that higher planting densities and/or narrower row spacing decreases the ascorbic acid content in *Capsicum*,[282] *Chrysanthemum coronarium*,[283] grape,[284] Kew pineapple,[285] tomato,[286] *Ziziphus mauritiana*,[287] and peas (Figure 10). Also, the observations that in apples there is an inverse relationship between the number of leaves per fruit and the ascorbic acid content of the fruits[288] and pruning of apples[289] and *Ziziphus mauritiana* (ber) trees[287, 290] and removal of part of the leaves in tomato[291] all of which increased the ascorbic acid content in these plants further support the view that less crowding (density) and the resulting improved light distribution within the plant canopy can increase the ascorbic acid of plants. Other experimental data, as presented below, however, seem to contradict this simplistic line of thought.

For example, snap beans planted at exactly the same planting densities were found to contain different ascorbic acid, depending on the distances between the plants and between the rows. Thus snap beans planted on narrow rows (27.9 cm apart) were found to contain 20% more ascorbic acid than beans planted on wider rows (55.9 cm apart).

After being canned, however, beans obtained from the two different row spacings contained similar ascorbic acid concentration.[292] These results are difficult to interpret because the total number of plants per unit surface was the same in all treatments and thus the mutual shading of the plants and their competition for light should have been the same under both planting conditions unless narrower rows gave a special three-dimensional structure to the canopy so that it improved the light interception by the crop as a whole. Also, the reports that higher planting densities may increase the ascorbic acid in tomato[293] and eggplant[294] point to factors other than light interception that might play a role in determining the amount of vitamin C in plants under certain conditions.

Changes in planting density may change other plant qualities without affecting their vitamin content. This has been documented in an experiment on the effect of planting density on the quality of fruits of mandarin trees planted at 4, 5, or 6 m between the rows and 2, 3, or 4 m within the rows. It was noted that close planting of trees (4 × 2 m, 1250 trees/ha) produced the highest yield of good-quality fruits with the best flavor score without having any adverse effect on their vitamin C content.[295] Plant density was also reported to have no effect on the ascorbic acid content of *Capsicum* grown in plastic tunnels.[296]

Carotene content in carrots seems to be also affected by the planting density. This conclusion is based on the only report of its kind known to us according to which carrots planted at higher densities stayed smaller, had a stump rather than pointed shape, ripened faster, and showed a faster rate of carotene accumulation than those grown at lower planting densities.[297]

In summary, based on the limited and sometime inconsistent reports available, it appears that increased planting density may affect the content of some vitamins under some circumstances; the mechanism involved, however, is not well understood.

B. Tree Vigor and Fruit Vitamin

Information as to the effect of tree vigor on the vitamin content of the fruits is very scarce. The only information known to the author is that of Murneek and Wittwer[298] indicating that under conditions of equivalent crop size, apples grown on weaker trees (as judged by the amount of their shoot growth and foliage) contained higher ascorbic acid concentration than those grown on more vigorous trees (Table 6). Recently Ray and Munsi[299] reported that litchi fruits grown on low-yielding trees contained higher ascorbic acid content than those grown on high-yielding trees.

TABLE 6
Effect of Tree Vigor on the Ascorbic Acid in the Golden Delicious and Jonathan Varieties of Apples

	Ascorbic acid (mg / 100 g)	
Tree vigor	Golden Delicious (light crop)	Jonathan (young tree)
Weak tree	7.5	6.3
Vigorous tree	6.3	4.1

From Murneek, A. E. and Wittwer, S. H., *Proc. Am. Soc. Hortic. Sci.*, 51, 97, 1948.

TABLE 7
Effects of Crop Size (Yield) and Variety on the Ascorbic Acid of Apples

	Ascorbic acid (mg / 100 g)	
Variety	Light crop	Heavy crop
Winesap	11.8	11.0
York	7.6	6.3
Jonathan	6.3	4.3

From Murneek, A. E. and Wittwer, S. H., *Proc. Am. Soc. Hortic. Sci.*, 51, 97, 1948.

C. Higher Yield, Lower Vitamins

Many agronomic practices, such as the use of fertilizers, plant protection and growth regulator chemicals, irrigation, and so forth, are mainly used to maximize the plant yield, and in cases to improve other culinary and cosmetic qualities. Whether these practices also improve the vitamin content of plants is discussed in chapters 5 and 7–9. Here some selected cases are reviewed to illustrate the possible relationships between the magnitude of yield increase and the quality of products as far as their vitamin contents are concerned.

Apples grown on trees bearing a light crop tend to be higher in ascorbic acid than those grown on trees with a heavy crop[298] (Table 7). In peaches, increasing the nitrogen fertilizer increased the leaf content of nitrogen, so there was a positive relationships between the leaf nitrogen and yield of peaches per tree. There was, however, a negative relationship between the yield and the ascorbic acid content in peaches (Figure 11).[301]

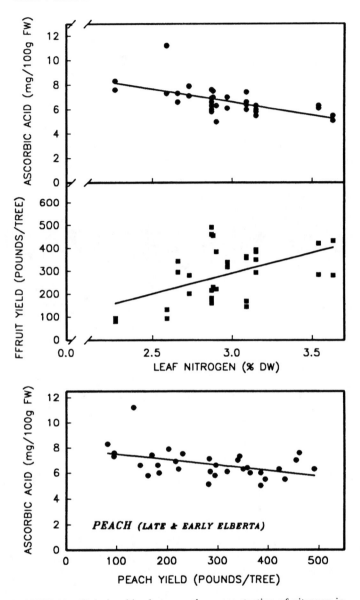

FIGURE 11. Relationships between the concentration of nitrogen in the leaves (increased due to increased use of nitrogen fertilization), yield, and the ascorbic acid concentration in peaches.[301]

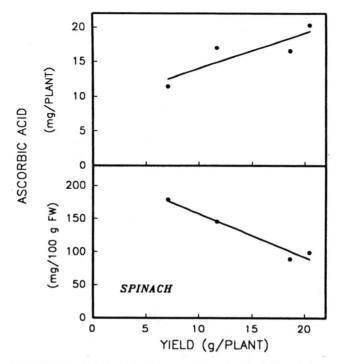

FIGURE 12. Relationships between the spinach yield and the ascorbic acid concentration in the leaves or the total ascorbic acid per plant. Data are averages of all N and Ca treatments of Wittwer, S. H., Schroeder, R. A., and Albrecht, W. A., *Soil Sci.*, 59, 329, 1945.

In spinach, Wittwer et al.[302] noted that application of nitrogen fertilizer increased spinach yield considerably but had an adverse effect on the concentration of ascorbic acid in the leaves. Thus, although the total ascorbic acid in each plant increased, the concentration of ascorbic acid in their leaves, however, decreased as the yield was increased (Figure 12).

In mandarin and lemon fruits, application of fertilizers was noted to increase the fruit yield, but the higher the yield, the lower was their vitamin C content.[303] Also, a comparison between the yield and the vitamin C content of two mandarin varieties showed that the variety (Mudkhed Seedless Santra) that normally produced higher yield also had lower ascorbic acid than the variety (Nagpur Santra) that produced lower yield.[304] Also, the observation that those varieties of peas producing higher yield tend to be lower in ascorbic acid than the lower-yielding varieties (Figure 13) indicates that higher-yielding varieties may be inferior to lower-yielding varieties as far as their vitamin C is con-

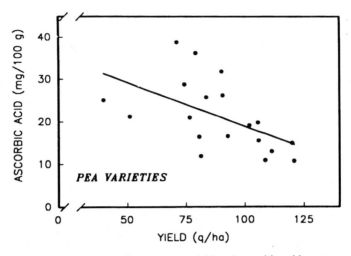

FIGURE 13. Relationships between yield and ascorbic acid concentration in 18 "main-season" varieties of peas.[305]

cerned. In western Canada, significant negative correlation was noted between the wheat yield per acre and the thiamin content in the kernels.[306]

Finally, the yield of plants grown in different geographical and climatic conditions could be very different. But whether their vitamin content would also be different is not clear. Based on the data of Hårdh et al.,[307] however, it seems that there is no clear relationships between the yield and the ascorbic acid content of several vegetables grown in very different parts of Norway (Table 8). For some further data of Hårdh and co-workers on the yield and vitamin C content of plants grown in different parts of Scandinavian countries see Tables 19 and 20 in chapter 4.

Further examples of the relationships between yield and the concentration of vitamins are given in the following tables and figures:

Vitamin	Chapter	Figure	Table
Vitamin C	4	7	
Vitamin C	5	2, 3	2, 15, 20
Vitamin C	7	1, 2	4
Carotene	5		2, 7, 19
Carotene	7		4
Thiamin	5	5	
Riboflavin	5	5	
Riboflavin	7	1	3
Folic acid, biotin			
Niacin and pantothenic acid	5		8
Vitamin E	7		3

TABLE 8

Yield, Ascorbic Acid, and Carotene Concentrations in Three Vegetables Grown in Different Geographical Locations in Norway

Location	Potato		Tomato			Sweet pepper		
	Tuber weight (g)	Ascorbic acid (mg / 100 g DW)	Yield (kg / m²)	Ascorbic acid (mg / 100 g FW)	Carotene (mg / 100 g DW)	Yield (kg / m²)	Ascorbic acid (mg / 100 g FW)	Carotene (mg / 100 g DW)
Kvithamar (63°28'N)	51	12	21	22	7	9	183	153
As (59°40'N)	158	17	29	18	8	4	144	164

Compiled and rounded off data from tables 18, 19, and 20 of Hårdh, J. E., Persson, A. R., and Ottosson, L., *Acta Agric. Scand.*, 27, 81, 1977.

A tradeoff between the plant yield and a number of quality and sensory properties of fruits and vegetables has been reported.[300] Information presented here seems to also point to a tradeoff between yield and the concentration of some vitamins. The nutritional consequences of these relationships need to be investigated.

D. Larger Size: Lower Vitamins

1. Ascorbic Acid

Plants, even when they are genetically as uniform as possible, do not produce parts (leaves, roots, seeds, fruits) of exactly the same size, even when they are grown under the most uniform root and atmospheric conditions practically possible. Furthermore, different varieties of the same plant may often produce fruits or vegetables of different size and shape. The question is whether different size fruits or vegetables have the same or different vitamin contents.

Available information summarized in Table 9 shows that at the same stage of ripening, larger fruits or vegetables usually have less ascorbic acid than smaller ones. This is not only the case when comparisons are made between different varieties (Figures 14 and 15),[68, 340] but also holds true among different-sized fruits and vegetables of the same variety (Figures 16–21).[35, 298, 319, 323, 326, 349]

The relationships between the size and the vitamin concentration may or may not be linear. In ripe tomatoes, for example, very small fruits grown on the same plant may have disproportionally higher ascorbic acid than larger fruits. Thus fruits weighing less than 10 g may have 50% more vitamin C than those weighing 30 g and more.[339] In the potato, however, medium-sized tubers appear to have the highest ascorbic acid concentration as compared with too small or too large potatoes (Figure 21).[349]

In lemon (Figure 16)[326] and grapefruit[325] (Figure 17),[323] there exists a negative linear relationship between the fruit size and its ascorbic acid content. In the tangerines grown in Thailand, however, although larger fruit tended to be much higher in percentage of juice, no consistent relationship was found between the fruit's size and the ascorbic acid concentration in its juice.[345]

2. Other Vitamins

In contrast to ascorbic acid, the concentration of most other vitamins tends to be higher in the larger-sized fruits or vegetables than in the smaller ones (Table 9). In carrots, for example, carotene concentration in the roots increases as the roots mature and at same time become larger in size[140, 141] (see page 250). It is thus recommended that farmers sow their carrots in the spring as soon as possible to increase the chances of larger and better-colored carrots at harvest time. Delayed

TABLE 9
Relative Vitamin Concentrations in Fruits and Vegetables of Different Size

Plant	Variety	Effect	Ref.
		Ascorbic acid	
Apples	SV[a]	Small > large	295, 308–312
Asparagus spear[b]	SV	Small > large	313
Blueberry	DV	Small > large	68
Brussels sprout	SV	Small > large	314, 315
Cabbage	DV	Small > large	316, 317
Cauliflower	SV	Small > large	320
Cucumber	DV	Small > large	321
Currant (black)	SV	Small > large	49
Currant (black)	DV	Small > large	322
Gooseberry	SV	Small > large	49
Grapefruit	SV	Small > large	323–325
Lemon	SV	Small > large	326, 327
Lime (Persian)	SV	Small > large	328
Mandarin	?	Small > large	329
Mango	?	Small > large	330
Muskmelon	DV	Small > large	331
Oranges	SV	Small > large	332, 333
Peas	DV	Small > large	35
Peas	SV	Small > large	334, 335
Petunia (leaf)	SV	Small > large	123
Sea buckthorn	?	Small > large	336
Strawberry	SV	Small > large	337, 338
Tomato	SV	Small > large	52, 256, 339
Tomato	DV	Small > large	63, 340–342
Zucchini	?	Small > large	343
Guava	DV	Large = small	344
Tangerine	SV	Large = small	345
Turnip greens	SV	Large = small	346
Sweet potato	DV	Large > small	347
		Carotene	
Carrots	DV	Small > large	38
Peas	SV	Small > large	335
Tomato	SV	Large = small	277
Purslane leaves	SV	Large > small	348
Sweet potato	DV	Large > small	347
Turnip greens	SV	Large > small	346
		Folic acid	
Peas	SV	Large = small	335

TABLE 9 (continued)

Plant	Variety	Effect	Ref.
Niacin			
Peas	SV	Large = small	335
Sweet potato	DV	Large = small	347
Pantothenic acid			
Tomato	DV	Large > small	176
Sweet potato	DV	Large = small	347
Pyridoxine			
Gourd	DV	Large > small	176
Tomato	DV	Large > small	176
Peas	SV	Large = small	335
Riboflavin			
Gourd	DV	Large > small	176
Tomato	DV	Large > small	176
Turnip greens	SV	Large > small	346
Peas	SV	Large = small	335
Sweet potato	SV	Large = small	347
Thiamin			
Gourd	DV	Large > small	176
Peas	DV	Large > small	35, 335
Turnip greens	SV	Medium > small	346
Tomato	SV	Large > small	176

[a] SV or DV indicates that comparisons were made between the same or different varieties, respectively. A question mark (?) indicates that it is not fully clear whether some or different varieties were compared.
[b] Spears with small and large diameters were compared.

sowing was shown to progressively result in more immature carrots, which were lower in carotene.[140] Among different varieties of carrots, however, those varieties producing smaller roots tend to be slightly higher in carotene than those producing larger roots.[140]

In summary, depending on the kind of plant and vitamin under consideration, larger fruits or vegetagbles may be lower in some but higher in other vitamins. In some cases, however, vitamin concentration is not affected by the size of fruit or vegetable (Table 9). Whether the magnitude of the differences observed in the vitamin content of large

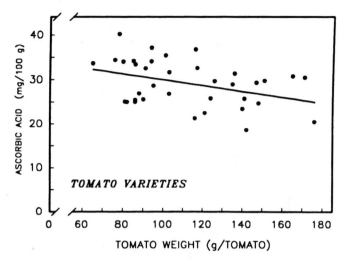

FIGURE 14. Relationships between the average weight of each tomato fruit and its ascorbic acid concentration in 36 different tomato varieties (18 variety plus 18 hybrids).[340]

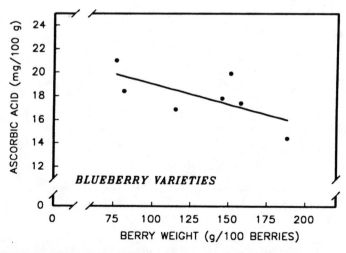

FIGURE 15. Relationships between the size of berry and its ascorbic acid concentration in different blueberry varieties.[68]

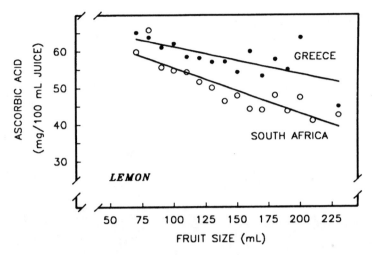

FIGURE 16. Relationships between the size and the ascorbic acid concentration in lemons cultivated in Greece or South Africa.[326]

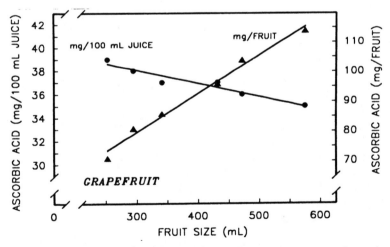

FIGURE 17. Relationships between the grapefruit and the concentration and total ascorbic acid in its juice. Grapefruit volume is calculated from the fruit diameter assuming a perfect sphere.[323]

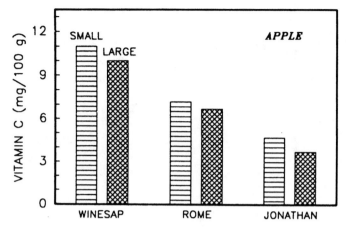

FIGURE 18. Relative ascorbic acid concentrations in small and large apples.[298]

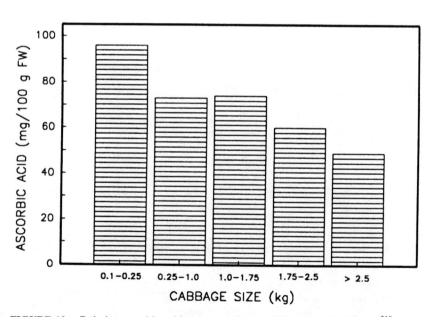

FIGURE 19. Relative ascorbic acid concentrations in different size cabbages.[319]

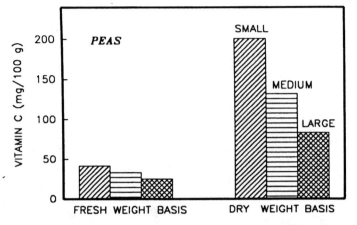

FIGURE 20. Relative ascorbic acid concentration in small, medium, and large peas.[35]

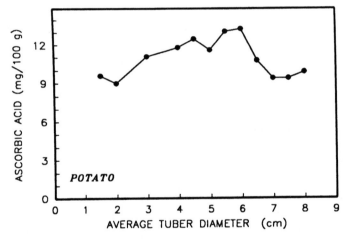

FIGURE 21. Relationships between average tuber diameter (average of length plus breadth) and the ascorbic acid concentration in the mature potatoes (variety Kufri Chandramukhi).[349]

and small fruits or vegetables is of any nutritional significance is not clear at this time. It should be emphasized, however, that although most of the differences noted in Table 9 are relatively small, there are cases where the differences observed are considerable. In peas, for example, the small (size 3) peas were noted to have about 32% higher ascorbic acid (171.1 vs. 129.4 mg/100 g), 34% higher carotene (1,530 vs. 1,144 µg/100 g), but 62%g lower thiamin (97 vs 1,592 µg/100 g) than the larger peas (size 5).[335]

With the above in mind, any decision on the part of breeders or farmers to produce plant varieties with larger fruits, for example, or on the part of the consumers in selecting larger fruits or vegetables from the supermarket shelf, is bound to be a tradeoff between different vitamins. As another example, one may mention the case of tomatoes, where smaller tomatoes may have higher ascorbic acid but would be lower in thiamin and riboflavin than larger tomatoes. Since tomatoes are a main source of ascorbic acid and not of thiamin and riboflavin in human nutrition, it may be concluded that if the culinary and price considerations are comparable, then the smaller tomatoes are a better buy than the larger ones. This argument, however, does not hold true for sweet potatoes where large roots are significantly higher in both ascorbic acid and total carotenoid than small ones.[347]

REFERENCES

1. Zubeckis, E., Ascorbic acid content of fruit grown at Vineland, Ontario, *A. R. Hortic. Exp. Stn. Prod. Lab.*, *Vineland*, 90, 1962; *Hortic. Abstr.*, 34, 2307, 1964.
2. Ferrão, A. M. A. B. da C. and Ferrão, J. E. M., Vitamin C in apricots—effects of variety and of processing, *Agricultura (Lisboa)*, 21, 29, 1964; *Hortic. Abstr.*, 35, 5185, 1965.
3. Trautner, K. and Somogyi, J. C., Änderungen der Zucker- und Vitamin-C Gehalte in Früchten während der Reifung, *Mitt. Geb. Lebensmittelunters. Hyg.*, 69, 431, 1978.
4. Koch, J. and Bretthauser, G., Über den Vitamin C Gehalt reifender Früchte, *Landwirt. Frorsch.*, 9, 51, 1956.
5. Zubeckis, E., Ascorbic acid content of fruit grown at Vineland, Ontario, *A. R. Hortic. Exp. Stn. Prod. Lab.*, *Vineland*, 90, 1962; *Hortic. Abstr.*, 34, 2307, 1964.
6. Stepanova, F. P. and Samorodova-Bianki, G. B., Nature of interaction between phenolic compounds, ascorbic acid and glutathione in the fruit of apple, *Tr. Prikl. Bot. Genet. Sel.*, 70(3), 94, 1981; *Hortic. Abstr.*, 54, 8892, 1984.
7. Joshi, S. M. and Seth, J. N., Physiological changes in the apple hybrid Chaubattia Princess during fruit development maturity and post harvest storage, *Prog. Hortic.*, 17(3), 221, 1985.
8. Joshi, S. M. and Divakar, B. L., Biochemical changes during fruit development and maturity in apple variety Esopus-Spitzenburg, *Prog. Hortic.*, 17(4), 304, 1985.
9. Harris, P. L. and Poland, G. L., Variations in ascorbic acid content of bananas, *Food Res.*, 4, 317, 1939.
10. Hayden, F. R., Heinze, P. H., and Wade, B. L., Vitamin content of snap beans grown in South Carolina, *Food Res.*, 13, 143, 1948.
11. Hibbard, A. D. and Flynn, L. M., Effect of maturity on the vitamin content of green snap beans, *Proc. Am. Soc. Hortic. Sci.*, 46, 350, 1945.
12. Flynn, L. M., Hibbard, A. D., Hogan, A. G., and Murneek, A. E., Effect of maturity on nutrients of snap beans, *J. Am. Diet. Assoc.*, 22, 415, 1946.
13. Gupta, A. K., Panwar, H. S., and Vashishtha, B. B., Growth and development changes in ber (*Zizyphus mauritiana* Lamk.), *Indian J. Hortic.*, 41, 52, 1984.
14. Bal, J. S. and Premsingh, Development physiology of ber (*Zizyphus mauritiana* Lam.) var. Umran. III. Minor chemical changes with reference to total phenolics,

ascorbic acid (vitamin C) and minerals, *Indian Food Packer*, 32(3), 66, 1978; *Food Sci. Technol. Abstr.*, 12(4), J447, 1980.

15. Rahman, F. M. M., Buckle, K. A., and Edwards, R. A., Changes in total solids, ascorbic acid and total pigment content of capsicum cultivars during maturation and ripening, *J. Food Technol.*, 13, 445, 1978.

16. Mudambi, S. R. and Rajagopla, M. V., Variation in vitamin C content of cashew apple with maturity, *J. Food Technol.*, 12, 555, 1977.

17. Sondhi, S. P. and Pruthi, J. S., Effect of variety/strain and stage of maturity on the quality of cashew apples, *Indian J. Hortic.*, 37, 270, 1980.

18. Roy, A. K. and Ojha, N. L., Ascorbic acid content in bottle gourd fruits during different ages of maturation and storage, *Natl. Acad. Sci. Lett. (India)*, 8(6), 175, 1985; *Chem. Abstr.*, 105, 169068a, 1986.

19. Zubeckis, E., Ascorbic acid in Veeport grape during ripening and processing, *Rep. Ont. Hortic. Exp. Stn. Prod. Lab.*, 1964, 114; *Hortic. Abstr.*, 36, 2696, 1966.

20. Webber, H. J., The vitamin C content of guavas, *Proc. Am. Soc. Hortic. Sci.*, 45, 87, 1944.

21. Srivastava, H. C. and Narasimhan, P., Physiological studies during the growth and development of different varieties of guavas (*Psidium guajava* L.), *J. Hortic. Sci.*, 42, 97, 1967.

22. Esteves, M. T. da C., Carvalho, V. D. de, Chitarra, M. I. F., Chitarra, A. B., and Paula, M. B. de, Characteristics of fruits of six guava (*Psidium guajava* L.), cultivars during ripening. II. Vitamin C and tannin contents, in *Anais do VII Congresso Brasileiro de Fruitcultura*, Vol. 2, Florianópolis, Brazil, 1984, 490; *Hortic. Abstr.*, 55, 7323, 1985.

23. Yusof, S. and Mohamed, S., Physiocochemical changes in guava (*Psidium guajava* L.) during development and maturation, *J. Sci. Food Agric.*, 38(1), 31, 1987.

24. Salmah, Y. and Suhaila, M., Physico-chemical changes in guava (*Psidium guajava* L.) during development and maturation, *J. Sci Food. Agric.*, 38(1), 31, 1987.

25. Abbas, M. F., Al-Niami, J. H. and Al-Ani, R. F., Some physiological characteristics of fruits of jujube (*Ziziphus spina-christi* L., Willd.) at different stages of maturity, *J. Hortic. Sci.*, 63(2), 337, 1988.

26. Okuse, I. and Ryugo, K., Compositional changes in the developing "Hayward" kiwi fruit in California, *J. Am. Soc. Hortic. Sci.*, 106(1), 73, 1981.

27. Nagy, S. and Wardowski, W. F., Effect of agricultural practice, handling, processing, and storage of fruits, in *Nutritional Evaluation of Food Processing*, Karmas, E. and Harris, R. S., Eds., Van Nostrand Reinhold, New York, 1988, chap. 4.

28. Ramadan, A. A. S. and Domah, M. B., Non-volatile organic acids of lemon juice and strawberries during stages of ripening, *Nahrung*, 30(7), 659, 1986.

29. Singh, A. and Abidi, A. B., Level of carbohydrate fractions and ascorbic acid during ripening and storage of litchi (*Litchi chinensis* Sonn) cultivars, *Indian J. Agric. Chem.*, 19(3), 197, 1986; *Chem. Abstr.*, 108, 219062d, 1988.

30. Tono, T. and Fujita, S., Determination of ascorbic acid by spectrophotometric method based on difference spectra. VI. Determination of vitamin C (ascorbic acid) in Satsuma mandarin fruit by difference spectral method and change in its content of the fruit during the development stage. *Nippon Shokuhin Kogyo Gakkaishi*, 32(4), 295, 1985; *Chem. Abstr.*, 103, 138711j, 1985.

31. Mowlah, G. and Itoo, S., Changes in pectic components, ascorbic acid, pectic enzymes, and cellulase activity in ripening and stored guava (*Psidium guajava*), *Nippon Shokuhin Kogyo Gakkaish*, 30(8), 454, 1983; *Chem. Abstr.*, 99, 193401g, 1983.

32. Selvaraj, Y. and Pal, D. K., Changes in the chemical composition of papaya (Thailand variety) during growth and development, *J. Food Sci. Technol.*, 19, 257, 1982.

33. Selvaraj, Y., Pal, D. K., Subramanyam, M. D. and Iyer, C. P. A., Changes in the chemical composition of four cultivars of papaya (*Carica papaya* L.) during growth and development, *J. Hortic. Sci.*, 57(1), 135, 1982.

34. Pruthi, J. S., Physiology, chemistry, and technology of passion fruit, *Adv. Food Res.*, 12, 203, 1963.

35. Heinze, P. H., Hayden, F. R., and Wade, B. L., Vitamin studies of varieties and strains of peas, *Plant Physiol.*, 22, 548, 1947.

36. Schroder, G. M., Satterfield, G. H., and Holmes, A. D., The influence of variety, size and degree of ripeness upon the ascorbic acid content of peaches, *J. Nutr.*, 25, 503, 1943.

37. Beckley, V. A. and Notley, V. E., The ascorbic acid content of sweet peppers, *J. Soc. Chem. Ind.*, 62, 14, 1943.

38. Pepkowitz, L. P., Larson, R. E., Gardner, J., and Owens, G., The carotene and ascorbic acid concentration of vegetable varieties, *Plant Physiol.*, 19, 615, 1944.

39. Simon, J., Ertrag und Vitamin C-Gehalt bei Paprika, *Bodenkultur*, 11, 208, 1960.

40. Michna, M., The formation of some chemical components in several pepper varieties of foreign origin in relation to the degree of fruit maturity, *Roczn. Nauk Rol.*, *Ser. A*, 91, 421, 1966; *Hortic. Abstr.*, 37, 3077, 1967.

41. Butkevič, S. T., Vitamins in sweet peppers, *Tr. Mold. Nauč Issled. Inst. Orošaem. Zemled. Ovoščev.*, 7(1), 55, 1967; *Hortic. Abstr.*, 37, 7124, 1967.

42. Ludilov, V. A., Possibilities of improving the quality of peppers with regard to content of ascorbic acid and P-active substances, *Kach. Ovoshch. Bakhchevykh Kul't.*, 50, 1981; *Chem. Abstr.*, 97, 88921n, 1982.

43. Khadi, B. M., Goud, J. V., and Patil, V. B., Variation in ascorbic acid and mineral content in fruits of some varieties of chili (*Capsicum Annuum* L.), *Qual. Plant.—Plant Foods Hum. Nutr.*, 37(1), 9, 1987.

44. Kitagawa, Y., Distribution of vitamin C related to the growth of some fruits. II. Tomato, pimento and strawberry, *J. Jpn. Soc. Food Nutr.*, 26(2), 139, 1973; *Food Sci. Technol. Abstr.*, 6(3), J384, 1974.

45. Trofimova, E. A., Dynamics of variation in the chemical composition of plum fruits during growth, *Nauchno-Teckhnicheskii Byulleten' Vsesoyuznogo Ordena Lenina Ordena Druzhby Narodov Nauchno Issledovarel'skogo Instituta Imeni N. I. Vavilova*, 134, 72, 1983; *Hortic. Abstr.*, 55, 2476, 1985.

46. Takebe, M. and Yoneyama, T., Plant growth and ascorbic acid. 1. Changes of ascorbic acid concentrations in the leaves and tubers of sweet potat (*Ipomea batatas* Lam) and potato (*Solanum tuberosum* L.), *Nippon Dojo Hiryogaku Zasshi*, 63(4), 447, 1992; *Chem. Abstr.*, 117, 190048v, 1992.

47. Sil'janova, Ju. I., The effect of boron and zinc on enzyme activity and accumulation of ascorbic acid in the fruit of cultivated rowan, *Tr. Kazansk. Sel'sk. Inst.*, 44, 31, 1964; *Hortic. Abstr.*, 36, 342, 1966.

48. Stino, K. R., Abdelfattah, M. A. and Nassar, H., Studies on vitamin C and oxalic acid concentration in spinach, *Agric. Res. Rev.*, 51(5), 109, 1973; *Hortic. Abstr.*, 46, 304, 1976.

49. Olliver, M., The ascorbic acid content of fruits and vegetables, *Analyst*, 63, 2, 1938.

50. Tombesi, L., Baroccio, A., Cervigni, T., Fortini, S., Tarantola, M., Venezian, M. E., Oxidase, catalase, carbonic anhydrase and peroxidase activity, and content of reduced glutathione and of ascorbic acid during the maturation of fruits and seeds, *Ann. Sper. Agrar.* (*Rome*), 6, 857, 1952; *Chem. Abstr.*, 47, 2835a, 1953.

51. Fryer, H. D., Ascham, L., Cardwell, A. B., Frazier, J. C., and Willis, W. W., Relation between stages of maturity and ascorbic acid content of tomatoes, *Proc. Am. Soc. Hortic. Sci.*, 64, 365, 1954.

52. Murneek, A. E., Maharg, L., and Wittwer, S. H., Ascorbic acid (vitamin C) content of tomatoes and apples, *Univ. Missouri, Agric. Exp. Stn. Res. Bull.*, 568, 1954.

53. Georgiev, H. P. and Balzer, I., Analytische Methoden zur Bestimmung der Reifungsstadien der Tomate, *Arch. Gartenbau*, 10, 398, 1962.

54. Dalal, K. B., Salunkhe, D. K., Boe, A. A., and Olson, L. E., Certain physiological and biochemical changes in the developing tomato fruit (*Lycopersicon esculentum* Mill.), *Food Sci.*, 30, 504, 1965.

55. Twomey, D. G. and Ridge, B. D., Note on L-ascorbic acid content of english early tomatoes, *J. Sci. Food Agric.*, 21, 314, 1970.

56. Hobson, G. E. and Davies, J. N., The tomato, in *The Biochemistry of Fruits and Their Products*, Vol. 2, Hulme, A. C., Ed., Academic Press, London, 1971, 440.

57. Voronina, M. V., Some characteristics of changes in the peroxidase activity and ascorbic acid content in the tissues of greenhouse tomatoes, *Tr. Prikl. Bot. Genet. Sel.*, 45(1), 209, 1971; *Hortic. Abstr.*, 43, 5362, 1973.

58. Burge, J., Mickelsen, O., Nicklow, C., and Marsh, G. L., Vitamin C in tomatoes: comparison of tomatoes developed for mechanical or hand harvesting, *Ecol. Food Nutr.*, 4, 27, 1975.

59. Kader, A. A., Stevens, M. A., Albright-Holten, M., Morris, L. L., and Algazi, M., Effect of fruit ripeness when picked on flavor and composition in fresh market tomatoes, *J. Am. Soc. Hortic. Sci.*, 102(6), 724, 1977.

60. Tahir, M. A. and Elahi, M., Effect of growth on pectinesterase, vitamin C, and protein content of tomato, *Pakistan J. Sci. Res.*, 30, 8, 1978.

61. Kanesiro, M. A. B., Faleiros, R. R. S., and Nascimento, V. A., Effect of fertilizer treatments on the vitamin C content of tomatoes, *Cientifica*, 6(2), 225, 1978; *Hortic. Abstr.*, 49, 5105, 1979.

62. López-Andréu, F. J., Lamela, A., Esteban, R. M., and Collado, J. G., Evolution of quality parameters in the maturation stage of tomato fruit, *Acta Hortic.*, 191, 387, 1986.

63. Mochizuki, T., Kamimura, S., Variation in vitamin C content of fruits in tomato lines with the high-pigment gene, *Bull. Veg. Ornament. Crops Res. Stn., B. Morioka*, 6, 1, 1986; *Plant Breeding Abstr.*, 56, 9187, 1986.

64. Reder, R., Speirs, M., Chochran, H. L., Hollinger, M. E., Farish, L. R., Geiger, M., McWhirter, L., Sheets, O. A., Eheart, J. F., Moore, R. C., and Carolus, R. L., The effects of maturity, nitrogen fertilization, storage and cooking, on the ascorbic acid content of two varieties of turnip greens. *South. Coop. Series Bull.*, 1, 1, 1943.

65. Valentinova, N. I. and Valentinov, V. A., Mineral compounds and free amino acids in watermelons, *Izv. Vyssh. Uchebn. Zaved. Pishch. Tekhnol.*, 5, 132, 1983; *Chem. Abstr.*, 100, 21750t, 1984.

66. Al-Hijiya, M. N., Ali, S. H., and Al-Delaimy, K. S., Proximate composition and the effect of freezing on ascorbic acid content of *Ziziphus* (Spina christi) fruit, *Iraqi J. Sci.*, 30(1), 73, 1989; *Chem. Abstr.*, 111, 113995g, 1989.

67. Zilva, S. S., Kidd, F., and West, C., Ascorbic acid in the metabolism of the apple fruit, *New Phytol.*, 37, 345, 1935.

68. Matzner, F., Über den Vitamin-C-Gehalt der Kulturheidelbeeren, *Erwerbsobstbau*, 7(6), 105, 1965.

69. Smith, F. G. and Walker, J. C., Relation of environmental and heredity factors to ascorbic acid in cabbage, *Am. J. Bot.*, 33, 120, 1946.

70. Tanusi, S., and Yamamoto, M. Changes of vitamin C content in the fruits of *Citrus sudachi* during maturation, *Tokushima Bunri Daigaku Kenkyu Kiyo*, 24, 69, 1981; *Chem. Abstr.*, 96, 67459g, 1982.

71. Reid, M. S., Heatherbell, D. A., and Pratt, H. K., Seasonal patterns in chemical composition of the fruit of *Actinidia chinensis*, *J. Am. Soc. Hortic. Sci.*, 107(2), 316, 1982.

72. Udoessien, E. I. and Iflon, E. T., Chemical studies on the unique and ripe fruits of *Dennettia tripatela* (pepper fruit), *Food Chem.*, 13, 257, 1984.

73. Yamaguchi, M., Howard, F. D., Luh, B. S., and Leonard, S. J., Effect of ripeness and harvest dates on the quality and composition of fresh canning tomatoes, *Proc. Am. Soc. Hortic. Sci.*, 76, 560, 1960b.

74. Hernández-Medina, E. and Vélez-Santiago, J., Response of acerola (*Malpighia puniciofolia* L.) to the application of lime and foliar sprays of magnesium and minor elements, *Proc. Carib. Reg. Am. Soc. Hortic. Sci.*, 4, 20, 1960; *Hortic. Abstr.*, 31, 5547, 1961.

75. Guadarrama, A., Some chemical changes during ripening of Barbados cherry (*Malpighia punicifolia*) fruits, *Riv. Facultad Agron. Univ. Central de Venezuela*, 13(1/4), 111, 1984; *Hortic. Abstr.*, 56, 896, 1986.

76. Itoo, S., Aiba, M., and Ishihata, K., Ascorbic acid content in acerola fruit from different production regions in relation to degree of maturity and its stability during processing, *Nippon Shokuhin Kogyo Gakkaishi*, 37(9), 726, 1990; *Chem. Abstr.*, 114, 184032p, 1991.

77. Wills, R. B. H., Lim, J. S. K., and Greenfield, H., Changes in chemical composition of "Cavendish" banana (*Musa acuminata*) during ripening, *J. Food Biochem.*, 8, 69, 1984.

78. Eheart, J. F., Moore, R. C., Speirs, M., Cowart, F. F., Cochran, H. L., Sheets, O. A., McWhirter, L., Geiger, M., Bowers, J. L., Heinze, P. H., Hayden, F. R., Mitchell, J. H., and Carolus, R. L., Vitamin studies on lima beans, *South. Coop. Series Bull.*, 5, 1, 1946.

79. Eheart, J. F., Wakeley, J. T., Speirs, M., Cowart, F. F., Miller, J., Heinze, P. H., Kanapaux, M. S., Sheets, O. A., McWhirter, L., Geiger, M., and Moore, R. C., Effect of different planting dates, bean maturity, and location on the vitamin content of lima beans, *South. Coop. Series Bull.*, 12, 5, 1951.

80. Kmiecik, W., Lisiewska, Z., and Jaworska, G., Content of vitamin C in raw, frozen and canned broad beans as a function of cultivar and degree of ripeness, *Roczniki Panstwowego Zakladu Higieny*, 41(1/2), 17, 1990; *Food Sci. Technol. Abstr.*, 24(3), J120, 1992.

81. Singh, B. P., Singh, S. P., and Chauhan, K. S., Certain chemical changes and rate of respiration in different cultivars of ber during ripening, *Haryana Agric. Univ. J. Res.*, 11(1), 60, 1981; *Food Sci. Technol. Abstr.*, 13(11), J1782, 1981.

82. Slepetys, J., Biologicla characteristics of buckthorn, *Frangula alnus* Mill, grown in the Lithuanian SSR. 3. Content of ascorbic acid, tannic and pectin compounds, sugars and anthracene compounds in the fruits, *Liet. TSR Mokslu Akad. Darb., Ser. C.*, 3, 44, 1984; *Chem. Abstr.*, 102, 42979f, 1985.

83. Leclerc, J. and Miller, M. L., Evolution des teneurs minérales et vitaminiques de la carotte au cours de la croissance et selon le mode de stockage, *Agrochimica*, 36, 19, 1992.

84. Matzner, F., Vitamin-C-Gehalt in Früchten der "Schattenmorelle" *Erwerbsobstbau*, 18(6), 83, 1976.

85. Erickson, L. C., The general physiology of citrus, in *The Citrus Industry*, Vol. II. Reuther, W., Bachlor, L. D., and Webber, H. J., Eds., University of California, Divsion of Agricultural Science, Berkeley, Calif., 1968, 86.

86. Sheets, O. A., Prementer, L., Wade, M., Gieger, M., Anderson, W. S., Peterson, W. J., Rigney, J. A., Wakeley, J. T., Cochran, F. D., Eheart, J. F., Young, R. W., and Massey, P. H., Jr., The nutritive value of collards, *South. Coop. Series Bull.*, 39, 5, 1954.

87. Omueti, O., Ojomo, O. A., Ogunyanwo, O., and Olafare, S., Biochemical composition and other characteristics of maturing pods of vegetable cowpea (*Vigna unguiculata*), *Exp. Agric.*, 22, 25, 1986.

88. Gouny, P. and Vangheesdaele, G., Notes on the biochemical changes occurring in black currant fruits during their development, *Ann. Physiol. Vég.*, 2, 269, 1960; *Hortic. Abstr.*, 32, 4482, 1962.

89. Hårdh, J. E., Factors affecting the vitamin C content of black currants, *Maataloust. Aikakausk.*, 36, 14, 1964; *Hortic. Abstr.*, 34, 6469, 1964.

90. Vestrheim, S., Ascorbic acid in black currants, *Meld. Nor. Landbrukshoegsk.*, 44(18), 1, 1965; *Hortic. Abstr.*, 36, 500, 1966.

91. Shirochenkova, A. I., Minaeva, V. G., and Zhanaeva, T. A., Activity and iso-enzyme composition of ascorbic acid oxidase and peroxidase in ripening and stored black currants, *Fiziol. Biokhim. Kul't. Rast.*, 17(6), 609, 1985; *Hortic. Abstr.*, 56, 2277, 1986.

92. Sawaya, W. N., Khatchadourian, H. A., Khalil, J. K., Safi, W. M., and Al-Shalhat, A., Growth and compositional changes during the various developmental stages of some Saudi Arabian date cultivars, *J. Food Sci.*, 47, 1489, 1982.

93. Damanski, A. F. and Topalović-Avramov, R., Vitamin C in figs during the vegetative period, *Acta Pharm. Jugoslav.*, 1, 37, 1951; *Chem. Abstr.*, 46, 4610c, 1952.

94. Gigliotti, A., Ascorbic acid in the maturation cycle of Merlot and Raboso Piave grapes, *Riv. Viticolt. Enol.*, 27(1), 15, 1974; *Chem. Abstr.*, 81, 101902y, 1974.

95. Metcalfe, E., Rehm, P., and Winters, J., Variations in ascorbic acid content of grapefruit and oranges from the Rio Grange valley of Texas, *Food Res.*, 5, 233, 1940.

96. Semochkina, L. G., Biologically actin substances in jujube fruits, *Subtrop. Kul't.*, 6, 12, 1988; *Chem. Abstr.*, 111, 150557m, 1989.

97. Baraniak, B., Bubicz, M., and Bochniarz, M., Effect of herbicides on carotene, α-tocopherol and L-ascorbic acid content in fodder kale, *Parmiet Pulawski*, 77, 143, 1982; *Chem. Abstr.*, 100, 81142f, 1984.

98. Fuke, Y. and Matsuoka, H., Changes in content of pectic substances, ascorbic acid and polyphenols, and activity of pectinestrase in kiwi fruit during growth and ripening after harvest. Studies on constituents of kiwi fruit cultures in Japan. Part. II. *J. Jpn. Soc. Food. Sci. Technol.*, 31(1), 31, 1984; *Hortic. Abstr.*, 54, 4349, 1984.

99. Bagaturiya, N. Sh., Mekhashishvili, V. P., Pkhakadze, R. R. and Dzhobava, T. S., Composition of lemon leaves and fruits, *Pishch. Prom-st. (Moscow)*, 5, 57, 1989; *Chem. Abstr.*, 111, 56107t, 1989.

100. Sanchez Conde, M. P., Variation in the mineral content of lettuce (*Lactuca sativa*) during the growth period, *An. Edafol. Agrobiol.* 43(1/2), 327, 1984; *Hortic. Abstr.*, 57, 8446, 1987.

101. Alessandrini, M., Ramos, A., Alvarez, E., and Col, Y., Important changes during ripening of Persian limes, *Cult. Trop.*, 9(1), 85, 1987; *Food Sci. Technol. Abstr.*, 20(12), J34, 1988.

102. Underhill, S. J. R. and Critchley, C., The physiology and anatomy of lychee (*Litchii Chinensis* Sonn.) pericarp during fruit development, *J. Hortic. Sci.*, 67(4), 437, 1992.

103. Basu, N. M., Ray, G. K., and De, N. K., The possible relationship of carotene, vitamin C, total acidity, pH, and sugar content of different varieties of mangoes during their green and ripe conditions, *J. Indian Chem. Soc.*, 24, 355, 1947; *Chem. Abstr.*, 42, 5579b, 1948.

104. Siddappa, G. S. and Bhatia, B. S., Tender green mangoes as a source of vitamin C, *Indian J. Hortic.*, 11, 104, 1954.

105. Askar, A., El-Tamimi, A., and Raouf, M., Constituents of mango fruit and their behaviour during growth and ripening, *Mit. Rebe.-Wein.-Obstbau und Fruecheverwert*, 22(2), 120, 1972; *Food Sci. Technol. Abstr.*, 4(11), J1792, 1972.

106. van Lelyveld, L. J., Ascorbic acid content and enzyme activities during maturation of the mango fruit and their association with bacterial black spot, *Agroplantae*, 7, 51, 1975b.

107. Morga, N. S., Lustre, A. O., Tunac, M. M., Balagot, A. H., and Soriana, M. R., Physico-chemical changes in Philippine *Carabao* mangoes during ripening, *Food Chem.*, 4(3), 225, 1979.

108. Kalra, S. K. and Tandon, D. K., Ripening-behaviour of "Dashehari" mango in relation to harvest period, *Scientia Hortic.*, 19, 263, 1983.

109. Vazquez-Salinas, C. and Lakshminarayana, S., Compositional changes in mango fruit during ripening at different storage temperatures, *J. Food Sci.*, 50(6), 1646, 1985.

110. Ali, Z. M., Lazan, H., and Chik, C. Z. C., Effect of heat treatment on the biochemical changes and activity of polyphenol oxidase during ripening of Harumanis and Malgoa mango fruits, *Proc. Malays. Biochem. Soc. Conf.*, 12, 61, 1986; *Chem. Abstr.*, 107, 216406n, 1987.

111. Mahani Binti Ibrahim, S. and Sani, H. A., Study of peroxidase enzyme activity and determination of the reducing sugar and ascorbic acid content in several cultivars of mango, *Proc. Malays Biochem. Soc. Conf.*, 12th, 67, 1986; *Chem. Abstr.*, 107, 235114g, 1987.

112. Sahni, C. K. and Khurdiya, D. S., Physicochemical changes during ripening in Dashehari Chausa, Neelum, and Amrapali mango, *Indian Food Packer*, 43(1), 36, 1989; *Chem. Abstr.*, 111, 20931t, 1989.

113. Hollinger, M. E., and Colvin, D., Ascorbic acid content of okra as affected by maturity, storage, and cooking, *Food Res.*, 10, 255, 1945.

114. Harding, P. L., Winston, J. R., and Fisher, D. F., Seasonal changes in the ascorbic acid content of juice of Florida oranges, *Proc. Am. Soc. Hortic. Sci.*, 36, 358, 1939.

115. Eaks, I. L., Ascorbic acid content of citrus during growth and development, *Bot. Gaz. (Chicago)*, 125, 186, 1964.

116. Higazi, A. M., Elhagah, M. H., and Elnagar, S. Z., Evaluation of some citrus varieties growtn at the Delta, ARE. I. Physical and chemical components, *Minufiya J. Agric. Res.*, 5, 349, 1982; *Hortic. Abstr.*, 55, 1547, 1985.

117. Damptey, H. B., Studies of preharvest metabolism of ascorbic acid in *Citrus sinensis*, Osbeck, *Ghana J. Chem.*, 1(1), 76, 1989; *Chem. Abstr.*, 114, 58953b, 1991.

118. Medawara, M. R., Notizen über Vitamin C in der Pflanze, *Phyton (Austria)*, 2, 193, 1950.

119. Morrison, M. H., The vitamin C and thiamin contents of quick frozen peas, *J. Food Technol.*, 9, 491, 1974.

120. Selman, J. D. and Rolfe, E. J., Studies on the vitamin C content of developing pea seeds, *J. Food Technol.*, 14, 157, 1979.

121. Sistrunk, W. A., Peach quality assessment: fresh and processed, in *Evaluation of Quality of Fruits and Vegetables*, Pattee, H. E., Ed., AVI Publishing, Westport, Conn., 1985, 1.

122. Sandhu, S. S. and Dhillon, B. S., Relation between growth pattern, endogenous growth hormones and metabolites in the developing fruit of Sharbati peach, *Indian J. Agric. Sci.*, 52(5), 302, 1982; *Chem. Abstr.*, 96, 214441a, 1982.

123. Yamamura, H. and Yamane, N., Occurrence of black satin on fruit skin (black spots) in relation to ascorbic acid content in pericarp tissues of Japanese persimmon, *Shimane Daigaku Nogakubu Kenkyu Hokoku*, 21, 18, 1987; *Chem. Abstr.*, 110, 21237n, 1989.

124. Hanson, G. P., Thorne, L., and Jativa, C. D., Vitamin C—a natural smog resistance mechanism in plants? *Lasca Leaves*, 20, 6, 1970; *Hortic. Abstr.*, 41, 1756, 1971.

125. Sinha, S. K. P., and Dogra, J. V. V., Variation in the level of vitamin C, total phenolics and protein in *Phyllanthus niruri* Linn. during leaf maturation, *Natl. Acad. Sci. Lett. (India)*, 4(12), 467, 1981; *Chem. Abstr.*, 97, 212843e, 1982.

126. Adisa, V. A., Studies on nutritional requirements of the latent phase of *Ceratocystis paradox*—the pineapple soft rot pathogen in Nigeria, *Nahrung*, 27(10), 951, 1983.
127. Omuaru, V. O. T., Izonfuo, W. A. L., and Braide, S. A., Enzymic browning in ripening plantain pulp (*Musa paradisiaca* L.) as related to endogenous factors, *J. Food Sci. Technol.*, 27(4), 239, 1990.
128. Streighthoff, F., Munsell, H. E., Ben-dor, B. A., Orr, M. L., Calleau, R., Leonard, M. H., Ezekiel, S. R., Kornblum, R., and Koch, F. G., Effect of large-scale methods of preparation and the vitamin content of food. I. Potatoes, *J. Am. Diet. Assoc.*, 22, 117, 1946.
129. Augustin, J., McDole, R. E., McMaster, G. M., Painter, C. C., and Sparks, W. C., Ascorbic acid content in Russet Burbank potatoes, *J. Food Sci.*, 40, 415, 1975.
130. Shekhar, V. C., Iritani, W. M., and Arteca, R., Changes in ascorbic acid content during growth and short-term storage of potato tubers (*Solanum tuberosum* L.), *Am. Potato J.*, 55, 663, 1978.
131. Ishii, G., and Saijo, R., Effect of various cultural conditions on total sugar content, vitamin C content and β-amylase activity of Daikon radish root (*Raphanus sativus*, L.), *J. Jpn. Soc. Hortic. Sci.*, 55(4), 468, 1987; *Hortic. Abstr.*, 59, 9059, 1989.
132. Sakamura, F., Changes in some chemical constituents of raspberries, *Rubus sieboldi* Blume, and *Rubus parvifolius* Linn. during ripening, *Kaseigaku Zasshi*, 33(7), 366, 1982; *Chem. Abstr.*, 97, 195819g, 1982.
133. Rouhani, I., Khosh-Kuhi, M., and Bassiri, A., Changes in ascorbic acid content of developing rose hips, *J. Hortic. Sci.*, 51, 375, 1976.
134. Selvaraj, Y., and Pal, D. K., Changes in the chemical composition and enzymatic activity of two sapodilla (*Manilkara zapota*) cultivars during development and ripening, *J. Hortic. Sci.*, 59(2), 275, 1984.
135. Ingle, G. S., Khedkar, D. M., and Dabhade, R. S., Effect of growth regulators on ripeingin of sapota fruit (*Achras sapota* Linn), *Indian Food Packer*, 36(1), 72, 1982; *Chem. Abstr.*, 98, 84820b, 1983.
136. Oba, K., Changes in vitamin C contents and ascorbate oxidase activity of vegetables after cutting and washing, *Nippon Kasei Gakkaishi*, 41(8), 715, 1990; *Chem. Abstr.*, 114, 162712h, 1991.
137. Nadiradze, M., Dynamics of some vitamins in grains of the Kakhil-8 wheat, *Tr. Tbilis. Gos. Univ.*, 109, 71, 1965; *Chem. Abstr.*, 66, 112962j, 1967.
138. Simpson, K. L., Relative value of carotenoids as precursors of vitamin A, *Proc. Nutr. Soc.*, 42, 7, 1983.
139. Rodriguez-Felix, A., and Cantwell, M., Developmental changes in composition and quality of prickly pear cactus cladodes (nopalitos), *Plant Foods Hum. Nutr.*, 38(1), 83, 1988.
140. Booth, V. H. and Dark, S. O. S., The influence of environment and maturity on total carotenoids in carrots, *J. Agric. Sci.*, 39, 226, 1949.
141. Wolf, E., Einfluss der Wachstumsdauer und steigender Nährstoffgaben auf den Karotin- und Vitamin-C-Gehalt von Möhren und Sellerie, *Landwirt. Forschung*, 7(2), 19, 1955.
142. Engst, R., Aufnahme und Speicherung von Schädlingsbekämpfungsmitteln in Möhren, *Qual. Plant. Mater. Veg.*, 14, 305, 1967.
143. Sweeney, J. P., and Marsh, A. C., Effects of selected herbicides on provitamin A content of vegetables, *J. Agric. Food Chem.*, 19(5), 854, 1971.
144. Phan, C. T. and Hsu, H., Physical and chemical changes occurring in the carrot root during growth, *Can. J. Plant Sci.*, 53, 629, 1973.
145. Fritz, D. and Weichmann, J., Influence of the harvesting date of carrots on quality and quantity preservation, *Acta Hortic.*, 93, 91, 1979.

146. Michalik, H. and Bakowski, J., The content and distribution of carotenoids in carrots in relation to harvest time, *Biuletyn Warzywniczy*, 20, 417, 1977; *Hortic. Abstr.*, 49, 549, 1979.

147. Lee, C. Y., Changes in carotenoid content of carrots during growth and post-harvest storage, *Food Chem.*, 20(4), 285, 1986.

148. Arima, H. K. and Rodriguez-Amaya, D. B., Carotenoid composition and vitamin A value of commercial Brazilian squashes and pumpkins, *J. Micronutr. Anal.*, 4, 177, 1988.

149. Ramos, D. M. R. and Rodriguez-Amaya, D. B., Determination of vitamin A value of common Brazilian leafy vegetables, *J. Micronutr. Anal.*, 3, 147, 1987.

150. Morga, N. S., Lustre, A. O., Balagot, A. H., Tunac, M. M., and Soriano, M. R., Physiocochemical changes in immature Carabao mangoes during post-harvest storage, *Philipp. J. Food Sci. Technol.*, 6(1-2), 3, 1982; *Chem. Abstr.*, 100, 137617j, 1984.

151. Aina, J. O., Physico-chemical changes in African mango (*Irvingia gabonesis*) during normal storage ripening, *Food Chem.*, 36(3), 205, 1990.

152. Lester, G. E. and Dunlap, J. R., Physiological changes during development and ripening of "Perlita" muskmelon fruits, *Scientia Hortic.*, 26(4), 323, 1985.

153. Miller, E. V. and Winston, J. R., Seasonal changes in the carotenoid pigments in the juice of Florida oranges, *Proc. Am. Soc. Hortic. Sci.*, 38, 219, 1941.

154. Ikemefuna, J. and Adamson, I., Chlorophyll and carotenoid changes in ripening palm fruit, *Elaeis guineënsis*, *Phytochemistry*, 23(7), 1413, 1984.

155. Giri, J., Bhuvaneswari, V., and Tamilarasu, R., Evaluation of the nutritive content of five varieties of papaya in different stages of ripening, *Indian J. Nutr. Diet.*, 17, 319, 1980.

156. Pruthi, J. S. and Lal, G., Chemical composition of passion fruit (*Passiflora edulis*, Sims.), *J. Sci. Food Agric.*, 10, 188, 1959.

157. Scott, G. C. and Belkengren, R. O., Importance of breeding peas and corn for nutritional quality, *Food Res.*, 9, 371, 1944.

158. Kon, M. and Shimba, R., Accumulation of carotenoids in squash, *Nippon Kasei Gakkaishi*, 39(10), 1059, 1988; *Chem. Abstr.*, 110, 21207c, 1989.

159. Rinno, G., Die Beurteilung des ernährungsphysiologischen Wertes von Gemüse, *Arch. Gartenbau*, 13, 415, 1965b.

160. Al-Shaibani, A. M. H. and Greig, J. K., Effects of stages of maturity, storage, and cultivar on some quality attributes of tomatoes, *J. Am. Soc. Hortic. Sci.*, 104(6), 880, 1979.

161. Lee, T. C. and C. O. Chichester, The influence of harvest times on nutritional values, in *Nutritional Quality of Fresh Fruits and Vegetables*, White, P. L. and Selvey, N., Eds., Futura Publ., Mount Kisco, N.Y., 1974, 111.

162. Ezell, B. D., Wilcox, M. S., and Crowder, J. N., Pre- and post-harvest changes in carotene, total carotenoids and ascorbic acid content of sweet potatoes, *Plant Physiol.*, 27, 355, 1952.

163. Zscheile, F. P., Beadle, B. W., and Kraybill, H. R., Carotene content of fresh and frozen green vegetables, *Food Res.*, 8, 299, 1943.

164. Pritchard, M. K. and Becker, R. F., Cabbage, in Eskin, N. A. M., Ed., *Quality and Preservation of Vegetables*, CRC Press, Boca Raton, Fla., 1989, chap. 9.

165. Hjarde, W., Hellström, V., and Åkerberg, E., The content of tocopherol and carotene in red clover as dependent on variety, conditions of cultivation and stage of development, *Acta Agric. Scand.*, 13, 3, 1963.

166. Kamimura, S., Ohgi, T., Takahashi, M., and Tsukamoto, T., Supplementary effects of vitamin A on cows under different forage conditions, *Gokujutsu Hokoku-Kagoshima Daigaku Nogakubu* 42, 37, 1992; *Chem. Abstr.*, 117, 211412h, 1992.

167. Sharaf, A. and El-Saadany, S. S., Biochemical studies on guava fruits during different maturity stages, *Chem. Mikrobiol. Technol. Lebensm.*, 10(5-6), 145, 1987.

168. Woolfe, J. A., *Sweet Potato*, Cambridge University Press, Cambridge, 1992.

169. Peynaud, E. and Ribereau-Gayon, P., The grape, in *The Biochemistry of Fruits and Their Products*, Vol. 2, Hulme, A. C., Ed., Academic Press, London, 1971, 172.

170. Yamaguchi, M., Perdue, J. W., and MacGillivray, J. H., Nutrient composition of white rose potatoes during growth and after storage, *Am. Potato J.*, 37, 73, 1960.

171. Lefebvre, J. M. and Leclerc, J., The influence of some cultivation procedures upon mineral and vitamin content of vegetables grown in glass houses, *Qual. Plant.—Plant Foods Hum. Nutr.*, 23(1/3), 129, 1973.

172. Sistrunk, W. A., Gonzalez, A. R., and Moore, K. J., Green beans, in Eskin, N. A. M., Ed., *Quality and Preservation of Vegetables*, CRC Press, Boca Raton, Fla., 1989, chap. 6.

173. Hunt, C. H., and Bethke, R. M., The riboflavin of certain hays and grasses, *J. Nutr.*, 20, 175, 1940.

174. Speirs, M., Effect of environment on the riboflavin content of plants, *South. Coop. Series. Bull.*, 36, 42, 1954.

175. Eheart, M. S. and Sholes, M. L., Nutritive value of cooked, immature and mature cowpeas, *J. Am. Diet. Assoc.*, 24, 769, 1948.

176. Withner, C. L., B-vitamin changes during growth of cucurbit and tomato fruits, *Am. J. Bot.*, 36, 517, 1949.

177. Ournac, A., and Décor, M., Riboflavin in the grape during the course of its development, *Ann. Technol. Agric.*, 16, 309, 1967; *Hortic. Abstr.*, 38, 7348, 1968.

178. Vedrina-Dragojević, L., Momirović-Čuljat, J., and Balint, L., Dynamics of the biosynthesis of riboflavin in developing soybean seed, *J. Agron. Crop. Sci.*, 154, 73, 1985.

179. Avakyan, A. G., Gevorkyan, L. A., Avetisyan, S. V., and Tarosova, E. O., Vitamins of the B complex in capsicum and tomato fruits, *Biol. Zh. Arm.*, 40(7), 607, 1987; *Hortic. Abstr.*, 58, 2156, 1988.

180. Meiklejohn, J., The vitamin B_1 content of potatoes, *Biochem. J.*, 37, 349, 1943.

181. Gleim, E. G., Tressler, D. K., and Fenton, F., Ascorbic acid, thiamin, riboflavin, and carotene contents of asparagus and spinach in the fresh, stored, and frozen states, both before and after cooking, *Food Res.*, 9, 471, 1944.

182. Tarosova, E. O., Egiazaryan, A. G., Avetisyan, S. V., and Gevorkyan, L. A., Dynamics of accumulation of B group vitamins in tomato fruits, *Biol. Zh. Arm.*, 42(7), 663, 1989; *Chem. Abstr.*, 112, 74006j, 1990.

183. Tramontano, W., Ganci, D., Pennino, M., and Dierenfeld, E. S., Age dependent α-tocopherol concentration in leaves of soybean and pinto beans, *Phytochemistry*, 31(10), 3349, 1992.

184. Contreras-Guzmàn, E., Strong III, F. C., and da Silva, W. J., Fatty acid and vitamin E content of nutrimaiz, a sugary/opague-2 corn cultivar, *J. Agric. Food Chem.*, 30, 1113, 1982.

185. Feldheim, W. and Thomas, B., Vitamin E und Getride, *Ernährungsforsch*, 2, 97, 1957.

186. Gebauer, H., Vitamin E und B in Pflanzen, *Qual. Plant.*, 3-4, 381, 1958.

187. Daood, H. G., Blacs, P. A., Kiss-Kutz, N., Hajdú, F., and Czinkotal, B., Lipid and antioxidant content of red pepper, in *Biological Role of Plant Lipids*, Biacs, P. A., Gruiz, K., and Kremmer, T., Eds., Plenum Publishing Co., New York, 1989, 491.

188. Tendille, C., Gervais, C., and Gaborit, T., Variations in the content of quinone compounds and of α-tocopherol of some chlorophyll-containing plant tissues under the influence of factors altering the chlorophyll content, *Ann. Physiol. Vég.*, 8, 271, 1966; *Hortic. Abstr.*, 37, 7173, 1967.

189. Molina-Torres, J. and Martinez, M. L., Tocopherol and leaf age in *Xanthium strumarium*, L., *New Phytol.*, 118, 95, 1991.
190. Combs, S. B. and Combs, G. F., Jr., Varietal differences in the vitamin E content of corn, *J. Agric. Food Chem.*, 33, 815, 1985.
191. Brown, F., The tocopherol content of farm feeding-stuffs, *J. Sci. Food Agric.*, 4, 161, 1953.
192. Hardin, K., Ridlington, J., and Leklem, J., Increased vitamin B_6 content in ripening bananas, *FASEB J.*, 3, A668, 1989.
193. Ferland, G. and Sadowski, J. A., Vitamin K_1 (Phylloquinone) content of green vegetables: effect of plant maturation and geographical growth location, *J. Agric. Food Chem.*, 40, 1874, 1992.
194. Salunkhe, D. K. and Desai, B. B., Effect of agricultural practice, handling, processing, and storage on vegetables, in *Nutritional Evaluation of Food Processing*, Karmas, E. and Harris, R. S., Eds., Van Nostrand Reinhold, New York, 1988, chap. 3.
195. Kays, S. J., *Postharvest Physiology of Perishable Plant Products*, AVI Book, Van Nostrand Reinhold, New York, 1991.
196. Carpenter, K. J., Schelstraete, Vilcich, V. C., and Wall, J. S., Immature corn as a source of niacin for rats, *J. Nutr.*, 118, 165, 1988.
197. Jones, W. W., van Horn, C. W., Finch, A. H., Smith, M. C., and Caldwell, E., A note on ascorbic acid: nitrogen relationships in grapefruit, *Science*, 99, 103, 1944.
198. Ting, S. V. and Attaway, J. A., Citrus fruits, in *The Biochemistry of Fruits and Their Products*, Vol. 2, Hulme, A. C., Ed., Academic Press, London, 1971, 107.
199. Nilsson, F., Ascorbic acid in black currants, *Landbrukshoegsk. Ann.*, 35, 43, 1969; *Hortic. Abstr.*, 39, 6428, 1969.
200. Vajic, B., Der Vitamin C-Gehalt der Paprikafrüchte und ihre Bedeutung als Schutznahrungsmittel für die Volksernährung, *Z. Vitaminforschung*, 11, 42, 1941.
201. Lampitt, L. H., Baker, L. C., and Parkinson, T. L., Vitamin-C content of potatoes. I. Distribution in the potato plant, *J. Soc. Chem. Indust. (Lond.)*, 64, 18, 1945.
202. Augustin, J., Johnson, S. R., Teitzel, C., Toma, R. B., Shaw, R. L., True, R. H., Hogan, J. M., and Deutsch, R. M., Vitamin composition of freshly harvested and stored potatoes, *J. Food Sci.*, 43(5), 1566, 1978.
203. Baird, E. A. and Howatt, J. L., Ascorbic acid in potatoes grown in New Brunswick, *Can. J. Res.*, 26C(4), 433, 1948.
204. Lo Coco, G., Composition of Northern California tomatoes, *Food Res.*, 10, 114, 1945.
205. Malewski, W. and Markakis, P., Ascorbic acid content of developing tomato fruit, *J. Food Sci.*, 36, 537, 1971.
206. Graifenberg, A., and Tesi, R., Variations in the vitamin C content in commercial glasshouse tomato varieties and in F_1 hybrids, *Rivista Ortoflorofrutticoltura Italiana*, 55(1), 69, 1971; *Hortic. Abstr.*, 42, 1501, 1972.
207. Hadi, B. C. and Al-Samarrie, A. A., Effect of maturation stages on some physical and chemical properties of tomato fruits, *Iraqi J. Agric. Sci. "Zanco,"* 6(3), 7, 1988; *Food Sci. Technol. Abstr.*, 21(6), J53, 1989.
208. Gonzalez, A. R. and Brecht, P. E., Total and reduced ascorbic acid levels in *Rin* and normal tomatoes, *J. Am. Soc. Hortic. Sci.*, 103(6), 756, 1978.
209. Nakasone, H. Y., Miyashita, R. K., and Yamane, G. M., Factors affecting ascorbic acid content of the acerola (*Malpighia glabra* L.), *Proc. Am. Soc. Hortic. Sci.*, 89, 161, 1966.
210. Skard, O. and Weydahl, E., Ascorbic acid—vitamin C—in apple varieties, *Medl. Nor. Landbrukshoegsk.*, 30, 477, 1950.

211. Ketsa, S. and Poopattarangk, S., Growth, physiocochemical changes and harvest indices of small edible-podded peas (*Pisum sativum* L. var. macrocarpon), *Trop. Agric. (Guildford, U.K.)*, 68(3), 274, 1991; *Chem. Abstr.*, 115, 203483a, 1991.

212. Traversi, D., Leo, P. de, Dipierro, S. and Borraccino, G., Variations in ascorbic acid and dehydroascorbic acid contents in berries of table grape cultivar Italia during ripening. *Riv. Viticolt. Enol.*, 39(1), 18, 1986; *Hortic. Abstr.*, 56, 5151, 1986.

213. Yusof, S., Mohamed, S., and Abu Bakar, A., Effect of fruit maturity on the quality and acceptability of guava pureé, *Food Chem.*, 30(1), 45, 1988.

214. Ting, S. V., Nutritional labeling of citrus products, in *Citrus Science and Technology*, Nagy, S., Shaw, P. E., and Veldhuis, M. K., Eds., The AVI Publishing, Westport, Conn., 1977, 401.

215. Shukla, J. P. and Prasad, A., Changing pattern of jamun (*Syzgium cumunii* Skeels) fruit during growth and development. II. Change in biochemical indexes, *Indian J. Agric. Chem.*, 13(2), 141, 1980; *Chem. Abstr.*, 96, 119181g, 1982.

216. Hussein, F., Physiological studies on the growth and development of yellow passion fruits grown at Asswan/Egypt, *Beitrag Trop. Subtrop. Landwirt. Tropenveterinärmed.*, 10(2), 153, 1972; *Hortic. Abstr.*, 44, 4300, 1974.

217. Rao, K. P. G. and Sulladmath, U. V., Changes in certain chemical constituents associated with maturation of okra (*Abelmoschus esculentus* (L.) Moench), pods, *Veg. Sci.*, 4(1), 37, 1977; *Hortic. Abstr.*, 49, 4234, 1979.

218. Reddy, N. S. and Kumari, R. L., Effect of different stages of maturity on the total and available iron and ascorbic acid content of soybean, *Nutr. Rep. Int.*, 37, 77, 1988.

219. Kuliev, A. A. and Akhundov, R. M., Changes in ascorbic acid and catechin contents of *Zizyphus jujuba* fruit during ripening, *Uchenyw Zap. Azerb. Un-t. Ser. Biol. Nauk*, 3/4, 54, 1975; *Hortic. Abstr.*, 47, 4141, 1977.

220. Heinze, P. H., Kanapauk, M. S., Wade, B. I., Grimball, P. C., and Foster, R. L., Ascorbic acid content of 39 varieties of snap beans, *Food Res.*, 9, 19, 1944.

221. Fan Yung, A. F. and Pavlova, G. N., The dynamics of vitamins content in green peas, *Izvestiya Vysshikh Uchebnykh Zavedenii Pishchevaya Tekhnologiya*, 2, 91, 1975; *Food Sci. Technol. Abstr.*, 8(6), J963, 1976.

222. Khan, M. and Habibullah, and Ali, S. M., Carotene content of certain vegetables at different stages of growth, *Pakist. J. Sci. Res.*, 17, 95, 1965; *Hortic. Abstr.*, 37, 6778, 1967.

223. Khan, N. and Elahi, M., Studies on pigments and vitamin E at different stages of growth of some leguminosae plants, *Pakistan J. Sci. Res.*, 20(4, 5), 282, 1977.

224. McDowell, L. R., *Vitamins in Animal Nutrition*, Academic Press, San Diego, 1989.

225. Geddes, W. F. and Levine, M. N., The distribution of thiamin in the wheat plant at successive stages of kernel development, *Cereal Chem.*, 19, 547, 1942.

226. Kondo, H., Mitsuda, H., and Iwai, K., Thiamine synthesis in leaves of cereal crops, *J. Agr. Chem. Soc. Jpn.*, 24, 128, 1950–51; *Chem. Abstr.*, 46, 10305i, 1952.

227. Hunt, C. H., Rodrihuez, L. D., and Bethke, R. M., The effect of maturity on the niacin and pantothenic acid content of the stalks and leaves, tassels, and grain of four sweet corn varieties, *Cereal Chem.*, 27, 157, 1950.

228. Booth, V. H., The α-tocopherol content of forage crops, *J. Sci. Food Agric.*, 15, 342, 1964.

229. Booth, V. H., and Hobson-Frohock, A., The α-tocopherol content of leaves as affected by growth rate, *J. Sci. Food Agric.*, 12, 251, 1961.

230. Machlin, L. J., Ed., *Handbook of Vitamins*, 2nd Ed., Marcel Dekker, New York, 1991.

231. Erdman, J. W., Jr., and Klein, B. P., Harveting, processing, and cooking influences on vitamin C in foods, in *Ascorbic Acid: Chemistry, Metabolism and Uses*, Seib, P. A.

and Tolbert, B. M., Eds., Adv. Chem. Series No. 200, Am. Chem. Soc., Washington, D.C., 1982, 499.

232. House, M. C., Nelson, P. M., and Haber, E. S., The vitamin A, B, and C content of artificially versus naturally ripened tomatoes, *J. Biol. Chem.*, 81(3), 495, 1929.

233. Scott, L. E. and Kramer, A., The effect of storage upon the ascorbic acid content of tomatoes harvested at different stages of maturity, *Proc. Am. Soc. Hortic. Sci.*, 54, 277, 1949.

234. Sayre, C. B., Robinson, W. B., and Wishnetsky, T., Effect of temperature on the color, lycopene, and carotene content of detached and vine-ripened tomatoes, *Proc. Am. Hortic. Sci.*, 61, 381, 1953.

235. Verma, A. N., and others, Quality determination of tomato (*Lycopersicon esculentum* Mill.) during natural and artifical ripening, *Poona Agric. Coll. Mag.*, 60, 72, 1970; *Hortic. Abstr.*, 41, 6969, 1971.

236. Pantos, C. E. and Markakis, P., Ascorbic acid content of artifically ripened tomatoes, *J. Food Sci.*, 38, 550, 1973.

237. Betancourt, L. A., Stevens, M. A., and Kader, A. A., Accumulation and loss of sugar and reduced ascorbic acid in attached and detached tomato fruits, *J. Am. Soc. Hortic. Sci.*, 102(6), 721, 1977.

238. Kader, A. A., Morris, L. L., Stevens, M. A., and Albright-Holton, M., Composition and flavor quality of fresh market tomatoes as influenced by some postharvest handling procedures, *J. Am. Soc. Hortic. Sci.*, 103(1), 6, 1978.

239. Gould, W. A., *Tomato Production, Processing and Quality Evaluation*, 2nd Ed., AVI Publishing, Westport, Conn., 1983.

240. De Carvalho, V. D., de Souza, S. M. C., Chitarra, M. I. F., Cardoso, D. A. M., and Chitarra, A. B., The quality of attached and detached tomatoes of the cultivar Giganta Kada ripened on or off the plant, *Pesquisa Agropecuaria Brasileira*, 19(4), 489, 1984; *Hortic. Abstr.*, 55, 1221, 1985.

241. Ketsa, S. and Wongveerakhan, A., Ascorbic acid content at maturity stages in tomato (*Lycopersicon esculentum* Mill.) cultivars, *Thai J. Agric. Sci.*, 20(4), 257, 1987; *Hortic. Abstr.*, 58, 6754, 1988.

242. Enăchescu, G. Löbl, D., and Brezeanu, D., The chemical composition of late tomatoes ripened after picking, *Lucr. Şti. Inst. Cerc. Horti-Vitic.*, 7, 953, 1966; *Hortic. Abstr.*, 37, 3184, 1967.

243. Lee, Y. C., Effect of ripening methods and harvest time on vitamin content of tomatoes, *Korean J. Food Sci. Technol.*, 16(1), 59, 1984b.

244. Pal, D. K., Divakar, N. G., and Subramanyam, M. D., A note on the physico-chemical composition of papaya fruits ripened on and off the plant, *Indian Food Packer*, 34(6), 26, 1980; *Food Sci. Technol. Abstr.*, 13(12), J1877, 1981.

245. Corazzi, L., Azzi, A., and Usai, A., Ascorbic acid content of vegetables grown in unheated greenhouses, *Ind. Aliment.* (*Pinerolo, Italy*), 28(277), 1179, 1989; *Chem. Abstr.*, 113, 210412p, 1990.

246. Kopec, K., and Števliková, M., Ascorbic acid content of vegetable capsicum cultivars, *Zborník UVTIZ, Zahradnictví*, 4(1/2), 91, 1977; *Hortic. Abstr.*, 50, 1872, 1980.

247. Rinno, G. and Becker, M., Untersuchungen über den Einfluss einiger gemüsebaulicher Massnahmen auf den Vitamingehalt des Gemüses, *Arch. Gartenbau*, 13, 329, 1965.

248. Rozov, N. F., The characteristics of growing, and the chemical composition of head lettuce, borage and cress in the phyotron, *Izv. Timiryazevsk. S-kh. Akad.*, 5, 141, 1973; *Hortic. Abstr.*, 44, 7564, 1974.

249. Lukovnikova, G. A. and Boos, G. V., The chemical composition of vegetables grown under cover, *Vestn. S-Kh. Nauki*, 12(2), 22, 1967; *Hortic. Abstr.*, 37, 6775, 1967.

250. Jungk, A., Beeinflussung des Vitamin- und Mineralstoffgehaltes von Pflanzen durch Züchtung und Anbaumassnahmen, *Landwirt. Forsch. Sonderh.*, 32, 222, 1975.

251. Hårdh, J. E. and Hårdh, K., Effect of radiation, day-length and temperature on plant growth and quality: a preliminary report, *Hortic. Res.* (*Edinburgh*), 12, 25, 1972.

252. Prihod'ko, S. N. and Musat, I. K., Studies on prickly pears in the Ukraine, *Bjull Glav. Bot. Sada*, 56, 101, 1964; *Hortic. Abstr.*, 36, 3593, 1966.

253. Sawamura, M., Hattori, M., Yanogawa, K., Manabe, T., Akita, T., and Kusunose, H., Chemical constituents and sensory evaluation of Satsumas grown in vinyl-clad houses and in the open, *J. Agric. Chem. Soc. Jpn.*, 57(8), 757, 1983; *Hortic. Abstr.*, 54, 1436, 1984.

254. Ivanov, V., and Velchev, V., Interest in fresh strawberries out of season, *B'lgarski Plodove Zelenchtsi i Konservi*, 6, 18, 1976; *Food Sci. Technol. Abstr.*, 9(4), J501, 1977.

255. Currence, T. M., A comparison of tomato varieties for vitamin C content, *Proc. Am. Soc. Hortic. Sci.*, 37, 901, 1940.

256. Brown, A. P. and Moser, F., Vitamin C content of tomatoes, *Food Res.*, 6, 45, 1941.

257. Wokes, F. and Organ, J. G., Oxidizing enzymes and vitamin C in tomatoes, *Biochem. J.*, 37, 259, 1943.

258. Crane, M. B. and Zilva, S. S., The influence of some genetic and environmental factors on the concentration of L-ascorbic acid in tomato fruits, *J. Hortic. Sci.*, 25, 36, 1949.

259. Daskaloff, Chr. and Ognjanowa, A., Die Änderung des Heterosiseffektes in F_1-Tomatensorten beim Gewächshaus-und Freilandanbau, *Arch. Gartenbau*, 10, 193, 1962.

260. Snezhko, V. L. and Al-Vagab, K., The effects of varietal characteristics, growing conditions and the degree of ripeness on the content of some biological substances, *Fiziol. Biokhim. Kul't. Rast.*, 3(5), 517, 1971; *Hortic. Abstr.*, 42, 6267, 1972.

261. Usai, A., and Stacchini, A., Chemical changes of vegetable food due to various crop treatments. II. Tomatoes, *Rassegna Chimica*, 22(4), 125, 1970; *Food Sci. Technol. Abstr.*, 4(1), J14, 1972.

262. Gutiev, O. G., The ascorbic acid content of tomato plants, *Tr. Prikl. Bot. Genet. Sel.*, 49(2), 295, 1973; *Hortic. Abstr.*, 44, 6781, 1974.

263. Fritz, D. and Venter, F., Einfluss von Sorte, Standort und Anbaumassnahmen auf messbare Qualitätseigenschaften von Gemüse, *Landwirt. Forschung. Sonderh.*, 30(1), 95, 1974.

264. Prodan, G., Some relationships concerning the chemical composition of tomato fruit, *Lucrari Stinntifice, Institutul Agronomic "N Balcescu,"* B, 14, 63, 1971; *Hortic. Abstr.*, 45, 2553, 1975.

265. Fritz, D., Habben, F. J., Reuff, B., and Venter, F., Die Variabilität einiger qualitätsbestimmender Inhalsstoffe von Tomaten, *Gartenbauwissenschaft*, 41(3), 104, 1976.

266. Zukowska, E., Assessment of the content of nutritious substances in the fruits of some varieties of greenhouse tomatoes, *Zeszyty Naukowe Akademii Rolniczej H. Kollataja Krakowie, Orgrodnictwo*, 11, 137, 1984; *Hortic. Abstr.*, 55, 3513, 1985.

267. Müller-Haslach, W., Arold, G., and Kimmel, V., Einfluss der Düngungsintensität auf die Qualität von Tomaten, *Bayerisches Landwirt. Jahresb., Sonderh.*, 63(1), 81, 1986.

268. Corzazzi, L., Porcu, M., and Usai, A., Uptake of lead from polluted soil by horticultural plants, *Riv. Merceol.*, 25(1), 33, 1986; *Chem. Abstr.*, 106, 4151e, 1987.

269. Kulikova, N. T., Content of ascorbic acid in green vegetables in polar region, *Sbornik Nauchnykh Trudov Poprikladnoi Bot. Genet. Sel.*, 107, 53, 1986; *Hortic. Abstr.*, 59, 1032, 1989.

270. Frazier, J. C., Ascham, L., Cardwell, A. B., Fryer, H. C., and Willis, W. W., Effect of supplemental lighting on the ascorbic acid concentration of greenhouse tomatoes, *Proc. Am. Soc. Hortic. Sci.*, 64, 351, 1954.

271. Novoderžkina, Ju. G., The effect of microelements in increasing the vitamin C synthesis in tomatoes, *Fiziol. Rast.*, 7, 121, 1960; *Hortic. Abstr.*, 31, 805, 1961.

272. Shattuck, V. I., Kakuda, Y., Shelp, B. J., and Kakuda, N., Chemical composition of turnip roots stored or intermittently grown at low temperature, *J. Am. Soc. Hortic. Sci.*, 116(5), 818, 1991.

273. Hårdh, J. E., Der Einfluss der Umwelt nördlicher Breitengrade auf die Qualität der Gemüse, *Qual. Plant.—Plant Foods Hum. Nutr.*, 25(1), 43, 1975.

274. Benoit, F. Ceustermans, N., Rouchaud, J., and Vlassak, K., Influence of a plastic cover on the quality of carrots and cabbage lettuces, *Rev. Agric.*, 37(2), 211, 1984; *Hortic. Abstr.*, 54, 7214, 1984.

275. Shigeyama, T. and Murakami, H., On the quality of papaw fruit and the utilization of unripe fruit in warm regions of Japan, *Bull. Fac. Agric. Miyazaki*, 10, 295, 1965; *Hortic. Abstr.*, 35, 8914, 1965.

276. Smith, L. L. W. and Morgan, A. F., The effect of light upon the vitamin A activity and the carotenoid content of fruits, *J. Biol. Chem.*, 101, 43, 1933.

277. Ellis, G. H. and Hamner, K. C., The carotene content of tomatoes as influenced by various factors, *J. Nutr.*, 25, 539, 1943.

278. Speirs, M., Effect of environment on the thiamine content of plants, *South. Coop. Series. Bull.*, 36, 48, 1954.

279. Madsen, E., The effect of CO_2 concentration on the content of ascorbic acid in tomato leaves, *Ugeskrift Agronomer*, 116(28), 592, 1971; *Hortic. Abstr.*, 42, 1483, 1972.

280. Kimball, B. A. and Mitchell, S. T., Effect of CO_2 enrichment, ventilation, and nutrient concentration on the flavor and vitamin content of tomato fruit, *HortScience*, 16(5), 665, 1981.

281. Ito, T., Photosynthetic activity in vegetable plants and its horticultural significance. VI. Plant growth, yield and quality of tomatoes grown in a fan-ventilated plastic house, *J. Jpn. Soc., Hortic. Sci.*, 41(1), 51, 1972; *Hortic. Abstr.*, 45, 1092, 1975.

282. Sinha, M. M., Effect of closer spacing and higher nutritional doses with and without gibberellic acid on yield and quality in chillies (*Capsicum annuum* L.), *Prog. Hortic.*, 7(1), 41, 1975; *Hortic. Abstr.*, 46, 5795, 1976.

283. Yang, S. B., Park, K. W., and Chiang, M. H., The effect of fertilizer application, spacing and sowing date on the growth and quality of *Chrysanthemum coronarium* L.), *Abstracts of Communicated Papers (Horticulture Abstracts)*, Korean Soc. for Hortic. Sci., 7(1), 72, 1989; *Hortic. Abstr.*, 59, 8284, 1989.

284. Amarjeet Singh and Daulta, B. S., Studies on the effect of cane and spacing levels in different cultivars of grapes (*V. vinifera* L.). III. Quality attributes, *Haryana J. Hortic. Sci.*, 19(1-2), 27, 1990; *Hortic. Abstr.*, 61, 3579, 1991.

285. Mustaffa, M. M., Influence of plant population and nitrogen on fruit yield, quality and leaf nutrient content of Kew pineapple, *Fruits*, 43(7-8), 455, 1988; *Hortic. Abstr.*, 59, 8769, 1989c.

286. Fehér, B., Effect of nutrient supply and plant density on tomato fruit composition, *Kertgazdaság*, 11(3), 29, 1979; *Hortic. Abstr.*, 51, 4654, 1981.

287. Yadava, L. S. and Godara, N. R., Effect of planning distance and severity of pruning on physico-chemical characters of ber (*Ziziphus mauritiana* Lamk.) cv. Umran, *Haryana J. Hortic. Sci.*, 16(1-2), 45, 1987; *Hortic. Abstr.*, 58, 8362, 1988.

288. Bhambota, J. R. and Shrestha, A. B., Effect of leaf area on the chemical composition of apple fruit, *J. Res. (Ludhiana)*, 2, 21, 1965; *Hortic. Abstr.*, 37, 4443, 1967.

289. Poniedzialek, W., Nosal, K., Porebski, S., and Sobolewska, A., Effect of summer pruning on tree growth and fruit yield and quality in the apple cultivar Spartan, *Zesyty Naukowe Akademii Rolniczej im. Hugona Kollataja Krakowie, Ogrodnictwo*, 220, 35, 1987; *Hortic. Abstr.*, 58, 7232, 1988.

290. Bajwa, G. S., Sandhu, H. S., and Bal, J. S., Effect of different pruning intensities on growth, yield and quality of ber (*Ziziphus mauritiana* Lamk.), *Haryana J. Hortic. Sci.*, 16(3-4), 209, 1987; *Hortic. Abstr.*, 59, 2629, 1989.

291. Libik, A., Starzecki, W., and Dudek, Z., Effect of leaf number reduction on the growth and yield of greenhouse tomatoes, *Zeszyty Naukowe Akademii Rolniczej im. Hugona Kollataja Krakowi, Ogrodnictwo*, 210(15), 57, 1987; *Hortic. Abstr.*, 58, 342, 1988.

292. Drake, S. R. and Silbernagel, M. J., The influence of irrigation and row spacing on the quality of processed snap beans, *J. Am. Soc. Hortic. Sci.*, 107, 239, 1982.

293. Pandita, M. L. and Bhatnagar, D. K., Effect of nitrogen, phosphorus and spacing on fruit quality of tomato cultivar HS-102, *Haryana Agric. Univ. J. Res.*, 11(1), 8, 1981; *Hortic. Abstr.*, 51, 9455, 1981.

294. Gutierrez, L. E., Minami, K., Camargo, T. P., and Mantovani, W., The effect of spacing on the contents of ascorbic acid and soluble carbohydrates in eggplant, *Anais Escola Superior Agricultura "Luiz de Queiroz,"* 33, 259, 1976; *Hortic. Abstr.*, 49, 4246, 1979.

295. Pirtskhalaishvili, S. K. and Tsereteli, G. A., The planting density factor in a young mandarin orchard, *Subtrop. Kul't.*, 3, 94, 1986; *Hortic. Abstr.*, 57, 1501, 1987.

296. Dobrzanska, J. and Michalik, H., Effect of pruing and spacing on the yield and quality of *Capsicum* grown in plastic tunnels, *Biuletyn Warzywniczy*, 28, 49, 1985; *Hortic. Abstr.*, 57, 5596, 1987.

297. Banga, O. and De Bruyn, J. W., Selection of carrots for carotene content. III. Planting density and ripening equilibrium of the roots, *Euphytica*, 5, 87, 1956.

298. Murneek, A. E. and Wittwer, S. H., Some factors affecting ascorbic acid content of apples, *Proc. Am. Soc. Hortic. Sci.*, 51, 97, 1948.

299. Ray, D. P. and Munis, P. S., A note on qualitative parameters and its association with leaf and soil nutrients in litchi (*Litchi chinensis* Sonn.), *Orissa J. Hortic.*, 18(1-2), 80, 1990; *Hortic. Abstr.*, 62, 9490, 1992.

300. Pattee, H. E., Ed., *Evaluation of Quality of Fruits and Vegetables*, AVI Publishing, Westport, Conn., 1985.

301. Wittwer, S. H., and Hibbard, A. D., Vitamin C–nitrogen relations in peaches as influenced by fertilizer treatment, *Proc. Am. Soc. Hortic. Sci.*, 49, 116, 1947.

302. Wittwer, S. H., Schroeder, R. A., and Albrecht, W. A., Vegetable crops in relation to soil fertility. II. Vitamin C and nitrogen fertilizers, *Soil Sci.*, 59, 329, 1945.

303. Maršanija, I. I., The effect of long term application of fertilizers on the vitamin C content in mandarin and lemon fruit, *Tr. Suhum. Opyt. Stan Efirno-mas. Kul'tur*, 9, 49, 1970; *Hortic. Abstr.*, 41, 9797, 1971.

304. Lavekar, K. B., Ballal, A. L., Rane, D. A., and Warke, D. C., Study of seasonal variation in physicochemical composition of Mudkhed Seedless Santra in comparison with Nagpur Santra (*Citrus reticulata*, Blanco), *J. Maharashtra Agric. Univ.*, 3(3), 224, 1978; *Hortic. Abstr.*, 50, 4703, 1980.

305. Jaiswal, S. P., Kaur, G., Kumar, J. C., Nandpuri, K. S., and Thakur, J. C., Chemical constituents of green pea and their relationships with some plant characters, *Indian J. Agric. Sci.*, 45(2), 47, 1975.

306. Hoffer, A., Alcock, A. W., and Geddes, W. F., The effect of variations in Canadian spring wheat on the thiamine and ash of long extraction flours, *Cereal Chem.*, 21, 210, 1944.

307. Hårdh, J. E., Persson, A. R., and Ottosson, L., Quality of vegetables cultivated at different latitudes in Scandinavia, *Acta Agric. Scand.*, 27, 81, 1977.

308. Batchelder, E. L. and Overholser, E. L., Factors affecting the vitamin C content of apples, *J. Agric. Res. (Washington, D.C.)*, 53(7), 547, 1936.

309. Matzner, F., Über den Gehalt und die Verteilung des Vitamin C in Äpfeln, *Erwerbsobstbau*, 4(2), 27, 1962.

310. Matzner, F., Über den Trockensubstanz-, Säure- und Vitamin-C-Gehalt in den Früchen der Sorten "Freiherr von Berlepsch" und "Roter Berlepsch," *Erwerbsobstbau*, 8(11), 208, 1966.

311. Martin, D., Ascorbic acid in Tasmanian apples, *Fld. Stat. Res. CSIRO Div. Plant Ind.*, 5(1), 45, 1966; *Hortic. Abstr.*, 37, 4444, 1967.

312. Martin, D., Vitamin C in apples, *N.Z.J. Agric.*, 116(6), 71, 1968; *Hortic. Abstr.*, 39, 313, 1969.

313. Drake, S. R., Nelson, J. W., Powers, J. R., and Early, R. E., Fresh and processed asparagus quality as influenced by field grade, *J. Food Qual.*, 2, 149, 1978c.

314. Woyke, H., The effect of head size on vitamin C, dry matter and total sugar content in brussels sprouts, *Biuletyn Warzywniczy*, 12, 267, 1971; *Hortic. Abstr.*, 42, 7697, 1972.

315. Horbowicz, M. and Bakowski, J., Effects of biological factors on the chemical composition of Brussels sprouts. Part II. The influence of head diamter on the chemical composition of Brussels sprouts, *Biuletyn Warzywniczy*, 32, 171, 1988; *Hortic. Abstr.*, 59, 6510, 1989.

316. Janes, B. E., The relative effect of variety and environment in determining the variations of per cent dry weight, ascorbic acid, and carotene content of cabbage and beans, *Pro. Am. Soc. Hortic. Sci.*, 45, 387, 1944.

317. Poole, C. F., Grimball, P. C., and Kanapaux, M. S., Factors affecting the ascorbic acid content of cabbage lines, *J. Agric. Res. (Washington, D.C.)*, 68(8), 325, 1944.

318. Branion, H. D., Roberts, J. S., Cameron, C. R., and McCready, A. M., The ascorbic acid content of cabbage, *J. Am. Diet. Assoc.*, 24, 101, 1948.

319. Schuphan, W., *Mensch und Nahrungspflanze*, Dr. Junk Verlag, Den Haag, 1976.

320. Jaiswal, S. P., Singh, J., Karus, G., and Thakur, J. C., Some chemical constituents of cauliflower curd and their relationship with other plant characters, *Indian J. Agric. Sci.*, 44(11), 726, 1974.

321. Bombasov, I. I., Ivanova, E. I., Toshcheva, M. N., and Bocharov, V. N., The size and chemical composition of cucumbers, *Kartofel'i Ovoshchi*, 4, 27, 1976; *Food Sci. Technol. Abstr.*, 9(4), J568, 1977.

322. Kel't, K. and Parksepp, I., Differences in the content of nutrients between groups of black currant berries of different sizes, *Nauchnye Tr. Estonskii Nauchnoissledovatel'skii Inst. Zemledeliya Melioratsii*, 56, 152, 1985; *Hortic. Abstr.*, 56, 136, 1986.

323. Long, W. G., Harding, P. L., and Soule, M. J., Jr., The ascorbic acid concentrations of grapefruit of different sizes, *Proc. Fla. St. Hortic. Soc.*, 70, 17, 1957.

324. Smith, P. F., Quality measurements on selected size of Marsh grapefruit from trees differentially fertilized with nitrogen and potash, *Proc. Am. Soc. Hortic. Soc.*, 83, 316, 1963.

325. Aparicio, J., Escriche, A. J., Artes, F., and Marín, J. G., A study of grapefruit (*Citrus paradisi* Macf.) cv. "Marsh seedless," Physicochemical characterization according to circumference in the Murcia region, *Fruits (Paris)*, 45(5), 489, 1990.

326. McDonald, R. E. and Hillebrand, B. M. Physical chemical characteristics of lemons from several countries, *J. Am. Soc. Hortic. Sci.*, 105(1), 135, 1980.

327. Escriche, A., Aparicio, J., Marín, J. G., Artés, F., and Nieves, M., Relationship between the chemical parameters of fruit quality and fruit equatorial diameters in

lemon (*Citrus lemon*, Burn.) cultivar Fino, *ITEA, Producción Vegetal*, 20(83), 47, 1989; *Hortic. Abstr.*, 61, 5432, 1991.

328. Hatton, T. T., Jr. and Reeder, W. F., Ascorbic acid concentrations in Florida-grown "Tahiti" (Persian) limes, *Proc. Tropical Region, Am. Soc. Hortic. Sci.*, 15, 89, 1971; *Hortic. Abstr.*, 43, 8101, 1973.

329. Beridze, Z. A., The effect of different rates of potassium on the chemical composition and mechanical and biochemical characteristics of mandarin fruit of different classes, *Subtrop. Kul't.*, 3, 111, 1988; *Hortic. Abstr.*, 59, 1591, 1989.

330. Singh, K. K. and Chadha, K. L., Factors affecting the vitamin C content of mango, *Punjab Hortic. J.*, 1, 171, 1961; *Hortic. Abstr.*, 33, 6223, 1963.

331. Kaur, G., Lal, T., Nandpuri, K. S., and Sharma, S., Varietal-cum-seasonal variation in certain physicochemical constituents in muskmelon, *Indian J. Agric. Sci.*, 47(6), 285, 1977.

332. Sites, J. W. and Camp, A. F., Producing Florida citrus for frozen concentrate, *Food Technol.*, 9, 361, 1955.

333. Kefford, J. F., The chemical constituents of citrus fruits, *Adv. Food Res.*, 9, 285, 1959.

334. Todhunter, E. N., and Sparling, B. L., Vitamin values of garden-type peas preserved by frozen-pack method. I. Ascorbic acid (vitamin C), *Food Res.*, 3, 489, 1938.

335. Lee, C. Y., Massey, L. M., Jr. and Van Buren, J. P., Effects of post-harvest handling and processing on vitamin contents of peas, *J. Food Sci.*, 47, 961, 1982.

336. Yao, Y., Tigerstedt, P. M. A., and Joy, P. Variation of vitamin C concentration and character correlation between and within natural sea buckthorn (*Hippophae rhamnoides* L.), populations, *Acta Agric. Scand., Section B, Soil Plant Sci.*, 42(1), 12, 1992.

337. Slate, G. L. and Robinson, W. B., Ascorbic acid content of strawberry varieties and selections at Geneva, New York in 1945, *Proc. Am. Soc. Hortic. Sci.*, 47, 219, 1946.

338. Anstey, T. H. and Wilcox, A. N., The breeding value of selecting inbred clones of strawberries with respect to their vitamin C content, *Sci. Agric.*, 30(9), 367, 1950.

339. Hallsworth, E. G. and Lewis, V. M., Variation of ascorbic acid in tomatoes, *Nature (Lond.)*, 154, 431, 1944.

340. Brown, G. B. and Bohn, G. W., Ascorbic acid in fruits of tomato varieties and F_1 hybrids forced in the greenhouse, *Proc. Am. Soc. Hortic. Sci.*, 47, 255, 1946.

341. Reynard, G. B. and Kanapaux, M. S., Ascorbic acid (vitamin C) content of some tomato varieties and species, *Proc. Am. Soc. Hortic. Sci.*, 41, 298, 1942.

342. Barooah, S. and Mohan, N. K., Correlation study between fruit size and ascorbic acid content in tomato (*Lycopersicon esculentum* Mill), *Curr. Res.*, 5, 82, 1976.

343. Kmiecik, W. and Lisiewska, Z., Vitamin C content in four cultivars of zucchuni as related to fruit size and harvest date, *Folia Hortic.*, 1(2), 27, 1989; *Hortic. Abstr.*, 61, 7956, 1991.

344. Du Preez, R. J. and Welgemoed, C. P., Variability in fruit characteristics of five guava selections, *Acta Hortic.*, 275(1), 351, 1990.

345. Ketsa, S., Effect of fruit size on juice content and chemical composition of tangerine, *J. Hortic. Sci.*, 63(1), 171, 1988.

346. Speirs, M., Miller, J., Whitacre, J., Brittingham, W. H., and Kapp, L. C., Effect of leaf size on moisture, ascorbic acid, carotene, thiamine, and riboflavin content of turnip greens, *South. Coop. Series Bull.*, 10, 18, 1951.

347. Lanier, J. J. and Sistrunk, W. A., Influence of cooking method on quality attributes and vitamin content of sweet potatoes, *J. Food Sci.*, 44(2), 374, 1979.

348. Simaan, F. S., Cowan, J. W., and Sabry, Z. I., Nutritive value of Middle Eastern foodstuffs. I. Composition of fruits and vegetables grown in Lebanon, *J. Sci. Food Agric.*, 15, 799, 1964.

349. Misra, J. B. and Chand, P., Relationship between potato tuber size and chemical composition, *J. Food Sci. Technol.*, 27(1), 63, 1990.

Chapter 7

ENVIRONMENTAL STRESS FACTORS
AND POLLUTANTS

I. MOISTURE STRESS (CROP IRRIGATION)

Irrigation of plants is a must in many parts of the world in order to produce a crop at all or to increase its yield. Plant parts (leaves or fruits) grown under water shortage conditions are often smaller and have a less appealing texture than larger, more turgid, "juicier" plants grown with adequate water. This is in part due to the relatively higher percentage of water (expressed as percentage of a plant's dry material) in the plants grown with adequate water.

Plants grown under water stress may have a different taste and flavor as compared with well-watered plants.[1] In certain cases these changes may be in a positive direction. For example, best-flavored tomatoes are reported to be those grown under dryland conditions.[2] Whether irrigation (or water stress) alters the synthesis and/or concentration of various plant vitamins has not been extensively studied.

In the following, first some of the observations made under natural conditions (rainy versus dry seasons, for example) and then some of the experimental data obtained under controlled water-supply conditions are reviewed. Interpretation of observations made under natural conditions is, however, very difficult since rainy climatic conditions are usually accompanied by lowered duration and/or intensity of sunshine and/or changes in temperature.

A. Ascorbic Acid

Rainy climatic conditions are noted to decrease the ascorbic acid content in turnip greens,[3] rose hips,[4] onions,[5] feijoa fruits,[6] and black currant,[7] as compared with dry (and thus sunnier and warmer) conditions. In the case of black currant, for example, fruits grown in hot, dry years were found to contain more than twice the ascorbic acid of those grown in wet years (247 vs. 110 mg/100 g, respectively).[7]

Reder et al.,[3] in experiments designed to study the effect of fertilizer and environmental conditions on the ascorbic acid content of turnip greens, noted that the ascorbic acid content was higher (190 mg/100 g) in plants grown at the location Experiment (Virginia), where the average daily rainfall was low and 49% of the days were sunny, than in the plants grown in the location Blacksburg (Georgia) (128 mg/100 g), which had the highest rainfall and only 27% of the days were sunny. The authors concluded that the ascorbic acid in this vegetable was

TABLE 1
Effect of Supplemental Irrigation of Field-Grown Turnip on the Ascorbic Acid and Carotene Concentrations in Turnip Greens[20]

	Ascorbic acid (mg / 100 g DW)[a]			Carotene (mg / 100 g DW)[a]		
	1950	1951	1952	1950	1951	1952
Control	1,235	1,152	1,343	79	61	51
Irrigated	1,257	1,088	1,389	78	77	54

[a] Values rounded.

inversely related to the amount of rainfall and directly related to the amount of sunshine.

Experiments conducted under controlled conditions have shown that increased water supply to the plants may reduce the ascorbic acid concentration in cabbage,[8,9] cauliflower,[10] celeriac,[9] cucumber,[11] muskmelon,[12] radish,[13] snap beans,[14] and tomato[8,15–17] but may increase its content in potato.[18] These observations indicate that the degree of water supply per se may affect the concentration of ascorbic acid in the plants. Whether part of this effect is due to changes in the relative amount of water (as percentage of dry matter) in the plants grown under different water-supply conditions is not clear.

Subjecting turnip plants to lower water supply was noted to lower the water content and increase the percent dry weight and the ascorbic acid content on the fresh weight basis, but decrease the ascorbic acid content on the dry weight basis, in their leaves.[19] The higher ascorbic acid, on the fresh weight basis, in the leaves of plants grown on drier soil was thus attributed to the lower percentage of water in these leaves, which were also tougher, had a wilted look, and were of poor quality as fresh vegetable.[19] In contrast, Sheets et al.[20] noted that irrigation does not affect the ascorbic acid of turnip greens when expressed on the dry weight basis (Table 1). Water deficiency stress (simulated by addition of polyethylene glycol to the rooting media) was noted to increase the ascorbic acid in spinach[21] but decrease it in wheat[22] and *Vinga* seedlings.[23]

The observations that ascorbic acid content in tomato and cucumber[24] and in tomato[25] was higher in the wet-cool years than in dryer years, on one hand, and the report that abundant rainfall during the last 15 days of ripening increases the ascorbic acid content in black currants,[26] along with the observation that irrigated snap beans contained 3.5% higher ascorbic acid content (18.4 vs. 14.9 mg/100 g) than those subjected to water deficiency,[27] indicate that under certain conditions ascorbic acid accumulation in plants is favored by more water and/or cooler climatic conditions.

Reports on the effect of water supply on the ascorbic acid content of fruits grown on trees are also inconsistent, a situation that may be due to the greater depth of tree roots as compared with that of annual plant and therefore to the relative lesser susceptibility of trees to short-term variations in soil water supply. For example, although moderate water stress was noted to increase the vitamin C content of Washington Navel oranges in Egypt,[28] in Texas, however, irrigation of trees was found to only occasionally affect the ascorbic acid concentration in the juice of grapefruit and oranges. It was thus concluded that fruits from groves that greatly differ in their water management will probably differ very little in their ascorbic acid content.[29] Also, irrigation of mango orchards was found to have only a slight effect on the vitamin C content of the fruits.[31]

In contrast, irrigation was noted to decrease[32] or increase[33] the content of ascorbic acid in apples, decrease its content in the Hamlin and Pineapple oranges, but increase it in Valencia oranges.[34] In the northern part of India, guavas harvested during the rainy seasons (July to October) were noted to have up to three times higher ascorbic acid concentration than those grown during the winter months.[35] Because of differences in the harvesting season, obviously factors other than just the water supply, such as light and temperature conditions, might have also played a role in the vitamin C content of guava. For the effect of seasonal variation of plant vitamins, see chapter 4. These results, along with the observation that the depth of the water table had little effect on the ascorbic acid content of oranges in Egypt,[30] support the view that irrigation of fruit trees has little or no effect on the vitamin C content of their fruits.

Reports on the effect of method of irrigation (furrow, sprinkler, drip, etc.) on the vitamin content of plants are inconsistent. For example, although sweet peppers irrigated with sprinkler and furrow irrigation methods did not differ in their vitamin C content,[36] other investigations point to the possible effect of method of irrigation on the plant vitamins. For example, the ascorbic acid content of snap beans irrigated by sprinkler was found to be significantly higher than in those irrigated by the rill method.[37] Cevik et al.[38] compared the effect of sprinkler, drip, and trickle irrigation on the ascorbic acid content of tomatoes grown in plastic-covered greenhouses in spring and in autumn and noted that spring tomatoes irrigated with drip and trickle methods had higher ascorbic acid concentration than those irrigated by the sprinkler method. Also, lettuce irrigated with furrow irrigation was noted to have more ascorbic acid than those irrigated by sprinkler system.[39]

The observations that (a) a regular and even wetting of soil, especially a day or two prior to harvest, may increase the ascorbic acid content in currant, strawberry, and raspberry but excessive moisture (downpours) on the eve of harvest could reduce the ascorbic acid content in currant and strawberries[40] and (b) large quantity of water shortly before harvest, by either rain or irrigation, reduces the vitamin C content of sweet peppers[36] indicate that along with the dilution effect, the rate of moisture entry into the soil and thus soil aeration may also play a role in the ascorbic acid content in the plant tops.

Irrigation can also affect numerous other plant quality parameters in the plants, which may not necessarily go hand in hand with the changes occurring in their vitamin content. In Israel, for example, a less frequent irrigation of orchards during the summer months resulted in smaller fruits with thicker peels and a higher number of oranges per tree. The juice of the fruits, however, contained higher sugar, TSS/acid ratio, and ascorbic acid concentration.[41]

Finally, one should note that the vitamin content of fruits and vegetables may drastically decrease if they are subjected to severe water loss (wilting) after they have been harvested. As an example, Akpapunam[42] noted a strong decrease in the carotenoid content of several Nigerian vegetables when subjected to wilting by exposure to sun at 35°C for four hours.

B. Carotene

Carrots grown under low soil moisture are known to have a better color,[43] which is in agreement with the observation that rainy climatic conditions[5,44] and increased soil moisture (under controlled conditions) decrease the carotene content in carrots.[43,45-48]

In other plants, however, the situation is less clear. For example, in snap beans, heavy irrigation was noted to decrease the carotene content of the pods, and the trends were the same regardless of whether the data were expressed on the dry or fresh weight basis.[14] On the other hand, liberal water supply and higher rainfalls were noted to increase the carotene content in corn and soybean foliage.[43] In Kansas, the carotene content in pasture grass was noted to be low during the hot summer months. The carotene content, however, increased after the fall rains.[43] Sheets et al.[20] did not find a consistent effect of irrigation on carotene content (expressed on the dry weight basis) in turnip greens during a three-year experiment; in some years, however, irrigated plants contained more carotene than nonirrigated plants (Table 1). Also, in broccoli no effect of water supply on the carotene content could be found.[49] Under controlled water supply conditions, subjecting the plants to water stress did not change the total carotenoids in *Portulacaria afra*[50] but increased the amount of β-carotene in *Digitalis lanata*.[51]

Part of the above inconsistencies on the effect of water supply and/or weather (rainfall) conditions on the carotene content of plants might be due to the confounding effect of soil water on soil temperature. Drier soils tend to be warmer than the more moist soils under similar conditions. Since temperature is known to exert a pronounced effect on the carotene content in several plants (chapter 4), it is thus conceivable that effect of soil water content on the carotene content of plants may be partly due to its effect on the soil temperature, which may be different at different times of the year and in different geographical locations. This point was noted by Nortjé and Henrico,[48] who irrigated the soils when 20, 40, 60, and 80% of the available water was depleted in the 0–30 cm depth and measured the soil temperature and carotene content of carrots grown on these soils. They noted that carrots grown in the least frequently irrigated plots (80% treatment) contained significantly more β-carotene than those grown with more frequent irrigation (20% treatment) (62 vs. 38 mg/100 g, respectively). These authors, however, noted that at 1400 h, soil temperature measured at 10-cm depth was on the average 4.5°C higher in the least frequently as compared to the most frequently irrigated plots, an observation that led them to conclude that the reduced carotene content in carrots brought about by high soil irrigation is the result of the lower temperature of the wetter soils. This is an important observation, which deserves more detailed follow-up studies.

Finally, Leclerc et al.[39] reported that lettuce grown with furrow-irrigation contained 14% more β-carotene than those grown with sprinkler method (1.22 vs. 1.07 mg/100 g, respectively). Whether this observation was also due to differences in soil temperature caused by different methods of water application to the plants remains to be seen.

C. Other Vitamins

Reports to the direct effect of irrigation on the content of other plant vitamins is very scarce and often contradictory. Indirect observations, however, indicate that the degree of water supply may also affect the concentration of some other vitamins in some plants. For example, cool weather with increasing rainfall during the week following blossoming was found to increase the thiamin content in the wheat kernels.[52] Also, maize grown in "normal" years had higher thiamin than those grown in "dry" years.[53] Wheat grown in a normal year contained higher thiamin, riboflavin, and pantothenic acid than wheat grown in a wet year but oats grown in a normal year contained less riboflavin and slightly more niacin than oats grown in a wet year.[53] Finally, the observation that irrigation did not affect the content of riboflavin and niacin in Bengal gram (*Cicer arietinum*)[54] indicates that the effect of water supply on the plant content of vitamins other than ascorbic acid and carotene is far from clear.

D. Vitamins and Tolerance to Drought

Water stress is believed by some to damage the plants by means of increased production of free radicals in the plant cells subjected to water shortage.[55,56] If so, then an increased production of vitamins such as E, C, and carotene (i.e., vitamins with free radical scavenging properties) should help the plants to reduce the destructive effects of water stress. If this hypothesis is correct, then water deficiency should increase the concentration of these vitamins in the plants. As noted above, this has been observed for vitamin C in some but not all cases. For vitamin E, however, the experimental results seem to be relatively more consistent.

For example, tocopherol content of soybean leaves was noted to increase as the amount of rainfall decreased.[57] This is consistent with the reports of Tanaka et al.[58], who showed that subjecting spinach to water deficiency stress increases the content of α-tocopherol in the leaves. Price and Hendry[59] studied the effect of drought on ten different grass species and noted that drought led to production of superoxide radical anions in the plants, and as the drought continued there was a rise in the peroxidation of membranes and loss of control over the uptake of transition metals, especially that of Fe, into the cells. Drought stress was also noted to cause an increase in the concentration of radical scavenging vitamin α-tocopherol, an increase in glutathione, and a decline in the ascorbic acid in the plant tissues. It was thus concluded that drought-tolerant species have the capacity to accumulate tocopherol while species with low tolerance to drought stress seem to have the capacity to accumulate glutathione.

E. Summary

Experimental data available seem to indicate that irrigation, a necessity in many areas of the world, may increase or decrease the content of some vitamins in plants. The magnitude of these changes, however, appears to be relatively small when compared with that of other factors covered in chapters 3–6 and thus may not have any nutritional consequences for the vitamin supply to the consumers. This is especially true if one considers the other and often positive changes known to occur in the other quality parameters in plant foods when they are irrigated. Thus, if a plant vitamin happens to be reduced as a result of irrigation, then one is faced with a tradeoff between yield, taste, and vitamin content. For tomatoes, for example, the case seems to be relatively clear since dry growing conditions, which may reduce their yield, improve their flavor as well as their vitamin C content. Although growing tomatoes with less water may be something that home and hobby gardeners may want to consider, for a commercial farmer, however, the economic consequences of lower tomato yield produced under water-limited conditions need to be taken into account. As for the role of vitamins in the physiology of plants subjected to drought stress, more information is needed before any clear picture can emerge.

II. VITAMINS AND TOLERANCE TO LOW-TEMPERATURE STRESS

Effect of temperature on the vitamin content of plants is covered in chapter 4. Here the emphasis is on the possible role of vitamins in the tolerance of plants to low-temperature stress.

Plants are differently susceptible to low (below the optimal) temperatures. The mechanism for cold tolerance in plants is not well understood although membranes are believed to be one of the sites where the temperature damage occurs.[1] One view is that the low-temperature damage may be mitigated by the free radicals formed in the tissues during the stress conditions. Some authors are of the opinion that the activity of free radical scavenging enzymes[60,61] and the amount of vitamin C in the cold-acclimated tissues[62] may play an important role in the degree of plants' tolerance to low temperature. Some reports seem to support this view.

Medawara,[63] for example, planted spinach from April to November at 15-day intervals in Austria and noted that maximum vitamin C concentration (150–200 mg/199 g) occurred in the leaves of those plants that grew during the cold months of November and December and at temperatures barely above 0°C or even under 0°C. This indicates a possible relationship between the temperature under which plants are grown and their vitamin C concentration. Although no cause-and-effect relationships could be deduced from this observation, other reports, however, seem to point to such a relationship. For example, exposure of wheat to low temperature (5°C) for 12 h was noted to raise the ascorbic acid content in the wheat coleoptile. Exposure to 0 or −5°C, however, decreased the ascorbic acid, by a much lesser extent in the cold-hardy than in the cold-susceptible variety.[64]

That the degree of cold hardiness in different winter wheat varieties and the content of ascorbic acid in their leaves might be related was also investigated by Andrews and Roberts.[65] They noted that the coleoptiles of wheat germinated at the hardening temperature of 1.5°C contained much higher ascorbic acid than those of wheat germinated at higher temperatures of 5, 10, or 20°C. When germinated at the hardening temperature of 1.5–3°C, a cold-hardy variety (Kharkov 22 M.C.) contained higher ascorbic acid than a less hardy variety (Alexander)

(93.9 vs. 71.1 mg/100 FW, respectively). When germinated at higher temperatures, however, varieties did not differ much in their ascorbic acid content. Also, artificial feeding of seedlings with ascorbic acid, by irrigating the germination substrate with ascorbic acid solution, was found to increase the ascorbic acid content in the leaves and at the same time increase the cold hardiness of the emerged seedlings. For example, 50% of the seedlings fed with a 1 mol/l solution of ascorbic acid during germination survived the freezing temperature of $-10°C$ for 8 h, while in the control seedlings all the leaves froze down to their base.[65] These data, along with other observations that cold- and frost-hardy varieties of grapevine contain more vitamin C than the less resistant varieties[66-68] and that apples grown in steppe regions with very severe winters were richer in vitamin C than the same varieties grown in milder climatic conditions,[40] support the view that ascorbic acid may play a role in the cold hardiness of some plants.

Higher ascorbic acid content and the degree of cold hardening do not seem to hold true in all cases. In lemon trees, for example, hardened leaves during winter months were noted to contain less ascorbic acid than the dehardened leaves during spring and summer months.[69] Whether part of this observation is due to possible differences in the amount of sunshine during winter and summer months was not mentioned by these authors.

From the data reported on the changes in plant vitamins during different seasons and the effect of temperature on some plant vitamins (chapter 4), it appears that in many cases the content of ascorbic acid is higher in the plants grown at lower temperatures or during colder months of the year, which also supports the view that cold hardiness and ascorbic acid content might be related. More research, however, is needed to elucidate the role of ascorbic acid in the cold tolerance of plants and its nutritional consequences for the vitamin C supply to those people living in the colder regions of the world, some of whom may depend solely on the locally produced fruits and vegetables for their daily diet.

III. SOIL SALINITY

One-third of the world's 230×10^6 hectares of irrigated land is estimated to be threatened by soil salinity.[70] Although the effect of salinity on plant growth has been extensively investigated[1,70] and subjecting the plant to salinity (or increasing the electrical conductivity) in their rooting medium has been found to improve the taste and other culinary factors in carrots,[72] melons,[73] and tomatoes,[74-77] information on the effect of salinity on the vitamin content of plants is very limited.

A. Ascorbic Acid

Reports on the effect of salinity on the ascorbic acid content of plants are not consistent. For example, Okasha et al.[78] reported that soil salinity decreased the size of the strawberry fruits but did not have any significant effect on their ascorbic acid content. Also, in radish[79] and pepper[80] salinity did not have any consistent effect on their content of ascorbic acid. In contrast, salinity was shown to decrease the concentration of ascorbic acid in the leaves of peanut[82] and cabbage,[79] and in the fruits of tomato[83] and okra.[84] Part of these inconsistencies might be due to the effect of increased salinity on the relative water content in the plant tissues, since Walker et al.[81] noted that in *Capsicum*, NaCl salinity increased the ascorbic acid content of fruits if expressed on the fresh weight basis, but appeared either to have no effect or even to decrease it if expressed on the dry weight basis.

It is not certain whether the adverse effects of NaCl salinity on the ascorbic acid, in the cases observed, are brought about by the high concentrations of Na and/or Cl ions. A recent work by Albu-Yaron and Feigin[85] points to some interactions between Cl and NO_3 on the ascorbic acid content in tomato juice, which not only influences the vitamin C content in fresh tomato juice but may also affect the corrosion of cans and the degree of ascorbic acid retention in the canned juice during storage. For example, increasing the NO_3 concentration in the growing medium from 1 to 100 mmol/l, and keeping the Cl concentration at the constant level of 5 mmol/l, increased the ascorbic acid in the fresh juice from 0.50 to 0.70 mmol/l. Also, increasing the Cl concentration from 5 to 110 mmol/l, and keeping the NO_3 concentration at a constant level of 20 mmol/l, increased the ascorbic acid in the juice from 0.48 to 0.70 mmol/l, indicating that at these concentration ranges the Cl (as well as NO_3) ions may have positive effects on the vitamin C content of tomato.

Chloride and sulfate fertilizers appear to affect the vitamin C concentration in strawberry and leaves of celery and grape differently. Thus, strawberry fruits[86] and leaves of celery and grape[87] contained less ascorbic acid when plants were supplied with Cl-containing rather than SO_4-containing fertilizers. In the range of 150–450 mg fertilizer/pot, however, Cl-containing fertilizer slightly increased while the same amount of SO_4-containing fertilizers reduced the ascorbic acid in the celery leaves considerably. Cl-containing potassium fertilizer was also noted to slightly increase the ascorbic acid content in sour cherries when compared with the trees fertilized with a SO_4-containing fertilizer.[88] These observations indicate that at higher concentration ranges, a situation that may occur under chloride or sulfate salinity conditions in the soils, Cl salts might be relatively less detrimental to the ascorbic acid content of fruits and vegetables than the sulfate salts. More research, however, is needed to verify these findings under field conditions.

B. Other Vitamins

NaCl has been reported to reduce the concentration of carotene in the leaves of radish, cabbage, lettuce,[79] and tomato.[89] Although Cl is proposed to be detrimental to carotene synthesis in carrots,[90] the unequivocal proof for the relative roles of Na and Cl in carotene synthesis by the plants grown under NaCl salinity conditions is lacking. In contrast to the higher plants, synthesis of β-carotene by a marine alga (*Dunaliella bradawil*) is strongly increased at higher levels of NaCl salinity.[91] No information could be found on the effect of salinity on the content of other plant vitamins.

IV. MINERAL TOXICITY AND DISPOSAL OF SEWAGE SLUDGE

In some areas of the world, sewage sludge and pig slurries are spread on agricultural lands as fertilizers because of their content of some plant nutrients. These materials, however, depending on their source, may also contain numerous heavy metals such as Zn and Cu,[92,93] which are added to pig ration to promote their growth.[94,95] Thus repeated application of these organic wastes to soil may gradually result in the accumulation of high amounts of some heavy metals in the top layers of soil[94,95] where plant roots normally grow. Although various food-chain implications[96] and health risks aspects[97] of spreading sewage sludges on agricultural lands have been investigated, and the effects of various heavy metals on some physiological processes in plants are known,[94,99] information as to the effect of disposal of sewage sludge on agricultural lands on the plant vitamins is scarce.

A. Effect of Toxic Levels of Inorganic Substances on Plant Vitamins

Lyon and Beeson[100] studied the effect of high levels of B, Cu, Fe, Mn, Mo, and Zn on the concentration of several vitamins in turnip greens and tomatoes under sand culture conditions (Table 2). They noted that although the toxic levels of all elements, with the exception of Fe, decreased the dry weight of plants, the only effect observed on the plant vitamins was that of boron, whose toxic levels in the nutrient solution (15.5 ppm) increased the thiamin concentration in turnip greens by 39% compared with that in plants supplied with 0.5 ppm (normal level) of boron. Niacin concentration was increased by the increased levels of B, Mn, and Mo and riboflavin concentration by Mn and Zn. Ascorbic acid concentration, however, decreased by increased

TABLE 2
Effect of Toxic Levels of Several Micronutrients on the Vitamin Content of Turnip Greens and Tomato Fruits

| Nutrient | Turnip greens | | | | Tomato fruits |
	Thiamin	Niacin	Riboflavin	Ascorbic acid	Ascorbic acid
B	Increase	Increase		Decrease	Decrease
Mn		Increase	Increase	Decrease	Decrease
Mo	Increase	Increase			Decrease
Zn			Increase		
Cu				Decrease	Increase

Compiled data from Lyon, C. B. and Beeson, K. C., *Bot. Gaz.* (*Chicago*), 109, 506, 1948.

levels of B, Mn, and Cu in turnip greens and by increased levels of B, Mn, Mo, and Cu in tomato fruits (Table 2). The extent of changes in vitamin concentration was by no means small. For example, increasing the concentration of Cu in the nutrient solution was noted to increase the ascorbic acid content of tomato fruits by 60%.[100] Other reports also indicate that the toxic levels of Cu may affect the ascorbic acid content of different plants differently. For example, although toxic levels of Cu did not change the ascorbic acid content in lettuce, cabbage, and spinach, they reduced that in the Chinese cabbage and garden cress.[101]

Adedeji and Fanimokun[102] studied the effect of 0, 1, 2, 4, 8, and 16 mg/l of $CuSO_4$ in nutrient solution on the growth and ascorbic acid content of two leafy vegetables (*Celosia argentea* and *Amaranthus dubius*) in Nigeria and noted that plants grew more vigorously at 2 mg/l of Cu but became chlorotic at higher Cu levels. As the leaf concentration of Cu increased, however, a significant decrease in their content of ascorbic acid was noted. In the *A. dubius*, for example, plants grown with 16 mg/l Cu contained almost half as much ascorbic acid as the control plants (47.8 vs. 96.6 mmol/ml).

Stadelamnn et al.[108] studied the effect of soil Cd on the yield and concentration of some vitamins in Italian ryegrass, radish, and spinach and noted that although addition of up to 64 mg Cd/kg soil increased the Cd content of Italian ryegrass by 153 times (0.54 vs. 83.1 mg/kg DW in the control and treated plants, respectively) and affected the yield to a slight extent, it did not change the concentration of riboflavin in the plant. The concentration of α-tocopherol was decreased and that of β-carotene was increased, but only by a nonsignificant extent. In radish, small addition of Cd to the soil increased the yield of tubers and

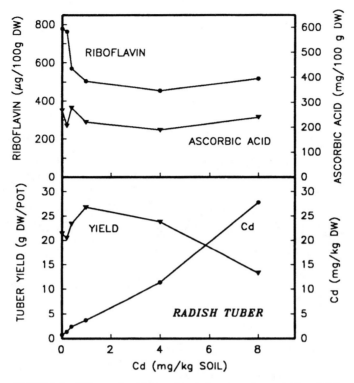

FIGURE 1. Effects of addition of cadmium to soil on yield and Cd and vitamin concentrations in the radish tuber.[108]

decreased the riboflavin concentration in the tubers; higher levels of Cd, however, decreased the yield but did not decrease the content of riboflavin any further. Ascorbic acid concentration in the tubers was apparently not affected by the Cd (Figure 1). In spinach, addition of Cd to soil progressively increased the Cd content of leaves, decreased the yield, and decreased the ascorbic acid in a nonlinear manner (Figure 2).

B. Effect of Organic Wastes on Plant Vitamins

Effect of sewage sludge on the ascorbic acid content of potato was investigated by Mondy et al.[109] They applied high amounts of an anaerobically digested municipal-industrial sludge (100–120 kg dry sludge/ha) for a five-year period to a Mardin channery silt loam soil in Binghamton, New York, and after one year of planting corn on these fields, they planted potato for the next two consecutive years and measured the yield and chemical composition of tubers. Their results

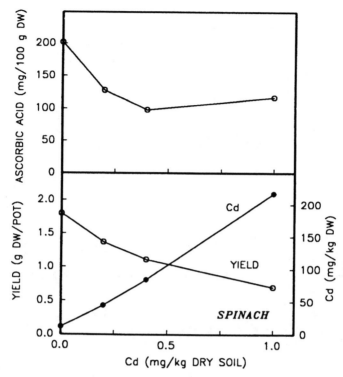

FIGURE 2. Effects of addition of cadmium to soil on yield and Cd and ascorbic acid concentrations in spinach leaves.[108]

show that although in the first year potato tubers had a slightly higher ascorbic acid content than the control plants, in the second year, however, when tuber generally had much higher ascorbic acid content than the first year, the trend was reversed, so tubers on sludge-treated soils contained less ascorbic acid than those grown in the control plot. Decrease in the ascorbic acid content of potato was also observed by others when fields were fertilized with municipal-industrial sewage sludge (applied at a rate of 300 kg N/ha),[110] or with sludge (effluent solids) at the rate of 500–2000 tons/ha.[111]

Based on an eleven-year experiment, Stadelmann et al.[105] reported that application of 2–5 tons of sewage sludge or pig slurry to slightly acid, low humus, and sandy loam soil increased the yield and the concentrations of Zn and Mn in the tubers of celeriac (turnip-rooted celery) (*Apium graveolens L. var. rapaceum*), but the concentrations of Cu and Cd in the tubers did not show any consistent relationships with the treatments employed. Sewage sludge treatment did not considerably change the concentrations of riboflavin and α-tocopherol in the celeriac

TABLE 3
Effects of 11 Years of Soil Application of Mineral or Organic Fertilizers on the Yield, Nitrate, Heavy Metal and Vitamin Concentrations in Celeriac (*Apium graveolens* L. var. *rapaceum*) Tubers

							Vitamins	
Treatment[a]	Yield[b]	NO_3[c]	Cu[c]	Zn[c]	Cd[c]	Mn[c]	B_2[d]	E[d]
Control	0.6	380	9	70	1.3	61	126	89
MF	18.8	672	9	52	0.7	41	100	100
SS1	26.5	732	11	80	1.0	26	115	87
SS2	28.2	961	11	73	0.8	15	98	86
PS1	14.9	621	8	80	1.1	76	99	74
PS2	15.9	1,078	5	123	1.2	103	93	92

[a] MF = mineral fertilizer; SS1 and SS2 = 2 and 5 tons of sewage sludge/ha/year, respectively; PS1 and PS2 = 2 and 5 tons of pig slurry/ha/year, respectively.
[b] Ton FW/ha.
[c] mg/kg DW.
[d] Vitamin concentrations are expressed as the percentage of that found in the plants fertilized with mineral fertilizer.

Rounded data from Stadelmann, F. X., Frossard, R., Furrer, O. J., Lehmann, L., and Moeri, P. B., *VDLUFA-Schriftenreihe, Kongressband* 1987, 23, 857, 1988

tubers. Pig slurry, however, reduced the tocopherol content of the tubers (Table 3).

Maync and Venter[100, 107] noted that sludge application increased the leaf concentrations of Cd, Zn, and Cu, did not affect the carotene content of lettuce, but significantly increased the ascorbic acid content in lettuce and kohlrabi leaves. Sludge application reduced the plant yield by a much higher extent than increasing the plant vitamin (Table 4).

Contamination of soil with Hg and Pb was noted to decrease the vitamin C content of cabbage.[112] Addition of chromium-containing petroleum-drilling mud to soil, however, was noted to increase the ascorbic acid in the leaves of cucumber and corn seedlings.[113]

Irrigating plants with the polluted water from the river Ulhas in India was noted to reduce the content of vitamin C in *Asteracantha longifolia* and vitamins B_1 and B_2 in *Phylanthus niruri*.[103] Kiss et al.[104] reported that water polluted with motor oil and 2, 4-D reduced the growth of barley, asparagus, and cucumber seedlings, and a mixture of 2, 4-D and motor oil slightly increased the ascorbic acid content in cucumber. It thus appears that the use of polluted waters for irrigation of plants may affect their vitamin contents, a subject matter that

TABLE 4
Effects of Soil Application of Sewage Sludge on the Yield (g FW / pot), Ascorbic Acid, and Carotene Concentrations (mg / 100 g FW) in Lettuce and Kohlrabi

Sludge (ton DW / ha)	Lettuce			Kohlrabi leaves	
	Yield	Carotene	Ascorbic acid	Yield	Ascorbic acid
0	204[a]	2.9	24.7	233	121
62	186	3.0	24.7	235	132
123	158	3.2	28.3	214	140
246	128	3.3	31.1	198	133
492	89	3.6	30.7	161	129
LSD	15	0.5	2.4	25.7	5.5

[a] Values rounded.

Compiled from tables 3, 5, and 6 of Maync, A. and Venter, F., *Gartenbauwissenschaft*, 46(2), 79, 1981 and tables 2 and 4 of Maync, A. and Venter, F., *Gartenbauwissenschaft*, 48(4), 145, 1983.

deserves much attention in the light of increasing degree of soil and water pollution especially in the developing countries.

In summary, the limited data available on the effect of soil contamination with various pollutants on the vitamin content of plants indicate that some plant vitamins, in particular vitamin C, are relatively responsive to the state of soil pollution. In the light of the increasing concern in many countries about the ways and means of disposing of municipal sludge and animal manure on the agricultural soils, more information is needed to establish any adverse effect of such practices on plant vitamins and/or other aspects of plant, and eventually human, health.

V. AIR POLLUTANTS

Air and soil pollution are becoming of increasing concern for ecologists, environmental protection agencies, and the general public. Although the effects of air[114-118] and soil[92, 98, 99] pollution on the growth of plants (mostly forest trees) have been the subject of numerous studies, very little information is available about whether they have any effect on the vitamin content of edible plants. In recent years, ascorbic acid and vitamin E, because of their antioxidant and free radical scavenging properties, are being increasingly investigated for their possible role in protecting plants against air-polluting gases such as ozone and SO_2.

A. Ozone

Collected experimental data on the effect of ozone on the concentration of vitamins in plants are not very consistent with respect to the effect of ozone on a given vitamin in different plants or different vitamins in a given plant (Table 5). Moreover, vitamin content in different plants seems to react differently to ozone exposure (Figure 3), which makes any comparison of different data very difficult. Here some selected reports are discussed in more detail.

Senger et al.,[128] for example, reported that the variations observed in the ascorbic acid content of spruce needles subjected to various ozone treatments are within the natural variability and are not due to the ozone treatment. Barnes,[119] however, noted that exposure of pine needles to 5 pphm (parts per hundred million) of ozone for 11 weeks increased the concentration of ascorbic acid in pine seedlings. In the needles treated for 20 weeks, however, the concentration of ascorbic acid (although still higher than the control) was not statistically different from that in the control plants.

Lee[157] studied the effect of ozone on the ascorbic acid content in the leaves of ozone-resistant and ozone-susceptible soybean genotypes and noted that: (a) the highest ascorbic acid content coincided with the time of highest concentration of photochemical oxidants in the air, (b) under low ozone environment (charcoal-filtered air), leaves of ozone-susceptible and ozone-resistant genotypes both showed rhythms in their ascorbic acid content, (c) under ozone stress environment, however, the rhythm variation in the ascorbic acid occurred only in the ozone-resistant genotype, and (d) leaves of the ozone-resistant genotype produced more ascorbic acid than leaves of the susceptible genotype. It was thus suggested that ascorbic acid, because of its free radical scavenging property, may protect the leaf cells from injury by ozone or other oxyradical products.

A possible relationship between the ascorbic acid content and the tolerance to ozone was also observed in different spinach varieties by Tanaka et al.[133, 158] These authors noted that although fumigation of spinach with 0.1 ppm ozone reduced the ascorbic acid content of the leaves by a slight extent, increasing the concentration of ozone to 0.5 ppm drastically reduced the leaf ascorbic acid within a few hours. Varieties were differently sensitive to ozone fumigation, and among the six varieties tested, those that contained higher ascorbate and glutathione were also more tolerant to ozone fumigation. Varities, however, did not differ in the activity of the enzymes that are presumed to protect the plants against free radicals.[133]

Duration of ozone exposure, as well as the rate by which different plants react to ozone treatment, may be among the reasons for the

TABLE 5
Effect of Various Air-Polluting Gases and Dusts on the Vitamin Concentration in Plants

Pollutant	Change	Plant	Ref.
		Ascorbic acid	
Dust (cement)	Increase	Lupine (foliage)	137
Dust (cement)	Increase	Oats (foliage)	137
Dust (cement)	Increase	Lettuce	137
NO_2	Increase	Radish root	156
Ozone	Increase	Cabbage	125
Ozone	Increase	Corn	125
Ozone	Increase	Spruce	126
SO_2	Increase	Forest trees	140
SO_2	Increase	Spruce	135
Ozone	No effect	Lettuce	125
Ozone	No effect	Peas (leaf)	127
Ozone	No effect	Spruce needle	128
Ozone	Variable	Pine needle	119
Air pollution	Decrease	*Bougainvillea spectabilis*	120
Air pollution	Decrease	Fig	121
Air pollution	Decrease	*Laurus nobilis*	121
Air pollution	Decrease	*Pyracantha coccinea*	121
Air pollution	Decrease	Several plants	122
Air pollution	Decrease	Avenue trees	123
Dust (Petro-coke)	Decrease	*Phaseolus aureus*	136
Dust + SO_2	Decrease	*Mangifera indica*	138
Dust + SO_2	Decrease	*Citrus medica*	138
Freeway gases	Decrease	Parsley	124
NH_3	Decrease	Several trees	155
NO_2	Decrease	Radish (leaf)	156
SO_2	Decrease	Beans (leaf)	141
SO_2	Decrease	Clover (ladino)	142
SO_2	Decrease	Larch needle	143
SO_2	Decrease	Lucerne	144
SO_2	Decrease	Mustard	150
SO_2	Decrease	Maize	150
SO_2	Decrease	*Phaseolus radiata*	150
SO_2	Decrease	Pine needle	143
SO_2	Decrease	Rice leaf	145, 146
SO_2	Decrease	Spruce needle	126, 147–148
SO_2	Decrease	Wheat (leaf)	151
SO_2	Decrease	*Vinga radiata*	152
Ozone	Decrease	Beans (leaf)	129
Ozone	Decrease	Faba beans (leaf)	130
Ozone	Decrease	Grape (leaf)	134
Ozone	Decrease	Potato	131
Ozone	Decrease	Spinach	132, 133
Ozone	Decrease	Tomato	125

TABLE 5 (continued)

Pollutant	Change	Plant	Ref.
		Carotene	
Dust (cement)	Increase	Lettuce	137
Dust (cement)	Increase	Lupine (foliage)	137
Dust (cement)	Increase	Oats (foliage)	137
Ozone	No effect	Cabbage	125
Ozone	No effect	Carrots	125
Ozone	No effect	Lettuce	125
Ozone	No effect	Tomato	125
Air pollution	Decrease	*Bougainvillea spectabilis*	120
Dust	Decrease	*Gardenia*(leaf)	139
Dust	Decrease	*Coccinia* (leaf)	139
NH_3	Decrease	Several trees	155
Ozone	Decrease	Faba beans (leaf)	130
Ozone	Decrease	Corn	125
Ozone	Decrease	Spinach	132
SO_2	Decrease	Beans (leaf)	141
SO_2	Decrease	Spinach	153
		Niacin	
Ozone	Increase	Carrots	125
Ozone	Increase	Strawberry	125
SO_2	Increase	Tomato (leaf)	154
Ozone	No effect	Lettuce	125
SO_2	Decrease	Lettuce	154
SO_2	Decrease	Peas (leaf)	154
		Pyridoxine	
SO_2	No effect	Lettuce	154
SO_2	No effect	Peas (leaf)	154
		Riboflavin	
Ozone	No effect	Lettuce	125
Ozone	No effect	Strawberry	125
		Thiamin	
Ozone	Increase	Cabbage	125
Ozone	Increase	Lettuce	125
Ozone	No effect	Carrots	125
Ozone	No effect	Strawberry	125
Ozone	Decrease	Tomato	125
SO_2	Decrease	peas (leaf)	154
		Vitamin E	
Ozone	Increase	Spruce	128, 135
Ozone	Increase	Grape (leaf)	134
SO_2	Increase	Spruce	135
SO_2	Increase	Fir	135

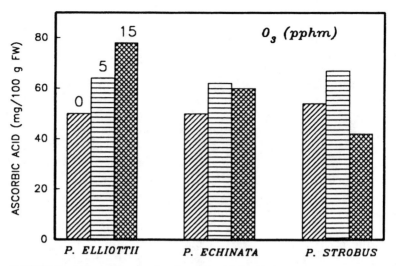

FIGURE 3.   Effect of five weeks of growth with increasing concentration of O_3 (pphm = parts per hundred million) on the ascorbic acid concentration in the needles of three different pine species.[119]

seemingly inconsistent reports on the effect of ozone on plant vitamins (Table 5). For example, Barnes[119] cities experiments in which a mere 2.5 h of exposing petunias to a relatively high (40 pphm) concentration of ozone was enough to increase the ascorbic acid concentration in their leaves. For pine seedlings, however, 11 weeks of exposure to low (5 pphm) ozone was needed to increase the ascorbic acid content in the needles. Opopol and Kushnir[160] noted that a continuous 30-min exposure of apples to ozone first reduced their content of ascorbic acid. After 60 min of exposure to ozone, however, the ascorbic acid content rose to a level even higher than that of the control fruits. These observations, along with those of Kuno,[161] that prolonged exposure (5 h/day for 10 days) of spinach to a low concentration (0.025 ppm) of ozone reduced the ascorbic acid concentration in the leaves of ozone-sensitivie varieties but not in the ozone-tolerant varieties, point to the possible differences in the response time required by different plants to set in motion the enzymatic mechanisms necessary for an increased rate of free radical scavenging in their tissues.

B. Sulfur Dioxide

Exposure to SO_2 was noted to decrease the concentration of ascorbic acid and niacin but increase that of vitamin E in the leaves of several forest trees and some other plants (Table 5). For example, exposure of larch (*Larix decidua*) and pine (*Pinus silvestris*) to SO_2 significantly reduced the concentration of ascorbic acid in needles. This, however, took place long before any visible symptom of injury (due to fumigation of plants with SO_2) could be observed in the needles.[143] Also, exposure of spruce (*Picea abies*) needles to sublethal concentrations of SO_2 under natural forest conditions in Austria was found to reduce the content of ascorbic acid and dehydroascorbic acid and increase the activity of ascorbic acid oxidase.[148] Also, Materna[147] reported that fumigation of five-year-old spruce trees for a period of six years to a total of 2000 h to low concentrations of SO_2 (< 1 mg m^{-3} air) significantly decreased the concentration of ascorbic acid in the needles at some, but not all, of the sampling dates.

Despite the above reports, the cause-and-effect relationships between the occurrence of forest dieback, concentration of air-polluting gases in the atmosphere, and the concentration of antioxidant vitamins in the plants remain to be proved.[162] For example, Osswald et al.[163] noted that in the East Bavarian mountains in Germany, the bleached (damaged or chlorotic) needles or the bleached segments of otherwise green needles actually contain more ascorbic acid than the fully green needles or the green segments of the partly chlorotic needles, respectively. The bleached needle segments of two- and three-year-old needles showed the largest increase of 45–350% in their ascorbic acid compared with the green needles or the green segments of the partly chlorotic needles. This raised the question as to the role of ascorbic acid as an antioxidant in protecting the spruce needles from bleaching, presumably caused by the air pollutants. In other words, why should the needle segments be bleached although they contained higher ascorbic acid than the apparently healthy green needles? In a cell suspension culture, Messner and Berndt[164] exposed spruce cells to SO_2, ozone, and H_2O_2 and noted that the chlorophyll content of the cells was reduced by a much lesser degree than that of the ascorbic acid and thus concluded that "...the increased ascorbic acid content in needles of damaged conifers is probably not an induced protection mechanism against oxidants."

What about the edible plants? Available reports are not very consistent but point to the strong difference in the way different plants react to SO_2 pollution. For example, SO_2 fumigation was noted to decrease the ascorbic acid content in the leaves of mustard (*Brassica nigra*) and *Phaseolus radiatus* (both relatively sensitive to SO_2) and maize (relatively tolerant to SO_2). Plants, however, reacted differently. It took one week of SO_2 fumigation to reduce the ascorbic acid concentration in the leaves of sensitive plants (*B. nigra* and *P. radiata*), but it took six weeks of SO_2 fumigation before any marked reduction in the ascorbic

TABLE 6

Effect of 2 h / Day Fumigation of 7-Day-Old Plants with SO_2 for 1–6-Week Periods on the Percent Reduction in the Ascorbic Acid Content[a] of the Leaves as Compared with Control Plants[150]

SO_2 ($\mu g\ m^{-3}$)	Weeks		
	1[b]	2	6
Brassica nigra			
79	1.5	4.5a	8.8a
131	3.3a	7.0a	12.3a
262	5.5a	8.4a	15.7a
Pinus radiata			
79	1.8	3.2a	10.8a
131	3.5a	5.2a	15.6a
262	5.0a	7.8a	21.7a
Zea mays			
79	0.0	1.2	2.8
131	0.5	1.8	3.3b
262	0.5	2.2	3.9b

[a] Values rounded.
[b] For each plant and fumigation duration combination, values followed by letters a and b are significantly different from the control plant at 0.01 and 0.5 levels, respectively.

acid could be detected in the leaves of SO_2-resistant corn plants (Table 6). Fumigation of tomato plants with SO_2 for up to ten weeks did not change the ascorbic acid content of tomato fruits on the fresh weight basis. If the data were expressed on the dry weight basis, however, then fumigation seemed to have caused a slight, but significant, decrease in the ascorbic acid content of the fruits.[165]

The effect of SO_2 on the concentrations of vitamins B_1, B_6, and niacin in the plant leaves was noted to depend on the plant species (peas, lettuce, or tomato) and the plant age at the time of fumigation. For example, when 14-, 26-, and 43-day-old tomato plants were fumigated each for 12 days, the younger plants (now 26 days old) showed a higher concentration of thiamin while the older plants (now 36 and 55 days old) both showed decreased thiamin content when compared with the same-age control plants. The concentration of niacin, however, increased in all tomato plants, even more so if they were older at the time of fumigation. In lettuce and peas, however, as soon as plants become necrotic due to fumigation, their content of thiamin shows a definite decline.[154]

C. Other Urban Pollutions and Dust

Are the fruits and vegetables produced in the vicinity of large metropolitan cities, with increased air pollution, lower or higher in vitamins than those grown in more urban areas? Available information indicates that general urban pollution can, in fact, alter the vitamin content of plants. For example, fruits grown in the industrial and urban area of the city of Split (former Yugoslavia) were found to have lowered levels of ascorbic acid (Table 5).[121] Parsley plants placed at the edge of a freeway in Switzerland were noted to have significantly ($p < 0.01$) less β-carotene and ascorbic acid in their leaves than plants located just 200 m away from the freeway, an effect that was attributed to the car exhaust gases.[124]

In Finland, Huttunen and Karhu[166] compared the ascorbic acid content of blueberries (*Vaccinium myrtillus L.*) and lingonberries (*Vaccinium vitis-idaea L.*) grown in urban and rural forest sites during two years and found the lowest ascorbic acid content in the blueberries grown in the urban (traffic polluted) sites of southern Finland and the lowest ascorbic acid in the lingonberries grown in the vicinity of industrial sites. In general, pollution was noted to have a greater effect on the ascorbic acid content of blueberries than on lingonberries.

Deposition of dust on plant surfaces is common in the dry parts of the world and/or the vicinity of some industries, such as cement factories, that do not have efficient filter facilities. In the Indian subcontinent, for example, suspended particulate matter was noted to make up 40–45% of the total air pollutants, and thus some perennial crops such as fruit trees may be covered with a thick layer of dust particles during the major part of the year![139] Estimations have shown that dust deposition on fruit trees may be in the range of 0.5–5 mg cm^{-2}; leaves of mango and fig may accumulate up to 8 mg cm^{-2} of dust during the dry seasons.[139] Although the effect of such heavy coverage of plant surfaces with dust on the vitamin content of plants has not been well investigated, the information available seems to indicate that dust, depending on its nature, may increase or decrease the concentration of ascorbic acid and carotenoids in different plants (Table 5). Although the detrimental effect of dust deposition on plant vitamins may primarily be due to reducing the amount of light reaching the plant tissues, its reported positive effect on plant vitamins is much harder to explain unless under certain conditions dust particles may serve as a source of some nutrients for some plants.[139]

In summary, although various air pollutants are reported to change the concentration of various vitamins in plants, the role of vitamins in the tolerance of plants against air-polluting gases is not certain. It is recommended that the studies on the effect of air pollution on the concentration of vitamins in plants be extended to include food plants, especially in the light of their possible effect on the quality of various fruits and vegetables grown in the vicinity of metropolitan areas.

REFERENCES

1. Hale, M. G. and Orcutt, D. M., *The Physiology of Plants under Stress*, Wiley-Intersci Publ., New York, 1987.
2. Stevens, M. A., Tomato flavor: effects of genotype, cultural practices, and maturity at picking, in *Evaluation of Quality of Fruits and Vegetables*, Pattee, H. E., Ed., AVI Publishing, Westport, Conn., 1985, chapter 14.
3. Reder, R., Ascham, L., and Eheart, M. S., Effect of fertilizer and environment on the ascorbic acid content of turnip greens, *J. Agric. Res. (Washington, D.C.)*, 66(10), 375, 1943.
4. Strasburger, M., The effect of atmospheric precipitation on the content of vitamin C in the fruits of wild rose in Nałęczowie, *Roczn. Sek. Dendrol. Polsk. Towarz. Bot.*, 19, 95, 1965; *Hortic. Abatr.*, 36, 7102, 1966.
5. Primak, A. P. and Litvinenko, M. V., Effect of growing conditions on the qualitative composition of some vegetables, *Kach. Ovoshchn. Bachchevykh Kul't.* 132, 1981; *Chem. Abstr.*, 97, 88756n, 1982.
6. Kvarachelija, M. S., The effect of weather on the quality and productivity of feijoa fruits, *Bjull. Vses. Inst. Rasten. Vavilova*, 12, 65, 1968; *Hortic. Abstr.* 40, 2285, 1970.
7. Shirochenkova, A. I., Minaeva, V. G., and Zhanaeva, T. A., Activity and iso-enzyme composition of ascorbic acid oxidase and peroxidase in ripening and stored black currants, *Fiziol. Biokhim. Kul't. Rast.*, 17(6), 609, 1985; *Hortic. Abstr.*, 56, 2277, 1986.
8. Krynska, W., Effect of irrigation on early cabbage and tomatoes grown on sloping ground, *Zeszyty Problemowe Postepow Nauk Rolniczych*, 181, 103, 1976; *Food Sci. Technol. Abstr.*, 9(10), J1504, 1977.
9. Jablónska-Ceglarek, R., Effects of irrigation, manuring with FYM, mineral fertilizers and green manure crops on the nutritional value of late head cabbage and celeriac, *Biuletyn Warzywniczy*, 25, 109, 1981; *Hortic. Abstr.*, 53, 7756, 1983.
10. Fritz, P. D., Käppel, R., and Weichmann, J., Vitamin C in gelagertem Blumenkohl, *Landwirt. Forsch.*, 32(3), 275, 1979.
11. Kryńska, W., Kawecki, Z., and Piotrowski, L., The effect of fertilization, irrigation and cultivar on the quality of fresh, sour and pickled cucumbers, *Zeszyty Naukowe Akademii Rolniczo-Technicznej Olsztynie, Rolnictwo*, 15, 109, 1976; *Hortic. Abstr.*, 47, 4552, 1977.
12. Wells, J. A. and Nugent, P. E., Effect of high soil moisture on quality of muskmelon, *HortScience*, 15(3), 258, 1980.
13. Park, K. W. and Fritz, D., Effects of fertilization and irrigation on the quality of radish (*Raphanus satiuvs L.*) grown in experimental pots, *Acta Hortic.*, 145, 129, 1984.
14. Janes, B. E., The effect of varying amounts of irrigation on the composition of two varieties of snap beans, *Proc. Am. Soc., Hortic. Sci.*, 51, 457, 1948.
15. Dastane, N. G., Kulkarni, G. N., and Cherian, E. C., Effects of different moisture regimes and nitrogen levels on quality of tomato, *Indian J. Agron.*, 8, 405, 1963.

16. Rudick, J., Kalmar, D., Geizenberg, C., and Harel, S., Low water tensions in defined growth stages of processing tomato plants and their effects on yield and quality, *J. Hortic. Sci.*, 52(3), 391, 1977.

17. Kunavin, G. A., and Braun, V. A., Irrigation and weed control in tomato cultivation, *Vestn. S-kh. Nauki Kaz.*, 11, 46, 1987; *Chem. Abstr.*, 108, 126618z, 1988.

18. Somers, G. F. and Beeson, K. C., The influence of climate and fertilizer practices upon the vitamin and mineral content of vegetables, *Adv. Food Tech.*, 1, 291, 1948.

19. Hunter, A. S., Kelly, W. C., and Somers, G. F., Effects of variations in soil moisture tension upon the ascorbic acid and carotene content of turnip greens, *Agron. J.*, 42, 96, 1950.

20. Sheets, O, Permenter, L., Wade, M., Anderson, W. S., and Gieger, M., The effect of different levels of moisture on the vitamin, mineral and nitrogen content of turnip greens, *Proc. Am. Soc. Hortic. Sci.*, 66, 258, 1955.

21. Watanabe, Y., Yoneyama, M., and Shimada, N., Effects of water stress treatment on sugars, vitamin C and oxalate contents of spinach, *Nippon Dojo Hiryogaku Zasshi*, 58(4), 427, 1987; *Chem. Abstr.*, 107, 233359s, 1987.

22. Sehtiya, H. L. and Srivastava, A. K., Effect of some regulants on water potential and metabolism of leaves of wheat (*Triticum aestivum L.*) seedling under induced water stress, *Indian J. Plant Physiol.*, 28(3), 215, 1985.

23. Mukherjee, S. P. and Choudhuri, M. A., Implications of water stress-induced changes in the levels of endogenous ascorbic acid and hydrogen peroxide in *Vinga* seedlings, *Physiol. Plant.*, 58, 166, 1983.

24. Boos, G. V. and Kolyukaev, Y. B., The physiological-biochemical characteristics of cucumbers and tomatoes grown in plastic houses, *Tr. Prikl. Bot. Genet. Sel.*, 55(2), 209, 1975; *Hortic. Abstr.*, 46, 10285, 1976.

25. Berényi, M., The effect of irrigation on the composition of processing tomatoes, *Duna-tisza Közi Mezőgazd. Kisérl. Int. Bull.*, 5, 47, 1970; *Hortic. Abstr.*, 41, 9177, 1971.

26. Ermakov, A. I. and Samorodova-Bianki, G. B., Variability in the chemical composition of black currant fruits, *Tr. Prikl. Bot. Genet. Sel.*, 37(1), 105, 1965; *Hortic. Abstr.*, 36, 499, 1966.

27. Drake, S. R., Silbernagel, M. J., and Dyck, R. L., The influence of irrigation, soil preparation and row spacing on the quality of snap beans, *Phaseolus vulgaris*, *J. Food Qual.* 7, 59, 1984.

28. Abdel-Messih, M. N. and El-Nokrashy, M. A. Effect of different soil moisture levels on growth, yield and quality of Washington navel oranges, *Agricultural Research Review*, 55(3), 47, 1977; *Food Sci. Technol. Abstr.*, 12(11), J1535, 1980.

29. Cruse, R. T., Wiegand, C. L., and Swanson, W. A., The effects of rainfall and irrigation management on citrus juice quality in Texas, *J. Am. Soc. Hortic. Sci.*, 107(5), 767, 1982.

30. Minessy, F. A., Barakat, M. A., and El-Azab, E. M., Effect of water table on mineral content, root and shoot growth, yield and fruit quality in "Washington Navel" orange and "Balady" mandarin, *J. Am. Soc. Hortic. Sci.*, 95(1), 81, 1970.

31. Azzouz, S., El-Nokrashy, M. A., and Dahshan, I. M., Effect of frequency of irrigation on tree production and fruit quality of mango, *Agric. Res. Rev.*, 55(3), 59, 1977; *Food Sci. Technol. Abstr.*, 12(11), J1551, 1980.

32. Iancu, M., Sotiriu, D., and Popescu, C. V., Research on the influence of soil moisture content on the physical properties, chemical composition and storage quality of apples, *Licrările Ştiinţifice Inst. Cercetări Pentru Pomicultură Piteşti*, 4, 195, 1975; *Hortic. Abstr.*, 47, 10157, 1977.

33. Semenenko, M. P., Effect of irrigation and soil type on content of minerals and vitamin C in fruits of lower Volga region, *Tr. Vses. Semin. Biol. Aktiv. (lech.)* *Veshchestvam Plodov Yagod*, 4*th*, 234, 1970; *Chem. Abstr.*, 81, 76925y, 1974.

34. Sites, J. W. and Camp, A. F., Producing Florida citrus for frozen concentrate, *Food Technol.*, 9, 361, 1955.

35. Chauhan, R., Kapoor, A. C., and Gupta, O. P., Note on the effect of cultivar and season on the chemical composition of guava fruits, *Haryana J. Hortic. Sci.*, 15(3-4), 228, 1986.

36. Špaldon, E. and Pevná, V., Vitamin C and dry matter contents in sweet pepper as affected by growing conditions, *Sborn. vys. Sk. Pol'nohosp. v Nitre, Agron. Fak.*, 5, 209, 1961; *Hortic. Abstr.*, 33, 7367, 1963.

37. Çevik, B., Kirda, C., and Dinç, G., Effect of some irrigation systems on yield and quality of tomato grown in a plastic covered greenhouse in the south of Turkey, *Acta Hortic.*, 119, 333, 1981.

38. Drake, S. R. and Silbernagel, M. J., The influence of irrigation and row spacing on the quality of processed snap beans, *J. Am. Soc. Hortic. Sci.*, 107, 239, 1982.

39. Leclerc, J., Miller, M. L., Joliet, E., Thicoipe, J. P. and Despujols, J., Lutte contre la nécrose et teneurs en minéraux et vitamines de la laitue batavia, *Agrochimica*, 36, 108, 1992.

40. Matusis, I. I. and Jurova, G. G., Relationships between ascorbic acid accumulation in top and small fruits and some meteorological factors, *Izv. Sib. Otd. Akad. Nauk SSSR, No. 5, Ser. Biol.-Med. Nauki*, 1, 76, 1968; *Hortic. Abstr.*, 40, 557, 1970.

41. Mantell, A., Meirovitch, A., and Goell, A., The response of Shamouti orange on sweet lime rootstock to different regimes in the spring and summer (Progress report, 1973-1974), *Special publication, Agric. Res. Organization, Bet Dagan*, 58, 16 pp., 1976; *Hortic. Abstr.*, 48, 1766, 1978.

42. Akpapunam, M. A., Effect of wilting, blanching and storage temperatures on ascorbic acid and total carotenoids content of some Nigerian fresh vegetables, *Qual. Plant.—Plant Foods Hum. Nutr.*, 34, 177, 1984.

43. Fernández, M. Del C. C., Effect of environment on the carotene content of plants, *South. Coop. Series Bull.*, 36, 24, 1954.

44. Michalik, H. and Bakowski, J., The content and distribution of carotenoids in carrots in relation to harvest time, *Biuletyn Warzywniczy*, 20, 417, 1977; *Hortic. Abstr.*, 49, 549, 1979.

45. Banga, O., De Bruyn, J. W., van Bennekom, J. L., and van Keulen, H. A., Selection of carrots for carotene content. V. The effect of the soil moisture content, *Euphytica*, 12, 137, 1963.

46. Fritz, D. and Venter, F., Einfluss von Sorte, Standort und Anbaumassnahem auf messbare Qualitätseigenschaften von Gemüse, *Landwirt, Forschung, Sonderh.*, 30(1), 95, 1974.

47. Moussa, A. G., Geissler, T., and Markgraf, G., Ertrag, Qualität und Nitratgehalt von Speisemöhre und Spinat bei steigender Stickstoffdüngung im Modellversuch, *Arch. Gartenbau*, 34(1), 45, 1986.

48. Nortjé, P. F. and Henrico, P. J., The influence of irrigation interval on crop performance of carrots (*Daucus carota* L.) during winter production, *Acta Hortic.*, 194, 153, 1986.

49. Massey, P. H., Jr., Eheart, J. F., Young, R. W., and Mattus, G. E., The effect of soil moisture, plant spacing, and leaf pruning on the yield and quality of broccoli, *Proc. Am. Soc. Hortic. Sci.*, 81, 316, 1962.

50. Guralnick, L. J. and Ting, I. P., Physiological changes in *Portulacaria afra* (L.) Jacq. during a summer drought and rewatering, *Plant Physiol.* 85(2), 481, 1987.

51. Stuhlfauth, T., Steuer, B., and Fock, H. P., Chlorophylls and carotenoids under water stress and their relation to primary metabolism, *Photosynthecia*, 24(3), 412, 1990.

52. O'Donnell, W. W. and Bayfield, E. G., Effect of whether, variety, and location upon thiamin content of some Kansas-grown wheats, *Food Res.*, 12, 212, 1947.

53. Hunt, C. H., Rodriguez, L. D., and Bethke, R. M., The environmental and agronomical factors influencing the thiamine, riboflavin, niacin, and pantothenic acid content of wheat, corn, and oats, *Cereal Chem.*, 27, 79, 1950.

54. Udayasekhara Rao, P. and Belavady, B., Effect of different agronomic practices on the protein and vitamin content of Bengal Gram (*Cicer arietinum*), *Indian J. Nutr. Dietet.*, 17, 6, 1980.

55. Bowler, C., Van Montagu, M., and Inzé, D., Superoxide dismutase and stress tolerance, *Annu. Rev. Plant Physiol.*, 43, 83, 1992.

56. Quartacci, M. F. and Navari-Izzo, F., Water stress and free radical mediated changes in sunflower seedlings, *J. Plant Physiol.*, 139, 621, 1992.

57. Burger, O. J. and Hauge, S. M., Relation of manganese to the carotene and vitamin contents of growing crop plants, *Soil Sci.*, 72, 303, 1951.

58. Tanaka, K., Masuda, R., Sugimoto, T., Omasa, K., and Sakaki, T., Water-deficiency-induced changes in the contents of defensive substances against active oxygen in spinach leaves, *Agric. Biol. Chem.*, 54(10), 2629, 1990.

59. Price, A. H., and Hendry, G. A. F., Stress and the role of activated oxygen scavengers and protective enzymes in plants subjected to drought, *Biochem. Soc. Trans.*, 17, 493, 1989.

60. Nakagawara, S. and Sagisaka, S., Increase in enzyme activities related to ascorbate metabolism during cold acclimation in poplar twigs, *Plant Cell Physiol.*, 25(6), 899, 1984.

61. Kuroda, H., Sagisaka, S., and Chiba, K., Seasonal changes in peroxide-scavenging system of apple trees in relation to cold hardiness, *Engei Gakkai Zasshi*, 59(2), 399, 1990; *Chem. Abstr.*, 114, 139883f, 1991.

62. Potapova, M. N., Biological (and biochemical) features of the common barberry, *Puti Adapt. Rast. Introd. Sev.*, 63, 1982; *Chem. Abstr.*, 99, 137013v, 1983.

63. Medawara, M. R., Notizen über Vitamin C in der Pflanze, *Phyton* (*Austria*), 2, 193, 1950.

64. Stikic, R. and Jevtic, M., Effects of low temperature on ascorbic acid content and osmotic water potential in different wheat cultivars, *Arh. Poljopr. Nauke*, 45(160), 427, 1984; *Chem. Abstr.*, 104, 31815q, 1986.

65. Andrews, J. E. and Roberts, D. W. A., Association between ascorbic acid concentration and cold hardening in young winter wheat seedlings, *Can. J. Bot.*, 39, 503, 1961.

66. Tarasashvii, K. M., Gvamichava, N. E., Kezeli, T. A., and Beridze, A. G., Significance of the ascorbic-acid glutathione system in frost-resistance of grapevines, *Soobshch. Akad. Nauk Gruz. SSR*, 107(3), 585, 1982; *Chem. Abstr.*, 98, 86345z, 1983.

67. Kazeli, T. A., Gvamichava, N. E., Tarashvili, K. M., Piranishvili, N. S., Kotaeva, D. V., and Chkhubianishvili, E. A., Redox metabolism in grapevines in relation to frost resistance, *Fiziol. Morozoustoich. Vinograd. Lozy*, 48, 1986; *Chem. Abstr.*, 106, 135378g, 1987.

68. Kezeli, T. A. and Beridze, A. G., Structural and histochemical changes in one-year grapevine shoots in relation to frost resistance, *Fiziol. Morozoustoich. Vinograd. Lozy* 84, 1986; *Chem. Abstr.*, 106, 135380b, 1987.

69. Alvarez, M. R., Candela, M. E., and Sabater, F., Ascorbic acid content in relation to frost hardiness injury in citrus lemon leaves, *Cryobiology*, 23, 263, 1986; *Chem. Abstr.*, 105, 39607q, 1986.

70. Epstein, E., Norlyn, J. D., Rush, D. W., Kingsbury, R. W., Kelley, D. B., Gunningham, G. A., and Wrona, A. F., Saline culture of crops: a genetic approach, *Science*, 210, 399, 1980.

71. Staples, R. C. and Toennissen, G. H., *Salinity Tolerance in Plants*, Wiley, New York, 1984.

72. Simon, P. W., Carrot flavor: effects of genotype, growing conditions, storage, and processing, in *Evaluation of Quality of Fruits and Vegetables*, Pattee, H. E., Ed., AVI Publishing, Westport, Conn., 1985, 315.

73. Mizrahi, Y. and Pasternak, D., Effect of salinity on quality of various agricultural crops, *Plant Soil*, 89, 301, 1985.

74. Clay, D. W. T. and Hudson, J. P., Effect of high levels of potassium and magnesium sulphates on tomatoes, *J. Hortic. Sci.*, 35, 85, 1960.

75. Mizrahi, Y. and Shoshana (Malis), A., and Zohar, R., Salinity as a possible means of improving fruit quality in slow-ripening tomato hybrids, *Acta Hortic.*, 190, 223, 1986.

76. Marks, M. J., Use of sodium chloride for tomatoes grown in solution culture, *J. Sci. Food Agric.*, 50, 423, 1990.

77. Cornish, P. S., Use of high electrical conductivity of nutrient solution to improve the quality of salad tomatoes (*Lycoperscion esculentum*) grown in hydroponic culture, *Aust. J. Exp. Agric.*, 32, 513, 1992.

78. Okasha, K. A., Helal, R. M., Khalaf, S., and El-Hifny, I. M., Effect of some soil types on fruit quality of Tioga strawberry cultivar, *Egypt J. Hortic.*, 13(1), 71, 1986.

79. Kim, C. M., Effect of saline and alkaline salts on the growth and internal components of selected vegetable plants, *Physiol. Plant.* 11, 441, 1958.

80. Sonneveld, C. and van der Burg, A. M. M., Sodium chloride salinity in fruit vegetable crops in soilless culture, *Neth. J. Agric. Sci.*, 39, 115, 1991.

81. Walker, R. R., Hawker, J. S., and Törökfalvy, E., Effect of NaCl on growth, ion composition and ascorbic acid concentration of *Capsicum* fruit, *Scientia Hortic.*, 12, 211, 1980.

82. Latha, V. M., Satakopan, V. M., and Jayasree, H., Salinity-induced changes in phenol and ascorbic acid content in groundnut (*Arachis hypogea*) leaves, *Curr. Sci.*, 58(3), 151, 1989.

83. Ponomareva, S. A. and Kubuzenko, S. N., Tomatoes under saline conditions, *Kartofel' Ovoshchi*, 10, 24, 1984; *Hortic. Abstr.*, 56, 8934, 1986.

84. Kirti Singh, Mangal, J. L., and Gupta, U. S., Compositional changes in okra (*Abelmoschus esculentus* (L) Moench) fruits as affected by sodium salts, *Prog. Hortic.*, 7(4), 19, 1976; *Hortic. Abstr.*, 47, 6599, 1977.

85. Albu-Yaron, A. and Feigin, A., Effect of growing conditions on the corrosivity and ascorbic acid retention in canned tomato juice, *J. Sci. Food. Agric.*, 59, 101, 1992.

86. Kramer, S., Untersuchungen über die Wirkung chloridhaltiger Kalidünger auf Ertrag, Ascorbinsäueregahlt der Früchte und vegetative Entwicklung der Erdbeere, *Arch. Gartenbau*, 11, 175, 1963.

87. Siegel, O. and Bjarsch, H. J., Über die Wirkung von Chlorid-und Sulfationen auf den Stoffwechsel von Tomaten, Sellerie und Reben. II. Der Einfluss auf die Trockensubstanz-und kohlenhydratbildung sowie den Ascorbinsäuregehalt, *Gartenbauwissenschaft*, 27, 103, 1962.

88. Matzner, F., Vitamin-C-Gehalt in Früchten der "Schattenmorelle," *Erwebsobstbau*, 18(6), 83, 1976.

89. Salama, F. M., Khodary, S. E. A., and Heikal, M., Effect of soil salinity and IAA on growth, photosynthetic pigments, and mineral composition of tomato and rocket plants, *Phyton* (*Austria*), 21, 177, 1981; *Hortic. Abstr.*, 52, 2997, 1982.

90. Freeman, J. A. and Harris, G. H., The effect of nitrogen, phosphorus, potassium and chlorine on the carotene content of the carrot, *Sci. Agric.*, 31, 207, 1951.

91. Ben-Amotz, A. and Avron, M., On the factors which determine massive β-carotene accumulation in the halotolerant *Dunaliella bardawil*, *Plant Physiol.*, 72, 593, 1983.

92. L'Hermite, P., Ed., Processing and use of organic sludge and liquid agricultural wastes, Reidel Publishing, Dordrecht, 1986.

93. Forster, C. F. and Senior, E., Solid wastes, in *Environmental Biotechnology*, Forster, C. F., Ed., Ellis Horwood, New York, 1987, Chap. 5.

94. L'Hermite, P. and Dehandtschutter, J., *Copper in Animal Wastes and Sewage Sludge*, Reidel Publishing, Dordrecht, 1981.

95. Klessa, D. A., Golightly, R. D., Dixon, J., and Voss, R. C., The effects upon soil and herbage Cu and Zn of applying pig slurry to grassland, *Res. Develop. Agric.*, 2(3), 135, 1985.

96. Page, A. L., Logan, T. J., and Ryan, J. A., *Land Application of Sludge*, Lewis Publishers, Chelsea, Mich., 1987.

97. Bitton, G., Damron, B. L., Edds, G. T., and Davidson, J. M., *Sludge—Health Risks of Land Application*, Ann Arbor Science Publisher, Ann Arbor, Mich., 1980.

98. Foy, C. D., Chaney, R. L., and White, M. C., The physiology of metal tolerance in plants, *Annu. Rev. Plant Physiol.*, 29, 511, 1978.

99. Shaw, A. J., *Heavy Metal Tolerance in Plants: Evolutionary Aspects*, CRC Press, Boca Raton, Fla., 1990.

100. Lyon, C. B. and Beeson, K. C., Influence of toxic concentration of micro-nutrient elements in the nutrient medium on vitamin concept of turnips and tomatoes, *Bot. Gaz.* (*Chicago*), 109, 506, 1948.

101. Livdane, B. and Ozolina, G., Activity of ascorbate oxidase and content of ascorbic acid in plant leaves differently supplied with copper, *Regul. Rosta Metab. Rast.*, 221, 1983; *Chem. Abstr.*, 100, 208540s, 1984.

102. Adedeji, F. O. and Fanimokun, V. O., Copper deficiency and toxicity in two tropical leaf vegetables (*Celosia argentea* L. and *Amaranthus dubius* Mart. ex Thell), *Environ. Exper. Bot.*, 24(1), 105, 1984.

103. Salgare, S. A. and Andhyaraujina, K. B., Effect of polluted water of Ulhaw River on the vitamin content of its vegetation—III. *J. Recent Adv. Appl. Sci.*, 4(1), 570, 1989; *Chem. Abstr.*, 113, 197332r, 1990.

104. Kiss, J., Fugedi, K. K., and Meszaros, M. H., Effect of irrigation water polluted with different chemicals on cultivated plants. II. Effect of motor-oil and sodium salt of 2, 4-D, *Tiscia*, 18, 9, 1983; *Chem. Abstr.*, 102, 108094x, 1985.

105. Stadelmann, F. X., Frossard, R., Furrer, O. J., Lehmann, L., and Moeri, P. B., Wirkung und Nachwirkung langjähriger hoher Klärschlamm- und Schweine-güllegaben auf die Qualität von Knollensellerie, *VDLUFA-Schriftenreihe*, *Kongress-band 1987*, 23, 857, 1988.

106. Maync, A. and Venter, F., Quality of vegetables as affected by high amounts of sewage sludge application. I. Influence of increasing sewage sludge application on components affecting quality of head lettuce, *Gartenbauwissenschaft*, 46(2), 79, 1981.

107. Maync, A. and Venter, F., Quality of vegetables as affected by high amounts of sewage sludge application. II. Influence of increasing sewage sludge application on components affecting quality of kohlrabi, *Gartenbauwissenschaft*, 48(4), 145, 1983.

108. Stadelmann, F. X., Frossard, R., and Moeri, P. B., Einfluss von Cadmium auf Ertrag, Physiologische Eigenschaften und Qualität von Ital. Raygras, Rettich und Spinat, *VDLUFA-Schriftenreihe*, *Kongressband 1985*, 16, 575, 1986.

109. Mondy, N. I., Naylor, L. M., and Phillips, J. C., Quality of potatoes grown in soils amended with sewage sludge, *J. Agric. Food Chem.*, 33, 229, 1985.

110. Dolinko, V. V., Reznik, A. P., Kachina, S. N., Bazarskii, Ya. E., Degodyuk, E. G., Garmatyuk, G. A., Chegrinets, G. Ya., Golyuga, V. A., Sidorov, V. N., and Odukha, V. I., Fertilizer from sewage sludge and pulping waste, *Bum. Prom-st.*, 9, 10, 1990; *Chem. Abstr.*, 115, 7527d, 1991.

111. Kaznachel, R. Ya., Solomko, G. I., and Konoko, L. N., Chemical composition of potatoes grown on plots fertilized with effluent solids, *Gigiena Sanitariya*, 36(9), 107, 1971; *Food Sci. Technol. Abstr.*, 4(4), J605, 1972.

112. Surkova, G. V., Hygienic standards for mercury and lead mixtures in soil, *Gig. Sanit.*, 4, 11, 1987; *Chem. Abstr.*, 107, 2149g, 1987.

113. Nemeth, M. and Meszaros, M., The effect of chromium-containing petroleum drilling muds on cultivated plants, *Bot. Kozl.*, 74-75(3-4), 427, 1987; *Chem. Abstr.*, 113, 151375d, 1990.

114. Dugger, M., Ed., *Air Pollution Effects on Plant Growth*, ACS Symposium series, No. 3, American Chemical Society, Washington, D.C., 1974.

115. Troyanowsky, C., Ed., *Air Pollution and Plants*, VCH Verlagsgesellschaft, Weinheim, Germany, 1985.

116. Hutchinson, T. C. and Meema, K. M., *Effects of Atmospheric Pollutants on Forest, Wetland and Agricultural Ecosystems*, Springer-Verlag, Berlin, 1987.

117. Schulte-Hostede, S., Darrall, N. M., Blank, L. W., and Wellburn, A. R., *Air Pollution and Plant Metabolism*, Elsevier Appl. Sci., London, 1988.

118. Treshow, M. and Anderson, F. K., *Plant Stress from Air Pollution*, Wiley, Chichester, 1989.

119. Barnes, R. L., Effect of chronic exposure to ozone on soluble sugar and ascorbic acid content of pine seedlings, *Can. J. Bot.*, 50, 215, 1972.

120. Saxena, L. M., Air pollution induced changes in foliar pigment and ascorbic acid content in *Bougainvillea spectabilis* Willd. and *Cassia siamea* Linn., *Acta Ecol.*, 7(1), 20, 1985; *Hortic. Abstr.*, 58, 7938, 1988.

121. Bacic, T. and Gligorijevic, S., Effect of air pollution on ascorbic acid content in fruits of some mediterranean trees, *Acta Biol. Med. Exp.* 14(2), 151, 1989; *Chem. Abstr.*, 115, 77652q, 1991.

122. Bessnova, V. P., Kozyukina, Zh. T., and Lyzhenko, I., The effect of technogenic conditions on the ascorbic acid and glutathione concentration in leaves of different plants, *Ukr. Bot. Zh.* 46(3), 83, 1989; *Chem. Abstr.*, 112, 41626h, 1990.

123. Krishnamurthy, R., Babu, R., Anbazhagan, M., Banerjee, G., and Bhagwat, K. A., Physiological studies on the leaves of some avenue trees, exposed to chronic low levels of air pollutants, *Indian For.*, 112(6), 503, 1986; *Chem. Abstr.*, 106, 135354w, 1987.

124. Flückiger, W. and Flückiger-Keller, H., Veränderungen im Gehalt an β-Carotin und Vitamin C in der Petersilie im nahbereich einer Autobahn, *Qual. Plant.—Plant Foods Hum. Nutr.*, 28(1), 1, 1978.

125. Pippen, E. L., Potter, A. L., Randall, V. G., Ng, K. C., Reuter III, F. W., and Morgan, A. I. Jr., Effect of ozone fumigation on crop composition, *J. Food Sci.*, 40, 672, 1975.

126. Bermadinger, G., Guttenberger, H., and Grill, D., Physiology of young Norway spruce, *Environ. Pollut.* 68(3-4), 319, 1990.

127. Melhorn, H., Tabner, B. J., and Wellburn, A. R., Electron spin resonance evidence for the formation of free radicals in plants exposed to ozone, *Physiol. Plant.*, 79, 377, 1990.

128. Senger, H., Osswald, W., Senser, M., Greim, H., and Elstner, E. F., Gehalt an Chlorophyll und den Antioxidantien Ascorbat, Glutathion und Tocopherol in Fichtennadeln (*Picea abies* [L.] Karst.) in Abhängigkeit von Mineralstoffenährung, Ozon und saurem Nebel, *Forstwissenschaftlichen Centralblatt*, 105, 264, 1986.

129. Guri, A., Variation in glutathione and ascorbic acid content among selected cultivars of *Phaseolus vulgaris* prior to and after exposure to ozone, *Can. J. Plant Sci.*, 63(3), 733, 1983.

130. Agrawal, M., Nandi, P. K., and Rao, D. N., Responses of *Vicia faba* plants to ozone pollution, *Indian J. Environ. Health*, 27(4), 318, 1985; *Hortic. Abstr.*, 56, 6968, 1986.

131. Enshina, A. N. and Voitik, N. P., Influence of regular ozone-treatment of potatoes and vegetables on their chemical composition, *Vopr. Pitan.*, 6, 61, 1989; *Chem. Abstr.*, 112, 117580m, 1990.

132. Sakaki, T. and Kondo, N., Destruction of photosynthetic pigments in ozone-fumigated spinach leaves, *Kokuritsu Kogai Kenkyusho Kenkyu Hokoku*, 28, 31, 1981; *Chem. Abstr.*, 96, 63874r, 1982.

133. Tanaka, K., Suda, Y., Kondo, N., and Sugahara, K., O_3 tolerance and the ascorbate-dependent H_2O_2 decomposing system in chloroplasts, *Plant Cell Physiol.*, 26(7), 1425, 1985.

134. Krieg, F., Einfluss der Nährstoffversorgung auf die Ozonafälligkeit der Weinrebe (*Vitis vinifera* L. cv. Pinot noir), Ph.D. thesis, institute of plant sciences, Swiss federal institute of technology, Zürich, Switzerland, 1992.

135. Mehlhorn, H., Seufert, G., Schmidt, A., and Kunert, K. J., Effect of SO_2 and O_3 on production of antioxidants in conifers, *Plant Physiol.*, 82, 336, 1986.

136. Prasad, B. J., and Rao, D. N., Growth responses of *Phaseolus aureus* plants to petro-coke pollution, *J. Exp. Bot.*, 32(131), 1343, 1981; *Chem. Abstr.*, 96, 194615e, 1982.

137. Berzina, A., Effect of cement dust on the growth and chemical composition of cultivated plants and pine seedlings, *Zagryaz. Prir. Sredy Kal'tsiisoderzh. Pyl'yu*, 73, 1985; *Chem. Abstr.*, 103, 87056q, 1985.

138. Agrawal, M. and Agrawal, S. B., Phytomonitoring of air pollution around a thermal power plant, *Atmos. Environ.*, 23(4), 763, 1988; *Chem. Abstr.*, 111, 11876b, 1989.

139. Das, T. M., Effect of deposition of dust particles on leaves of crop plants on screening of solar illumination and associated physiological processes, *Environ. Pollution*, 53, 421, 1988.

140. Osswald, W. F. and Elstner, E. F., Comparative studies on spruce diseases in sulfur dioxide exposed forest areas and unpolluted areas, *Spez. Ber. Kernforschungsanlage Juelich*, Juel-Spez-369, 54, 1986; *Chem. Abstr.*, 106, 143247d, 1987.

141. Sandmann, G. and Gamez Gonzales, H., Peroxidative processes induced in bean leaves by fumigation with sulfur dioxide, *Environ. Pollut.*, 56(2), 145, 1989.

142. Murray, F., Some responses of ladino clover (*Trifolium repens* L. cv. Regal) to low concentrations of sulfur dioxide, *New Phytol.*, 100(1), 57, 1985.

143. Keller, Th. and Schwager, H., Air pollution and ascorbic acid, *Eur. J. Forest Pathol.*, 7, 338, 1977.

144. Murray, F., Changes in growth and quality characteristics of lucerne (*Medicago sativa* L.) in response to sulfur dioxide exposure under field conditions, *J. Exp. Bot.*, 36(164), 449, 1985.

145. Nandi, P. K., Agrawal, M., and Rao, D. N., Sulfur dioxide-induced enzymatic changes and ascorbic acid oxidation in *Oryza sativa*, *Water, Air, Soil Pollut.*, 21(1-4), 25, 1984; *Chem. Abstr.*, 100, 133758h, 1984.

146. Agrawal, S. B., Agrawal, M., Nandi, P. K., and Rao, D. N., Effects of sulfur dioxide and quinalphos singly and in combination on the metabolic functions and growth of *Oryza sativa* L., *Proc. Indian Natl. Sci. Acad., Part B*, 53(2), 169, 1987.

147. Materna, J., Einfluss niedriger Schwefeldioxydkonzentrationen auf die Fichte, *Mitteil. der Forstlichen Bundesversuchsanstalt (Wien)*, 97, 219, 1972.

148. Grill, D., Esterbauer, H., and Welt, R., Einfluss von SO_2 auf das Ascorbinsäuresystem der Fichtennadeln, *Phytopath Z.*, 96, 361, 1979.

149. Keller, T., Effect of winter fumigation with sulfur dioxide on physiological processes of spruces, *Mitt. Forstl. Bundesversuchsanst.* (*Wien*), 137(1), 1981; *Chem. Abstr.*, 96, 212027w, 1982.

150. Varshney, S. R. K. and Varshney, C. K., Effect of SO_2 on ascorbic acid in crop plants, *Environ. Pollution* (*Ser. A*), 35(4), 285, 1984.

151. Davieson, G., Murray, F., and Wilson, S., Effects of sulfur dioxide and hydrogen fluoride, singly and in combination, on growth and yield of wheat in open-top chambers, *Agric. Ecosyst. Environ.* 30(3-4), 317, 1990.

152. Singh, N., and Rao, D. N., Effects of sulfur dioxide on injury and foliar concentrations of pigments, ascorbic acid and sulfur in *Vinga radiata* (L), *I. Environ. Biol.*, 9 (Suppl. 1), 107, 1988.

153. Shimazaki, K., Sakaki, T., Kondo, N., and Sugahara, K., Active oxygen participation in chlorophyll destruction and lipid peroxidation in SO_2-fumigated leaves of spinach, *Plant Cell Physiol.*, 21(7), 1193, 1980.

154. Unzicker, H. J., Jäger, J. H., and Steubing, L., Einfluss von SO_2 auf den Vitamingehalt von Pflanzen, *Angew. Botanik*, 49, 131, 1975.

155. Agrawal, S. B. and Agrawal, M., Relative susceptibility of certain tree plants to gaseous ammonia, *Pollut. Res.*, 7(1-2), 1, 1988; *Chem. Abstr.*, 110, 226811s, 1989.

156. Erval'd, M. A. and Bychkova, Z. N., The effect of nitrogen dioxide on some physiological processes in plants, *Sbornik Nauchnykh Rabot, Ryazanskii Sel'skokhozyaistvennyi Institut*, 27(1/3), 125, 1971; *Hortic. Abstr.*, 42, 7703, 1972.

157. Lee, E. H., Plant resistance mechanisms to air pollutants; rhythms in ascorbic acid production during growth under ozone stress, *Chronobiol. Int.*, 8(2), 93, 1991.

158. Tanaka, K., Suda, Y., Kondo, N., and Sugahara, K., Ozone tolerance and substances protecting against active oxygen toxicity, *Kokuritsu Kogai Kenkyusho Kenkyu Hokoku*, 82, 167, 1985; *Chem. Abstr.*, 103, 82970y, 1985.

160. Opopol, N. I. and Kushnir, G. V., Toxicological hygenic aspects of the use of ozone during fruit storage, *Izv. Akad. Nauk Mold. SSR, Ser. Biol. Khim. Nauk*, 6, 58, 1985; *Chem. Abstr.*, 104, 205682m, 1986.

161. Kuno, H., Characteristics of spinach leaf injury by photochemical oxidants and mechanisms of ozone resistance in spinach cultivars. IV. Effect of low ozone levels on enzymes and substances participating in the detoxication of active oxygen, *Taiki Osen Gakkaishi*, 24(4), 259, 1989; *Chem. Abstr.*, 112, 2299j, 1990.

162. Madamanchi, N. R., Hausladen, A., Alscher, R. G., Amundson, R. G., and Fellows, S., Seasonal changes in antioxidants in red spruce (*Picea rubens* Sarg.) from three field sites in the northern United States, *New Phytol.*, 118, 331, 1991.

163. Osswald, W. F., Senger, H., and Elstner, E. F., Ascorbic acid and glutathione contents of spruce needles from different locations in Bavaria, *Naturforsch*, 42c, 879, 1987.

164. Messner, B. and Berndt, J., Ascorbic acid and chlorophyll content in cell cultures of spruce (*Picea abies*): changes by cell culture conditions and air pollutants, *Z. Naturforsch*, 45c, 614, 1990.

165. Lotstein, R. J., Davis, D. D., and Pell, E. J., Quality of tomatoes harvested from plants receiving chronic exposure to sulfur dioxide, *HortScience*, 18(1), 72, 1983.

166. Huttunen, S. and Karhu, M., Effect of air pollution and other factors on ascorbic acid content of blueberries and lingoberries, *Angew. Botan.*, 60, 277, 1986.

Chapter 8

EFFECT OF PESTS AND PATHOGENS

I. ACCELERATED VITAMIN LOSS

Plants attacked by various pests may have lower content of some vitamins, especially vitamin C (Table 1). This is apparently due to an accelerated rate of vitamin loss (over that in the healthy tissues), which sets in when plant tissues are attacked by foreign organisms.[43]

Infection of apples,[44] for example, with pathogenic fungi was noted to strongly accelerate the loss of ascorbic acid in the fruits (Table 2). Infection of oranges with *Aspergillus flavus*[9] or with *Curvularia lunata*[45] reduced their ascorbic acid in some cases by as much as 55% after 12 days of infection.[45] Mango infected with *Botryodiplodia* lost almost 100% of its ascorbic acid during a 10-day period, while the healthy fruits lost only 32.6% of their vitamin.[14] Infection of tomato plant and tomato fruits with *Alternaria solani* reduced the ascorbic acid content of the fruits by 30–40%.[6] Infection of tomato fruits with *Nigrospora oryzae*, *Phoma exigua*, *Rhizoctonia solani*, or *Stemphylium vesicarium* reduced the ascorbic acid content of the fruits to zero or near-zero levels.[46] In sweet potatoes, infection of roots with soft rot fungus (*Rhizopus stolonifer*) reduced the carotene by half (50 vs. 24 mg/100 g) and the ascorbic acid to 2/3 (32 vs. 20 mg/100 g in the healthy and infected plants, respectively).[47]

Virus infection can also alter the concentration of vitamin C in plants. In radish, for example, the concentration of ascorbic acid increases in the leaves and in the roots as the plant matures. The rate of increase, however, was noted to be much slower in the plants infected with radish mosaic virus.[48] A reduction of 79 and 66% was noted in the lettuce leaves infected with leaf mosaic virus (LMV) at the 5–7 leaf stage or heading stage, respectively.[49]

Some pathogens may also increase the concentration of some plant vitamins (Table 1). For example, mosaic virus was noted to increase the ascorbic acid concentration in the tubers of 13 of 17 potato varieties one month after storage.[35] Also, infection of tomato plant with tobacco mosaic virus (TMV) was found to increase the ascorbic acid content in its stem and leaves.[38] Infection of wheat with stem rust disease (*Puccinia graminis*) increased the concentration of pantothenic acid and niacin in the seedlings, especially as pustules developed. Since rust urediospores were found to contain 7–10 times more vitamins than the plant tissues, it was thus concluded that the higher concentrations of

TABLE 1
Effect of Pathogenic Organisms on the Concentration of Vitamins in Plant

Organism	Effect	Plant	Ref.
	Ascorbic Acid		
Cytospora	Increase	Peaches (leaf)	1
Glomus fasciculatus	Increase	Chili	2
Meloidogyne incognita	Increase	Tomato root	3
Puccinia arachidis	Increase	Peanut	4
Acremonium recifei	Decrease	Tomato	5
Alternaria solani	Decrease	Tomato	6
Aspergillus flavus	Decrease	Apples	7
A. flavus	Decrease	Bananas	8
A. flavus	Decrease	Oranges	9, 10
A. flavus	Decrease	Pear	11
A. flavus	Decrease	Tomato	5
A. niger	Decrease	Lime	12
A. niger	Decrease	Tomato	5
A. parasiticus	Decrease	Apples	7
A. parasiticus	Decrease	Oranges	9, 10
A. parasiticus	Decrease	Pear	11
Botryodiplodia theobromae	Decrease	Bananas	8
B. theobromae	Decrease	Oranges	13
B. theobromae	Decrease	Mango	14
B. theobromae	Decrease	Mosambi	15
B. theobromae	Decrease	Guava	16
Colletotrichum gloeosporioides	Decrease	Guava	16
Curvularia lunata	Decrease	Pointed gourd	17
Drechslera australiense	Decrease	Tomato	18
Fusarium	Decrease	Cotton	19
F. chlamydosporium	Decrease	Tomato	5
F. equiseti	Decrease	Guava	16
F. equiseti	Decrease	Pointed gourd	17
F. equiseti	Decrease	Tomato	5
F. oxysporium	Decrease	Bananas	8
Geotrichum candidum	Decrease	Round gourds	20
G. candidum	Decrease	Tomato	5
Helminthosporium speciferum	Decrease	Bananas	21
Macrophoma mantegazziana	Decrease	Citrus (leaf)	22
Pseudopeziza rubiae	Decrease	*Rubia cordifolia*	23
Phoma exigua	Decrease	Alma (*Phyllanthus emblica*)	24
Phomopsis citri	Decrease	Oranges	13
P. mangiferae	Decrease	Mango	24
P. viticola	Decrease	Grape	25
Phytophthora infestans	Decrease	Tomato	26
Rhizoctonia solani	Decrease	Guava	16
Rhizopus stolonifer	Decrease	Peaches	27
Sclerotinia sclerotiorum	Decrease	Cauliflower	28
Sclerotium rolfsii	Decrease	Tomato	29
Thielaviopsis paradoxa	Decrease	Oranges	13
Uncinula necator	Decrease	Grape	30

TABLE 1 (continued)

Organism	Effect	Plant	Ref.
Fusarium moniliforme	No effect	Pineapple	31
Ustilago scitaminea	No effect	Sugarcane	32
chlorotic spot virus	Decrease	Peanut (leaf)	33
Cucumber mosaic virus	Decrease	Cape gooseberry	34
Mosaic virus	Increase	Potato	35
Tobacco mosaic virus (TMV)	Increase	Tomato	38
Striate mosaic virus	No effect	Sugarcane	36
TEV virus	Decrease	Tabasco pepper root	37
Virus	Decrease	Apples (leaf)	39

Biotin

Organism	Effect	Plant	Ref.
Puccinia striiformis	Increase	Wheat (leaf)	41
P. graminis	Increase	Wheat (leaf)	41

Carotene

Organism	Effect	Plant	Ref.
Callosobruchus maculantus	Decrease	Cowpea	42

Niacin

Organism	Effect	Plant	Ref.
Aspergillus flavus	Increase	Wheat (spring)	40
Fusarium	Decrease	Cotton	19

Pantothenic acid

Organism	Effect	Plant	Ref.
Aspergillus flavus	Increase	Wheat kernel	40
Puccinia striiformis	Increase	Wheat (leaf)	41
P. graminis	Increase	Wheat (leaf)	41

Pyridoxine

Organism	Effect	Plant	Ref.
Aspergillus flavus	Increase	Wheat kernel	40

Riboflavin

Organism	Effect	Plant	Ref.
Aspergillus flavus	Increase	Wheat kernel	40
Callosobruchus maculantus	Decrease	Cowpea	42

Thiamin

Organism	Effect	Plant	Ref.
Aspergillus flavus	Decrease	Wheat kernel	40
Puccinia striiformis	Increase	Wheat (leaf)	41
P. graminis	Increase	Wheat (leaf)	41
Callosobruchus maculantus	Decrease	Cowpea	42

TABLE 2
Ascorbic Acid Content of Apples (Variety Kesari)
Infected with *Aspergillus niger*[44]

Incubation period (day)	Ascorbic acid (mg / 100 g pulp)	
	Healthy	Infected
0	2.08	—
3	2.00	0.79
6	2.00	0.43
9	1.8	0.00
12	1.7	0.00

vitamins in the infected plants were mainly due to the high vitamin concentration in the urediospores.[50]

Also some insects appear to alter the synthesis and/or translocation of vitamins in plants. For example, infection of wheat with aphids (*Sitobion avenae*) was noted to increase the niacin and thiamin content of the wheat flower.[51] By considering the relative distribution of thiamin and niacin in the different parts of the wheat kernel (chapter 2), this observation seems to indicate that aphid infestation had either increased the synthesis and/or transport of these vitamins from the leaves to the kernels or had altered the relative distribution of vitamins between different parts of the wheat kernel.

From a single known report of its kind, it appears that some microorganisms employed for the biological control of insects may also affect the content of plant vitamins. Jacob et al.[52] reported that the spraying of peach trees with the bacterial preparation Thuricide and Dipel (containing the *Bacillus thuringiensis*) to control the *Grapholitha molesta* increased the ascorbic acid content in the fruits. This observation is hard to interpret unless the mere presence of this microorganism on (or in) the plant tissues, although not patheogenic to the tree itself, triggered the plant to produce more ascorbic acid.

II. VITAMIN CONTENT AND SUSCEPTIBILITY TO PESTS AND PATHOGENS

In analogy to the currently controversial question in human medicine, one may also raise the question of whether there is any relationship between the plants' content of certain vitamins (such as vitamin C) and the degree of their resistance to various stress factors, including pathogenic organisms. This, we think, is a justified question since the infection of plants with fungi is noted to increase the production of oxygen free radicals in the infected tissues,[53] and an increased peroxidation of lipids has been noted to occur in several instances in the plants infected with various pathogens.[54] In this context one may ask: (a) What is the pathological significance of the often-observed accelerated loss of vitamin C in the infected plant tissues? and (b) Does a higher content of antioxidant vitamins in a plant tissue make it less susceptible to attack by microorganisms? Vidhyasekaran,[55] in a review of this subject, noted that ascorbic acid is needed for the development of cyanide-resistant respiration, a kind of respiration that is characteristic of resistant reactions in plants. The observation, however, that in some cases high ascorbic acid content in plants apparently suppresses the development of disease symptoms while in other cases it may result in enhanced susceptibility of host to pathogens[55] points to a host-pest specificity in this respect.

Infection of tomato with nematode (*Meloidogyne incognita*), for example, was found to reduce the ascorbic acid content of the fruits. Interestingly, however, the variety resistant to nematode was noted to retain more (lose less) of its original ascorbic acid than the variety susceptible to the nematode.[56] This might indicate that the ability of a plant to maintain a given level of vitamin C in its cells and its ability to resist the attack by nematode may be related. Such a relationship was also observed in the susceptibility of mango fruits to bacterial black spot infection caused by *Erwinia mangiferae*. It was noted that bacterial infection reduced the vitamin C content of fruits. The variety Sensation, which contained a higher concentration of ascorbic acid than the other three varieties tested, was also more tolerant to the bacterial infection. It was therefore concluded that an ascorbic acid concentration of 40–50 mg/100 g or higher in fruits renders them more tolerant (less susceptible) to the necrosis caused by the bacterial infection.[57,58]

Rattan and Saini[59] studied the relationships between the ascorbic acid content of five tomato cultivars differing in their susceptibilty to *Phytophthora parasitica* and noted that high fruit ascorbic acid (31–34 mg/100 g) and disease resistance were somehow associated since fruits of the cultivars susceptible to this fungi had lower ascorbic acid (17–22 mg/100 g). Also, Plakhova[60] reported that rose cultivars resistant to rust disease caused by *Phramidium butleri* contained higher ascorbic acid in their young leaves than the less resistant cultivars.

In contrast to the above, some plants may be attracked more by some organisms if they are high in certain vitamins. Thus Singh et al.[61] noted a positive correlation between the ascorbic acid content of tomato fruits and the degree of damage they received by the fruit borer (*Heliothis armigera* Hubner). Huang and Huang[62] reported that rice varieties containing less ascorbic acid in their leaves are actually more resistant to the blast diseases caused by *Pyricularia oryzae*. Also, Reddy and Khare[4] observed that under healthy conditions, peanut variety Jyothi, which is susceptible to rust disease (caused by *Puccinia arachidis*), has a relatively higher concentration of ascorbic acid in its leaves than the variety IGG-1697, which is resistant to this rust. When plants were infected with the fungi, however, the ascorbic acid was found to increase in the leaves of the susceptible variety and to decrease in the leaves of the resistant variety when compared with the corresponding healthy plants. Also, wheat varieties susceptible to rust disease were reported to be especially high in several vitamins (names not mentioned) as compared with the resistant varieties. It was postulated that the high content of some vitamins at the flowering stage creates a favorable condition for stem rust development.[63]

Some of the inconsistencies in the higher or lower vitamin content in the disease-infected plants as compared with the healthy counterparts may be due to the time after infection when measurements were made. For example, Milo and Smith[64] noted that infection of beans with TMV resulted in an initial rise in the leaf ascorbic acid followed by a rapid reduction prior to development of the lesions in the leaves.

III. SUMMARY

The above-cited reports, along with the observation that ascorbic acid sprayed onto tomato plants decreases the rate of penetration of root knot nematodes and their multiplication[65] and reduces the infectivity of cowpea banding virus,[66] appear to point to some possible role(s) that ascorbic acid, and/or other vitamins, may play in the plant's defense mechanism against some organisms. Thus those plants that could maintain a higher ascorbic acid in their tissue when infected with a foreign organism may be more resistant to the attack than others, as shown in some instances cited above. Finally, the observation made by Peresypkin et al.[67] might shed further light on the possible role of ascorbic acid in plants. They noted that when cultivars of peas resistant to *Peronospora pisi* are attacked by this fungus, ascorbic acid accumulates in the tissues surrounding the infection point, a situation that apparently does not take place when susceptible cultivars are attacked by this fungus.

The question whether ascorbic acid may be among the compounds that plants appear to employ in their first lines of defense against the intruding pathogens is intriguing. Further research is, however, needed to test this observation. Along this line of thought it is worth noting that there appears to be some relationship between mineral-nutrition status of plants, their susceptibility to various insects and microorganisms, and

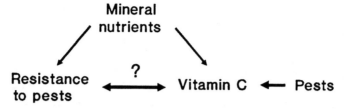

FIGURE 1. Relationships between the supply of mineral nutrients and the ascorbic acid content in plants and the resistance of plants to pests and pathogens.

the content of ascorbic acid in the plants. Thus nitrogen and potassium fertilizer are often noted to respectively decrease and increase the incidence of plants' attack by various plant pests (insects and various microorganisms).[68-72] At the same time, the numerous observations that some mineral nutrients, especially nitrogen and potassium, decrease and increase the ascorbic acid content in many different plants, respectively (chapter 5), point to possible interrelationships between some mineral nutrients, vitamin C content, and susceptibility for attack by some plant pests (Figure 1). Although the parallelism between the effect of mineral nutrients on susceptibility of attack by pests and pathogens, on the one hand, and the respective effects on the ascorbic acid content in plants does not prove any cause-and-effect relationships by itself, the overwhelming number of reports in this area are hard to ignore. Further research, however, is needed to establish an unequivocal relationship between the vitamin C content and the susceptibility to pests and pathogens in plants. In this respect, information obtained in animals and in plants may prove to complement each other for a better understanding of the mechanisms involved.

REFERENCES

1. Zavarzin, V. I., *Cytospora* and the desiccation of peach trees, *Sadovodstvo*, 3, 25, 1965; *Hortic. Abstr.*, 35, 7532, 1965.
2. Bagyaraj, D. J . and Sreeramulu, K. R., Preinoculation with VA mycorrhiza improves growth and yield a chilli transplanted in the field and saves phosphatic fertilizer, *Plant Soil*, 69(3), 375, 1982.
3. Kannan, S., Studies in nematode infected root-knots of the tomato plant, *Indian J. Exp. Biol.*, 6, 153, 1968; *Hortic. Abstr.*, 39, 5038, 1969.
4. Reddy, P. N. and Khare, M. N., Physiology of groundnut rust disease: changes in total sugars, phenols, ascorbic acid, peroxidase and phenol oxidase, *J. Oilseeds Res.*, 5, 102, 1988.
5. Oladiran, A. O. and Iwu, L. N., Changes in ascorbic acid and carbohydrate contents in tomato fruits infected with pathogens, *Plant Foods Hum. Nutr.*, 42, 373, 1992.

6. Reda, F., Mousa, O. M., Sejiny, M. J., and Nawar, L. S., Responses of tomato cultivars to infection with *Alternaria solani* in relation to growth and chemical composition in their fruits, *Egypt. J. Phytopathol.*, 17(2), 83, 1985; *Chem. Abstr.*, 107, 55887c, 1987.

7. Sinha, K. K. and Singh, A., Chemical changes in apples due to aflatoxin producing aspergilli, *J. Food Sci. Technol.*, 24(1), 44, 1987; *Chem. Abstr.*, 107, 114398u, 1987.

8. Singh, H. N. P., Prasad, M. M., and Roy, A. K., Sugar and vitamin level in chinia variety of banana under pathogenesis, *Natl. Acad. Sci. Lett.* (*India*), 14(12), 459, 1991; *Chem. Abstr.*, 117, 68765t, 1992.

9. Singh, A., and Sinha, K. K., Biochemical changes in musambi fruits innoculated with species of aflatoxin producing aspergilli, *Curr. Sci.*, 51(17), 841, 1982; *Chem. Abstr.*, 97, 195968e, 1982.

10. Sinha, K. K. and Singh, A., Biochemical changes in orange fruits during infestation with *Aspergillus flavus* and *Aspergillus parasiticus*, *Natl. Acad. Sci. Lett.* (*India*), 5(5), 143, 1982; *Chem. Abstr.*, 98, 195860g, 1983.

11. Sinha, K. K., and Singh, A., Chemical changes in pear fruits due to aflatoxin producing aspergilli, *Indian Phytopathol.*, 37(3), 545, 1984.

12. Reddy, B. C., Reddy, P. V., and Raju, D. G., Post-infection changes in acid lime fruits caused by *Aspergillus niger*, *Indian Phytophtol.*, 37(1), 185, 1984.

13. Gautam, S. P., Rajak, R. K., and Malaviya, N., Changes in "vitamin C" content of citrus fruits during pathogenesis by *Phomopsis citri*, *Botryodipoldia theobromae* and *Thielaviopsis paradoxa*, *Comp. Physiol. Ecol.*, 10(1), 26, 1985; *Chem. Abstr.*, 103, 85249f, 1985.

14. Srivastava, M. P. and Tandon, R. N., Effect of *Botryodiplodia* infection on the vitamin C content of mango fruit, *Sci. Cult.*, 35, 419, 1966; *Hortic. Abstr.*, 37, 3907, 1967.

15. Srivastava, M. P. and Tandon, R. N., Changes in ascorbic acid content of mosambi fruit induced by *Botryodipolodia theobromae*, *Can. J. Plant Sci.*, 48, 337, 1968; *Hortic. Abstr.*, 38, 8513, 1968.

16. Adisa, V. A., Metabolic changes in post-infected guava fruits, *Fitopatol. Bras.*, 8(1), 81, 1983; *Chem. Abstr.*, 100, 65193t, 1984.

17. Prasad, A. K. and Roy, A. K., Biochemical changes in pointed gourd fruit under pathogeneis, *Indian Bot. Rep.*, 5(1), 83, 1986; *Chem. Abstr.*, 106, 81739e, 1987.

18. Kapoor, I. J. and Tandon, R. N., Post-infection changes in ascorbic acid content of tomato fruits caused by *Drechslera australiense*, *Curr. Sci.*, 38, 397, 1969.

19. Polyanskaya, L. A., Variation in the vitamin content of cotton plant infected with Fusarium wilt, *Izv. Akad. Nauk Turkm. SSR, Ser. Biol. Nauk*, 2, 75, 1970; *Chem. Abstr.*, 73, 73937c., 1970.

20. Sumbali, G. and Mehrotra, R. S., Post-infection chemical changes in round gourds infected with *Geotrichum candidum*, *Indian J. Mycol. Plant Pathol.*, 12(1), 48, 1983.

21. Prasad, M. M., Post-infection changes in vitamin C content of banana fruits, *Cur. Sci.*, 46(6), 197, 1977; *Hortic. Abstr.*, 48, 919, 1978.

22. Giorbelidze, A. A., Effect of fungus causing gummosis in citrus, *Macrophoma mantegazziana* (Penz.) Berl. & Vogl., on some physiological processes in lemon, *Subtrop. Kul't.*, 3, 143, 1985; *Plant Breeding Abstr.*, 56, 3189, 1986.

23. Nagraja, T. G. and Thite, A. N., Some physiological changes in leaves of *Rubia cordifolia* Linn. under pathogenesis, *Biovigyanam*, 14(2), 107, 1988; *Chem. Abstr.*, 112, 4809z, 1990.

24. Reddy, S. M. and Laxminarayana, P., Post-infection changes in ascorbic acid contents of mango and alma caused by two fruit-rot fungi, *Curr. Sci.* 53(17), 927, 1984.

25. Arya, A., and Lal, B., Biochemical changes in grapes infected wtih *Phomopsis viticola*, *J. Plant Sci. Res*, 2(1–4), 53, 1986; *Chem. Abstr.*, 107, 195060c, 1987.

26. Gladilovič, B. R. and Drel, R. I., The effect of late blight infection on the chemical composition of tomatoes, *Zap. Leningr. Sel'.-hoz. Inst.*, 100, 155, 1965; *Hortic. Abstr.*, 36, 3141, 1966.

27. Singh, R. S. and Prashar, M., Studies on metabolic changes in amino acid, sugars, total acidity and vitamin C in peach fruit infected with *Rhizopus stolonifer*, *Indian Phytopathol.*, 37(2), 334, 1985.

28. Sharma, S. L., Sharma, R. C., and Sharma, I., Biochemical changes in cauliflower infected by *Sclerotinia sclerotiorum* (Lib.) de Bary, *J. Res. (Punjab Agric. Univ.)*, 22(4), 679, 1985; *Chem. Abstr.*, 106, 30165g, 1987.

29. Prasada, B. K., Sinha, T. S. P., and Shanker, U., Biochemical changes in tomato fruits caused by *Sclerotium rolfsii*, *Indian J. Mycol. Plant Pathol.*, 17(3), 316, 1987.

30. Ghure, T. K. and Shinde, P. A., Effects of powdery mildew disease caused by *Uncinula necator* (Schew.) Burr on the quality of grape berries, *J. Maharashtra Agric. Univ.*, 12(3), 400, 1987; *Chem. Abstr.*, 108, 201907y, 1988.

31. Chalfoun, S. M. and de Carvalho, V. D., Effect of *Fusarium* infection on physical, chemical and physicochemical characteristics of Smooth Cayenne cv. pineapples, *Hamm Wopecuaria Brasileira*, 17(7), 1031, 1982; *Food Sci. Technol. Abstr.*, 16(3), J497, 1984.

32. Padmanaban, P., Alexander, K. C., and Shanmugam N., Some metabolic changes induced in sugarcane by *Ustilago scitaminea*, *Indian Phytopathol.*, 41(2), 229, 1988.

33. Sreenivasulu, P., and Nayudu, M. V., Influence of groundnut chlorotic spot virus (GCSV) on ascorbic acid metabolism of *Arachis hypogaea*, *Indian Phytopathol.*, 35(1), 133, 1982.

34. Joshi, R. D., Dubey, L. N., and Gupta, A. K., Changes in ascorbic acid, carbohydrate and nitrogen contents of cape gooseberry fruits infected with cucumber mosaic virus, *Sci. Cult.*, 43(1), 40, 1977; *Food Sci. Technol. Abstr.*, 9(7), J1048, 1977.

35. Smith, A. M. and Gillies, J., The distribution and concentration of ascorbic acid in the potato (*Salanum tuberosum*), *Biochem. J.*, 34, 1312, 1940.

36. Sreenivasulu, P., Raju, B. C., and Nayudu, M. V., Effect of striate mosaic virus infection on nitrogen fraction, phenols, ascorbic acid and oxidative enzymes of sugarcane (*Saccharum officinarum* L.) leaves, *Indian J. Microbiol.*, 21(4), 351, 1981.

37. Ghabrial, S. A. and Pirone, T. P., Physiology of tobacco etch virus-induced wilt of Tabasco peppers, *Virology*, 31, 154, 1967; *Hortic. Abstr.*, 37, 5101, 1967.

38. Matkovics, B., Szabó, L., and Varga, Sz. L., Study of host-parasite interaction in tomato plants, *Acta Biol. Szeged.* 27(1–4), 17, 1981.

39. Tuleuov, Zh. T., Effect of latent viruses on apple leaves, *Vestn. S-kh. Nauki Kaz.*, 11, 50, 1988; *Chem. Abstr.*, 110, 151452q, 1989.

40. Kao, C. and Robinson, R. J., *Aspergillus flavus* deterioration of grain: its effect on amino acides and vitamins in the whole wheat, *J. Food Sci.*, 37, 261, 1972.

41. Andreev, L. N., Role of vitamins in the resistance of plants to rust, *Abh. Akad. Wiss. DDR, Abt. Math., Naturwiss., Tech.*, 341, 1982; *Chem. Abstr.*, 99, 102412x, 1983.

42. Etokakpan, O. U., Eka, O. U., and Ifon, E. T., Chemical evaluation of the effect of pest infestation on the nutritive value of cowpeas *Vinga unguiculata*, *Food Chem.*, 12(3). 149, 1983.

43. Adisa, V. A., The influence of molds and some storage factors on the ascorbic acid content of oranges and pineapple fruits, *Food Chem.*, 22, 139, 1986.

44. Bisen, P. S., Effect of fungal infection in vitamin C content of apples, *Curr. Sci.*, 43(19), 625, 1974.

45. Agarwala, G. P. and Verma, K. S., Changes in the ascorbic acid contents in orange fruits (*Citrus reticulata* Linn) infected with *Curvularia lunata* (Wakk.) Boedjin, *Natl. Acad. Sci. Lett. (India)*, 4(10), 401, 1981; *Chem. Abstr.*, 97, 3717c, 1982.

46. Reddy, S. M., Kumar, B. P., and Reddy, S. R., Vitamin C changes in infected fruits of tomato, *Indian Phytopathol.*, 33(3), 511, 1980; *Chem. Abstr.*, 96, 196688e, 1982.

47. Thompson, D. P., Phenols, carotene and ascorbic acid of sweet potato roots infected with *Rhizopus stolonifer*, *Can J. Plant Sci.*, 59, 1177, 1979.

48. Upadhyaya, P. P., Influence of radish mosaic virus infection on ascorbic acid content of radish (*Raphanus sativus*), *Curr. Sci.*, 46(18), 647, 1977.

49. Fegla, G.I., Shawkat, A. L. B., and Ramadan, N. A., Effect of infection date of lettuce mosaic virus on seed transmission, vegetative growth, and certain content of lettuce plants, *Iraqi J. Agric. Sci. "Zanco,"* 1(1), 91, 1983; *Food Sci. Technol. Abstr.*, 17(3), J215, 1985.

50. Husain, M., Hobbs, C. D., and Futrell, M. C., Effect of infection by *Puccinia graminus tritici* on the vitamin B_6, niacin, and pantothenic acid contents of wheat plants, *Phytopathology*, 54, 502, 1964.

51. Lee, G., Stevens, D. J., Stokes, S., and Wratten, S. D., Duration of cereal aphid population and the effects on wheat yield and bread making quality, *Ann. Appl. Biol.*, 98, 169, 1981.

52. Iacob, M., Matei, I., Voica E., and Vladu, S., The influence of some treatments against *Grapholitha molesta* Busck, applied on the irrigated sands close to the Jiu River, on the quality and quantity of peach fruits, *An. Inst. Cercet. Prot. Plant.*, *Acad. Stiinte Agric. Silvice*, 16, 385, 1981; *Chem. Abstr.*, 99, 153830q, 1983.

53. Chai, H. B. and Doke, N., Superoxide anion generation: a response of potato leaves to infection with *Phytophtora infestans*, *Phytopathol*, 77(5), 645, 1987.

54. Bowler, C., Van Montagu, M., and Inzé D., Superoxide dismutase and stress tolerance, *Annu. Rev. Plant Physiol.*, 43, 83, 1992.

55. Vidhyasekaran, P., *Physiology of Disease Resistance in Plants*, Vol. II, CRC Press, Boca Raton, Fla., 1988, chapter 4.

56. Goswami, B. K., Raychaudhuri, S. P., and Paul Khuranna, S. M., Changes in vitamin C content of fruit in nematode resistant and susceptible tomato varieties with infection of nemotode and/or tobacco mosaic virus, *Z. Pflanzenkrankheiten Pflanzenschutz*, 78, 355, 1971.

57. Van Lelyveld, L. J., Bacterial black spot in mango (*Mangifera indica* L.) fruits. Ascorbic acid and the hypersensitive reaction as a means of resistance, *Agroplantae*, 7, 45, 1975.

58. Van Lelyveld, L. J., Ascorbic acid content and enzyme activities during maturation of the mango fruit and their association with bacterial black spot, *Agroplantae*, 7, 51, 1975.

59. Rattan, R. S. and Saini, S. S., Association of fruit rot resistance with ascorbic acid content in tomato (*Lycopersicon esculentum* Mill.), *Vegetable Sci.*, 6(1), 54, 1979; *Hortic. Abstr.*, 51, 7841, 1981.

60. Plakhova, T. M., On some biochemical characteristics of rose cultivars differing in resistance to rust, *Byulleten' Vsesoyuznogo Ordena Lenina Rastenievodstva Imeni N. I. Vavilova*, 38, 82, 1974; *Hortic. Abstr.*, 45, 6799, 1975.

61. Singh, D., Singh, S., Bajaj, K. L., Kaur, G., and Gill, C. K., Influence on physical and biochemical factors on the incidence of fruit borer (*Heliothis armigera* Hubner), *J. Res. (Punjab Agric. Univ.)*, 19(1), 31, 1982; *Chem. Abstr.*, 98, 86437f, 1983.

62. Huang, X. and Huang, H., Rice resistance to blast disease. I. Biochemical cause of seedling resistance, *Xiamen Daxue Xuebao, Ziran Kexueban*, 25(4), 461, 1986; *Chem. Abstr.*, 106, 15887a, 1987.

63. Andreev, L. N., Filippov, V. V., Khisarova, L. Ts., Content of vitamins in leaves of wheat plants characterized by varietal resistance to stem rust, *Vestn. Akad. Nauk Kaz. SSR*, 4, 59, 1976; *Chem. Abstr.*, 85, 43776e, 1976.

64. Milo, G. E., Jr., and Santilli, V., Changes in the ascorbate concentration of Pinto bean leaves accompanying the formation of TMV-induced local lesions, *Virology*, 31, 197, 1967; *Hortic. Abstr.*, 37, 4930, 1967.

65. Al-Sayed, A. A. and Montasser, S. A., The role of ascorbic and glutamic acids in controlling the root knot nematode *Meloidogyne javanica*, Egypt. *J. Phytopathol.*, 18(2), 143, 1986; *Chem. Abstr.*, 108, 89411p, 1988.

66. Prakash, J. and Sadruddin, Effect of different pH, incubation period and reducing agents on the cowpea banding mosaic virus and inhibitory activity of three phenolic acids, *Bangladesh J. Bot.*, 12(1), 50, 1983; *Chem. Abstr.*, 101, 67668w, 1984.

67. Peresypkin, V. F., Kirik, N. N., and Koshevskii, I. I., Accumulation of ascorbic acid in pea cultivars differing in their resistance to *Peronospora*, *Vestn. Sel'sk. Nauki*, 8, 25, 1977; *Hortic. Abstr.*, 48, 3505, 1978.

68. Dale, D., Plant-mediated effects of soil mineral stresses on insects, in *Plant Stress-Insect Interactions*, Heinricks, E. A., Ed., Wiley Intersci. Publ., New York, 1988, 35.

69. International Potash Institute, *Fertilizer Use and Plant Health*, Int. Potash Inst., Bern, Switzerland, 1976.

70. Huber, D. M. and Arny, D. C., Interactions of potassium with plant disease, in *Potassium in Agriculture*, Munson, R. D., Ed., ASA-CSSA-SSSA, Madison, Wis., 1985, 467.

71. Huber, D. M., The use of fertilizers and organic amendments in the control of plant diseases, in Pimentel, D., Ed., *CRC Handbook of Pest Management in Agriculture*, 2nd Ed., Vol. I, CRC Press, Boca Raton, Fla., 1991, 405.

72. Perrenoud, S., *Potassium and Plant Health*, Int. Potash Inst., Bern, Switzerland, 1990.

Chapter 9

PLANT PROTECTION CHEMICALS

Industrialized agriculture often requires the use of numerous chemical compounds such as herbicides, fungicides, insecticides, nematicides, and growth regulators for various purposes during the plants' course of growth and postharvest period in storage. For example, some apple orchards may be sprayed as many as 20 times during a growing season.[1] Some fruits and vegetables may even be treated with up to eight or more pesticides applied 6, 10, 20, or more times during a given growing season.[2] Whether such a repeated application of various substances on plants and the residues they may leave in the plant and/or soil affect the vitamin content of plants has been very little documented,[3-7] especially by scientists in the Western countries where the bulk of these chemicals are produced and used. The scientists in the former Soviet Union or Eastern block countries, however, seem to have paid much more attention to this subject, and thus the majority of available data on this subject are published in the non-English journals from which only their abstracts are accessible to English-speaking scientists. Unfortunately, based on the abstracts alone, one is not able to gain detailed information as to the methodology and analytical methods employed and, more important, the magnitude of the effects observed. It is thus hard to make a definitive comment about these works except by citing them (Table 1). This certainly does not do justice to the vast number of apparently extensive research programs conducted in this field by these scientists and the valuable information that has been collected. Here only some of the reports for which more detailed information was available are reviewed.

From more than 450 cases summarized in Table 1, and despite some contrasting effects found by different authors for the effect of a given compound on the concentration of a given vitamin in a given plant, it can easily be seen that in the majority of cases, the use of various plant protection chemicals either does not affect, or increases, the content of some vitamins in some of the plants tested. The number of cases in which the use of a plant protection chemical was noted to decrease plant vitamins is relatively small. As noted above, due to the limited space available, the subject of residues of plant protection chemicals in the soil or plants will not be treated here; readers can refer to other sources for this information.[283-286]

Finally, very little information is available about whether the altered vitamin content in plants (directly or indirectly) exposed to various plant protection chemicals is due to an increased concentration of these

TABLE 1
Effect of Plant Protection Compounds on the Vitamin Concentration in the Edible Parts of Plants Unless Otherwise Indicated

Compound[a]	Effect[b]	Plant	Ref.
		Ascorbic Acid	
2,4-D	I	Beans (bush)	3
2,4-D	I	Beans (green)	8
2,4-D	V	Buckwheat (leaf)	9
2,4-D	I	Carrots (cells)	10
2,4-D	I	Eggplant	11
2,4-D	I	Grapefruit	12, 13
2,4-D	I	Mango	14-16
2,4-D	I	Pepper	11
2,4-D	I	Strawberry	17
2,4-D	N	Tangerine	18
2,4-D	I	Tomato	3, 11, 19
2,4-D	D	Tomato	20
2,4-D	D	Beans (leaf)	21
2,4,5T[g]	I	Tomato	3
2,4,5-T	I	Grapefruit	12
2,4,5-T	I	Mango	15, 16, 22
2,4,5-T	D	Tomato	20
2,4,5-T	N	Tangerine	18
2,4,5-T	N	Ber	23
Acifluorfen[h]	D	Cucumber	24
Acifluorfen	D	Mustard (leaf)	25
Alar® (Daminozide)[g]	N	Grape	26
Alar	N	Tomato	27
Alar	N	Cherry	77, 78
Aldicarb (Temlik 10G®)[i]	I	Potato	28
Aminol Forte[g]	I	Oranges	30
Antitranspirant	D	Brinjal	29
Atrazine[h]	N	Apples	31
Atrazine	V	Citrus	32
Atrazine	I	Tomato	33
Benlate® (benomyl)[f]	I	Apples	34
Benlate	I	Cucumber (leaf)	35
Benlate	I	Strawberry	17
Benzyladenine[i]	I	Tomato	37
BIF-2 (Tribifos)[*,f]	I	Cauliflower	38
BIF-2[g]	I	Sweet pepper	39
BIF-36[g]	I	Sweet pepper	39
Binuron[*,h]	I	*Rosa canina*	40
Biomycin[an]	I	Tomato	41
Bordeaux mixture[f]	I	Pepper	42
Bordeaux mixture	I	Tomato	42
Captan[f]	I	Apples	43
Captan	I	Strawberry	17, 44
Caragard®[h]	D	Mandarin	45

TABLE 1 (continued)

Compound[a]	Effect[b]	Plant	Ref.
		Ascorbic Acid	
Carbendazim (Derosal®)[f]	I	Apples	34
Carbendazim	I	Strawberry	79
Carbofuran (Furadan®)[i]	I	Strawberry	17
Carbofuron	I	Chili	48
Carbofuron	I	Potato	49
Carbofuron	N	Faba beans	50
CCC (Cycocel®, chlormequat chloride, TUR*)[g]	I	Apples	51
CCC	D	Apples	59
CCC	D	Black-fruit *Sorbus*	60
CCC	I	Carrots	53
CCC	N	Cauliflower	38
CCC	I	Cucumber	54
CCC	I	Mandarin	55
CCC	D	Okra	61
CCC	I	*Phyllanthus urinaria*	52
CCC	I	Potato	56
CCC	D	Raspberry	62
CCC	D	Strawberry	62
CCC	I	Tomato	54, 57
CCC	N	Tomato	27, 63
CCC	I	Tomato (leaf)	58
CCC + kinetin[g]	I	Tomato	64, 65
CCC + Camposan®	I	Red raspberry	66
CCC + metal chelates	I	Tomato	67
Chloramben (Amiben®)[h]	I	Cucumber	68
Chlordane[i]	I	Strawberry	17
Chlorflurenol[g]	I	Red pepper	69
Chlorothalonil[f]	N	Tomato	70
Chlorthiamid[h]	I	Gooseberry	71
Chlorthiamid	N	Black currant	71
Cidial® (phenthoate)[i]	I	Peaches	85
CIPA[g]	I	Grapefruit	12
Cloprop* (Fruitone*, 3-CPA)[g]	I	Pineapple	72
Cotoran®(fluometuron)[h]	I	*Rosa canina*	40
Cresacin*[g]	N	Tomato	73
Cuprosan®,[f]	I	Grape	74
Dacthal®,[H]	D	Onion	75
Dalapon[h]	D	Potato	76
Demeton-S-methyl (Meta-Isosystox)[i]	I	Potato	28
Dextramine-N*,[g]	N	Tomato	73
Dextrel*,[g]	I	Red raspberry	80
Dextrel	I	Tomato	54

TABLE 1 (continued)

Compound[a]	Effect[b]	Plant	Ref.
		Ascorbic Acid	
Dextrel	I	Cucumber	54
Dextrel	D	Cucumber	81
DDT[i]	I	Citrus fruit	82
DDT + lindane[i]	N	Apples	31
Dichlormid[th]	D	*Ipomea hederacea*	83
Dicryl*,[h]	I	Cucumber	84
Dicotex® + linuron[h]	D	Potato	86
Difoset* (EBF-5)[g]	I	Cauliflower	38
Dimethoate (Cygon®)[i]	N	Apples	31
Dimethoate	I	Black currant	87
Dimethoate	I	Strawberry	17
Dimethoate	N	Tomato	70
Dimilin® (diflubenzuron)[i]	I	Peaches	85
Dinoseb (Nixol)[h]	I	Potato	28
Dipel + Lannate®,[i]	I	Peaches	85
Diphenamid[h]	N	Sweet potato	89
Diquat (Reglone®)[h]	N	Potato	28
Diquat	N	Potato	90
Diquat	I	Potato	91
Dithane M-45® (mancozeb)[f]	D	Potato	92
Dymid® (diphenamid)[i]	I	Strawberry	17
EBF-2[g]	D	Sweet pepper	39
Enolofos EC-50*,[j]	I	Potato	91
Ergostim*,[g]	N	Tomato	70
Ethephon (ethylene, Ethrel®, Camposan®)[g]	D	*Achras sapota*	107
Ethylene	I	Bean sprout	114
Ethephon	I	Cauliflower	38
Ethephon	N	Cherry	77, 78
Ethephon	D	Clementine	101
Ethephon	N	Currant	109
Ethephon	I	Cucumber	54
Ethephon	I	Lemon	93
Ethephon	I	Litchi	94
Ethephon	D	Mandarin	102
Ethephon	N	Mango	110
Ethephon	D	Onion	103
Ethephon	I	Oranges	95
Ethephon	D	Papaya	96, 104
Ethephon	D	Pepper	105
Ethephon	D	Pomegranate	106
Ethephon	I	Pumpkin	96
Ethephon	D	Pumpkin	97
Ethephon	N	Tangerine	18

TABLE 1 (continued)

Compound[a]	Effect[b]	Plant	Ref.
		Ascorbic Acid	
Ethephon	I	Tomato	54, 98-100, 111, 112
Ethephon	D	Tomato	108
Ethephon	N	Tomato	113
Fenitrothion[i]	D	Cabbage	46
Fermate*,[f]	I	Mango	115
Fluchloralin[h]	I	Onion	116
Fluchloralin	I	Tomato	117
Formothion[i]	N	Mandarin	118
GA Gibberellic acid)[g]	V	Apples	143, 144
GA	I	Ber	119
GA	D	Celery	141, 142
GA	I	Cherry	77, 78
GA	I	Currant (black)	120
GA	N	Grape	266
GA	I	Grape	121
GA	I	Guava	122
GA	I	Leek	123
GA	I	Loquat	124
GA	I	Mango peel	125
GA	I	Mango	126
GA	I	Mung bean sprout	127
GA	I	Oranges	128
GA	I	Papaya	129
GA	I	Potato	130, 131
GA	I	*Rosa rugosa*	132
GA	D	*Rosa canina* (leaf)	132
GA	I	Soybean sprouts	133
GA	I	Strawberry	134
GA	N	Tangerine	18
GA	I	Tomato	135-139
GA	N	Tomato	140
Guthion®,[i]	I	Strawberry	17
Hydrel*,[g]	I	Tomato	54
Hydrel	I	Cucumber	54
Hydrel	D	Cucumber	81
Hydrel	D	Mandarin	102
Hydrel	D	Tomato	147, 148
IAA (indoleacetic acid)[g]	I	Chili	149
IAA	I	Potato	130
IAA	I	Soybean sprouts	133
IAA + Benzyladenine[g]	I	Soybean sprouts	150
IBA (indolebutyric acid)[g]	I	Carrots	151
Indolepropionic acid[g]	I	*Capsicum*	136
Indolepropionic acid	I	Tomato	136
Isoproturon[h]	I	Onion	116

TABLE 1   (continued)

Compound[a]	Effect[b]	Plant	Ref.

Ascorbic Acid

Compound[a]	Effect[b]	Plant	Ref.
Ivin[*,g]	I	Cucumber	152
Ivin	I	Tomato	54
Karathane® (dinocap)[f]	D	Gooseberry	88
Karathane[f]	I	Cucumber (leaf)	35
Karbatox 75®,[i]	D	Cabbage	153
Karbatox 75	D	Kale	153
Kartolin[*,g]	I	Beet root	154
Kernite® (fosamine)[g]	I	Tomato	155
Lead arsenate[g]	I	Citrus	128, 156
Lead arsenate	D	Citrus	157
Lindane (benzene hexachloride)[i]	I	Tomato	36
Lindane	I	Plant cell	158
Linuron (Afalon®)[h]	N	Cabbage	159
Linuron[h]	I	Carrots	160
Linuron	D	Carrots	161
Linuron	N	Carrots	162
Linuron + prometryne + propazine[b]	D	Carrots	163
Linuron	D	Potato	164
Malathion (carbophos)[i]	I	Cabbage	47
Malathion	N	Mandarin	118
Maleic hydrazide[g]	I	Mango	22
Maleic hydrazide[g]	D	Oranges	166
Maleic hydrazide[h]	N	Potato	165
Mancozeb[f]	N	Tomato	70
Maneb[f]	N	Potato	167
Mesoranil® (aziprotryne)[h]	I	Cabbage	168
Mesoranil	I	Onion	169
Mesoranil	D	Kale (fodder)	170
Mesoranil + linuron	I	Onion	171
Metasystox®,[i]	I	Chili	48
Methabenzthiazuron[h]	I	Tomato	117
Methyl bromide[i]	I	Tomato	172
Methyl bromide	D	Tomato	173
Methyl Topsin®,[f]	I	Apples	174
Metoxuron[h]	I	Potato	28, 91
Metribuzin (Sencor®)[h]	I	Potato	175, 176
Metribuzin	I	Tomato	177
Mugan[*,?]	I	Mung bean sprouts	178
Murfotox®,[i]	I	Peaches	85
β-Naphthoxyacetic acid[i]	I	*Capsicum*	136
β-Naphthoxyacetic acid	I	Tomato	137
NAA (Naphthalene acetic acid)[g]	N	Ber	23
NAA	I	Cabbage	179
NAA + Zn	I	Cabbage	183

TABLE 1 (continued)

Compound[a]	Effect[b]	Plant	Ref.
	Ascorbic Acid		
NAA	I	Chili	180
NAA	I	Litchi	181
NAA	N	Grape	26
NAA	I	Mango	15, 16, 126
NAA	I	Peaches	182
NAA	N	Tangerine	18
Nimrod® (bupirimate)[f]	I	Cucumber	184
Nitiran* (Nitran?)[h]	N	Cabbage	185
Nitrofor[Y®, h]	I	Tomato	177
Oil sprays[i]	D	Oranges	186
Oryzalin (Dirimal®)[h]	D	Currant (black)	187
Oryzalin	N	Sweet potato	89
Oryzalin + chloramben	N	Sweet potato	89
Oxyfluorfen[h]	D	Mustard (leaf)	25, 188
Oxytetracycline*, [an]	I	Tomato	41
P-Chlorophenoxyacetic acid[g]	I	Beans (snap)	189
P-Chlorophenoxyacetic acid	I	Tomato	19
Paraquat[h]	D	Mustard (leaf)	25
Paraquat	D	Spinach	190, 191
Paraquat	I	Tomato	117
Parathion[i]	I	Citrus	36
Parathion	N	Beans (green)	192
Pendimethalin[h]	I	Onion	116
Penicillin[an]	I	Barley	193
Permethrin[i]	N	Tomato	70
Permethrin + tetramethrin[i]	I	Tomato	70
Phorate[i]	I	Potato	49
Phosphamide (Phosphamidon?)[7]	N	Potato	90
Plictran® (cyhexatin)[i]	I	Cucumber	194
Polyram 80®, [f]	D	Strawberry	195
Pirimor G[j]	N	Potato	167
Prometryne[h]	I	Carrots	160, 196
Prometryne	N	Carrots	162
Prometryne	D	Onion	198
Prometryne	I	Potato (leaf)	197
Propazine[h]	N	Carrots	162
Propyzamide[h]	V	Lettuce	199
Putrescine dichloride[g]	I	*Litchi chinesis*	200
Pyramin® (chloridazon)[h]	I	Beet	201
Quinomethionate[f]	I	Currant (black)	202
Ramrod® (propachlor)[h]	I	Kale (fodder)	170
Ramrod[h]	N	Cabbage	159
Ramrod	D	Mandarin	45
Ramrod	D	Onion	203
Ramrod + mesoranil®, [h]	I	Onion	171

TABLE 1 (continued)

Compound[a]	Effect[b]	Plant	Ref.
Ascorbic Acid			
Ramrod + Dacthal[®,h]	I	Onion	171
Reglone[®] (diquat dibromide)[h]	D	Mandarin	45
Reglone	D	Onion	103
Reglone	N	Potato	90
Riboflavin[vi]	I	Grape	121
Ridomil[®] (metalaxyl)[f]	D	Potato	92, 164
Ridomil M2[f]	N	Potato	167
Roundup[®] (glyphosate)[h]	I	Grape	204
Rovral[®] (iprodione)[f]	I	Strawberry	79
Rubigan[®] (fenarimol)[f]	D	Cucumber	205
Rubigan	D	Tomato	205
Saifos[*,h]	N	Potato	90
Semeron[®] (desmetryn)[h]	I	Cabbage	206
Semeron	I	Kale (fodder)	170
Semeron	D	Mandarin	45
Semeron + Ramrod[®]	D	Cabbage	203
Semeron + Treflan[®]	D	Cabbage	203
Carbaryl (Sevin[®])	N	Apples	31
Sevin	D	Cabbage	46
Sevin	I	Cabbage	47
Siapton-Ap[®,?]	N	Tomato	70
Simazine (Gesatop[®])[h]	I	Apples	207, 208
Simazine	D	Apples	212
Simazine	N	Apples	31
Simazine	V	Citrus	32
Simazine	I	Currant (red)	145
Simazine	I	Currant (black)	210
Simazine	I	Gooseberry	146
Simazine	I	Lettuce	211
Simazine	D	Raspberry	213
Simazine + amitrole[h]	I	Apples	31
Simazine + atrazine[h]	I	Apples	209
Simazine + atrazine	I	Currant	209
Simazine + atrazine	I	Grape	209
Simazine + dalapon[h]	D	Raspberry	214
SR-5[g]	I	Sweet pepper	215
SR-5	I	Beet root	154
Streptomycine[an]	I	Tomato	216
Sumilex[®] (procymidone)[f]	I	Strawberry	79
Sodium sulfanilate[?]	D	Peanut (leaf)	217
Thiabendazole[f]	I	Apples	34
Thiobencarb[h]	N	Tomato	117
Thiodan[®,i]	I	Strawberry	17
Thiodan	D	Cucumber	194
Thiodicarb[i]	D	Tomato	218
Thiram[f]	I	Pepper	219

TABLE 1   (continued)

Compound[a]	Effect[b]	Plant	Ref.

Ascorbic Acid

Compound[a]	Effect[b]	Plant	Ref.
Thuricide® + Lannate®,[i]	I	Peaches	85
Topsin® (thiophanate-methyl)[f]	I	Cucumber (leaf)	35
Trakephon® (buminafos)[h]	D	Onion	103
Treflan® (trifluralin)[h]	N	Cabbage	159
Treflan	I	Kale (fodder)	170
Treflan	D	Mandarin	45
Treflan	N	Red peppers	220
Treflan	I	Tomato	177, 221
Treflan + simazine	I	Cabbage	222
Triacontanol[g]	N	Strawberry	223
Triacantanol	I	Tomato	224
Triazine[h]	I	Tomato	199
Trichlorfon (Chlorofos)[i]	I	Cabbage	47, 225
Trichlorfon	N	Tomato	70
Ultracid (methidathion)[i]	I	Peaches	85
Utal* + sitrin*,[h]	I	Potato	226
Vincolozolin[f]	N	Tomato	70
Zineb[f]	I	Apples	43,227
Zineb (Sineb?)[f]	N	Potato	90
Zolone® (phosalone)[i]	I	Apples	228
Zolone	I	Peaches	85

Biotin

Compound[a]	Effect[b]	Plant	Ref.
2,4,5-T[g]	I	Apricots	278

Carotene

Compound[a]	Effect[b]	Plant	Ref.
2,4-D[h]	I	Strawberry	17
2,4-D[g]	D	Beans (leaf)	229
2,4-D	D	Buckwheat (leaf)	9
2,4,5-T[g]	I	Persimmon	36
Alar®[g]	I	Guava	230
Aldicarb[i]	I	Potato	49
Aldrin[i]	N	Carrots	231
Atrazine[h]	I	Corn (leaf)	197
Azide*,[f]	N	Sweet potato	233
Benlate® (benomyl)[h]	N	Lettuce	234-236
Bentazone[h]	I	Radish (leaf)	237
Bromophos (Nexion®)[i]	I	Carrots	238, 239
Carbofuron[i]	N	Potato	49
CCC (TUR*)[g]	I	Tomato	64
CCC	I	Tomato (leaf)	240
CCC + kinetin[g]	I	Tomato	65

TABLE 1   (continued)

Compound[a]	Effect[b]	Plant	Ref.
		Carotene	
CCC + metal chelates	I	Tomato	67
CDEC[?]	N	Spinach	241
Chloramben (Amiben®)[h]	N	Squash (Hubbard)	241
Chloramben	I	Carrots	3, 160, 232
Chloramben	I	Squash (Butternut)	241
Chlorfenvinphos (Birlane®)[i]	I	Carrots	38, 239, 242
Chlorflurenol[g]	D	Red pepper	69
Chlorofos[i]	I	Carrots	243
Chloroxuron[h]	D	Carrots	244
Chlorpropham (CIPC)[h]	I	Squash (butternut)	241
Chlorpropham	I	Carrots	241
Chlorpropham	I	Lettuce	234-236
Chlorpropham	N	Squash (Hubbard)	241
Cotoran (fluometuron)[h]	I	*Rosa canina*	40
CPTA[h]	D	Carrots (leaf)	245
CPTA	D	Carrots	246
CPTA	D	Gourd (ornamental)	247
CPTA	D	Pepper	248
Dacthal®,[h]	D	Onion (green)	249
DBCP (Nemagon®)[n]	I	Carrots	250, 251
DBCP	I	Citrus	251
DDT[i]	I	Carrots	243
Dichlobenil (Casoron®)[h]	D	Currant (leaf)	252
Didin*,[?]	D	Lettuce	253
Dieldrin[i]	D	Carrots	254
Dieldrin	N	Carrots	231
Dinocap (Karathane)[f]	D	Gooseberry	88
Dinoseb[h]	I	Squash (butternut)	241
Dinoseb	N	Squash (Hubbard)	241
Diphenamid[h]	N	Sweet potato	89
Diuron[h]	D	Apples (leaf)	255
EDB, W-85[n]	I	Carrots	250
EDB	I	Carrots	251
EDB	I	Citrus	251
Endothal[h]	N	Spinach	241
Ethephon (ethylene)[g]	I	Carrots	256
Ethephon	I	Tomato	99, 257, 258
Ethephon	N	Tomato	111, 112
Fluchloralin[h]	N	Tomato	117
Fonofos (Dyfonate®)[i]	I	Carrots	238, 239
GA[g]	D	Carrots	141, 259, 260
GA	I	Carrots (leaf)	260
GA	D	Celery	142
GA	D	Mango	125
GA	I	Mango	126
GA	D	Tomato	139

TABLE 1   (continued)

Compound[a]	Effect[b]	Plant	Ref.
Carotene			
Heptachlor[i]	I	Carrots	243
Heptachlor	D	Carrots	254
IBA[g]	I	Carrots	151
Iprodione (Rovral®)[f]	I	Lettuce	234-236
Lindane (γ-HCH)[i]	D	Carrots	261, 262
Lindane	I	Carrots	243
Linuron[h]	N	Carrots	162
Linuron	D	Carrots	244, 263
Linuron	D	Peas (leaf)	264
Linuron	I	Carrots	160, 232, 241
Linuron + monolinuron[h]	I	Carrots	238
Linuron + monolinuron	I	Carrots	239
MCPA[h]	D	Barley (leaf)	265
Mesoranil®,[h]	D	Kale (fodder)	170
Metafos[i]	I	Carrots	243
Me-Br + chloropicrin[l]	D	Carrots	254
Metoxuron[h]	D	Carrots	238, 239
NAA[g]	I	Mango	126
Neburon[h]	D	Carrots	266
Nemagon®,[n]	I	Carrots	267, 268
Nemagon	I	Sweet corn kernel	267
Nitrofen[h]	D	Carrots	244
Norflurazon (SAN 9789)[h]	D	Daffodil flower	269
Norflurazon	D	Tulip tree flower	269
Oryzalin[h]	N	Sweet potato	89
Oryzalin	D	Currant (black, leaf)	187
Oryzalin + chloramben[b]	N	Sweet potato	89
Oxyflurofen[h]	D	Mustard (leaf)	188
Paeraquat[h]	N	Tomato	117
Pendimethalin[h]	N	Tomato	117
Pendimethalin + thiobencarb[h]	I	Tomato	117
Phorate[i]	N	Potato	49
Polychloropinene[i]	I	Carrots	243
Polychlorocamphene[i]	I	Carrots	243
Prometryne (Gesagard®)[h]	I	Carrots	160, 232
Prometryne[h]	N	Carrots	162
Prometryne	D	Carrots	196
Prometryne	I	Potato (Leaf)	197
Prometryne	I	Sunflower (leaf)	197
Propazine[f]	I	Carrots	160
Propazine	N	Carrots	162
Propyzamide (Kerb®)[h]	I	Lettuce	234, 236
Propyzamide	N	Lettuce	235
Ramrod (propachlor)[h]	D	Kale (fodder)	170
Ramrod®	D	Carrots	270
Ramrod	D	Onion (green)	249

TABLE 1 (continued)

Compound[a]	Effect[b]	Plant	Ref.
		arotene	
Rogor® (dimethoate)[i]	I	Carrots	243
Semeron®,[h]	D	Kale (fodder)	170
Sevin®,[i]	I	Carrots	243
Simazine[h]	I	Apples (leaf)	255
Simazine	I	Bermudagrass	271
Simazine	I	Corn (leaf)	197
Simazine	D	Cabbage	272
Simazine	D	Barley (leaf)	273
Solan*,[h]	D	Tomato	198
SR-5[g]	I	Sweet pepper	215
Telone®,[i]	I	Carrots	3, 250, 267, 268
Telone	I	Sweet corn kernel	267
Terbacil[h]	D	Apples (leaf)	255
Thiobencarb[h]	N	Tomato	117
Treflan®,[h]	D	Kale (fodder)	170
Vinclozolin (Ronilan®)[f]	N	Lettuce	234-236
		Folic acid	
Maneb[f]	N	Potato	167
Phosphamide[i]	N	Potato	90
Pirimor G®,[i]	N	Potato	167
Reglone®,[h]	N	Potato	90
Rimodil M2®,[f]	N	Potato	167
Saifos*,[h]	N	Potato	90
Sodium sulfanilate?	D	Peanut (leaf)	217
Zineb (Sineb?)[f]	N	Potato	90
		Niacin	
2,4-D[g]	D	Beans (leaf)	229
2,4,5-T[g]	I	Apricots	278
Afalon®,[h]	N	Beans	276
Aretit®,[h]	N	Beans	276
Chloramben[h]	I	Cucumber	68
Dalapon[h]	D	Grape	280
Dinocap (Karathane®)[f]	N	Gooseberry	88
Gesagard®,[h]	N	Beans	276
Karbatox 75®,[i]	D	Cabbage	153
Karbatox 75	D	Kale	153
Nexoval*,[h]	N	Beans	276
Radokor* (simazine)[h]	D	Grape	279
Ramrod®,[h]	N	Beans	276

TABLE 1 (continued)

Compound[a]	Effect[b]	Plant	Ref.
Pantothenic acid			
2,4-D[g]	I	Beans (leaf)	229
2,4,5-T[g]	I	Apricots	278
Dalapon[h]	D	Grape	280
Radokor* (simazine)[h]	I	Grape	280
Pyridoxine			
2,4-D amine[h]	D	Rye	282
Maneb[f]	N	Potato	167
Pirimor G[®,i]	N	Potato	167
Ridomil M2[®,f]	N	Potato	167
Riboflavin			
2,4-D[g]	D	Beans (leaf)	229
Afalon[®] (linuron)[h]	N	Beans	276
Aretit[®] (dinoseb acetate)[h]	N	Beans	276
Dinocap (Karathane[®])[f]	D	Gooseberry	88
Ethephon[g]	N	Tomato	111
Gesagard[®,h]	N	Beans	276
Nexoval[®h]	N	Beans	276
Ramrod[®h]	N	Beans	276
Thiram[f]	I	Pepper	219
Tsilt*,[f]	D	Barley	277
Thiamin			
2,4-D[g]	D	Beans (leaf)	229
2,4-D[h]	D	Wheat	274
2,4-D amine[h]	D	Wheat	275
2,4-D amine	D	Rye	275
Afalon[®h]	N	Beans	276
Aretit[®,h]	N	Beans	276
Chloramben[h]	N	Cucumber	68
Chlorothalonil[f]	N	Tomato	70
Dimethoate[i]	N	Tomato	70
Dinocap (Karathane[®])[f]	D	Gooseberry	88
Ergostim*,[g]	N	Tomato	70
Ethephon[g]	N	Tomato	111
Gesagard[®,h]	N	Beans	276
Mancozeb[f]	N	Tomato	70
Maneb[f]	N	Potato	167

TABLE 1   (continued)

Compound[a]	Effect[b]	Plant	Ref.
Nexoval*,[h]	N	Beans	276
Permethrin[i]	N	Tomato	70
Pirimor G ®,[i]	N	Potato	167
Ramrod®,[h]	N	Beans	276
Ridomil M2®,[f]	N	Potato	167
Siapton-Ap*,[g]	N	Tomato	70
Thiram[f]	I	Pepper	219
Trichlorofon[i]	N	Tomato	70
Tsilt*[f]	D	Barley	277
Vinclozolin[f]	N	Tomato	70

Tocopherol

Chlorfenson[ac]	I	Beans (leaf)	281
DDT[i]	I	Beans (leaf)	281
Dichlone (Phygon®)[?]	I	Beans (leaf)	281
Mesoranil®,[h]	D	Kale (fodder)	170
Ramrod®,[h]	D	Kale (fodder)	170
Semeron®,[h]	D	Kale (fodder)	170
Treflan®,[h]	D	Kale (fodder)	170

Summary[c]

	No effect	Decrease	Increase
Ascorbic acid	65	74	222
Carotene	27	45	79
Thiamin	18	6	0
Riboflavin	6	3	1
Niacin	6	5	2
Pantothenic acid	0	1	3
Tocopherol	0	4	3
Folic acid	7	1	0
Pyridoxine	3	1	0
Biotin	0	0	1

* Superscripts next to the compounds indicate the purpose for which each compound has been used (as noted in the original publication or as classified in the Kidd, H. and James, D. R. Eds., *European Directory of Agrochemical Products*, Volumes 1–3, Royal Society of Chemistry, Thomas Graham House, Cambridge, England, 1990). Abbreviations used are: ac = acaricide; an = antibiotic; f = fungicide; g = growth regulators; h = herbicide; i = insecticides; n = nematicides; vi = vitamin; ? = not clear or hard to classify; * = not clear whether the name is a registered trade name or a common name.

[b] D = decrease; I = increase; N = no effect; V = variable effects at different concentrations of compound.

[c] Total number of cases (cited above) where the concentration of vitamins was found to be not affected, decreased, or increased by the use of various plant protection chemicals.

compounds (or their metabolites) within the plant tissues. Circumstantial evidence, however, indicates that this is the most plausible explanation for the observations made. It suffices, however, to note that some chemicals, such as lindane, were found to be absorbed by plants such as carrots to an extent that made them unsuitable for the preparation of baby foods.[261]

I. HERBICIDES

Application of herbicides to soils can change the concentration of numerous plant vitamins not only in the target plants, but also in nontarget plants; the effects, however, strongly depend on the kinds of soil, plant, herbicide, and vitamin under consideration (Table 1). For example, the herbicide paraquat was noted to decrease the ascorbic acid content in spinach[190, 191] and mustard leaves[25] but increase the ascorbic acid content of tomatoes.[117] The results obtained by different investigators could also be contradictory. For example, use of the herbicide prometryne in carrot fields was noted to increase the ascorbic acid in carrots by some[160, 196] and have no effect by others.[162] Also the herbicide simazine was noted to increase,[207, 208] decrease,[212] or have no effect[31] on the ascorbic acid content in apples. The herbicide linuron was noted to have no effect,[162] decrease,[261, 262] or increase[160, 232, 241] the carotene content in carrots. The reason for these apparently contradictory results may be due to differences in cultural (and climatic) conditions, amount of herbicides used, and properties of soils in which plants were grown.

Soil properties are known to influence the rates of herbicide movements into the soil profile and thus could affect the rate of its absorption by the plant roots located at different depths. This in turn might not only affect the outcome of herbicide action on the target plants, but might also affect their absorption by the nontarget plants. For example, soil application of simazine and atrazine to young orange orchards was noted to change the ascorbic acid content in th leaves 6–7 months after the application, the extent of which depended on the kind of soil and the amount of herbicide used. Thus, although in the heavy soil simazine increased the ascorbic acid content of the leaves considerably, in the light soil, however, it had the opposite effect. In the heavy soil, the effect of atrazine on the ascorbic acid was reversed at different application rates: low rate increased and high rate decreased the ascorbic acid contents in the orange leaves. No explanation was given for the differences in the effects of herbicides on the leaf concentration of ascorbic acid content in trees grown in different soils.[32]

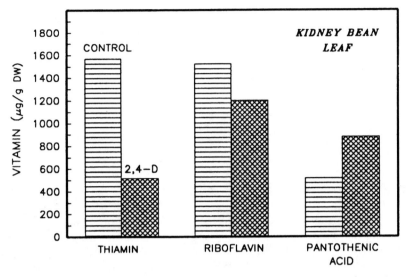

FIGURE 1. Effect of 2,4-D (applied to the base of kidney bean leaf) on the concentration of three vitamins in the leaves six days later.[229]

Herbicides may differently affect the concentration of vitamins in different plant organs. Thus in red kidney beans, for example, application of 50 μl of a 0.1% solution of 2,4-D to the base of primary leaves reduced the concentrations of thiamin, riboflavin, nicotinic acid, and carotene but increased that of pantothenic acid in the leaves just six days after the treatment (Figure 1). In the plant's stem, however, the concentrations of most vitamins tested were increased by the 2,4-D application.[229]

Effect of herbicides on plant vitamins may also depend on the concentration of the active ingredient used. For example, spraying buckwheat leaves with 2,4-D (10 μg/plant) increased the ascorbic acid content of the leaves. Higher concentrations (1 mg/plant), however, initially increased the leave's ascorbic acid, but this was followed by a decrease with the passage of time.[9] Also, the effect of triazine herbicide on plant vitamins seems to be dependent on the concentration used; low concentrations of triazine decreased while higher concentrations increased the vitamin C in tomato.[199]

Herbicides may differ in their effects on the concentration of a given vitamin (Table 1). For example, soil application of linuron was noted to increase while metoxuron reduced the total carotene content of carrots.[238] Soil applications of herbicides Kerb 50® (propyzamide) and CIPC (chlorpropham) in lettuce fields increased the total carotene content of the leaves by 23 and 48%, respectively. Treatments did not change the relative ratios of carotene stereoisomer.[234, 235] Manteiga kale grown on farms using herbicides was found to contain lower β-carotene than that grown on the so-called "natural farms."[287]

As noted above, the effect of herbicides on plant vitamins may not be entirely due to a possible uptake of the intact compound or its metabolite by the plant since in some cases the mere removal of the weeds by mechanical methods was noted to also alter the concentration of some vitamins in the plants. For example, mechanical removal of weeds by hoeing was found to increase the β-carotene concentration in lettuce by more than two times in 1987 (15.0 vs. 7.3 μg/g FW, respectively) and more than five times in 1986 (20.8 vs. 3.6 μg/g FW, respectively). Weed infestation reduced plant yield, but the reduction in carotene concentration was two to four times higher than the reduction in yield. It was thus postulated that weeds may actually interfere with carotene synthesis in lettuce leaves.[268] No explanation was given as to how this interference may take place.

In a "normal" year, lettuce contained similar β-carotene content whether grown with mechanical weeding method or by the use of the herbicides alachlor and pendimethalin. In a year with adverse (rainy) weather, however, plants grown in herbicide-treated soils contained 21–32% less carotene, supposedly because of more favorable conditions for the absorption of these herbicides by the lettuce roots in the rainy years.[268] Whether leaching of herbicide in the rainy year played any role in the effects observed was not mentioned.

That mere removal of weeds may also affect the plant vitamins was also noted by Mohammed and Ali,[221] who found that hand weeding of tomato plots not only increased fruit yield, but also increased the ascorbic acid content of the fruits; hand-cultivated fruits had higher ascorbic acid content than plants treated with the herbicides Treflan® and Sencor® (Figure 2). Also, Sanyal et al.[289] reported that the vitamin C content of bananas increased when fields were weeded by hand. Furthermore, a comparison between the vitamin content in the plants subjected to hand weeding or treated with herbicides indicates that at least part of the effect of herbicides seems to be merely due to the removal of competing weeds. Increasing the amount of herbicides, however, increased the concentration of vitamin C in the fruits (Figure 3).[289]

II. FUNGICIDES AND INSECTICIDES

Numerous reports indicate that many fungicides and insecticides (usually applied directly to the plants) may alter the concentration of various vitamins, the magnitude and direction of which depend on the environmental conditions, kind of plant, type of compound, and vitamin

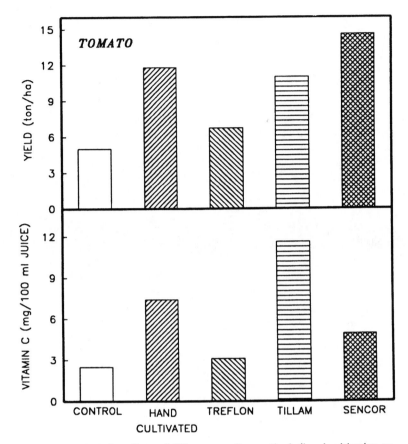

FIGURE 2. Relative effects of different weeding methods (hand cultivation or the use of three herbicides) on the tomato yield and vitamin C concentration in the fruits. Control plots were left without weeding.[221]

under study (Table 1). For example, the insecticide carbofuron was noted to increase the ascorbic acid content of strawberry,[17] chili,[48] and potato[49] but had no effect on the ascorbic acid content of faba beans.[50] The insecticide dimethoate was found to have no effect on the ascorbic acid content of apple[31] and tomatoes[70] but increased the ascorbic acid content of black currants.[87] Also, spraying orange trees with lead arsenate to combat the Mediterranean fruitfly was found to reduce the ascorbic acid concentration by two-thirds or even half in the juice of treated fruits as compared to that in the untreated fruits.[157] Spraying the same compound on grapefruits in Florida to reduce their acidity, however, increased the ascorbic acid concentration in the fruits.[128]

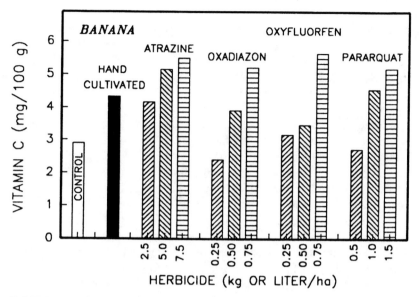

FIGURE 3. Effect of different weeding methods (by hand or by the use of herbicides) on the vitamin C concentration in bananas. Control plants were not weeded.[289]

Most fungicides appear to increase the ascorbic acid content in plants. This has been noted in the use of Benlate® in apple[34] and strawberry;[17] captan in apple[43] and strawberry;[17,44] carbendazim in apple;[34] Cuprosan® in grape; macozeb in potato;[92] fermate in mango;[115] and Nimrod® in cucumber.[184] Fungicides may also increase the carotene content in some plants. This has been noted in the use of iprodione in lettuce[234-236] and propazine in carrot.[160]

Little information is available on the effect of insecticides and fungicides on the concentrations of other plant vitamins such as thiamin, riboflavin, niacin, pantothenic acid, tocopherol, folic acid, pyridoxine, and biotin (Table 1). Information available, however, indicates that insecticides and fungicides may have no effect or may decrease the concentration of some of these plant vitamins (Table 1).

The carotene content of plants may be affected by a whole range of insecticides and fungicides (Table 1). For example, soil application of the insecticides bromophos, chlorfenvinphos, and fonofos was found to increase the concentration of total carotenes in carrots by up to 21% (without changing the relative proportion of various carotene stereoisomers). Soil application of pesticides also shortened the time carrots needed to reach their maximum carotene content by 7–14 days.[238] The fungicide iprodione sprayed on lettuce was reported to increase the total carotene content of the leaves by 36%,[234] but, the fungicides vinclozolin and benomyl did not affect the leaf carotene content.[234, 235]

The effect of pesticides on plant vitamins may depend on the concentration of active chemical used. This has been observed in apples, where a 25% reduction (from the usual amounts) in the use of several pesticides was noted to considerably increase the concentration of ascorbic acid in the fruits.[290]

In some cases plant protection chemicals may not affect the vitamin content of fruits at the time of their harvest but may affect the rate of vitamin loss after fruits have been harvested, i.e., during their postharvest periods. For example, spraying black currants with Bordeaux mixture or thiram to combat the *Septoria* leaf spot did not affect the ascorbic acid content of the fresh fruits but increased the rate of ascorbic acid loss during the fruit processing.[291] In contrast, spraying snap beans with 400 ppm of *p*-chlorophenoxyacetic acid four days before their harvest reduced the rate of ascorbic acid loss in the harvested pods so that four days after their harvest, treated pods contained almost 40% more ascorbic acid than the untreated pods (11.9 versus 51.7% decrease in the treated and untreated pods, respectively).[189]

Soil properties and the kind of plant protection chemicals may strongly interact for their effect on plant vitamins long after the plants have been harvested. For example, carrots grown on clay soil contaminated with dieldrin lost their carotene at a slower rate during storage. The rate of carotene loss was, however, higher if soil was contaminated with phenitrothion (fenitrothion) or with methyl bromide. When carrots were grown on humus soil, loss of α- and β-carotene in carrots was reduced by dieldrin but increased by phenitrothion or heptachlor contamination. On sandy soil, however, loss of both α- and β-carotene was reduced by dieldrin.[292] Retardation of vitamin loss during postharvest storage has also been observed for the vitamin C in lettuce by triacontanol.[293]

Information on the effect of plant protection chemicals on the bioavailability of plant vitamins is scarce. Based on the only report of its kind known to us, it appears that the spraying of plants with the fungicide tsilt not only decreased the amount of riboflavin and thiamin in the grains of barley, but also reduced the bioavailability of these vitamins (to rats fed on these grains), especially when the plants were grown with high levels of nitrogen fertilizer.[277]

High levels of nitrogen fertilizer are known to make plants more susceptible to attack by various insect pests, pathogenic fungi, and viruses.[294-297] This in turn would increase the need for plant protection chemicals to protect the plant against pathogens and insects. In this connection, the above information deserves particular attention espe-

TABLE 2
Effect of Soil Application of Nematicides on the Carotene
Concentrations in Carrots and Maize Seeds

Treatment	(Gal/acre)	β-Carotene (μg / 100 g) Carrots	Carotene (μg / 100 g) Maize seed
Control		4,927	1,526
Telone	10	5,881**[a]	1,720*
	20	6,270**	1,810*
	30	7,315**	1,918**
Nemagon	1	5,668**	1,652*
	2	6,182**	1,703*
	3	6,650**	1,821*

[a] * and ** indicate significantly different from the control by 0.05 and 0.01 levels, respectively.

Compiled from tables 2 and 3 of Salunkhe, D. K., Wu, M., Wu, M. T., and Singh, B. *J. Am. Soc. Hortic. Sci.*, 96, 357, 1971.

cially as the increased use of plant protection chemicals may prove to reduce the availability of plant vitamins to a considerable extent.

III. NEMATICIDES

Available information indicates that the treatment (fumigation) of soils with various nematicides may also affect the vitamin content of plants grown in them (Table 1). For example, soil fumigation with methyl bromide was noted to decrease the ascorbic acid concentration in tomato.[298] Bajaj and Mahajan,[289] however, reported that the use of nematicides significantly increased the ascorbic acid but decreased the concentration of β-carotene in tomato. In contrast, soil application of the nematicides Telone® and Nemagon® was found to significantly increase the β-carotene in carrots and in maize seeds (Table 2).[267, 268]

IV. PLANT GROWTH REGULATORS

Growth regulators are an indispensable part of industrialized agriculture and are used for various purposes such as improving fruit set, inducing thinning, enhancing or delaying fruit ripening, inducing seedlessness, facilitating fruit abscission for mechanical harvesting, and increasing the size and color of fruits.[300] Information available on whether use of these compounds would adversely affect the content of vitamins in plants is summarized in Table 1. It can be seen that the effect of these compounds on plant vitamins, similar to other

TABLE 3

**Effect of Gibberellin and CCC on the Ascorbic Acid Concentration
in Satsuma Mandarins Taken from the Interior and Exterior
of Trees[302]**

	Ascorbic acid (mg / 100 g FW)[a]			
	Exterior fruits		Interior fruits	
Treatment	Flavedo	Juice	Flavedo	Juice
Control	173.6a	29.1a	91.5a	24.7ab
GA	158.2b	29.7a	76.9b	26.0a
CCC	142.2c	30.3a	80.5ab	22.4c

[a] Values within each column followed by different letters are significant at % level.

compounds discussed above, strongly depends on the nature of the chemical, plant, and vitamin under consideration.

Among the growth regulators, the effects of 2, 4-D, Cyocel, ethephon, gibberellic acid (GA), and naphthalene acetic acid (NAA) on plant vitamins have been the subject of more studies than any other compound. Most growth regulators seem to increase the plant vitamins. This has been noted for the effect of 2,4-D on the ascorbic acid content of beans,[3,8] eggplant,[11] grapefruit,[12,13] mango,[14-16] pepper,[11] and strawberry;[17] for the effect of CCC on apple,[51] carrots,[53] cucumber,[54] mandarin,[55] potato,[56] and tomato;[54,57] for the effect of GA on celery,[141,142] cherry,[77,78] currant,[120] grape,[121] guava,[122] leek,[123] loquat,[124] mango,[126] orange,[128] papaya,[129] and potato;[130,131] and for the effect of NAA on cabbage,[179] chili,[180] litchi,[181] mango,[15,16,126] and peaches.[182]

The effect of growth regulators on plant vitamins may differ in different plants and may not be consistent in all cases (Table 1). For example, ethephon was noted to have no effect on the ascorbic acid content in cherry,[77,78] currant,[109] mango,[110] and tangerine;[18] decrease the ascrobic acid content of clementine,[101] mandarin,[102] onion,[103] papaya,[96,104] pepper,[105] pomegranate,[106] pumpkin,[18] and tomato;[108] and increase the ascorbic acid content in cucumber,[54] lemon,[93] litchi,[94] oranges,[95] pumpkin,[96] and tomatoes.[54,98-100] Also, gibberellic acid was noted to decrease the carotene content of carrots,[141,259,260] celery,[142] mango,[125] and tomato[139] and increase the carotene content in carrot leaves[260] and in mango.[126]

The effect of growth regulators on plant vitamins may depend on the kind of tissue under consideration. For example, application of gibberellin and CCC to fruits of Satsuma mandarin was noted to strongly reduce the ascorbic acid in the flavedo part of the fruit and had relatively little effect on the ascorbic acid content in the fruit's juice (Table 3). With this in mind, some selected reports are discussed in more detail to show the interactions with temperature, time course of effect, and effects on the vitamins during storage.

Treating potato tubers with CIPC [Isopropyl *N*-(3-chlorophenyl) carbamate] to inhibit sprouting during storage was reported to significantly reduce the ascorbic acid content of the tubers when measured after two months of storage at 5°C. When tubers were stored at 25°C, however, the differences in the ascorbic acid concentration of treated and control tubers were not consistent in the two varieties tested.[301]

Length of time after application when the effect of growth regulators on plant vitamin can be observed may vary from days to months. For example, application of 2,4,5-T to apricots was noted to significantly increase the concentrations of niacin, biotin, and pantothenic acid in the fruits within a few days of application.[278] On the other hand, presowing soaking of the peas in a 10-ppm solution of gibberellic acid for 2 h was noted to increase the ascorbic acid content in the peas produced (by sowing the treated seeds) by threefold (37.4 vs. 113.4 mg/ 100 g in control and treated plants, respectively),[136] an observation that is hard to interpret.

Ethephon (ethrel) is a compound that decomposes on or within the plant tissues and releases ethylene, a plant growth regulator often used to ensure uniform ripening in several fruits and vegetables.[300, 303] For processing tomatoes, for example, the application of ethephon results in a 100% ripening of the fruits for a once-over mechanical ripening, while in the untreated fields only 80–85% may be ripe at the time of harvest.[303] Available data indicate that the use of ethrel on tomato and various other crops such as citrus, papaya, and mango may change their ascorbic acid content; the direction of change, however, seems to depend on the plant and the experimental conditions (Table 1).

Growth regulators may also affect the rate of vitamin loss in the plants after their harvest. For example, Mitchell et al.,[189] in a greenhouse study, sprayed snap beans with *p*-chloropheoxyacetic acid four days prior to their harvest and measured the ascorbic acid content of the harvested beans four days after their harvest. They noted that the ascorbic acid in the untreated pods of beans decreased by 37.5% but in the treated pods, it decreased by only 11.7%. Thus, four days after harvest, treated beans contained 78% more ascorbic acid than the untreated control plants (19.6 vs. 11.0 mg/100 g, respectively). Under field conditions, the effect of spraying beans with *p*-chloropheoxyacetic acid was even more pronounced. Thus, four days after harvest, ascorbic acid content in the treated pods had decreased by 31.8% while that in the treated pods had decreased by 3.6% only. Nine days after storage, the treated pods contained 76% more ascorbic acid than those from the untreated plants. Spraying of apples with Alar during the summer months was found to reduce the rate of vitamin C loss during the fruit's storage.[304] Retardation of vitamin loss during postharvest storage has been reported in mango fruits treated with GA,[305] in peaches treated with CCC,[306] in grapes treated with Alar® and kinetin,[307] and in potatoes treated with maleic hydrazide and piperidinoacetanilide-HCl.[308] Based on these limited data, it appears that some growth-regulating compounds have a long-lasting effect on the content of some vitamins in some fruits and vegetables.

Cyocel®, another growth regulator used for different purposes in different crops,[300] can also change the content of plant vitamins. Interestingly, the number of cases where its use was noted to increase the content of some vitamins in plants outweighs the cases where no effect or a decrease in plant vitamins has been noted (Table 1).

Finally, the effect of a growth regulator on plant vitamins may depend on its concentration. For example, application of GA at 25, 50, and 100 ppm to apple trees (5–6 weeks after the fall of petals) was noted to increase the weight of apples retained on the trees from 141.9 in the control trees to 142.7, 146.5, and 160.6 in the treated fruits, respectively. The ascorbic acid concentration in the harvested apples was also dependent on the GA spray and ranged from 7.05 in the control fruits to 8.38, 7.20, and 5.12 mg/100 g in the treated fruits, respectively. This shows that although 25 ppm GA significantly increased the ascorbic acid content of apples, higher GA concentrations progressively decreased the ascorbic acid content of the fruits. Thus although the apples sprayed with 100 ppm GA were ca. 13% heavier (160.6 vs. 141.9 g), they had 38% less ascorbic acid (5.12 vs. 7.05 mg/100 g) than the control (untreated) fruits.[144]

V. USE OF CHEMICALS TO INCREASE PLANT VITAMINS

Should the chemicals that are (inadvertently) noted to increase plant vitamins be intentionally used to increase the concentration of vitamins in plants? This question is raised only because the use of "chemical stimuli" or "bioregulators" for the purpose of increasing plant carotenoids has been brought up by some authors.[309] The use of bioregulators to increase plant vitamins is proposed to be advantageous for tree crops because of the slowness and great expense of breeding new varieties high in vitamins and the difficulty in breeding into a single variety all of the desired traits.[309] As an example of success in this field, it was noted that immersion of immature citrus fruits for 30 s in a 5000-ppm solution of a bioregulator CPTA (2-[4-chloropheylthio]-triethylamine hydrochloride) could increase the provitamin A content in citrus fruits. Also, the provitamin A content of tree-ripe Navel oranges was apparently increased from 100 IU/100 g DW to 1500 IU/100 g and that of tree-ripe grapefruit from 21 IU/100 g to 1900 IU/100 g, i.e., by a factor of 90 times! The ability of this bioregulator to increase carotenoid biosynthesis has been also noted for a variety of other plants, including carrots, sweet potatoes, apricots, prunes, peaches, and mycelia of *Blakeslea trispora*.[309, 310] Even application of two mutagents to tomatoes was noted to increase their vitamin C content.[311] Whether use of these or similar compounds to increase plant vitamins is a nutritionally sound practice remains to be seen, particularly since the adverse effects of chemical residues may outweigh any beneficial effect they may produce by increasing the plant vitamins.

VI. SUMMARY

Available experimental evidence indicates that plant protection chemicals applied to the soil (to combat soil pests or kill weeds) or to the plants directly (to combat various insect and microbial pests or cause physiological or cosmetic changes) could change the concentration of some vitamins under certain agronomic conditions. These observations, whose numbers are by no means insignificant (Table 1), provide indirect and circumstantial evidence that, apart from those plant protection chemicals directly applied to the target plant, also the chemicals applied to soil to kill weeds are apparently absorbed by the nontarget plants and thus they could alter the concentration of vitamins in these plants. That the effect of chemicals on plant vitamins could be detected several months after the application date (such as the effect of various herbicides on the vitamin content of fruits grown on trees, Table 1) is surprising, and no other plausible explanation could be found but the presumption that compounds are absorbed by the roots, transported to the fruits, and there and then alter the concentration of plant vitamins.

This postulation, however, needs to be experimentally proven and may not hold true in all cases. In the only experimental case known to me, however, no relationship could be established between the herbicide residues and the ascorbic acid in lettuce leaves.[199] This may lead to the assumption that either plant protection chemicals themselves are not responsible for the effects observed or their concentration (or that of their metabolites) is below the detection limit of the presently available analytical techniques. The fact that their application has been noted to change the plant vitamins in so many instance, however, points to the presence of the original compound and/or its metabolite in the plant cells at some concentration noticeable by the plant but not by presently used analytical methods.

In support of the above view is the observation made by Borodulina and Shishkina.[208] These investigators applied simazine on soils with two different levels of humus content and studied the ascorbic acid in the apple fruits grown on these soils. They noted that when simazine was applied to a dark-gray loamy soil high (5.8%) in humus, simazine remained in the top 10 cm of soil and no change was noted in the chemical composition of apples. However, when simazine was applied to a soil with lower humus (3.2%), it penetrated into the soil to a depth of 30 cm and also increased the ascorbic acid content in apples grown in this soil! This observation clearly indicates that the deeper penetration of simazine into the soil brought it into contact with the absorbing plant roots and thus caused the increase in the ascorbic acid content of the fruits. Whether such effects of herbicides were due to their effect on root physiology or whether the herbicides had to reach the target tissue, in this case the apple fruit, to induce the changes observed is not clear.

REFERENCES

1. Luckmann, W. H. and Metcalf, R. L., The pest-management concept, in *Introduction to Pest Management*, 2nd Ed., Metcalf, R. L., and Luckmann, W. H., Eds., John Wiley, New York, 1982, chapter 1.
2. Benbrook, C. M., What we know, don't know, and need to know about pesticide residues in food, in *Pesticide Residues and Food Safety*, Tweedy, B. G., Dishburger, H. J., McCarthy, J., and Murphy, J., Eds., ACS Symposium Series 446, American Chemical Society, Washington, D.C., 1991, 140.
3. Wu, M. T., Salunkhe, D. K., The use of certain chemicals to increase nutritional value and to extend quality in economic plants, *CRC Crit. Rev. Food Sci. Nutr.*, 4, 507, 1973.
4. Wu, M. T. and Salunkhe, D. K., The use of certain chemicals to increase nutritional value and to extend quality in economic plants, in *Storage, Processing, and Nutritional Quality of Fruits and Vegetables*, Salunkhe, D. K., Ed., CRC Press, Cleveland, Ohio, 1974, 79.
5. Karmas, E. and Harris, R. S., Eds., *Nutritional Evaluation of Food Processing*, Van Nostrand Reinhold, New York, 1988.
6. Eskin, N. A. M., Ed., *Quality and Preservation of Fruits*, CRC Press, Boca Raton, Fla., 1991.
7. Kays, S. J., *Postharvest Physiology of Perishable Plant Products*, AVI Book, Van Nostrand Reinhold, New York, 1991.
8. Rathore, V. S. and Wort, D. J., Growth and yield of bean plants as affected by 2,4-D micronutrient spray, *J. Hortic. Sci.*, 46, 223, 1971.
9. Wort, D. J., The application of sublethal concentrations of 2,4-D, and in combination with mineral nutrients, *World Rev. Pest Control*, 1(4), 6, 1962.
10. Earnshaw, R. A. and Johnson, M. A., Control of wild carrot somatic embryo development by antioxidants, *Plant Physiol.*, 85, 273, 1987.
11. Yakushkina, N. I. and Kravtsova, B. E., The influence of growth stimulants on the yield and quality of the fruit of several vegetables, *Dokl. Vsesoyuz. Akad. Sel'skokhoz. Nauk V.I. Lenina*, 22(2), 15, 1957; *Chem. Abstr.*, 51, 14028b, 1957.
12. Chundawat, B. S. and Randhawa, G. S., Effect of plant growth regulators on fruit set, fruit drop and quality of Saharanpur special variety of grapefruit (*Citrus paradisi* Macf.), *Indian J. Hortic.*, 29, 277, 1972.
13. Chundawat, B. S. and Randhawa, G. S., Effect of plant growth regulators on fruit set, fruit drop and quality of Foster and Duncan cultivars of grapefruit (*Citrus paradisi* Macf.), *Haryana J. Hortic. Sci.*, 2(1/2), 6, 1973; *Hortic. Abstr.*, 45, 7829, 1975.

14. Chadha, K. L. and Singh, K. K., Effect of NAA, 2,4-D and 2,4,5-T on fruit drop, size and quality of Langra mango, *Indian J. Hortic.*, 20, 30, 1963.

15. Veera, S. and Das, R. C., Effect of growth regulators on development and quality of fruits in mango (*Mangifera indica* L.), *S. Indian Hortic.*, 19(1/4), 29, 1971; *Hortic. Abstr.*, 44, 3574, 1974.

16. Singh, A. R., Effect of foliar sprays of nitrogen and growth regulators on the physico-chemical composition of mango (*Mangifera indica* L.), *Plant Sci. (India)*, 8, 75, 1976.

17. Cheng, B. T., Effect of pesticides on the quality of strawberries, *J. Chinese Agric. Chem. Soc.*, 27(3), 373, 1989.

18. Singh, M. I. and Chohan, G. S., Effect of various plant growth regulators on granulation in citrus cv. Dancy tangerine, *Indian J. Hortic.*, 41, 221, 1984.

19. Singh, S. N. and Choudhury, B., Effect of various plant regulators and their methods of application on quality of tomato fruits, *Indian J. Hortic.*, 23, 156, 1966; *Hortic. Abstr.*, 38, 3595, 1968.

20. Mehrotra, O. N., Garg, R. C., and Singh, I., Growth, fruiting and quality of tomato (*Lycopersicon esculentum* Mill) as influenced by growth regulators, *Prog. Hortic.*, 2(1), 57, 1970; *Hortic. Abstr.*, 42, 1482, 1972.

21. Panič, M. and Franke, W., Der Einfluss von 2,4-D auf den Ascorbinsäuregehalt und die Aktivität der Ascorbinsäureoxidase in Bohnen- und Weizenblättern bei Blattapplikation, *Agnew. Botanik*, 45(1/2), 33, 1971; *Hortic. Abstr.*, 42, 1254, 1972.

22. Date, W. B., Preharvest treatment with growth regulators on ascorbic acid content of mangoes, *Sci. Cult.*, 26(5), 227, 1960.

23. Bal, J. S., Singh, S. N., Randhawa, J. S., and Jawanda, J. S., Effect of growth regulators on fruit drop, size and quality of Ber (*Zizyphus mauritiana* Lamk.), *Indian J. Hortic.*, 41, 182, 1984.

24. Kenyon, W. H. and Duke, S. O., Effects of acifluorfen on endogenous antioxidants and protective enzymes in cucumber (*Cucumis sativus* L.), *Plant Physiol.*, 79, 862, 1985.

25. Kunert, K. J., Herbicide-induced lipid peroxidation in higher plants: the role of vitamin C, in *Oxygen Radicals in Chemistry and Biology*, Bors, W., Saran, M., and Tait, D., Eds., Walter de Gruyter, Berlin, 1984, 383.

26. Tafazoli, E., Increasing fruit set in *Vitis vinifera*, *Scientia Hortic.*, 6, 121, 1977.

27. Mohammed, A. K., Ashour, N., and Abdul-Hadi, A. I., Effect of different planting dates and foliar spray with Cycocel and Alar on yield characteristics of tomato, *Mesopotamia J. Agric.*, 16(1), 34, 1981.

28. Rouchaud, J., Moons, C., Detroux, L., Haquenne, W., Seutin, E., Nys, L., and Meyer, J. A., Quality of potatoes treated with selected insecticides and potato-haulm killers, *J. Hortic. Sci.*, 61(2), 239, 1986.

29. Prakash, M., Ramachandran, K. and Nagarajan, M., Influence of antitranspirants on yield and quality of brinjal, *South Indian Hortic.*, 40(2), 113, 1992; *Hortic. Abst.*, 63(5), 3490, 1993.

30. Dzhakeli, E. M., Sardzhveladze, G. P., and Kharebava, L. G., Effect of Aminol Forte on some characteristics of citrus crops, *Subtrop. Kul't.*, 2, 97, 1988; *Chem. Abstr.*, 109, 206647u, 1988.

31. Schubert, E., Einfluss ausgewählter Insektizide und Herbizide auf die Fruchtqualität des Apfels, *Nahrung*, 18(5), 557, 1974.

32. Goren, R. and Moneselise, S. P., Some physiological effects of triazines on citrus trees, *Weeds*, 14, 141, 1966.

33. Kanaujia, J. P., Mohan, N., Pandey, A. K., and Pandey, P. K., Effect of atrazine on growth, ascorbic acid and chlorophyll content of tomato, *Prog. Hortic.*, 15(3), 178, 1983; *Hortic. Abstr.*, 55, 4496, 1985.

34. Cano, M. P., De la Plaza, J. L., and Muñoz-Delgado, L., Effects of several postharvest fungicide treatments on the quality and ripening of cold-stored apples, *J. Agric. Food Chem.*, 37, 330, 1989.

35. Rzaeva, S. I., Dusting of cucumber seeds with fungicides for protection from powdery mildew, *Khim. Sel'sk. Khoz.*, 12, 35, 1981; *Chem. Abstr.*, 96, 64095z, 1982.

36. NAS (National Academy of Science), *Principle of Plant and Animal Pest Control, Vol. 6, Effect of Pesticides on Fruit and Vegetable Physiology*, National Academy of Science, Publication 1698, Washington, D.C., 1968.

37. Reda, F., Mousa, O. M., Sejiny, M. J., and Nawar, L. S., Effect of benzyladenine on resistance of tomato cultivars to early-blight, *Egypt. J. Phytopathol.*, 18(2), 89, 1986; *Chem. Abstr.*, 108, 89375v, 1988.

38. Shishov, A. D., Matevosyan, G. L., Sutulova, V. I., and Ivanova, S. I., Effect of physiologically active substances on the growth, development, and yield of cauliflower, *Fiziol. Biokhim. Kul't. Rast.*, 22(4), 360, 1990; *Chem. Abstr.*, 113, 147196d, 1990.

39. Sovetkina, V. E., Dymova, G. I., and Matevosyan, G. L., Phosphorylated benzimidazoles effect on greenhouse sweet peppers, *Agrokhimiya*, 7, 103, 1988; *Chem. Abstr.*, 109, 124336c, 1988.

40. Strelets, V. D., Bukina, N. V., Kochetkov, V. P., and Tsybul'ko, N. S., Herbicides in young *Rosa canina* plantings, *Khim-Farm. Zh.*, 16(11), 1365, 1982; *Chem. Abstr.*, 98, 48579w, 1983.

41. Beleva, L., The role of biomycin, oxytetracycline and chlornitromycin in increasing the resistance of tomatoes to the agent of bacterial canker, *Corynebacterium michiganense, Nauchni Trudove, Vissh Selskostopanski Institut "Georgi Dimitrov," Rasteniev'dstvo*, 23, 371, 1973; *Hortic. Abstr.*, 45, 1750, 1975.

42. Baida, T. A., Bordeaux mixture to control black bacterial spot and its effect on yield and quality of fruit in the nightshade family, *Zashch. Plodovykh Ovoshchn. Kul't.*, 141, 1982; *Chem. Abstr.*, 99, 117714w, 1983.

43. Šibkova, N., Fungicides and apple quality, *Zašč. Rast. Vred. Bolez.*, 11(1), 55, 1966; *Hortic. Abstr.*, 36, 6461, 1966.

44. Engst, R., Noske, R., and Voigt, J., Zum Einfluss von Pflanzenschutz und Schädlingsbekämpfungsmitteln auf einige Inhaltsstoffe von Obst und Gemüse. 1. Mitt. Beeinflussung von Äpfeln und Erdbeeren, *Nahrung*, 13(3), 249, 1969.

45. Ahmed, S. T., Use of herbicides in mandarin orchards in humid subtropical areas and their effects on weeds and on fruit quality, *Ratsion. Ispol'z. pestits., Udobr. Pochv Trop. Subtrop.*, 77, 1983; *Chem. Abstr.*, 100, 116376q, 1984.

46. Baicu, T., Stefan, A., and Ionescu, M., The effect of carbaryl and fenitrothion on cabbages and their efficiency in controlling *Mamestra brassicae, Analele Inst. Cercetări Pentru Protectia Plantelor*, 7, 267, 1969; *Hortic. Abstr.*, 42, 5885, 1972.

47. Germanova, V. I., Effect of insecticides on growth, yield and quality of cabbage, *Nauk. Pr.—Ukr. Sil's'kogospod. Akad.*, 134, 1981; *Chem. Abstr.*, 98, 67044j, 1983.

48. De, S. K. and Laloraya, D., Fertilizer-pesticide interaction effect on ascorbic acid content of *Capsicum annuum, Indian J. Agric. Chem.*, 16(1), 65, 1983; *Chem. Abstr.*, 100, 22055g, 1984.

49. Marwaha, R. S., Nematicides induced changes in the chemical constituents of potato tubers, *Plant Foods Hum. Nutr.*, 38, 95, 1988.

50. All, A. A., Antonius, G. F., Khattab, M. M., and Othman, M. A. S., Analysis of carbofuran residues and metabolites on faba beans (*Vicia faba*) in relation to quality related properties, *Alexandria Sci. Exch.* 9(3), 221, 1988; *Chem. Abstr.*, 110, 191414k, 1989.

51. Khaustovich, I. P., Use of TUR on apple trees, *Khim. Sel'sk. Khoz.*, 5, 31, 1991; *Chem. Abstr.*, 115, 87415d, 1991.

52. Dogra, J. V. V. and Sinha, S. K. P., Cyocel induced changes in vitamin C contents in *Phyllanthus urinaria* Linn. shoot, *Biol. Bull. India*, 5(1), 26, 1983; *Chem. Abstr.*, 99, 18018r, 1983.

53. Vytskaya, I. Ya. and Sergeeva, L. S., Effect of chlorocholine chloride and its derivatives on carrot yield, *Ovoshchevod. Sev.-Zapadn. Zone RSFSR*, Semenov, G. V., Ed., Leningrad, 1982, 55; *Chem. Abstr.*, 100, 134188j, 1984.

54. Zhukova, P. S., New growth regulators for tomatoes and cucumbers, *Vestsi Akad. Navuk BSSR, Ser. Sel'skagaspad. Navuk*, 4, 54, 1987; *Chem. Abstr.*, 108, 217703q, 1988.

55. Kalmykova, T. I., Effect of retardants on growth, physiological-biochemical, and generative processes in citrus plants, *Prog. Teknol. Plodovod. Vinograd.*, 92, 1982; *Chem. Abstr.*, 98, 193277a, 1983.

56. Shadeque, A. and Pandita, M. L., Effect of cyocel (CCC) as foliar spray on growth, yield and quality of potato (*Solanum tuberosum* L.), *J. Res.—Assam Agric. Univ.*, 3(1), 34, 1982; *Chem. Abstr.*, 101, 185968s, 1984.

57. Borisova, V. P. and Kolobkova, V. I., Application of TUR during raising of tomato transplants and its effect on the yield, *Tr. Dal'nevost. NII S.-Kh.*, 24, 130, 1978; *Hortic. Abstr.*, 50, 5269, 1980.

58. Emmerikh, F. D., Effect of chlorocholine chloride and presowing seed hardening on tomato resistance to negative temepratures, *Fiziol. Biokhim. Kul't. Rast.*, 14(5), 506, 1982; *Chem. Abstr.*, 98, 12881z, 1983.

59. Martines Cano, A., Lilov, D., and Baev, Kh., Effect of some growth regulators in retarding anthesis of the apple cultivar Goldspur, *Fiziol. Rast. (Sofia)*, 8(4), 33, 1982; *Chem. Abstr.*, 98, 174722a, 1983.

60. Kur'yata, V. G., Remenyuk, G. L., Knysh, V. M., and Prokopenko, L. S., Effect of chlorocholine chloride on the growth processes, yield and quality of black-fruit Sorbus, *Fiziol. Biokhim. Kul't. Rast.*, 16(6), 548, 1984; *Chem. Abstr.*, 102, 91314u, 1985.

61. Gowda, N. C., Effects of inter-row spacings and Cyocel on growth, yield and quality attributes of bhendi (*Abelmoschus esculentus* (L.) Moench.) cv. Pusa Sawani, *Thesis Abstracts, Haryana Agric. Univ.*, 9(4), 345, 1983; *Hortic. Abstr.*, 55, 1176, 1985.

62. Kur'yata, V. G., Remenyuk, G. L., and Prokopenko, L. S., Effect of chlorocholine chloride on the growth, yield capacity and production quality of small-fruit crops, *Fiziol. Biokhim. Kul't. Rast.*, 17(4), 366, 1985; *Chem. Abstr.*, 103, 173905x, 1985.

63. Kerin, V. and Ivanov, P., Effect of the plant growth regulator chlorocholine chloride (CCC) on some chemical properties of tomatoes, *B"lgarski Plodove Zelenchutsi i Konservi*, 12, 24, 1979; *Food Sci. Technol. Abstr.*, 13(1), J70, 1981.

64. Budykina, N. P., Volkova, R. I., Drozdov, S. N., Klykova, V. V., and Derusov, V. S., Use of chlorocholine chloride and kinetin to accelerate ripening of tomatoes in the northern zone of hothouse vegetable production, *Khim. S-Kh. Khoz.*, 2, 38, 1984; *Chem. Abstr.*, 100, 169852y, 1984.

65. Drozdov, S. N., Volkova, R. I. and Klykova, V. V., Possible use of synthetic growth regulators for increasing tomato productivity in the northern zone of hothouse vegetable growing, *Termorezist. Prod. S-Kh. Rast.*, 119, 1984; *Chem. Abstr.*, 103, 49657d, 1985.

66. Kur'yata, V. G., Sogur, L. N., and Dabizhuk, T. M., Combined application of chlorocholine chloride and Camposan M to red raspberries: physiological background, *Fiziol. Biokhim. Kul't. Rast.*, 20(3), 296, 1988; *Chem. Abstr.*, 109, 33799q, 1988.

67. Volkova, R. I., Budykina, N. P., Kurets, V. K., Seliverstova, L. A., and Rudakova, G. Ya., Application of chelates in the cultivation of vegetables in greenhouses in northern climates, *Khim. Sel'sk. Khoz.*, 1, 45, 1987; *Chem. Abstr.*, 106, 175211d, 1987.

68. Hankin, L. and Hill, D. E., Yield losses and increases in certain vitamins in cucumbers treated with the herbicide 3-amino-2,5-dichlorobenzoate and urea fertilizer, *Soil Sci.*, 134(3), 193, 1982.

69. Lukasik, S., Achremowicz, B., Kulpa, D., and Fraczek, T., Effect of chlorflurenol on the yield and technological value of red pepper (*Capsicum annuum* L.), *Acta Aliment*, 12(2), 101, 1983.

70. Decallonne, J. R. and Meyer, J. A., Investigation on the influence of commonly used pesticides and growth-promoting chemicals on the quality and composition of tomatoes, *Med. Fac. Landbouww. Rijksuniv. Gent.*, 49/3b, 969, 1984.

71. Byast, T. H. and Hance, R. J., Effect of chlorthiamid on the ascorbic acid content of black currants and gooseberries, *Weed Res.*, 12, 272, 1972.

72. Soler, A., The use of Fruitone 3 CPA as a growth regulator on pineapple (Smooth Cayenne) in the Ivory Coast, *Fruits*, 40(1), 31, 1985; *Hortic. Abstr.*, 55, 4867, 1985.

73. Zhukova, P. S. and Anikhovskaya, T. E., Use of retardants and growth regulators on tomato seedlings, *Vestsi Akad. Navuk BSSR, Ser. Sel'skagaspad. Navuk*, 2, 92, 1990; *Chem. Abstr.*, 114, 57458g, 1991.

74. Bychenko, N. I. and Bychenko, I. I., The effect of some fungicides on vine development and productivity, *Vinodel. Vinograd. SSSR*, 32(1), 36, 1972; *Hortic. Abstr.*, 42, 7606, 1972.

75. Shestopalova, N. A., Application of dacthal in irrigated onions, *Khim. Sel'sk. Khoz.*, 12(3), 53, 1974; *Hortic. Abstr.*, 44, 8652, 1974.

76. Chetverikova, Z. N., Effect of herbicides on weed infestation, yield and quality of potatoes in the wooded steppe area of the Tyumen region, *Nauchnye Tr., Omskii Ordena Lenina Sel'skokhozyaistvennyi Institut imeni S.-M.-Kirova*, 148, 109, 1976; *Food Sci. Technol. Abstr.*, 10(9), J1327, 1978.

77. Drake, S. R., Proebsting, E. L., Jr., and Nelson, J. W., Influence of growth regulators on the quality of fresh and processed "Bing" cherries, *J. Food Sci.*, 43, 1695, 1978.

78. Drake, S. R., Proebsting, E. L., Jr., Carter, G. H., and Nelson, J. W., Effect of growth regulators on the ascorbic acid content, drained weight and color of fresh and processed "Rainer" cherries, *J. Am. Soc. Hortic. Sci.*, 103(2), 162, 1978.

79. Gancheva, I., Effect of the fungicides Rovral, Sumilex and Derosal used in the control of strawberry gray rot (*Botrytis cinerea* Pers.) on some plant physiological processes and on the biochemical composition of the fruits, *Gradinar. Lozar. Nauka*, 19(6), 30, 1982; *Chem. Abstr.*, 99, 1722t, 1983.

80. Kur'yata, V. G., Dabizhuk, T. M., and Lobov, V. R., Physiological basis for the use of dextrel as a growth retardant, *Fiziol. Biokhim. Kul't. Rast.*, 23(1), 45, 1991; *Chem. Abstr.*, 114, 223452x, 1991.

81. Zhukova, P. S., Zabara, Yu. M., Pushkina, G. I., and Anikhovskaya, T. E., Effect of growth regulators on the yield and pickling quality of cucumbers, *Vestsi Akad. Navuk BSSR, Ser. Sel'skagaspad. Navuk*, 1, 50, 1990; *Chem. Abstr.*, 113, 128032f, 1990.

82. Bartholomew, E. T., Stewart, W. S., and Carman, G. E., Some physiological effects of insecticides on citrus fruits and leaves, *Bot. Gaz. (Chicago)*, 112, 501, 1950.

83. Fedtke, C. and Strang, R. H., Synergistic activity of the herbicide safner Dichlormid with herbicides affecting photosynthesis, *Z. Naturforsch.*, 45c, 565, 1990.

84. Zelenin, V. M., Effect of the herbicide dicryl on the yield and chemical composition of cucumbers, *Aktual. Vopr. Zashch. Rast.*, Mordvintsev, P. V., Ed., 1982, 76; *Chem. Abstr.*, 98, 193297g, 1983.

85. Iacob, M., Matei, I., Voica, E., and Vladu, S., The influence of some treatments against *Grapholitha molesta* Busck, applied on the irrigated sands close to the Jiu River, on the quality and quantity of peach fruits, *An. Inst. Cercet. Prot. Plant*, *Acad. Stiinte Agric. Silvice*, 16, 385, 1981; *Chem. Abstr.*, 99, 153830q, 1983.

86. Ivantsov, N. K., Effects of using herbicides over several tuber generations on the productivity of potatoes, *Khim. S-Kh. Khoz.*, 3, 40, 1984; *Chem. Abstr.*, 100, 204900m, 1984.

87. Cwiertniewska, E., Effects of the organophosphorus insecticides, dimethoate, thiometon, and methyl demeton-S, on the content of vitamin C and reducing sugars in the fruits of black currant. II. Effect of dimethoate on the vitamin C content of the black curant fruit, *Rocz. Panstw. Žakl. Hig.*, 24(2), 151, 1973; *Chem. Abstr.*, 79, 133470f, 1973.

88. Szajkowski, Z. and Gertig, H., The effect of spraying gooseberries and currants with Karathane 25 on the level of some vitamins in the fruits, *Bromatol Chem. Toksykol.*, 11(3), 297, 1978; *Hortic. Abstr.*, 49, 6666, 1979.

89. Hammett, L. K. and Monaco, T. J., Effect of oryzalin and other herbicide treatments on selected quality factors of sweet potatoes, *J. Am. Soc. Hortic. Sci.*, 107(3), 432, 1982.

90. Smirnova, E. V., Stepanova, E. N., and Konovalova, L. V., Effect of pesticides on the nutritive value of potatoes, *Khim. Sel'sk. Khoz.*, 14(9), 75, 1976; *Food Sci. Technol. Abstr.*, 9(3), J425, 1977.

91. Lisinska, G., Effect of various factors on the composition of potatoes and quality of potato crisps made therefrom, *Zeszyty Nauk. Akad. Rolniczej Wroclaw Rosprawy (Poland)*, 31, 1981; *Food Sci. Technol. Abstr.*, 16(11), J1930, 1984.

92. Kocourek, V., Hajslova, J., Darebnik, V., and Musil, J., Studies on the effect of ethylenebisdithiocarbamate fungicides on the quality of potatoes, *Sb. UVTIZ. Potavin. Vedy*, 4(1), 15, 1986; *Chem. Abstr.*, 105, 148068k, 1986.

93. Gill, D. S., Jwa, M. S. B., Singh, R., and Brar, W. S., Effect of ethrel spraying on lemon fruit development (*Citrus lemonoides* L.), *Sci. Cult.*, 48(11), 404, 1982; *Chem. Abstr.*, 99, 18083h, 1983.

94. Sadhu, M. K., and Chattopadhyay, G., Effect of a post-harvest fruit dip in ethephon on the ripening of litchi fruits, *J. Hortic. Sci.*, 64(2), 239, 1989.

95. Shihab, K. H. and Abd-Al-Hadi, A. M., Use of ethrel for the ripening of Navel orange and its effect on fruit quality and composition, *J. Agric. Water Resour. Res.*, 4(1), 137, 1985; *Chem. Abstr.*, 103, 155749v, 1985.

96. Nagy, S. and Wardowski, W. F., Effect of agricultural practice, handling, processing, and storage of fruits, in *Nutritional Evaluation of Food Processing*, Karmas, E. and Harris, R. S., Eds., Van Nostrand, Reinhold, New York, 1988, chapter 4.

97. Shanmugavelu, K. G., Srinivasan, C., and Thamburaj, S., Effect of Ethrel (ethephon, 2-chloroethylphosphonic acid) on pumpkin (*Cucurbita moschata* Poir), *S. Indian Hortic.*, 21(3), 94, 1973; *Hortic. Abstr.*, 45, 4917, 1975.

98. Singh, S. K., and Singh, V. P., Effect of ethrel (2-chloroethyl phosphonic acid) on some parameters of fruit development and yield of tomato var. Pusa buby, *Natl. Acad. Sci. Lett. (India)*, 4(6), 235, 1981; *Chem. Abstr.*, 96, 117436v, 1982.

99. Arora, S. K., Pandita, M. L., and Singh, K., Effect on Ethrel on chemical composition of tomato (*Lycoperscion esculentum* Mill.) varieties HS-101 and HS-102, *Haryana Agric. Univ. J. Res.*, 13(2), 254, 1983; *Chem. Abstr.*, 101, 34430j, 1984.

100. Zhukova, P. S., and Anikhovskaya, T. E., Use of growth regulators on field tomatoes, *Khim. Sel'sk. Khoz.*, 4, 54, 1988; *Chem. Abstr.*, 109, 2387z, 1988.

101. Al-Jebori, K. H., and Al-Ani, A. M., Effect of ethrel on ripening of clementine (*Citrus reticulata*) and its effect on fruit quality, *J. Agric. Water. Resour. Res.*, 4(3), 221, 1985; *Chem. Abstr.*, 104, 33257h, 1986.

102. Dzhibladze, K. M., Effect of synthetic growth regulators on fruits ripening in the mandarin in postharvest treatment, *Subtrop. Kul't.*, 5, 123, 1983; *Chem. Abstr.*, 100, 84417k, 1984.

103. Kozlova, V. F., Effect of preharvest drying of the leaves of bulb onions on their preservation, *Khranenie Plodoovoshchn. Prod. Kartofelya*, 120, 1983; *Chem. Abstr.*, 101, 37350u, 1984.

104. Shanmugavelu, K. G., Rao, V. N. M., and Srinivasan, C., Studies on the effect of certain plant regulators and boron on papaya (*Carica papaya* L.), *S. Indian Hortic.*, 21(1), 19, 1973; *Hortic. Abstr.*, 44, 7180, 1974.

105. Graifenberg, A. and Giustiniani, L, Influence of ethrel on the uniform ripening of peppers for mechanical harvesting, *Rivista della Ortoflorofrutticoltura Italiana*, 64(1), 63, 1980; *Food Sci. Technol. Abstr.*, 13(8), J1212, 1981.

106. Shaybany, B. and Sharifi, H., Effect of pre-harvest applications of ethephon on leaf abscission, fruit drop and constituents of fruit juice of pomegranates, *J. Hortic. Sci.*, 48, 293, 1973.

107. Ingle, G. S., Khedkar, D. M., and Dabhade, R. S., Effect of growth regulators on ripening of sapota fruit (*Achras sapota* Linn), *Indian Food Packer*, 36(1), 72, 1982; *Chem. Abstr.*, 98, 84820b, 1983.

108. Elkner, K., Bakowski, J., and Babik, I., The effect of Ethrel on the chemical composition of tomatoes cultivated for processing, *Biuletyn Warzywniczy*, 27, 531, 1984; *Hortic. Abstr.*, 56, 6140, 1986.

109. Kaukovirta, E. and Murto, O., Effects of ethephon on fruit ripening and abscission in some black and red currant varieties cultivated in Finland, *J. Sci. Agric. Soc. Finland*, 48(2), 170, 1976; *Food Sci. Technol. Abstr.*, 9(1), J101, 1977.

110. Andam, C. J., Response of maturing "Carabao" mango fruits to pre-harvest ethephon application, *NSTA Technol. J.*, 8(3), 4, 1983; *Chem. Abstr.*, 100, 46963c, 1984.

111. Watada, A. E., Aulenbach, B. B., and Worthington, J. T., Vitamins A and C in ripe tomatoes as affected by stage of ripeness at harvest and by supplementary ethylene, *J. Food Sci.*, 41, 856, 1976.

112. Kader, A. A., Morris, L. L., Stevens, M. A., and Albright-Holton, M., Composition and flavor quality of fresh market tomatoes as influenced by some postharvest handling procedures, *J. Am. Soc. Hortic. Sci.*, 103(1), 6, 1978.

113. Lee, Y. C., Effect of Ethephon treatment on vitamin and mineral contents of fresh tomatoes, *Korean J. Food Sci. Technol.*, 15(4), 409, 1983; *Chem. Abstr.*, 100, 66885g, 1984.

114. Tajiri, T., Studies on cultivation and keeping quality of bean sprouts. VII. Improvement of bean sprouts production by the intermittent treatment of ethylene gas, *Nippon Shokuhin Kogyo Gakkaishi*, 29(10), 596, 1982; *Chem. Abstr.*, 98, 174720y, 1983.

115. Mustard, M. J. and Lynch, S. J., Effect of various factors upon the ascorbic acid content of some Florida-grown mangos, *Fla. Agr. Expt. Stn. Bull.*, 406, 1, 1945; *Chem. Abstr.*, 42, 5580b, 1948.

116. Raghav, M., Singh, A., and Srivastava, S., Biochemical studies on regulatory effects of herbicides in onion (*Allium cepa* L.), *J. Recent Adv. Appl. Sci.*, 2(1), 215, 1987; *Chem. Abstr.*, 107, 231321z, 1987.

117. Singh, A. B., Singh, A., Abidi, A. B., and Singh, R. P., Regulatory effect of herbicides on the composition of tomato fruits, *Sci. Cult.*, 53(6), 190, 1987.

118. Shawky, I. and Shaaban, A. M., Effect of insecticide sprays on fruit quality of Balady mandarin, *Egyptian J. Hortic.*, 1(1), 67, 1974; *Hortic. Abstr.*, 44, 9011, 1974.

119. Singh, A. K., Singh, B. P., and Ram, R. B., Effect of gibberellic acid on physical and chemical characters of ber fruit (*Ziziphus mauritiana*), *Bangladesh Hortic.*, 10(1), 47, 1982; *Hortic. Abstr.*, 54, 3061, 1984.

120. Karabanov, I. A. and Šubert, V. A., The after-effect of gibberellin on the ascorbic acid content in black currant berries in connection with mineral nutrition, *Fiziol. Rast.*, 17, 177, 1970; *Hortic. Abstr.*, 40, 8044, 1970.

121. Gvamichavas, N. E., Kezeli, T. A., Tarasashvili, K. M., Kikvidzf, M. V., Takaishvili, sT. K., and Piranishvili, N. S., The effects of gibberellin and riboflavin on growth and

fruiting of the grapevine, *Izv. Akad. Nauk Gruz. SSR, Ser. Biol.*, 14(6), 400, 1988; *Chem. Abstr.*, 110, 149753v, 1989.

122. Biswas, B., Ghosh, S. K., Ghosh, B., and Mitra, S. K., Effect of growth substances on fruit weight, size and quality of guava cv. L-49, *Indian Agric.*, 32(4), 245, 1988; *Chem. Abstr.*, 111, 110946f, 1989.

123. Furuta, M. and Aketagawa, T., Effect of gibberellin and N^6-benzyladenine on preservation of vegetables, *Niigata-kenshkuhin Kenkyusho Kenkyu Hokoku*, 17, 13, 1980; *Chem. Abstr.*, 96, 179711q, 1982.

124. Rao, S. N., Rao, C. S., and Rao, P. B., Effect of gibberellic acid on loquat (*Eriobotrya japonica* Lindl.), *J. Hortic. Sci.*, 38, 1, 1963.

125. Khader, S. E. S. A., Effect of preharvest application of GA_3 on postharvest behavior of mango fruits, *Scientia Hortic.*, 47(3-4), 317, 1991.

126. Rath, S. and Rajput, C. B. S., Effect of β-naphthoxyacetic acid and gibberellic acid on chemical composition of mango fruits, *Orissa J. Agric. Res.*, 3(2), 155, 1990; *Chem. Abstr.*, 114, 242711x, 1991.

127. Hee, P. B. and Suk, K. M., The effect of gibberellin on the content of vitamin C during the growth of mung bean sprouts, *Han'guk Yongyang Siklyong Hakhoechi*, 10(1), 117, 1981; *Chem. Abstr.*, 98, 142116q, 1983.

128. Kefford, J. F., The chemical constituents of citrus fruits, *Adv. Food Res.*, 9, 285, 1959.

129. Nakasone, H. Y., Papaya, in *CRC Handbook of Fruit Set and Development*, Monselise, S. P., Ed., CRC Press, Boca Raton, Fla., 1986, 277.

130. Kiryukhin, V. P., Use of GA and IAA for growing purposes, *Khim. sel'. Khoz.*, 7(3), 55, 1969; *Field Crop Abstr.*, 22, 3000, 1969.

131. Kumar, P., Alka, Rao, P., and Baijal, B. D., Effect of some growth regulators on plant growth, tuber initiation, yield and chemical composition of potato (*Solanum tuberosum* L.), *Pak. J. Bot.*, 13(1), 69, 1981; *Chem. Abstr.*, 96, 196644n, 1982.

132. Brugovitzky, E. and Popovici, G., The effect of gibberellic acid on the ascorbic acid content of two rose species, *Naturwissenschaften*, 53, 312, 1966; *Hortic. Abstr.*, 36, 7103, 1966.

133. Kim, S. O., Effect of growth regulators on growth and vitamin C biosynthesis during germination of soybeans, *J. Korean Soc. Food Nutr.*, 17(2), 115, 1988; *Food Sci. Technol. Abstr.*, 21(12), J149, 1989.

134. Mihteleva, L. A., Gibberellin and berry productivity, *Priroda*, 55(9), 105, 1966; *Hortic. Abstr.*, 37, 2435, 1967.

135. Avdonin, N. S. and Kuan, T. J., The effect of gibberellin on the yield, quality and metabolism of tomatoes and lettuce, *Vestn. S-Kh. Nauki*, 6(1), 32, 1961; *Hortic. Abstr.*, 31, 4665, 1961.

136. Srivastava, R. P. and Srivastava, K. K., Effect of presowing treatment with growth substances on the chemical composition of important vegetable crops. I. Ascorbic acid content in (a) tomato, (b) pea and (c) capsicum, *Sci. Cult.*, 30(6), 292, 1964.

137. Oza, A. M. and Rangnekar, Y. B., Effect of application of gibberellic acid on the ascorbic acid content of tomato (*Lycopersicon esculentum* Mill.), *Indian J. Agric. Sci.*, 39, 980, 1969.

138. Emmerikh, F. D., Treatment of tomato seedlings with gibberellin, *Agrokhimiya*, 1, 122, 1974; *Food Sci. Technol. Abstr.*, 8(3), J378, 1976.

139. Achremowicz, B., Lukasik, S., and Kulpa, D., Effect of some morphactins on the quality of tomato and *Capsicum* fruits, *Ann. Univ. Mariae Curie-Sklodowska, E. (Agricultura)*, 37, 277, 1982; *Hortic. Abstr.*, 56, 393, 1986.

140. Saleh, M. M. S. and Abdul, K. A., Effects of gibberellic acid and Cyocel on growth, flowering and fruiting of tomato *Lcopersicon esculentum* Mill. Plants, *Mesopotamia J. Agric.*, 15(1), 137, 1980.

141. Clijsters, H., Effect of gibberellin and fertilizes on the growth, mineral composition and contents of water, pigments and ascorbic acid of some horticultural plants, *Agricultura (Louvain)*, 9, 151, 1961; *Hortic. Abstr.*, 32, 6548, 1962.

142. Shibutani, S. and Kinoshita, K., Studies on the influence of gibberellin sprays on celery. (The influence of gibberellin on components of celery, such as chlorophyll, total carotene, vitamin C), *Sci. Rep. Fac. Agric. Okayama*, 27, 27, 1966; *Hortic. Abstr.*, 36, 6549, 1966.

143. Srivastava, R. P. and Agarwal, N. C., Effect of gibberellic acid sprays on the fruit drop, size, weight and chemical composition of Red Delicious apples, *Sci. Cult*, 32, 546, 1966; *Hortic. Abstr.*, 37, 4442, 1967.

144. Srivastava, R. P. and Agarwal, N. C., Effect of gibberellic acid on fruit crops. I. Apple, *Indian J. Hortic.*, 25, 170, 1968.

145. Kawecki, Z., Effect of various management systems on the growth and yield of red currants and chemical compositions of berries, *Zesz. Nauk. Akad. Roln.-Tech. Olsztynie, Roln.*, 26, 161, 1979; *Chem. Abstr.*, 91, 103627t, 1979.

146. Kawecki, Z., Wazbinska, J., and Kopytowski, J., Effect of increased amounts of Gesatop 50 on the growth and yield of two gooseberry cultivars, *Zesz. Nauk. Akad. Roln.-Tech. Olsztynie, Roln.*, 38, 151, 1983; *Chem. Abstr.*, 101, 224726h, 1984.

147. Sal'kova, E. G., Bulantseva, E. A., Zvyagintseva. Yu. V., and Zlobinskaya, O. I., Hydrel and Dihydrel as accelerators of ripening in hothouse tomatoes, *Khim. Sel'sk. Khoz.*, 2, 40, 1984; *Chem. Abstr.*, 100, 169853z, 1984.

148. Guseinov, Yu. A., Effect of Hydrel on the uniformity of ripening in tomatoes, *Khim. S-Kh. Khoz.*, 10, 40, 1987; *Chem. Abstr.*, 108, 2118s, 1988.

149. Mishra, R. S. and Khatai, M., Effect of growth substances on the ascorbic acid content of fresh green and ripe fruits of chili, *Indian J. Sci. Ind. Sect. A*, 3, 177, 1969; *Hortic. Abstr.*, 40, 6448, 1970.

150. Lee, S. H., and Chung, D. H., Studies on the effects of plant growth regulator on growth and nutrient composition in soybean sprout, *Hanguk Nonghwa Hakhoe Chi.*, 25(2), 75, 1982; *Chem. Abstr.*, 97, 210370z, 1982.

151. Maurya, K. R., Effect of IBA on germination, yield, carotene and ascorbic acid content of carrots, *Indian J. Hortic.*, 43(1/2), 118, 1986.

152. Zhukova, P. S. and Anikhovskaya, T. E., Use of growth regulators in cucumber cultivation, *Khim. Sel'sk. Khoz.*, 5, 28, 1991; *Chem. Abstr.*, 115, 87414c, 1991.

153. Maruszewska, M. and Gertig, H., Effect of "Karbatox 75 suspension" on content of some vitamins in cabbages, *Bromatol. Chem. Toksykol.*, 15(1/2), 19, 1982; *Food Sci. Technol. Abstr.*, 15(9), J1470, 1983.

154. Stepanova, Z. A., Effect of physiologically active substances on biochemical composition of table beet roots, in *Ispol'z. Regul. Rosta Polim. Mater. Ovoshchevod.*, Burmistrov, A. D., Ed., 1984, 29; *Chem. Abstr.*, 104, 47070d, 1986.

155. Wu, J. R., Shao, Y. L., Wei, A. F., Qian, L. H., and Yan, Z. X., Effects of krenite on increase of quality and quantity in tomato fruits, *Acta Hortic. Sinica*, 13(2), 107, 1986; *Hortic. Abstr.*, 56, 7865, 1986.

156. Li, H. C. and Hu, Y. C., Effects of lead arsenate sprays on the physiological functions and fruit quality of sweet oranges, *Yuan Yi Hsueh Pao*, 3(2), 129, 1964; *Chem. Abstr.*, 62, 7050c, 1965.

157. Nelson, E. M. and Mottern, H. H., Effect of lead arsenate spray on the composition and vitamin content of oranges, *Am. J. Public Health*, 22, 587, 1932.

158. Vittoria, A., Treatment with hexachlorocyclohexane and introduction of a concept of semiquantitative cytochemical analysis of total ascorbic acid, *Boll. Soc. Ital. Sper.*, 29, 461, 1953; *Chem. Abstr.*, 49, 4926i, 1955.

159. Dobrzanski, A., Effect of herbicides on some indices of the nutritional value of vegetables, *Przemysl Fermentacyjny i Owocowo Warzywny*, 22(6), 21, 1978; *Food Sci. Technol. Abstr.*, 11(7), J1288, 1979.

160. Kankanyan, A. G., Nazaryan, R. G., and Ananyan, A. M., Effect of herbicide on qualitative indexes of carrots, *Izv. Sel'skohoz. Nauk*, 14(1), 119, 1971; *Chem. Abstr.*, 75, 75192w, 1971.

161. Zhukova, P. S., Rogozhnikov, V. G., Dubenetskaya, M. M., and Voitik, N. P., Effect of linuron on some indexes of the chemical composition of carrots, *Vestsi Akad. Navuk BSSR, Ser. Sel'skagaspad Navuk*, 1, 67, 1982; *Chem. Abstr.*, 96, 176027k, 1982.

162. Galeev, N. A. and Galeev, R. R., Herbicides for carrot plantings, *Zashch. Rast. (Moscow)*, 11, 37, 1982; *Chem. Abstr.*, 98, 67041f, 1983.

163. Devochkin, F. A., and Evsyutina, Z. S., Characteristics of the growth and yield of carrots following mulching and chemical methods for weed control, *Dokl. TSKhA.*, 216, 100, 1976; *Chem. Abstr.*, 86, 26851d, 1977.

164. Ciecko, Z., and Nowak, G., Effect of pesticides on yield and quality of potato tubers, *Pestycydy (Warsaw)*, 3, 17, 1989; *Chem. Abstr.*, 114, 37772b, 1991.

165. Mukerjee, D. and Chava, N. R., Storage response of potatoes after pre-harvest application of maleic hydrazinde, *J. Indian Bot. Soc.*, 67, 103, 1988.

166. Chen, Z. S. and Lin, J. Q., Effect of controlling summer shoots with MH on growth and fruiting on citrus, *Fujian Agric. Sci. Technol.*, 1, 32, 1982; *Hortic. Abstr.*, 53, 2157, 1983.

167. Leth, T. and Kirknel, E., Nutrient content in potatoes after treatment with pesticides, *Publikation Statens Levnedsmiddelinstitut*, 124, 1986; *Food Sci. Technol. Abstr.*, 19(3), J90, 1987.

168. Zhukova, P. S. and Belova, V. I., Effectiveness of herbicides on white cabbage plantings with the application of various doses of potassium fertilizers, *Khim. Sel'sk. Khoz.*, 10, 36, 1983; *Chem. Abstr.*, 99, 208058b, 1983.

169. Zhukova, P. S. and Kharitonova, A. P.,,, Effect of herbicides on the composition of onions, *Vestsi Akad. Navuk BSSR, Ser. Sel'skagaspad. Navuk*, 1, 81, 1988; *Chem. Abstr.*, 109, 49954x, 1988.

170. Baraniak, B., Bubicz, M., and Bochniarz, M., Effect of herbicides on carotene, α-tocopherol and l-ascorbic acid content in fodder kale, *Pamiet Pulawski*, 77, 143, 1982; *Chem. Abstr.*, 100, 81142f, 1984.

171. Zhukova, P. S., Kharitonova, A. P., and Sil'vanovich, S. F., Herbicide content of onions in the ploughed soil layer in relation to mineral fertilization, *Vestsi Akad. Navuk BSSR, Ser. Sel'skagaspad. Navuk*, 2, 80, 1986; *Chem. Abstr.*, 105, 92873z, 1986.

172. van Wambeke, E., van Achter, A., and van Assche, A., Influence of repeated soil disinfection with methyl bromide on bromide content in different tomato cultivars and some qaulity criteria of these tomato fruits, *Med. Fac. Landbouww. Rijksuniv. Gent.*, 44, 895, 1979.

173. Weyns, J., Roucoux, P., and Decallonne, J., Experiemntal characterization of the influence of a few pesticides on the quality and composition of tomatoes, *Meded. Fac. Landbouwwet., Rijksuniv. Gent.*, 48(4), 1175, 1983.

174. Kotsev, K., Effect of some systemically acting fungicides on the content of dry matter, acids, sugars, and vitamin C in apple fruits, in *Systemfungiz. Int. Sympl.*, Lyr, H. and Polter, C., Eds., Akad. Verlag, Berlin, 1975, 259; *Chem. Abstr.*, 88, 59282p, 1978.

175. Mondy, N. I. and Munshi, C. B., Chemical composition of potato as affected by the herbicide, metribuzin; enzymatic discoloration, phenols and ascorbic acid content, *J. Food Sci.*, 53(2), 475, 1988.

176. Kasimov, Kh. A., Baratov, K. B., and Babaev, I. I., Effect of Sencor on the food value of potatoes, *Izv. Akad. Nauk Tadzh. SSR, Otd. Biol. Nauk.*, 1, 110, 1980; *Chem. Abstr.*, 93, 180848j, 1980.

177. Zhukova, P. S. and Rogozhnikov, V. G., Cultivation of tomatoes using herbicides in conjunction with agrotechnical methods, *Vestsi Akad. Navuk BSSR, Ser. Sel'skagaspad. Navuk*, 4, 82, 1983; *Chem. Abstr.*, 100, 81156p, 1984.

178. Alieva, Z. N. and Gasanov, T. G., Effect of Mugan on components of the respiratory chain in barley and mung bean seedlings, *Issled. Spetsifichnosti Deistviya Fiz.-Khim. Faktorov Biol. Ob'ekty*, 89, 1990; *Chem. Abstr.*, 115, 44156c, 1991.

179. Patil, V. S. and Patil, A. A., Effect of NAA on certain quality attributes of cabbage varieties, *J. Maharashtra Agric. Univ.*, 14(2), 232, 1989; *Hortic. Abstr.*, 62, 3900, 1992.

180. Patil, U. B., Sangale, P. B., and Desai, B. B., Chemical regulation of yield and composition of chili (*Capsicum annum* L.), fruits, *Curr. Res. Rep.*, 1(1), 39, 1985; *Chem. Abstr.*, 103, 137052h, 1985.

181. Sharma, S. B. and Dhillon, B. S., Effect of zinc sulfate and growth regulators on the quality of litchi fruits, *Indian Food Packer*, 39(2), 29, 1985; *Chem. Abstr.*, 106, 14592p, 1987.

182. Kaundal, G. S., Minhas, P. P. S., and Grewal, G. P. S., Chemical deblossoming of peach (*Prunus persica* Batsch) as an aid to quality improvement, *Sci. Cult.*, 55(10), 418, 1989.

183. Misra, S. G., Gupta, A. K., Varshney, M. L., and Sharma, K. N., Effect of naphthalene acetic acid and chelated zinc on vitamin C content and yield of cabbage (*Brassica oleracea var. capitata*), *Prog. Hortic.*, 16(1/2), 92, 1984; *Hortic. Abstr.*, 55, 8595, 1985.

184. Avetisyan, K. V. and Kostanyan, A. V., Effect of Nimrod on quality of cucumbers, *Biol. Zh. Arm.*, 38(7), 623, 1985; *Chem. Abstr.*, 103, 173888u, 1985.

185. Amirov, B. V., Manankov, M. E. and Saparov, A. S., Effectiveness of combined application of herbicide and fertilizers in cabbage cultivation, *Vestn. S-Kh. Nauki Kaz.*, 11, 34, 1990; *Chem. Abstr.*, 115, 87419h, 1991.

186. Sites, J. W. and Camp, A. F., Producing Florida citrus for frozen concentrate, *Food Technol.*, 9, 361, 1955.

187. Nikolova, G. and Ivanov, A., Effect of some soil herbicides on weeds, growth and yield of black currant, *Gradinar. Lozar. Nauka*, 20(5), 49, 1983; *Hortic. Abstr.*, 54, 6076, 1984.

188. Kunert, K. J., The diphenyl-ether herbicide Oxyflurofen: a potent inducer of lipid peroxidation in higher plants, *Z. Naturforsch.*, 39c, 476, 1984.

189. Mitchell, J. W., Ezell, B. D., and Wilcox, M. S., Effect of *p*-chlorophenoxyacetic acid on the vitamin C content of snap beans following harvest, *Science*, 109, 202, 1949.

190. Law, M. Y., Charles, S. A., and Halliwell, B., Glutathione and ascorbic acid in spinach (*Spinacia oleracea*) chloroplasts, *Biochem. J.*, 210, 899, 1983.

191. Schuphan, W. and Weinmann, W., The effect of parathion treatment of spinach on its valuable constituents and on the biological value of the leaf proteins, *Z. Pflanzenk.*, 71, 1964; *Hortic. Abstr.*, 34, 2957, 1964.

192. Hivon, K. J., Doty, D. M., and Quackenbush, F. W., Ascorbic acid and ascorbic acid oxidizing enzymes of green bean plants deficient in manganese, *Plant Physiol.*, 26, 832, 1951.

193. Sinha, B. K., Chaubey, V. K., and Chaturvedi, R. C., Influence of penicillin on growth and composition of barley, *Agric. Sci. Dig.*, 1(3), 182, 1981; *Chem. Abstr.*, 99, 171249m, 1983.

194. Aslanyan, G. Ts., Tatevosyan, A. E., and Aleksanyan, D. S., Effect of Thiodan and Plictran on the nutritional value of greenhouse cucumbers, *Biol. Zh. Arm.*, 36(8), 712, 1983; *Chem. Abstr.*, 99, 208110n, 1983.

195. Freeman, J. A., The control of strawberry fruit rot in coastal British Columbia, *Can. Plant. Dis. Survey*, 44, 96, 1964; *Hortic. Abstr.*, 35, 774, 1965.

196. Dubenetskaya, M. M., Patent, R. L., Voitik, N. P., Voinova, I. V., and Krasnaya, S. D., Study on the nutritive value of carrots grown with the use of the herbicide prometryn, *Vopr. Pitan.*, 6, 50, 1981; *Chem. Abstr.*, 96, 33571x, 1982.

197. Kurbatskii, N. Ya., Tsikin, Yu. E., and Yakovlev, A. P., Effect of syn-triazines on the content of pigments and vitamin C in plant leaves, *Pochvy Udobr.*, *Urozhai*, 225, 1976; *Chem. Abstr.*, 89, 124474q, 1978.

198. Sivtsev, M. V. and Kuznetsova, E. A., Carbohydrates, ascorbic acid, and glutathione in leaves of vegetables treated with herbicides, *Fiziol. Biokhim. Kul't. Rast.*, 9(4), 406, 1977; *Chem. Abstr.*, 87, 128445n, 1977.

199. Holtkamp, S., Untersuchungen zum Einfluss von Kreb 50 (Propyzamid) auf den Gehalt an Vitamin C in Kopfsalat (*Lactuca sativa* L.), *Nachrichenbl. Dtsh. Pflanzenschutzdienstes* (*Braunschweig*), 32(9), 133, 1980.

200. Mitra, S. K. and Sanyal, D., Effect of putrescine on fruit set and fruit quality of litchi, *Gartenbauwissenschaft*, 55(2), 83, 1990.

201. Zhukava, P. S., Kamzolova, O. I., Dubenetskaya, M. M., and Voitik, N. P., Effect of the herbicide pyramin on the quality of table beets, *Vestsi Akad. Navuk BSSR, Ser. Sel'skagaspad. Navuk*, 2, 56, 1982; *Chem. Abstr.*, 97, 70983z, 1982b.

202. Yakovleva, R. S., The use of Morestan in black currants, *Khimiya Sel'skom Khozyaistve*, 11(6), 42, 1973; *Hortic. Abstr.*, 44, 204, 1974.

203. Bairambekov, Sh. B., Effect of herbicides on the quality of vegetables *Dokl. TSKhA*, 256, 91, 1979; *Chem. Abstr.*, 94, 151478d, 1981.

204. Sevumyan, M. A., Oganisyan, R. V., Aslanyan, G. Ts., and Aleksanyan, D. S., Roundup testing and its toxicological characterization, *Izv. S-kh. Nauk*, 26(4), 44, 1983; *Chem. Abstr.*, 99, 117777u, 1983.

205. Avetisyan, K. V., Method for determining rubigan residues, their detoxification and effect on the quality of tomatoes and cucumbers, *Izv. S-Kh. Nauk*, 7, 40, 1986; *Chem. Abstr.*, 105, 220839g, 1986.

206. Dubenetskaya, M. M., Patent, R. L., Voitik, N. P., Bogdan, A. S., Enshina, A. N., and Karaseva, A. E., Effect of the herbicide Semeron on some indexes of the food value and harmlessness of cabbage, *Gig. Sanit.*, 7, 76, 1980; *Chem. Abstr.*, 93, 112463e, 1980.

207. Portnoi, M. I. and Balaban, V. D., Use of simazine in young non-fruit-bearing apple orchards, *Khim. Sel'sk. Khoz.*, 10, 28, 1984; *Chem. Abstr.*, 102, 19501k, 1985.

208. Borodulina, V. S. and Shishkina, E. E., The effect of simazine and atrazine on fruit quality in apples, *Khimiya Sel'skom Khozyaistve*, 11(4), 52, 1973; *Hortic. Abstr.*, 44, 119, 1974.

209. Oksenyuk, Yu. F., Use of herbicides in fruit growing in the maritime territory, *Tr. Dal'nevost. Nauchno-Issled. Inst. Sel'sk. Khoz.*, 13(2), 311, 1973; *Chem. Abstr.*, 83, 38676b, 1975.

210. Vosiljus, R., The use of simazine in black currants, *Materialy 5-oj Pribalt. Nauč. Konf. Zašč Rast.*, 157, 1965; *Hortic. Abstr.*, 38, 7251, 1968.

211. Shinohara, Y., Tanaka, K., Suzuki, Y., and Yamasaki, K., Growing conditions and quality of vegetables. I. Effect of nutrition and foliar spray treatment on the ascorbic acid content of leaf vegetables, *Jpn. J. Hortic. Sci.*, 47(1), 63, 1978; *Hortic. Abstr.*, 48, 10487, 1978.

212. Belobrov, A. V. and Yanchik, K. L., The yield and chemical composition of apples from simazine-treated orchards, *Sadovod. Vinograd. Vinodel. Mold.*, 5, 49, 1976; *Hortic. Abstr.*, 47, 4259, 1977.

213. Tarlapan, M. I., Grigel, T. A., and Muntyanu, R. G., Effect of simazin and complete fertilizer on the yield and quality of raspberries, *Khim. Sel'sk. Khoz.*, 13(9), 697, 1975; *Food Sci. Technol. Abstr.*, 8(3), J350, 1976.

214. Zueva, N. P., Simazine plus dalapon, *Sadovodstvo*, 5, 24, 1973; *Hortic. Abstr.*, 44, 1406, 1974.

215. Dymova, G. I., Effect of benzimidazole and its derivatives on yield, dynamics of fruit bearing, and biochemical composition of fruits in growth of sweet peppers in winter hot houses, *Ispol'z. Regul. Rosta Polim. Mater. Ovoshchevod.*, 23, 1984; *Chem. Abstr.*, 104, 64084d, 1986.

216. Beleva, L., The role of antibiotics in increasing the resistance of tomatoes to bacterial canker, *Corynebacterium michiganese*, *Nauchni Trudove*, Vissh Sel-skostopanski Institut "Georgi Dimitrov," Rasteniev"dstvo, 23, 357, 1973; *Hortic. Abstr.*, 45, 1749, 1975.

217. Zhang, L. H. and Lin, K. H., Mechanisms of selective action of sodium sulfanilate on plants, *Pestic. Biochem. Physiol.*, 32(1), 11, 1988.

218. Antonious, G. F., Abdel-All, A., and Youssef, M. M., Residual behavior of thiodi-carb in squash and tomatoes and its effect on ascorbic acid content of tomatoes, *Alexandria Sci. Exch.*, 9(2), 143, 1988; *Chem. Abstr.*, 110, 187767k, 1989.

219. Fel'dman, A. L. and Kobeleva, S. M., The effect of TMTD on the nutritional value of potatoes, sweet pepeprs and eggplants, *Himija Sel'Hoz.*, 8(3), 32, 1970; *Chem. Abstr.*, 41, 1413, 1971.

220. Rybachok, N. O. and Luchinina, E. G., Effect of herbicides on yield and quality of red peppers, *Konservnaya i Ovoshchesushil'naya Promyshlennost*, 9, 9, 1978; *Food Sci. Technol. Abstr.*, 11(7), J1251, 1979.

221. Mohammed, A. I. and Ali, A. Y., Effect of herbicides on yield and fruit quality of tomato, *Acta Hortic.*, 190, 191, 1986.

222. Kolesnikov, V. A. and Ignatov, V. A., Use of herbicides on transplanted white-headed cabbage, *Khim. Sel'sk. Khoz.*, 7, 41, 1984; *Chem. Abstr.*, 101, 124790f, 1984.

223. Osman, B. H. and Lundergan, C. A., Effect of triacontanol on yield and fruit composition of spring-harvested "Tangi" and "Dover" strawberries, *HortScience*, 20(1), 73, 1985.

224. Subbiah, S., Ramanathan, K. M., Francis, H. J., Effect of triacontanol on ascorbic acid content and yield of tomato var. Pusa Ruby, *Madras Agric. J.*, 67(11), 758, 1980; *Chem. Abstr.*, 96, 117447z, 1982.

225. Rjazancev, A. V. and Budazapov, V. C., The effect of trichlorofon on the yield and quality of white cabbage, *Tr. Permsk. sel'.-hoz. Inst.*, 39, 422, 1968; *Hortic. Abstr.*, 40, 943, 1970.

226. Martynyuk, I. V., Weed control in potatoes, *Vestn. S-kh. Nauki* (*Moscow*), 7, 135, 1990; *Chem. Abstr.*, 113, 128045n, 1990.

227. Talaš, A. I., The effect of fungicides on tree growth and fruit quality of the apple varieties Renet Simirenko and White Rosemary, *Himija sel'. Hoz.*, 7(4), 28, 1969; *Hortic. Abstr.*, 40, 2996, 1970.

228. Isin, M. M., Shanimov, Kh. I., and Kairova, G. N., Effect of phosalane on apple growth and yield, *Vestn. S-kh. Nauki Kaz.*, 12, 52, 1985; *Chem. Abstr.*, 104, 163697n, 1986.

229. Luecke, R. W., Hamner, C. L., and Sell, H. M., Effect of 2,4-dichlorophenoxyacetic acid on the content of thiamine, riboflavin, nicotinic acid, pantothenic acid and carotene in stems and leaves of red kidney bean plants, *Plant Physiol.*, 24, 546, 1949.

230. Singh, K. and Chauhan, K. S., Effect of pre-harvest application of calcium, potas-sium and alar on fruit quality and storage life of guava fruits, *Haryana Agric. Univ. J. Res.*, 12(4), 649, 1982; *Chem. Abstr.*, 99, 21111w, 1983.

231. Schuphan, W., Rückstände von Aldrin und Dieldrin in Wurzeln von Möhren (*Daucus carota* L) und ihr Einfluss auf den Biologischen Wert, *Z. Pflanzenkrank. Pflanzenschutz*, 67, 340, 1960.

232. Restuccia, G., Two years of research on chemical weed control in carrots grown during winter and spring, *Tecn. Agric.*, 22, 89, 1970; *Hortic. Abstr.*, 41, 6981, 1971.

233. Constantin, R. J., and Hernandez, T. P., Effect of Azide soil treatment on quality and yield of sweet potatoes, *HortScience*, 12(5), 457, 1977.

234. Rouchaud, J., Moons, C., and Meyer, J. A., The effects of selected herbicide and fungicide treatments on the carotenes and xanthophylls in lettuce, *J. Hortic. Sci.*, 60(2), 245, 1985.

235. Rouchaud, J., Moons, C., and Meyer, J. A., The effects of herbicide and fungicide treatments on the growth and provitamin A content of lettuce, *Pestic. Sci.*, 16, 88, 1985.

236. Rouchaud, J., Moons, C., and Meyer, J. A., Effects of pesticide treatment on the carotenoid pigment of lettuce, *J. Agric. Food. Chem.*, 32, 1241, 1984.

237. Buschmann, C., Grumbach, K. H., and Bach, T. J., Herbicides which inhibit photosynthesis II or produce chlorosis and their effect on production and transformation of pigments in etiolated radish seedlings (*Raphanus sativus*), *Physiol. Plant.*, 49(4), 455, 1980.

238. Rouchaud, J., Moons, C., Meyer, J. A., Effect of selected insecticides and herbicides on the carotene content of summer carrots, *Scientia Hortic.*, 19, 33, 1983.

239. Rouchaud, J., Moons, G., Gillet, J., and Meyer, J. A., Effects of pesticide treatments on the provitamin A content of freshly harvested and cool stored carrots, *Tests Agrochem. Cultiv.*, 6, 114, 1985; *Chem. Abstr.*, 106, 80252d, 1987.

240. Sukhareva, I. Kh. and Shevtsov, V. I., The use of preparation TUR in cucumbers and tomatoes growing under cover, *Khim. Sel'sk. Khoz.*, 12(8), 46, 1974; *Hortic. Abstr.*, 45, 3196, 1975.

241. Sweeney, J. P. and Marsh, A. C., Effects of selected herbicides on protvitamin A content of vegetables, *J. Agric. Food Chem.*, 19(5), 854, 1971.

242. Rouchaud, J., Moons, C., and Meyer, J. A., Effect of soil treatment with the insecticide chlofenvinphos and of covering of the culture with plastic film on the provitamin A content of early carrots, *J. Agric. Food Chem.*, 30, 1036, 1982.

243. Zatserkovskii, V. A., Zatserkovskaya, G. A., Changes in some indexes of chemical composition of carrot root crops grown in soils treated with insecticides, *Zakhist. Rosl.*, 23, 107, 1976; *Chem. Abstr.*, 89, 54735z, 1978.

244. Bhagat, P., Saimbhi, M. S., Sharma, B. N., and Paul, Y., Effect of some herbicides on the chemical composition of carrot root (*Daucus carota* L.), *J. Res. (India)*, 13(2), 202, 1976; *Hortic. Abstr.*, 47, 9520, 1977.

245. Rhodes, B. B. and Hall, C. V., Effect of CPTA, temperature and genotype on carotene synthesis in carrot leaves, *HortScience*, 10(1), 22, 1975; *Hortic. Abstr.*, 46, 3430, 1976.

246. Lee, C. Y., Changes in carotenoid content of carrots during growth and post-harvest storage, *Food Chem.*, 20(4), 285, 1986.

247. El-Sayed Osman, M., Mummery, R. S., and Valadon, L. R. G., Effect of CPTA and of nicotine on chlorophyll and carotenoid contents of an ornamental gourd, *Ann. Bot.*, 53(1), 21, 1984; *Hortic. Abstr.*, 54, 2655, 1984.

248. Simpson, D. J., Baqar, M. R., and Lee, T. H., Chemical regulation of plastic development. III. Effect of light and CPTA on chromoplast ultrastructure and carotenoids of *Capsicum annuum*, *Z. Pflanzenphysiol.* 82(3), 189, 1977.

249. Korzun, G. P. and Kulik, L. V., The effect of herbicidal weed control on the biochemical composition of onions, *Ovochivnitstvo Bashtannitstvo Resp. Mizhvid. Temat. Nauk. Zbornik*, 18, 11, 1974; *Hortic. Abstr.*, 45, 9479, 1975.

250. Emerson, G. A., Thomason, I. J., Paulus, A. O., Dull, G. G., and Snipes, J. W., Effects of soil fumigation on the quality and nutritive values of carrots, *Fed. Proc.*, 30, 584Abs, 1971.

251. Thomason, I. J., Castro, C. E., Baines, R. C., and Mankau, R., What happens to soil fumigants after nematode control? *Calif. Agric.* 25(9), 10, 1971.

252. Janković, R. and Bojić, M., Effect of different rates of dichlobenil on the chloroplast pigment content of red currant leaves, *Jugoslovensko Voćarstvo*, 17(65), 21, 1983; *Hortic. Abstr.*, 54, 6077, 1984.

253. Benoit, F. and Ceustermans, N., Qualitative aspects of tomatoes grown by NFT, *Boer en de Tuinder*, 90, 21, 1984; *Hortic. Abstr.*, 54, 9207, 1984.

254. Walkowska, A., Olejnik, D., and Urbanowicz, M., Content and stability of α- and β-carotene and some selected essential chemical components in carrot cultivated in insecticide-contaminated soil. Part II. α- and β-carotene content of carrot, *Bomatol. Chem. Toksykol.*, 15(1-2), 31, 1982; *Chem. Abstr.*, 97, 176547a, 1982.

255. Stoimenova, I., Study of some herbicides used in nursery of two-year-old apple trees, *Gradinar. Lozar. Nauka*, 13(4), 37, 1976; *Chem. Abstr.*, 86, 38505g, 1977.

256. Bewick, T. A., Binning, L. K., and Simon, P. W., Effect of ethephon on the carotene content of early planted carrots, *Acta Hortic.*, 201, 125, 1987.

257. Buescher, R. W. and Doherty, J. H., Color development and carotenoid levels in *rin* and *nor* tomatoes as influenced by ethephon, light and oxygen, *J. Food Sci.*, 43, 1816, 1978.

258. Paz, O., Janes, H. W., Prevost, B. A., and Frenkel, C., Enhancement of fruit sensory quality by post-harvest applications of acetaldehyde and ethanol, *J. Food Sci.*, 47(1), 270, 1982.

259. Linser, H. und Zeid, F. A., Reinprotein, Chlorophyll, Carotin und Kohlenhydrate bei *Daucus carota* im Verlauf der Vegetationsperiode des ersten Jahres unter dem Einfluss von Wachstumsregulatoren, *Z. Pflanzenernähr. Bodenk.*, 138(2), 181, 1975.

260. Neumann, K. H. and Schwab, B., Untersuchungen über den Einfluss von Giberellinsäurespritzungen auf den Ertrag, die Anatomie der Wurzel und die Karotinverteilung bei Karotten, *Z. Pflanzenerähr. Bodenk.*, 138(1), 19, 1975.

261. Engst, R., Blazovich, M., and Knoll, R., Über das Vorkommen von Lindan in Möhren und seinen Einfluss auf den Carotingehalt, *Nahrung.*, 11(5), 389, 1967.

262. Engst, R., Aufnahme und Speicherung von Schädlingbekäpfungsmitteln in Möhren, *Qual. Plant. Mater. Veg.*, 14, 305, 1967.

263. Beckmann, E. O. and Pestemer, W., The influence of herbicide treatment with different humus supplies on the yield and composition of carrots, *Landwirt. Forsch.*, 28(1), 41, 1974; *Hortic. Abstr.*, 46, 1318, 1976.

264. Milivojević, D. and Marković, D., Effects of linuron and monolinuron herbicides on pea chloroplasts, *Photosynthetica*, 23(3), 386, 1989.

265. Kleudgen, H. K., Veränderungen der Pigment- und Prenylchinongehalte in Chloroplasten von Gerstenkeimlingen nach Applikation des Wuchsstoffherbizids MCPA (4-Chlor-2-methylphenoxyessigsäure), *Z. Naturforsch.*, 34c, 106, 1979.

266. Nikiforova, E. L., Results of trials with herbicides to control *Stellaria media* in carrots and onions, *Himija Sel'.hoz.*, 4(3), 52, 1966; *Hortic. Abstr.*, 37, 7091, 1967.

267. Salunkhe, D. K., Wu, M., Wu, M. T., and Singh, B., Effects of Telone and Nemgon on essential nutritive components and the respiratory rates of carrot (*Daucus carota* L.) roots and sweet corn (*Zea mays* L.) seeds, *J. Am. Soc. Hortic. Sci.*, 96, 357, 1971.

268. Wu, M., Singh, B., Wu, M. T., Salunkhe, D. K., and Dull, G. G., Effect of certain soil fumigants on essential nutritive components and the respiratory rate of carrot (*Caucus carota* L.) roots, *HortScience*, 5, 221, 1970.

269. Hloušek-Radojčić, A. and Ljubešić, N., The development of daffodil chromoplasts in the presence of herbicides SAN 9789 and SAN 9785, *Z. Naturforsch.*, 43c, 418, 1988; *Hortic. Abstr.*, 58, 8983, 1988.

270. Pospíšilová, J. and Janyška, A., Side effects of the herbicides Liro CIPC 40 and Ramrod WP on onion seedlings, *Bull. Vyzkumny Šechitelsky Ustav Zelinářsky Olomouc*, 23/24, 3, 1979/1980; *Hortic. Abstr.*, 53, 5018, 1983.
271. Monson, W. G., Burton, G. W. and Wilkinson, W. S., Effect of N fertilization and simazine on yield, protein, amino-acid content, and carotenoid pigments of coastal Bermudagrass, *Agron. J.*, 63, 928, 1971.
272. Kalinin, F. L. and Ponomarev, G. S., Influence of simazine on the pigment and carbohydrate exchange of plants, *Ukr. Bot. Zh.*, 20(1), 52, 1963; *Chem. Abstr.*, 61, 15280g, 1964.
273. Kleudgen, H. K., Die Wirkung von Simazin auf die Bildung der plastidären Prenyllipide in Keimlingen von *Hordeum vulgare* L., *Z. Naturforsch.*, 34c, 110, 1979.
274. Davis, K. R., Peters, L. J., and Le Tourneau, D., Variability of vitamin content in wheat, *Cereal Foods World*, 29(6), 364, 1984.
275. Bogdan, A. S., Patent, R. L., and Dubenetskaya, M. M., Thiamine supply of animals fed rye and wheat grain grown with the use of herbicide from the group of chlorinated phenoxy acids, *Vopr. Pitan.*, 1, 61, 1980; *Chem. Abstr.*, 92, 145383m, 1980.
276. Koslowska, A., Effect of herbicides on yield, vitamin content and germination of bean seeds, *Howdowla Roslin, Aklimatyzacia i Nasinmictwo*, 20, 517, 1976.
277. Ivanov, O. N. and Stefanovich, E. A., Content and bio-availability of group B vitamins in grain, *Vestsi Akad. Navuk BSSR, Ser. Sel'skagaspad. Navuk*, 4, 40, 1989; *Chem. Abstr.*, 112, 215417h, 1990.
278. Catlin, P. B., Changes in some B vitamins associated with the responses of the Tilton apricot fruit to 2,4,5-trichlorophenoxyacetic acid, *Proc. Am. Soc. Hortic. Sci.*, 74, 174, 1959.
279. Gevorkyan, L. A., Avakyan, B. P., and Kalantarov, A. A., Effect of herbicides on the vitamin B group levels in grapes, *Izv. S-Kh. Nauk*, 23(10), 33, 1980; *Chem. Abstr.*, 94, 115563u, 1981.
280. Gevorkyan, L. A., Kalantarov, A. A., Sevumyan, M. A., and Danielyan, L. G., Effect of herbicides on group B vitamin levels in grapes, *Biol. Zh. Arm.*, 33(3), 334, 1980; *Chem. Abstr.*, 93, 162289y, 1980.
281. Blagonravova, L. N., Nilov, G. I., Avdoshina, E. G., The effect of organochlorine pesticides on the vitamin E content of plant leaves, *Byulleten Gosudarstvennogo Nikitskogo Botanicheskogo Sada*, 3(19), 57, 1972; *Hortic. Abstr.*, 44, 3962, 1974.
282. Dubenetskaya, M. M., Bogdan, A. S., and Patent, R. L., Effect of herbicides of the 2,4-D group on the content of vitamin B_6 in grain crops, *Zdravookhr. Beloruss.*, 5, 47, 1977; *Chem. Abstr.*, 87, 79572r, 1977.
283. Anonymus, *Pesticide Residue Problems in the Third World, A Contribution of the GTZ Residue Laboratory in Darmstadt and Its Foreign Activities*, TZ-Verlag, Rossdorf, Germany, 1979.
284. National Research Council, *Alternative Agriculture*, National Academy Press, Washington, D.C., 1989.
285. World Health Organization, *Our Planet, Our Health*, World Health Organization, Geneva, 1992, chapter 3.
286. Pimentel, D. and Hanson, A. A., *CRC Handbook of Pest Management in Agriculture*, 2nd ed., Volumes I-III, CRC Press, Boca Raton, Fla., 1991.
287. Mercadante, A. Z. and Rodriguez-Amayua, D. B., Carotenoid composition of a leafy vegetable in relation to some agricultural variables, *J. Agric. Food. Chem.*, 39(6), 1094, 1991.
288. Giannopolitis, C. N., Vassiliou, G., and Vizantinopoulos, S., Effect of week interference and herbicides on nitrate and carotene accumulation in lettuce, *J. Agric. Food Chem.*, 37, 312, 1989.

289. Sanyal, D., Sarkar, B., and Mitra, S. K., Herbicidal weed control in banana, *Pesticides*, 23, 29, 1989.

290. Babii, V. S. and Tsvetkova, A. G., Effect of apple trees of various pesticide concentration in restricted-volume spraying, *Khim. Sel'sk. Khoz.*, 10, 511, 1972; *Food Sci. Technol. Abstr.*, 6(1), J23, 1974.

291. Geard, I. D., Control of *Septoria* spot of black currants, *Tasmanian J. Agric.*, 28, 226, 1957; *J. Sci. Food Agric.*, 10, i-137, 1959.

292. Olejnik, D., Walkowska, A., and Urbanowicz, M., Content and stability of α- and β-carotenes and some selected essential chemical components in carrots cultivated in soil contaminated with insecticides. Part III. Stability of α- and β-carotenes of carrots, *Bromatol. Chem. Toksykol.*, 15(3), 139, 1982; *Chem. Abstr.*, 97, 214499w, 1982.

293. Patil, S., Chikkasubbanna, V., and Gowda, J. V. N., Effect of preharvest sprays of triacontanol on the storage life of lettuce (*Lactuca sativa* L.) cv. Great Lakes, *J. Food Sci. Technol.*, 26(3), 156, 1989.

294. Singh, R., Influence of nitrogen supply on host susceptibility to tobacco mosaic virus infection, *Phyton (Austria)*, 14(1-2), 37, 1970.

295. Leath, K. T. and Ratcliffe, R. H., The effect of fertilization on disease and insect resistance, in *Forage Fertilization*, Mays, D. A., Ed., American Society of Agronomy, Madison, Wis., 1974, 481.

296. Marschner, H., *Mineral Nutrition of Higher Plants*, Academic Press, London, 1986

297. International Potash Institute, *Fertilizer Use and Plant Health*, Int. Potash Inst., Bern, Switzerland, 1976.

298. Decallonne, J. R. and Meyer, J. A., A survey of a three years study on the effect of agrochemicals on the quality and composition of tomatoes, in *The Effects of Modern Production Methods on the Quality of Tomatoes and Apples*, Gormley, T. R., Sharples, R. O., and Dehandtschtter, J., Eds., Commission of the European Communities, Luxembourg, 1985, 5.

299. Bajaj, K. L. and Mahajan, R., Influence of some nematicides on the chemical composition of tomato fruits, *Qual. Plant.—Plant Foods Hum. Nutr.*, 27(3-4), 335, 1977.

300. Monselise, S. P., Ed., *CRC Handbook of Fruit Set and Development*, CRC Press, Boca Raton, Fla., 1986.

301. Ponnampalam, R. and Mondy, N. I., Effect of sprout inhibitor isopropyl *N*-(3-chlorophenyl) carbamate (CIPC) on phenolic and ascorbic acid content of potatoes, *J. Agric. Food Chem.*, 34, 262, 1986.

302. Izumi, H., Ito, T., and Yoshida, Y., Relationship between ascorbic acid and sugar content in citrus fruit peel during growth and development, *Jpn. Soc. Hortic. Sci.*, 57(2), 304, 1988.

303. Lürssen, K., Ethylene and agriculture, in *The Plant Hormone Ethylene*, Mattoo, A. and Suttle, J. C., Eds., CRC Press, Boca Raton, Fla., 1991, 315.

304. Ben, J. and Kropp, K., The effect of CCC and Alar on the chemical composition and storage quality of Jonathan apples, *Acta Hortic.*, 179(2), 825, 1986; *Hortic. Abstr.*, 57, 961, 1987.

305. Khader, S. E. S. A., Singh, B. P., and Khan, S. A., Effect of GA_3 as a post-harvest treatment of mango fruit on ripening, amylase and peroxidase activity and quality during storage, *Scientia Hortic.*, 36(3-4), 261, 1988.

306. Sud, G. and Nayital, R. K., Effect of CCC (cyocel) on the shelf life of peach cv. Opulent, *Indian Food Packer*, 43(4), 5, 1989; *Chem. Abstr.*, 112, 20154n, 1990.

307. Dhillon, B. S., Ladania, M. S., Bhullar, J. S., and Randhawa, J. S., Effect of plant growth regulators on the storage of Anab-E-Shahi grapes, *Indian J. Hortic.*, 42, 18, 1985.

308. Chan, M. T. and Karanov, E., Effect of maleic acid hydrazide (MAH) and piperidi-noacetanilide hydrochloride (PAH) on vitamin C content in stored potato tubers, *Fiziol. Rast. (Sofia)*, 9(2), 36, 1983; *Chem. Abstr.*, 99, 138356w, 1983.
309. Maier, V. P. and Yokoyama, H., Vitamin induction in fruits and vegetables with bioregulators, in *Nutritional Qualities of Fresh Fruits and Vegetables*, White, P. L. and Selvey, N., Eds., Futura Publ. Co., Mount Kisco, N.Y., 1974, 177.
310. Hsu, W. J., Yokoyama, H., and Coggins, C. W., Jr., Carotenoid biosnythesis in *Blakeslea trispora*, *Phytochemistry*, 11, 2985, 1972.
311. Novitskaya, E. I., The effect of chemical mutagenes on tomatoes, *Vestn. S-kh. Nauki*, 2, 27, 1973; *Hortic. Abstr.*, 43, 7769, 1973.

Chapter 10

EPILOGUE

I. AGRONOMIC AND NUTRITIONAL ASPECTS

Vitamins are among the essential components of foods for many animals, including human beings, and are produced mainly by plants. Thus any genetic and environmental factors that affect their concentration in the plants could conceivably have nutritional consequences for the consuming organisms. Data collected in this book show that:

a. The concentration of vitamins may strongly vary depending on the plant variety, in some cases by more than severalfold.

b. Large variations can occur in the concentration of a given vitamin in a given plant grown in different geographical locations around the world.

c. Numerous investigations have shown that excessive use of nitrogen fertilizers can decrease the ascorbic acid in many fruits and vegetables. Nitrogen fertilizers may, however, increase the concentration of other vitamins such as carotene and thiamin.

d. Plants grown in the greenhouse often have lower ascorbic acid than those grown in the field.

e. Organically grown plants generally have the same vitamin C content as those grown by conventional methods. Concentrations of vitamins B_1 and B_{12}, however, may be higher in plants grown with organic fertilizers.

f. Use of various agrochemicals such as pesticides and herbicides may alter the concentration of vitamins in plants, in many cases by increasing them.

g. When plants are attacked by foreign organisms (fungi, virus, insects, etc.), their content of some vitamins, especially ascorbic acid, decreases.

h. Both air and soil pollution can modify the concentration of vitamins in plants.

It is reasonable to assume that the natural or man-made variations in plant vitamins may have a special nutritional relevance in those situations were food diversity is low or nonexistent and the food consumed is primarily locally produced.

The variations in plant vitamins may present serious difficulties when it comes to nutritional labelling of plant products.[1] It is proposed that informing the public about the large differences that may exist in the vitamin content of fruits and vegetables, depending on which variety one chooses during shopping, may result in an increased demand for the vitamin-rich varieties compared to "better-looking" varieties. An increased demand for high-vitamin varieties may in turn provide a strong enough incentive for breeders, farmers, and distributors to work toward improving the content of vitamins in the fruits and vegetables we enjoy and need to consume.

From a practical point of view, the variations in plant vitamins may be roughly classified as (1) those brought about by natural factors over which one has little or no control and (2) those caused by man-made factors over which one might exert some kind of control if it is desired to do so.

1. Those brought about by natural factors:
   - Variations due to genetic differences
   - Variations caused by climatic factors such as amount of sunlight, temperature, and rainfall variations due to geographical location
   - Yearly fluctuations
   - Soil type
   - Pests
2. Those brought about by man-made factors:
   - Use of certain varieties
   - Use of fertilizers
   - Cultivation in greenhouses
   - Effects of air and soil pollution
   - Use of various plant protection chemicals

An example of a new emerging method to modify the vitamin content in plants—the use of an intermittent supply of nutrients to vegetables grown with various forms of nutrient solution culture (chapter 5)—needs to be emphasized. This is a promising technology, which, when its details have been worked out, could have two positive results: (1) increase in the vitamin C content and (2) decrease in the nitrate content of vegetables. Other areas deserving more research have been noted in different chapters.

Finally, the relatively large varietal differences, the yearly fluctuations in the vitamin content of various fruits and vegetables, and the effect of location of growth on plant vitamins indicate that the vitamin concentrations in food plants as published in vitamin handbooks and nutritional manuals[13–15] need to be interpreted with due caution.

II. PHYSIOLOGICAL ASPECTS (STRESS AND VITAMINS IN PLANTS)

In animals, ascorbic acid affects the toxicity and/or carcinogenicity of numerous organic and inorganic pollutants, many of which are ubiquitous in the air, water, and foods.[2-7] In this connection it is of interest that animals that cannot synthesize their own vitamin C often benefit from increased intake of vitamin C when subjected to toxic compounds or stressful conditions.[2-7]

Based on the material presented here, it appears that also in plants, various stress factors such as low temperature, water stress, use of agrochemicals, and attack by pests alter the vitamin content of plants. Whether these changes have any cause-and-effect relationships with the mechanism used by plants to defend themselves against these stress factors is not clear. Although spraying of higher plants such as lettuce, celery, spinach, petunia, and roses with ascorbic acid was noted to increase their resistance to smog or ozone exposure,[8-9] and supplementing the cyanobacterium *Nostoc muscorum* with ascorbic acid was noted to protect this organism against Cr and Pb toxicity,[10] an unequivocal proof for the direct role of vitamins in protecting higher plants against various stress factors remains to be seen.

Do vitamins play exactly the same role(s) in different organisms, for example humans and plants? This is an intriguing question that needs to be investigated. One of the reasons for this uncertainty is the fact that although the metabolic roles of various vitamins have been intensively investigated, the majority of experiments have been carried out by using microorganisms, animals, and humans as experimental material and astonishingly little by using plants. In this respect, one may ask whether the natural variations observed in the vitamin content of plants, some of which may be in the order of severalfold, have any significance for the physiology of different plant varieties or plants living under different conditions, or the variations observed are important only for the organisms consuming the plants and not for the plants themselves.

The roles supposedly played by antioxidant vitamins in human diseases, ranging from the common cold[11,12] to some forms of cancer and especially then tolerance to various stresses (chapter 1), on one hand, and the changes in the content of these vitamins in plants subjected to various stresses ranging from low temperature to various pathogenic organisems (chapters 7–9), on the other, may be indicative of similar physiological roles of these vitamins in plants and animals. This is an intriguing thought, which needs to be investigated.

REFERENCES

1. Leveille, G. A., Bedford, C. L., Kraut, C. W., and Lee, Y. C., Nutrient composition of carrots, tomatoes and red tart cherries, *Fed. Proc.*, 33(11), 2264, 1974.
2. Calabrese, E. J., *Nutrition and Environmental Health. The Influence of Nutritional Status on Pollutant Toxicity and Carcinogenicity*, Vol. 1, *The Vitamins*, Wiley-Interscience, New York, 1980.
3. Calabrese, E. J., *Nutrition and Environmental Health. The Influence of Nutritional Status on Pollutant Toxicity and Carcinogenicity*, Vol. 2, *Minerals and Macronutrients*, Wiley Interscience, New York, 1981.
4. Calabrese, E. J., Does exposure to environmental pollutants increase the need for vitamin C? *J. Environ. Path. Toxicol. Oncol.*, 5(6), 81, 1985.
5. Omaye. S. T., Tillotson, J. A., and Sauberlich, H. E., Metabolism of L-ascorbic acid in the monkey, in *Ascorbic Acid: Chemistry, Metabolism, and Uses*, Seib, P. A. and Tolbert, B. M., Eds., Adv. Chem. Series, 200, Am. Chem. Soc., Washington, D.C., 1982, 317.
6. Nakano, K. and Suzuki, S., Stress-induced changes in tissue levels of ascorbic acid and histamine in rats, *J. Nutr.*, 114, 1602, 1984.
7. Street, J. C. and Chadwick, R. W., Ascorbic acid requirements and metabolism in relation to organochlorine pesticides, *Ann. N.Y. Acad. Sci.*, 258, 132, 1975.
8. Freebairn, H. T., The prevention of air pollution damage to plants by the use of vitamin C spray, *J. Air Pollution Cont. Assoc.*, 10, 314, 1960.
9. Freebairn, H. T. and Taylor, O. C., Prevention of plant damage from air-borne oxidizing agents, *Proc. Am. Soc. Hortic. Sci.*, 76, 693, 1960.
10. Rai, L. C., and Raizada, M., Impact of chromium and lead on *Nostoc muscorum*: regulation of toxicity by ascorbic acid, glutathione, and sulfur-containing amino acids, *Ecotoxicol. Environ.* Safety, 15, 195, 1988.
11. Pauling, L., Evolution and the need for ascorbic acid, *Proc. Natl. Acad. Sci. (USA)*, 67(4), 1643, 1970.
12. Pauling, L., The significance of the evidence about ascorbic acid and the common cold, *Proc. Natl. Acad. Sci. USA*, 68(11), 2678, 1971.
13. Food and Agriculture Organization, *Food Composition Tables for the Near East*, FAO Food and Nutrition Paper, 26, FAO, Rome, 1982.
14. Dukec, J. A. and Atchley, A. A., *CRC Handbook of Proximate Analysis Tables of Higher Plants*, CRC Press, Boca Raton, Fla, 1986.
15. U.S. Department of Agriculture, *USDA Agricultural Handbook Series 8. Composition of Foods—Raw, Processed, Prepared*, Comparison of Nutrient Data Research, Human Nutrition Information Service, U.S. Department of Agriculture, Washington, D.C, 1976–1989.

INDEX

A

Abies numidica, tocopherol, maturity, 252
Acerola, ascorbic acid:
  maturity, 242, 247
  rootstocks, 63
Achras sapota (*see* sapota)
Acifluorfen, ascorbic acid, 336
Acremonium recifei, ascorbic acid, 324
Aesculus, ascorbic acid, 113
Afalon® (*see* linuron)
Aging (*see also* maturity), 239, 252
Air pollution, effect on vitamins, 305–313
Alar® (*see* daminozide)
Aldicarb, carotene, 343
Aldrin, carotene, 343
Aleuron layer, vitamins in, 30
Alfalfa:
  carotene, 165
  riboflavin, maturity, 244
  vitamins in sprouts, 32
Alma, ascorbic acid:
  pests, 324
  range in varieties, 46
Alocasia (*see* taro)
Alternaria solani, ascorbic acid, 323, 324
Altitude, ascorbic acid, 136
Amaranthus hybridus, ascorbic acid:
  light, 90
  loss by wilting, 10
Amaranthus dubius, ascorbic acid, 301
Amiben® (*see* Chloramben)
Aneurine (*see* thiamin)
Aphid (*see Sitobion avenae*)
Apium graveolens (*see* celeriac)
Apple:
  ascorbic acid:
    air pollution, 309
    altitude, 136
    biennial bearing, 21
    breeding for high vitamin, 64
    cold hardiness, 298
    crop size, 261
    fertilizers, 158, 161, 163, 171, 175
    in fruits from colder regions, 298
    geographical variations, 127
    growth regulators, 356–358

    light, 90, 94, 101–103
    low- and high-vitamin varieties, 43, 59
    maturity, 241, 247
    ozone, 309
    peel vs. flesh, 24, 28, 29
    pesticides, 349, 352, 353
    position on tree, 101–103
    postharvest loss, effect of:
      calcium, 175
      growth regulator Alar, 357
      urea, 171
      zinc, 178
    pruning, 259
    range in varieties, 44, 55
    rootstocks, 63
    size-vitamin relationship, 266, 270
    temperature, 104, 105
    tree vigor, 261
    water supply, 292
    yearly fluctuations, 120, 122
  biotin:
    peel vs. flesh, 25
    range in varieties, 53
  niacin:
    peel vs. flesh, 25
    range in varieties, 49
  pantothenic acid, range in varieties, 50
  pesticides, vitamins, 336–340, 342–344, 346, 349, 352, 353
  riboflavin:
    peel vs. flesh, 25
    range in varieties, 50
  thiamin, range in varieties, 51
  tocopherol:
    peel vs. flesh, 26
Apricots:
  ascorbic acid:
    altitude, 136
    maturity, 241
    range in varieties, 44
    rootstocks, 63
    yearly fluctuations, 120
  carotene:
    altitude, 137
    range in varieties, 48
    yearly fluctuations, 121

Apricots (*Continued*)
  niacin in varieties, 49
  pesticides, vitamins, 343, 346, 347
  thiamin in varieties, 51
Aretit ® (*see* dinoseb acetate)
Arisarum vulgare, ascorbic acid, 111
Armenia:
  ascorbic acid, potatoes, 136
  carotene, apricots, 137
Artichoke, Jerusalem, carotene, 165
Artificial ripening (*see* ripening method)
Ascorbic acid:
  air pollution, 305–312
  altitude, 136
  anticancer property, 7
  CCC, 337
  chloroplast, 26
  color of rose flower and hips, 61
  concentration gradients in plants, 21
  contribution from foods, 2
  conversion to dehydroascorbic acid, 4
  in different plant parts, 21
  in different plant varieties, 43
  diurnal fluctuations, 111
  early and late varieties, 43
  ethephon, 338
  ethylene, 338
  farmyard manure, 197
  fertilizers, 158–164
  flower pollination, 21
  fruit trees, rootstocks, 62, 63
  fungicides, 351
  genotypic differences, 43
  geographical variations, 126, 127, 131–134
  gradients in plant tissues, 22, 23
  greenhouse vegetables, 256, 258
  growth regulators, 355–359
  insecticides, 351
  irrigation (*see also* water supply)
  leaf:
    chloroplast, 26
    variegated, 89
  light, 89, 92, 94–97
  loss after harvest, 11
    bruising, 10
    calcium, role of, 175
    nitrogen role of, 169
    organic fertilizers, 198
    pesticides, 357
    shredding, 10

  temperature, 89
  wilting, 10
  maleic hydrazide, 340
  maturity, 239, 241–242
  manganese toxicity, 301
  mineral toxicity, 301
  molybdenum toxicity, 301
  naphthalene acetic acid (NAA), 341
  organic fertilizers, 187, 189
  ozone (*see* air pollution)
  peel vs. flesh, 24
  pericarp and placenta, 24, 25, 27
  pesticides, effects on, 336–343
  polar plants, 108
  Pollution, air (*see* air pollution)
  position of fruit on plant, 94, 101–102
  postharvest loss, (*see* loss after harvest)
  potato, varieties, 58
  rootstocks, 62, 63
  seed germination, 32
  seasonal fluctuations, 113
  sewage sludge, 302, 305
  size-vitamin relationship, 266–272
  shade (*see* light)
  soil water (*see* water supply)
  sprouts (*see* seed germination)
  spruce needle, 104
  sulphur dioxide (*see* air pollution)
  sunny vs. cloudy weather, 92
  supermarket and roadside stands, 10
  temperature, 104, 106–108, 131
  tolerance-vitamin relationship
    low temperature stress, 297
    of mango to *Erwinia mangiferae*, 327
    of peanut to *Puccinia arachidis*, 328
    of tomato to *Meloidogyne incognita*, 327
    of tomato to *Phytophthora parasitica*, 327
    of rice to *Pyricularia oryzae*, 328
    of rose to *Phramidium butleri*, 327
  translocation from other plant parts, 89
  water supply, 291
  yearly fluctuations, 120
Asparagus:
  ascorbic acid:
    geographical variations, 127
    gradients in, 22, 23
    light, 90
    peel vs. flesh, 24